SYSTEMS ENGINEERING

WILEY SERIES IN SYSTEMS ENGINEERING

Andrew P. Sage

SYSTEMS ENGINEERING

ANDREW P. SAGE
School of Information Technology and Engineering
George Mason University

A Wiley-Interscience Publication
JOHN WILEY & SONS, INC.
New York / Chichester / Brisbane / Toronto / Singapore

Copyright © 1992 by John Wiley & Sons, Inc.

Library of Congress Cataloging in Publication Data:

Sage, Andrew P.
    Systems Engineering/Andrew P. Sage
        p.   cm.— (Wiley series in systems engineering)
    "A Wiley-Interscience publication."
    Includes bibliographical references and index.
    1. Systems engineering.   I. Title.   II. Series.
    TA168.S18   1992
    620'.001'1—dc20   92-19523
    ISBN 0-471-53639-3   CIP

Printed in the United States of America

10 9 8 7 6 5 4 3 2

To LaVerne

Preface

This book discusses some fundamental considerations associated with the engineering of large-scale systems or systems engineering. We begin our effort by first discussing the need for systems engineering and then providing several definitions of systems engineering. We next present a structure describing the systems engineering process, the result of which is a life-cycle model for systems engineering processes. This is used to motivate discussion of the functional levels, or considerations, involved in a systemic process: systems science and operations research methods, systems methodology and design, and systems management. Although a number of methods-based discussions are found throughout the book, our focus is clearly on systems methodology, systems design, and systems management. We are much more directly concerned with the **process** of systems engineering than we are with the **product** of systems engineering. Of course, the two are strongly related; process leads to product. The best way to insure high-quality, trustworthy systems engineering products is through use of an appropriate process.

Following an introductory chapter on systems engineering, Chapter 2 provides a discussion of several life-cycle models for systems engineering processes. This leads to a detailed discussion of systems engineering methodology. Chapter 3 presents the important subject of systems management. These discussions are very significant in that systems engineering fundamentally is concerned with systems management or technical direction of efforts that leads to the design of a system to meet the needs of a client group.

Next we discuss configuration management, systems integration, and various operational methods associated with assuring the quality of a system. These approaches, although very important, are primarily directed at assuring system product quality. Chapter 4 is devoted to a fairly detailed discussion of operational and task-level system quality assurance through configuration management, audits and reviews, standards, and systems integration.

Of even greater importance than product quality is the subject of process quality. Most contemporary discussions of competitiveness offer convincing arguments that major improvements in productivity and quality are only accommodated through process-related approaches. We present an important chapter addressed specifically to the subject of strategic level quality assurance and management. This subject, also known as total quality management (TQM), is a very major issue for contemporary approaches at the process or systems management level of systems engineering.

This presentation of TQM, in Chapter 5, completes the first part of the book concerning systems methodology, design, and management. The second part of the text is devoted to a discussion of various systems design and management approaches, especially those concerned with system effectiveness evaluation and the role of humans in systems. Chapter 6 is a transition chapter that includes user and system level requirements identification, risk management, and associated systems engineering methods. Chapters 7, 8, and 9 provide a basis for system design and evaluation, and for meaningfully including human and organizational considerations in systems engineering efforts. To these ends, we discuss decision assessment, systems evaluation, cost–benefit and cost–effectiveness analysis, and cognitive ergonomics for system design for human interaction.

This text is written for both advanced undergraduates and beginning graduate students in systems engineering and such related areas as engineering management. It should also have value for other engineering areas that offer courses in systems design methodology and systems management. Prerequisites for the text are moderate. It is generally assumed that the reader has a fundamental background common to college juniors or seniors in engineering in the United States. This includes a knowledge of differential and integral calculus, differential equations, probability, and concepts associated with the modeling of dynamic systems. However, very little specific use of this sort of material is found here, even though lack of appreciation for these approaches will be a handicap.

Perhaps more importantly, some appreciation for the engineering design of large systems is assumed. The book is not specific to any particular engineering specialty. Doubtlessly, my experiences in software systems engineering, decision support systems engineering, information systems engineering, systems management of emerging technologies, and command-and-control systems engineering have influenced the presentation. However, advanced level courses in these areas are not at all needed for this presentation and what I discuss is as relevant to construction management and manufacturing as it is to software productivity.

The book is designed for a one-semester course for students in these areas, which I have used in my classes. Doubtlessly more material is contained here than can be covered easily in one semester, however. With some expansion (through case studies and projects) a two-semester course is entirely feasible. A great deal depends upon the background of the students entering the course. For those desiring a one-semester first course in systems engineering, Chapters 1 through 6 are suggested. Chapters 7, 8, and 9, especially when augmented with case studies and software packages for decision assessment and cost-effectiveness analysis, can easily comprise a second-semester effort. I have had success in using the book for a one-semester course for those who have had a basic introduction to systems engineering. In this case, Chapters 1, 2, and 3 can be treated as review material and Chapters 4 through 9 covered in depth.

The book should also be attractive to the many professionals in industries concerned with systems engineering, management, and technical direction related efforts. These include professionals in such diversified areas as project management, software engineering, information systems engineering, manufacturing, command and control, and defense systems acquisition and procurement. I hope that you enjoy reading *Systems Engineering* as much as I have enjoyed (finally) writing it.

My thoughts concerning systems engineering have evolved over a long period of time, longer than I sometimes easily admit. I hope I have provided reference to the large number of original source documents that have influenced my studies and research within the chapters of this text. Several individuals deserve particular note here. I suspect that the earliest influencer of my transitioning from communications engineering to control engineering to the modeling and simulation of large systems and ultimately to systems engineering was John E. Gibson. Jack had much to say about systems engineering that was very worthwhile and this has had a major influence in my thoughts here, and in my professional career. His untimely death in the summer of 1991 was a considerable loss to the entire systems engineering community. Other colleagues who provided much valuable fodder and advice included James L. Melsa, Bill Rouse, Chelsea C. White III, Jim Palmer, John Warfield, Wil Thissen, and Madan Singh.

Most importantly, I have had the good fortune to have been taught a great deal by a considerable number of doctoral students who assisted me mightily in gaining an understanding of some facets of systems engineering. These include Don Farris, Joe Cole, Bob Hawthorne, Craig Miller, David Rajala, Maurice Roesch, Adolfo Lagomasino, Bernard White, Steve Post, and Charlie Smith. Although there are many more, these are particularly noteworthy because our efforts together at Southern Methodist University, the University of Virginia, and George Mason University are related directly to some specific discussions in this text.

During latter portions of this effort, my research has been supported in part by the National Science Foundation under Grant EET-8820124 concerned with systems management of emerging technologies. Brien Benson provided much support during this research effort. Brien also greatly assisted in reading through portions of this manuscript in draft form and offered a number of suggestions for improvement. Thanks, Brien. Lisa Van Horn and her associates at John Wiley were excellent in copyediting and supervising the production process. George Telecki, also at Wiley, gave much editorial encouragement as well.

To these people, and all the many others, heartfelt thanks for your part in assisting me to obtain a very good education—in systems engineering.

ANDREW P. SAGE

Fairfax, Virginia
June 1992

Contents

SYSTEMS ENGINEERING

Chapter **1**

An Introduction to Systems Engineering

This book discusses some fundamental considerations associated with the engineering of large-scale systems, or *systems engineering*. We begin our effort with a chapter that provides an overview of efforts to follow. First, we discuss the need for systems engineering, and then provide several definitions of systems engineering. Next, we present a structure describing the systems-engineering process. The result of this is a *life-cycle model* for systems engineering processes. This model serves as a framework for discussion of the functional levels, or considerations, involved in a systemic process: systems science and operations research methods, systems methodology and design, and systems management. Although there will be a number of methods-based discussions throughout the book, our focus is clearly on *systems methodology*, *systems design*, and *systems management*.

1.1. CHARACTERISTICS OF SYSTEMS ENGINEERING

Systems engineering is a management technology. This statement requires further explanation. Technology is the result of, and represents the totality of, the organization, application, and delivery of scientific knowledge for the intended enhancement of society. This is a functional definition of technology as a fundamentally human activity. Associated with this definition is the fact that a technology inherently involves a purposeful human extension of one or more natural processes.

Management involves the interaction of an organization with the environment. Consequently, a management technology involves the interaction of science, the organization, and its environment. Figure 1.1 illustrates these conceptual interactions. Notably missing from this discussion is any mention

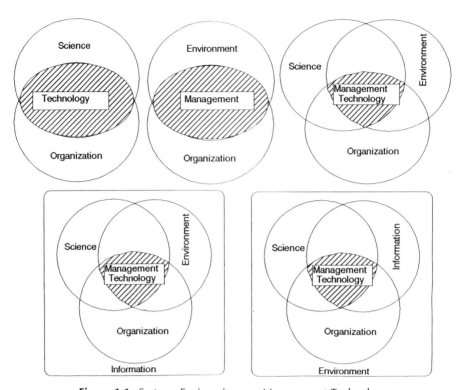

Figure 1.1. Systems Engineering as a Management Technology

of information. Information is the glue that enables the interactions shown in the figure and is, therefore, an elemental quantity, which we assume is capable of being acquired, represented, and used. Information is, in many ways, *"the"* essential ingredient in the systems engineering process. For this reason, it might be viewed as a background ingredient for systems engineering. Alternatively, we might consider information as the third ingredient and the environment as the background for systems engineering. Each of these perspectives is also shown in Figure 1.1.

We can think of a physical science basis, a management science basis, and an information science basis for technology. The physical science basis involves primarily matter and energy processing. The management science basis involves human and organizational concerns. In many ways, the information science basis is the most difficult, as information is not a truly fundamental quantity, but one that derives from the structure and organization inherent in the physical sciences and management sciences and the uses to which these sciences are put in the realization of operational systems.

This leads to the notion of physical systems design, management systems design, and information systems design. An information system and a physi-

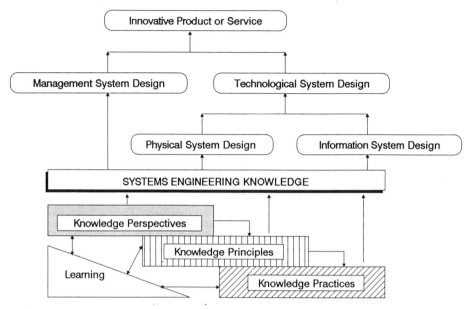

Figure 1.2. Systems Engineering in the Production of Innovative Products and Services

cal system also may support the management process. Technology involves information processing, matter, and energy processing. Therefore, we can group physical systems design and information systems design components into a single category called *technological systems design*. Figure 1.2 illustrates these interrelations. It also indicates that systems engineering knowledge [1] includes the following:

1. *Knowledge Perspectives:* Views held relative to future directions in the technological area under consideration, such that proactive and future-oriented system evolution is possible;
2. *Knowledge Principles:* Generally, formal problem solving approaches to knowledge, usually needed and employed in new situations and/or unstructured environments;
3. *Knowledge Practices:* The accumulated wisdom and experience that have led to the development of standard operating policies for well-structured problems.

These interact with one another to enable continual improvement in performance. It is on the basis of the appropriate use of these types of knowledge that we are able to accomplish the technological and management system designs that lead to an innovative product, process, service, or *systems*.

Generally, we use the term "system" to refer to products, processes, or services. However, it is often necessary to distinguish among these.

We continue our discussion and definition of systems engineering by indicating one possible structural definition of systems engineering. Systems engineering is management technology to assist and support policy making, planning, decision making, and associated resource allocation or action deployment. It accomplishes this through quantitative and qualitative formulation, analysis, and interpretation of the impacts of action alternatives with reference to the user's needs, values, and institutional perspectives.

The key words in this *structural definition of systems engineering* are formulation, analysis, and interpretation. In fact, all of systems engineering can be thought of as consisting of formulation, analysis, and interpretation of the various ingredients at what we call phases in the life cycle of a system. These activities

Formulation,

Analysis,

Interpretation

are very important and may also be thought of as aids to better understanding of system behavior. We may exercise these in a formal sense or in an *as if* or experientially based intuitive sense. They may be disaggregated into a larger number of related activities, as indicated later. These steps are the components comprising one dimension of a framework for systems engineering.

To resolve large-scale and complex problems or manage large systems, we must be able to deal with issues that involve the following:

1. Many considerations and interrelations;
2. Many different and perhaps controversial value judgments;
3. Knowledge from several disciplines;
4. Risks and uncertainties involving future events that are difficult to predict;
5. Fragmented decision-making structures;
6. Needs perspectives and value perspectives as well as technology perspectives;
7. Resolution of issues at the level of institutions and values as well as the level of symptoms.

To be truly useful, the professional practice of systems engineering must promote the development and use of aids for both technological and management systems design. These provide the support that allow clients to cope with multifarious large-scale issues. In addition to the many things that need to be done to accomplish a systems study, or to build and operationalize a large system, there are several potential pitfalls that need to be avoided. Some of these are as follows:

1. There may be an overreliance upon a specific analytic tool or technology that is advocated by a particular group.
2. There may be considerations of perceived needs and problem resolution activities only at the level of symptoms but not at the level of institutions or values.
3. There may be a failure to develop and apply an appropriate methodology for issue resolution, and associated system fielding, that allows:
   a. Identification of major issues;
   b. Identification and exposure of interactions among steps of the problem solution procedure;
   c. Identification of the structure of the situation at hand as an inherent and integral part of the problem-solving approach.
4. There may be a failure to involve the client, to the extent necessary, in the development of problem resolution, or system design, alternatives.
5. There may be a failure to consider the effects of cognitive biases that result from poor information-processing heuristics.
6. There may be a failure to identify a sufficiently robust set of options or alternative courses of action.
7. There may be a failure to identify risks associated with the costs and benefits (or effectiveness) of the system to be fielded. These risks may originate from the technologies themselves, from elements of the organization or the environment, or from considerations that affect the information that is being processed.
8. There may be a failure to properly relate the system that is designed and implemented to the cognitive style and behavioral constraints that affect the user of the system. In other words, system design for human interaction [2] is not given appropriate consideration.
9. Improper attention may be paid to the needs for reliability, availability, and maintainability. This leads to dysfunctional systems that have a shorter than desired life.
10. There may be a failure to integrate the system under development with the existing systems in the environment in which the new system is to operate. The needed proactive support from top management in the client or stakeholder group may not exist. Changes in the organization and environment that may result from implementation of the new system may not have been identified.

The result of these failures is that the objectives for the new system may not be identified or clearly understood, an immature set of alternatives may be considered, an ineffective set of tools for analysis and design may be used, the use of the system may only relieve symptoms as contrasted with resolution of the issue at hand, there may be inadequate financial and human resources to fully implement the system, and the impacts of the system upon the organizational culture and environment may produce unintended results. In other words, the formulation, analysis, or interpretation is defective.

1.2. THE EMERGENCE OF SYSTEMS ENGINEERING

Throughout history, the development of more sophisticated tools has often been associated with a decrease in dependence on human physical energy as a source of effort. Generally, this is accomplished by control of nonhuman sources of energy in an automated fashion. Often, these involve intellectual and cognitive effort. The Industrial Revolution [3] represented a major thrust in this direction.

In most cases, a new tool or machine makes it possible to perform a familiar task in a somewhat new and different way, typically with enhanced efficiency and effectiveness and sometimes with increased understandability as well. In a smaller number of cases, a new tool has made it possible to do something entirely new and different that could not be done before. Concerns associated with the *design* of tools such that they can be used efficiently and effectively has often been addressed on an implicit "trial and error" basis. When tool designers were also tool users, which was more often than not the case for the simple tools and machines of the past, the resulting designs were often good initially or soon evolved into good designs.

When physical tools, machines, and systems became so complex that it was no longer possible for a single individual to design them, and a design team was then necessary, a host of new problems emerged. This is the condition today. To cope with this, a number of methodologies associated with systems design engineering have evolved. Through these, it has been possible to decompose large design issues into smaller component subsystem design issues, design the subsystem, and then build the complete system as a collection of these subsystems.

Even so, problems remain. Just simply connecting together the individual subsystems often does not result in a system that performs acceptably, either from a technological efficiency perspective or from an effectiveness perspective. This has led to the relization that systems integration engineering and systems management throughout an entire system life cycle are necessary. Thus, contemporary efforts in systems engineering focus on tools and methods, on the system design methodology that promotes appropriate use of these tools, and on systems management approaches that permit the imbedding of design approaches within organizations and environments. The use of appropriate tools, as well as systems methodology and management constructs, enables *system design for more efficient and effective human interaction* [2].

1.3. FROM COMPUTER TECHNOLOGY TO INFORMATION TECHNOLOGY

Sometime around the middle of this century, the use of a new type of machine became widespread. This machine was fundamentally different from the usual combination of motors, gears, pulleys, and other physical components that assisted humans, perhaps in an automated fashion, in performing such physical tasks as pulling or even flying. Although this new machine could assist in performing functions associated with physical tasks (e.g., computing the optimal trajectory for an aircraft to move between two locations with minimum energy consumption and cost), it could also assist humans in a number of primarily *cognitive* tasks. These include planning, resource allocation, and decision making. Of course it can assist in many other productive support ventures, such as enterprise management.

It is doubtlessly correct to say that the modern stored-program digital computer is a product of the physical and material sciences, as its components are physical entities. It is also important, even more important for our efforts here, to note that the digital computer is intended to provide assistance to users through more appropriate use of information and associated knowledge. Thus, the purpose of the digital computer is more cognitive and intellectual support than it is physical support. It is perhaps somewhat confusing to dwell on the point that computers often support physical activity, such as flying an airplane, navigating a ship, or whatever. Computers are obviously used in this manner; it is also true that an enormous amount of computer effort supports purely symbolic activity, such as managing financial accounts, organizing commercial data bases, and searching data for underlying conceptual patterns. Perhaps the first use of computers was in World War II to break German radio transmission codes. This is certainly a symbolic rather than a physical activity. The real point here is that the computer, while a product of the physical and material sciences, is primarily a support tool for humans in the performance of cognitive and physical tasks.

The computer is an *information machine*, a *knowledge machine*, and a *cognitive support machine*. It has led to the growth of a new engineering area of endeavor, which involves information and knowledge technology. This new professional area is broadly concerned with efforts whose structure, function, and purpose are associated with the acquisition, representation, storage, transmission, and use of data that is of value for typically cognitive support, but which often results in human supervisory control of some sort of physical effort.

Associated with this are a plethora of information technology products and services that have the potential to affect profoundly the lives of each of us. Clearly, this is happening now, and the rate at which these changes will occur is surely going to accelerate. Information technology products and services based on computers and communications, such as telecommunications, command and control, automated manufacturing, electronic mail, and office automation, are common words today. Computer-aided everything will perhaps be a common term tomorrow. The results could be truly exciting: electronic access to libraries and shopping services; individualized and personalized systems of interactive instruction; individualized design of aids to the disabled and handicapped; and prediction and planning in business, agriculture, health, and education. This has led to a fundamental change in the way in which systems engineering is accomplished. Some now call the field *computer-aided systems engineering* [4].

Thus, we see that the physical and material science basis for engineering and technology is now augmented by an information science basis made possible primarily through the development of the modern microprocessor, and associated computers. Computer usage in systems engineering efforts has become ubiquitous. In general, computer systems may be used as

Imbedded system,
Mission critical system,
Information system

components within a larger system. Imbedded system components are essential: without them, the overall system is totally dysfunctional. If a critical system element is missing, the operation of the overall system will be dysfunctional. However, a missing information system element will not necessarily make system operation impossible. Nonetheless, information system elements may, if effectively and efficiently used, greatly increase the benefits of system operation while reducing the costs of system design and development.

1.4. THE NEED FOR AND DEFINITIONS OF SYSTEMS ENGINEERING

System management and integration issues are of major importance in determining the effectiveness, efficiency, and overall functionality of system designs. To achieve a high measure of functionality, it must be possible for a system design to be efficiently and effectively produced, used, maintained, retrofitted, and modified throughout all phases of a system design and development life cycle. This life cycle begins with conceptualization and identification of needs, moves through specification of system requirements through and with architectural structures, to the production of a system, to system installation and evaluation, and ends with operational implementation or deployment and associated evolutionary maintenance. We call this overall cycle the *systems engineering life cycle*.

In reality, there are many difficulties associated with the production of functional, reliable, and trustworthy systems of large scale and scope. There are many studies that indicate deficiencies in systems engineering products, such as the following

1. Large systems are expensive.
2. System capability is often less than promised and expected.
3. System delivery is often quite late.
4. Large-system cost overruns occur often.
5. Large-system maintenance is complex and error prone.
6. Large-system documentation is inappropriate and inadequate.
7. Large systems are often cumbersome to use, and system design for human interaction is generally lacking.
8. Individual subsystems often cannot be integrated.
9. Large systems often cannot be adapted to a new environment or modified to meet evolving needs.
10. Large-system performance is often unreliable.
11. Large systems often do not perform according to specifications.
12. System requirements often do not adequately capture user needs.
13. Unanticipated risks and hazards often materialize.
14. The system is of low quality.

Many potential difficulties emanate from these deficiencies. Among these are inconsistent, incomplete, or otherwise imperfect identification of user requirements for a system and associated technical specifications; system requirements that do not provide for change as user needs evolve over time; and poorly defined management structures for product design and delivery. These lead to products that are difficult to use, that do not solve the intended problem, that operate in an unreliable fashion, that are unmaintainable, and that as a result are not used. These same studies generally show that the major problems associated with the production of reliable systems have more to do with the *organization and management of complexity* than with direct technological concerns that affect individual subsystems and specific physical science areas.* A major objective of systems engineering is to improve effectiveness, including availability, reliability, maintainability, quality, and trustworthiness and at the same time to reduce costs.

These benefits are directly associated with the quality of systems design and management in the systems engineering process. A key way to improve systems engineering products is to improve systems engineering processes.

---

*Contemporary studies indicate, for example, that approximately 90% of the cost of many current systems is for software and roughly 65% of maintenance monies are spent for software maintenance; hardware concerns cannot be ignored, nor can the needed integration of hardware with software.

Figure 1.3. Three Fundamental Levels of Systems Engineering

This book has, therefore, a strong process orientation. We envision a hierarchy of performance levels for systems engineering efforts, as shown in Figure 1.3.

> *Systems engineering* is the design, production, and maintenance of trustworthy systems within cost and time constraints.

Our discussion of systems engineering is assisted by definitions of the structure, purpose, and function of systems engineering. Table 1.1 presents these.

Each of these definitions is important for our discussion. The functional definition of systems engineering points to our concern with the various tools and techniques that enable us to design systems. Often, these are *systems science and operations research* tools for analysis of systems themselves. Specific design of systems may involve a diversity of physical and material science tools, as well. Also, the functional definition notes our concern with a

TABLE 1.1   Definitions of Systems Engineering

Structural Definition   Systems engineering is management technology to assist clients through the formulation, analysis, and interpretation of the impacts of proposed policies, controls, or complete systems upon the perceived needs, values, institutional transactions of stakeholders.

Functional Definition   Systems engineering is an appropriate combination of theories and tools, carried out through use of a suitable methodology and set of systems management procedures, in a useful setting appropriate for the resolution of real-world problems that are often of large scale and scope.

Purposeful Definition   The purpose of systems engineering is information and knowledge organization that will assist clients who desire to develop policies for management, direction, control and regulation activities relative to forecasting planning, development, production and operation of total systems to maintain overall integrity and integration as related to performance and reliability.

combination of these tools. We denote the effort to obtain this combination *systems methodology and design*. Finally, the definition implies that we aim to accomplish this in a useful and appropriate setting. We use the term *systems management* to refer to the cognitive tasks necessary to produce a useful process from a systems methodology and design study. This product is an appropriate combination of systems science and operations research methods, with suitable technical direction and the leadership associated with systems management for resolving issues. Each of the three functional levels shown in Figure 1.3 is important. None can be safely neglected.

The structural definition focuses on our need for a framework for problem resolution. From a formal perspective at least, this consists of three fundamental steps:

Issue formulation,

Issue analysis,

Issue interpretation.

A detailed discussion of the framework that leads to these three steps and a number of smaller steps into which they may be decomposed can be found in the next chapter.

System design is a major subject for systems engineering.* Regardless of the way in which the design process is characterized and regardless of the type of process or system that is being designed, all characterizations will necessarily involve the following [5–9]:

1. *Formulation of the Design Problem:* The needs and objectives of a client group are identified, and potentially acceptable design alternatives, or options, are identified or generated.

2. *Analysis of the Alternative Designs:* The impacts of the identified design options are evaluated.

3. *Interpretation and Selection:* The impacts of the design alternatives are compared with one another. The needs and objectives of the client group are used as a basis for evaluation. The most acceptable alternative is selected for implementation or further study in a subsequent phase of the systems engineering life cycle.

Our model of the systems process, shown in Figure 1.4, is based upon this conceptualization. As we have indicated here, these three steps can be disaggregated into a number of others.

---

*Often, the term *systems design engineering* is used in a very broad context to indicate the entire systems engineering process. Our definition and interpretation of design is quite broad also, although not this all inclusive.

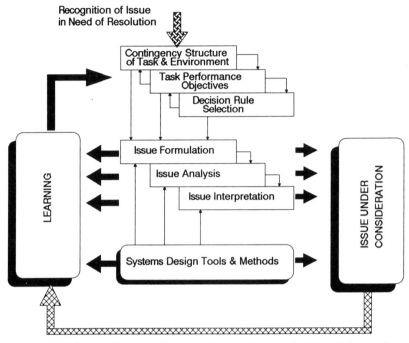

Figure 1.4. The Formal Process of Systems Engineering as a Problem Solving Effort

Without question, this is a formal model of the way in which design is accomplished. Within this formal framework, there is the need for much iteration as the actual process of analysis and evaluation unfolds. Also, this description does not emphasize the key role of information and information requirements determination. Although requirements determination must be the first phase in any useful systems engineering life cycle, there are requirements issues associated with the formulation step in all phases of the life cycle [10] and in all application areas. This role is strongly emphasized in software systems engineering [11, 12].

More importantly, this morphological* framework, in terms of phases of the design process and steps within these phases, does not emphasize the different types of information and different types of support that are needed within each step at the various phases. During the issue formulation step, the support that is needed tends to be of an affective, perceptive, or gestalt nature. Intuition-based experiential wisdom plays a most important role. During the analysis step, the need is typically for quantitative and algorithmic

---

*Morphology was originally used in a biological or chemical context to mean study of organic form. It is a relatively common word in systems engineering and simply denotes *form*.

support, typically through use of one of the formal methods of systems engineering, a few of which are described in later chapters. In the interpretation step, the effort shifts to a blend of the perceptive and the analytical.

Even when these realities are associated with a (morphological) framework for design, it still represents an incomplete view of the way in which people do, could, or should accomplish design, planning, or other problem-solving activities. The most that can be argued is that this framework is correct in an "as if" manner. It is a morphological box that consists of a number of phases and steps. There is also a third dimensional variable, effort level, and others could doubtlessly be identified as well.

A number of questions may be posed with respect to formulation, analysis, and interpretation that clearly indicate the role of values in every portion of a systems engineering effort. Although there are a vast number of specific questions that may be posed, there are a number of generic questions that are independent of the particular life-cycle phase being considered.

Issue formulation questions of importance in this regard are as follows:

What is the problem? The needs? The constraints? The alterables?

How do the stakeholder group and clients to the study bound the issue?

How does the systems engineering team bound the issue, and is this different from the way in which the stakeholder group bounds it?

What objectives are to be fulfilled?

What alternative options are appropriate?

How are the alternatives described?

What alternative environmental scenarios are relevant to the issue?

Analysis questions of importance are as follows:

How are pertinent problem or issue variables selected?

How is the issue formulation disaggregated for analysis?

What generic outcomes or impacts are relevant?

How are outcomes and impacts described across various societal sectors?

How are risks and uncertainties described?

How are ambiguities and other information imperfections described?

How are questions of planning period and planning horizon dealt with?

Interpretation concerns of importance in a systems engineering study are as follows:

How are values and attributes disaggregated and structured?

Does the technological effort involved in value and attribute structuring augment, replace, or obfuscate intuitive affect or experiential familiarity with the issue?

How are flawed judgment heuristics and cognitive information processing biases dealt with?

Are value perspectives altered by the systems engineering process itself?

A final question might be asked. *How is total issue resolution time at any specific phase in the systems engineering life cycle divided between the activities involved in formulation, analysis, and interpretation?* This is important because the allocation of resources to various systems engineering activities may reflect a number of value perspectives of the clients or stakeholders who support a study and the systems engineers who perform it.

The professional practice of systems engineering must be associated with an awareness of appropriate methods, tools, and techniques (systems science and operations research methods), appropriate design and development approaches (systems methodology and design), and appropriate human judgment at the product level and process level for organization of the systems engineering management process (systems management). Thus, the three functional components of systems engineering are each necessary to make effective resource allocations. There are other ingredients in the systems engineering environment and Figure 1.4 shows an expanded conceptual model of the complete systems engineering process.

The term *methodology* is sometimes misused, even in systems engineering. As we use it, a methodology is an open set of procedures for problem solving. Consequently, a methodology involves a set of methods, a set of activities, and a set of relations between the methods and the activities. To use a methodology we must have an appropriate set of methods. (Strictly speaking it is *not* correct to use methodology as a synonym for method. Sadly, it is not uncommon to do this, even in systems engineering.) Generally, the methods are provided by systems science and operations research and include a variety of qualitative and quantitative approaches from a number of disciplines. Associated with a methodology is a structured framework in which particular methods are employed for resolution of a specific issue.

Systems engineering efforts are also concerned with technical direction and management of the process of systems development, including identification of requirements, production of the system, and maintenance of the system. In adopting the process management technology of systems engineering and applying it, we become very concerned with making sure that correct systems are designed and not just that system design products are correct according to some potentially ill-conceived notions of what the system should do. Appropriate tools to enable efficient and effective error prevention and detection in the systems design process enhances the production of correct system design products.

To ensure that correct systems are designed requires that considerable emphasis be placed on the front end of the systems life cycle. In particular, there needs to be considerable emphasis on the accurate definition of a system, what it should do, and how people should interact with it before it is produced. In turn, this requires emphasis upon conformance to system requirements specifications and the development of standards to insure compatibility and integratability of system products. Such areas as documentation and communication are important. Thus, we see the need for the technical direction and management technology efforts that comprise the systems engineering process.

1.5. A LIFE-CYCLE METHODOLOGY FOR SYSTEMS ENGINEERING

The primary goal of systems engineering is the creation of a set of operational products to enable the accomplishment of desired tasks that fulfill identified needs of a client or stakeholder group. This is primarily and fundamentally an engineering task and a design task.

1.5.1. Systems Design

Design is the creative process through which system products, presumed to be responsive to client needs and requirements, are conceptualized or specified, implemented, and maintained. It is clear that this definition of design makes it inseparable from the overall life-cycle process of systems engineering. For even though specific design activities may not include such early systems engineering efforts as requirements identification or such late efforts as maintenance, an appropriate design must be responsive to and supportive of the other phases in the systems engineering life cycle. There are four primary ingredients in our definition of design, and they apply to software design as well as to the design of hardware and physical systems.

1. Design results in specifications or architecture for a product, process, or system.
2. Design is a creative process.
3. Design activity is conceptual in nature.
4. A successful design must be broadly responsive to client needs and requirements.

Good systems engineering practice requires that the systems engineer be responsive to each of these four ingredients for quality design efforts. The final ingredient requires of the client a set of needs and requirements for the desired product, process, or system. This information requirement serves as the input to the systems engineering process that leads to design. Systems engineering is creative, and it is a process that is conceptual and pragmatic in nature. The initial result of this creative, conceptual, and pragmatic process is information concerning the specifications or architecture for the product or service that will ultimately be manufactured, implemented, installed, or

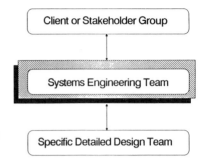

Figure 1.5. The Systems Engineering Team as a Broker and Management of Information

brought to fruition in some other way. A later result is the system itself and plans for its evolution over time.

In the detailed efforts that follow, we first discuss systems engineering methodology and systems engineering life cycles as necessary elements in the systems engineering process. This leads to a number of conceptual frameworks for the process of systems engineering. From this perspective, systems engineers provide an interface between the client or stakeholder group to whom an operational system will ultimately be delivered and a *detailed design* group who are responsible for specific system production and implementation. Figure 1.5 illustrates this view of systems engineering as an interface group that provides conceptual design and technical direction to enable the products of a detailed design group to be responsive to client needs. There are, of course, other views of systems engineers and systems engineering functions.

It is, hopefully, clear from our discussion of the closely interrelated nature of systems engineering processes and products that there are a variety of knowledge frameworks or perspectives that are necessary for systems engineering. These include, but are by no means limited to, the detailed design aspects of both hardware and software design. Along with other substantive issues, we must be concerned with characterization and understanding of the thought process of detail designers in organizing information about design, including the acquisition, representation, and use of this information. We must pay particular attention to the requirements for successful information support to designers and aids that support design processes for the production of high-quality, reliable systems. This concern naturally raises important issues about the knowledge base for design and how this knowledge base might best be employed in a system that supports decision and design processes. To accomplish this, both analytical and perceptual capabilities are required, and the knowledge base and model base used should be able to support a purposeful interplay of these capabilities. This may involve fundamental constructs in decision support systems engineering [13].

Clients or stakeholders and designers each have needs and requirements that must be satisfied by the results of a successful systems engineering effort.

Information requirements are, therefore, multifaceted. In a very real sense, determination of those requirements is the most important aspect of a successful systems engineering effort. A systems engineering product that fails to meet user requirements is likely to be seen as a failure, whatever its other merits.

1.5.2. Systems Engineering Methodology

There are many ways in which we can characterize systems engineering. We could describe systems engineering as an activity involving iterative hypothesis generation and test of alternatives or concepts. The hypothesis step involves primarily inductive skills, generally based upon experience, that enable generation of design alternatives. These are evaluated through the primarily deductive activity of evaluation or testing. An initial hypothesis is often not acceptable. When it is rejected, generally through evaluation, iteration back through the hypothesis generation step enables modification of the hypothesis, leading, hopefully, to a successful alternative.

More often than not, the hypothesis generation and option evaluation efforts are first conducted in a preliminary way to obtain several concepts that might work. Several potential options are identified and then subjected to at least a preliminary evaluation to eliminate clearly unacceptable alternatives. The surviving alternatives are next subjected to more detailed design efforts, and more complete architectures or specifications are obtained. Again, these are evaluated, and a final choice is made, which can be developed with detailed design testing and at least preliminary operational implementation. Once this has occurred, operational evaluation and test of the system can occur. The system design may be modified as a result of this evaluation, leading to an improved system and, ultimately, operational imple-

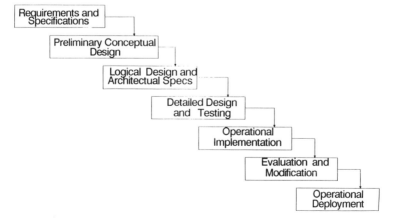

Figure 1.6. Systems Engineering Life-Cycle Phases

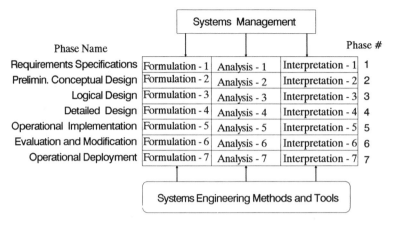

Figure 1.7. Methodological Framework for Systems Engineering

mentation. What we have, then, is a life cycle model of the systems engineering process that consists of a number of phases and a sequence of steps accomplished within each of the phases. Figure 1.6 illustrates a typical sequence of phases.

We have identified a phased system engineering life cycle that consists of seven phases:

1. Requirements and specifications identification.
2. Preliminary conceptual design.
3. Logical design and system architecture specification.
4. Detailed design, production, and testing.
5. Operational implementation.
6. Evaluation and modification.
7. Operational deployment.

These are sequenced in an iterative manner as shown in Figure 1.6. There are many descriptions of systems life cycles and associated methodologies and frameworks for system design and development [5, 8, 14], and we outline only one of them in any detail in this chapter. In general, the overall process is structured as in Figure 1.7, which illustrates an expansion of Figures 1.3 and 1.6 to accommodate the steps, and the phases that are described above. Systems methodology and design is the middle box in this figure and represents a two-dimensional morphological box.

1.6. PHASES IN THE LIFE CYCLE OF SYSTEMS ENGINEERING

The requirements and specification phase of the systems engineering life cycle has as its goal the identification of client or stakeholder needs, activities, and objectives for the functionally operational system. This phase should result in the identification and description of preliminary conceptual design considerations for the next phase. It is necessary to translate operational deployment needs into requirements specifications so that these needs may be addressed by the system design and development efforts. Thus, information requirements specifications are affected by, and affect each of the other design and development phases of the systems engineering life cycle.

As a result of the requirements specifications phase, there should exist a clear definition of design issues such that it becomes possible to make a decision concerning whether to undertake preliminary conceptual design. If the requirements specifications effort indicates that client needs can be satisfied in a functionally satisfactory manner, then documentation is typically prepared concerning specifications for the preliminary conceptual design phase. Initial specifications for the following three phases of effort are typically also prepared, and a concept design team is selected to implement the next phase of the life-cycle effort.

Preliminary conceptual system design typically includes, or results in, an effort to specify the content and associated architecture and general algorithms for the system product in question. The primary goal of this phase is to develop conceptualization of some sort of prototype that is responsive to the specifications identified in the previous phase of the life cycle. A preliminary conceptual design, one that is responsive to user requirements for the system and associated technical system specifications, should be obtained. Rapid prototyping of the conceptual design is clearly desirable for many applications as one way of achieving an appropriate conceptual design.

The desired product of this phase of activity is a set of detailed design and architectural specifications that should result in a useful system product. There should exist a high degree of user confidence so that a useful product will result from detailed design, or the entire design effort should be redone or abandoned. Another product of this phase is a refined set of specifications for the evaluation and operational deployment phases of the life cycle. In the third phase, these are translated into detailed representations in logical form such that system development may occur. A product, process, or system is produced in the fourth phase of the life cycle. This is not the final system design, but rather the result of implementation of the design that resulted from the conceptual design effort of the last phase. User guides for the product should be produced such that realistic operational test and evaluation can be conducted.

Evaluation of the detailed design and the resulting product, process, or system is achieved in the sixth phase of the systems engineering life cycle.*

---

*Depending upon the specific application being considered, the whole systems engineering life-cycle process could be called "design," "manufacturing," or some other appropriate designator, as we have already noted. System acquisition is an often used word to describe the entire systems engineering process and the resultant systems engineering product.

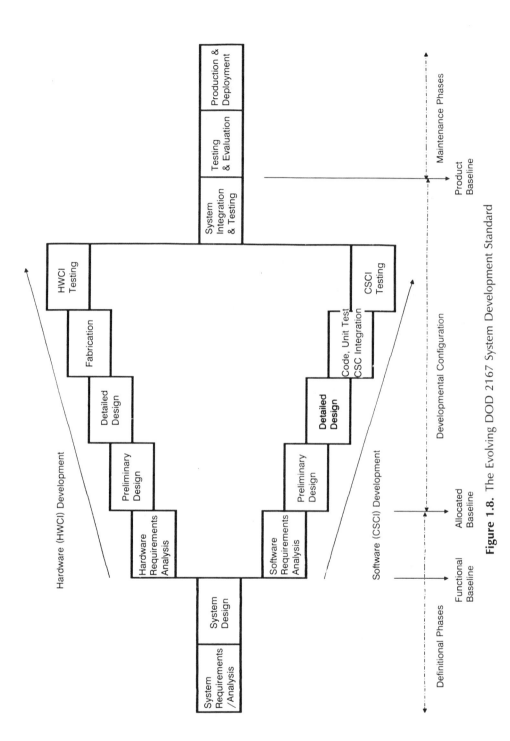

Figure 1.8. The Evolving DOD 2167 System Development Standard

Preliminary evaluation criteria are obtained as a part of requirements specifications and modified during the following two phases of the design effort. The evaluation effort must be adapted to other phases of the design effort such that it becomes an integral and functional part of the overall design process. Generally, the critical issues for evaluation are adaptations of the elements present in the requirements specifications phase of the design process. A set of specific evaluation test requirements and tests are evolved from the objectives and needs determined in the requirements specifications. These should be such that each objective measure and critical evaluation component can be measured by at least one evaluation test instrument.

If it is determined, perhaps through an operational evaluation, that the resulting systems product cannot meet user needs, the system life-cycle process reverts iteratively to an earlier phase, and the effort continues. An important by-product of system evaluation is determination of ultimate performance limitations for an operationally realizable system. Often, operational evaluation is the only realistic way to establish meaningful information concerning functional effectiveness of the result of a systems engineering effort. Successful evaluation is dependent upon having predetermined explicit evaluation standards.

The last phase of the systems life-cycle effort includes final acceptance and operational deployment. Maintenance and retrofit can be defined either as additional phases in life cycles, or as part of the operational deployment phase. Either is an acceptable way to define the system life cycle. There have been many proposed systems engineering life cycles. We discuss some of these throughout the book, especially in Chapters 2 and 3.

There are advantages to developing and accepting standards for the system life cycle. This, for example, is what the U.S. Department of Defense is attempting to do with its standard 2167 [15] for software design. Figure 1.8 illustrates both the hardware and the software phases of this evolving standard. Standards are very important in the systems engineering process, and will be discussed at appropriate points in this text.

1.7. OBJECTIVES FOR SYSTEMS ENGINEERING

There are many goals that could be stated for systems engineering. The following are among the most important such goals.

1. The systems engineering process should encompass all phases of the system life cycle, including transitioning between phases.
2. The systems engineering process should support problem understanding, and communication among all interested parties at all phases in the process.
3. The systems engineering process should enable capture of design and implementation needs for the systems engineering product early in the

life cycle, generally as part of the requirements specifications and conceptual design phases.

4. The systems engineering process and associated methodology should support both bottom-up and top-down approaches to systems design and development.

5. The systems engineering process should enable an appropriate mix of design, development, and systems management approaches.

6. The systems engineering process should support quality assurance of both the product and the process that leads to the product.

7. The systems engineering process should support system product evolution over time.

8. The systems engineering process should be supportive of appropriate standards and management approaches for configuration management.

9. The systems engineering process should support the use of automated aids for design that assist in the production of high-quality trustworthy systems.

10. The systems engineering process should be based upon a methodology that is teachable and transferable and one that makes the process visible and controllable at all life-cycle phases of development.

11. The systems engineering process should have appropriate procedures to enable definition and documentation of all relevant factors at each phase in the system life cycle.

12. The systems engineering process should support operational product functionality, revisability, and transitioning, both at the initial time of operational implementation and later at the time that a system is phased out of service or retired.

13. The systems engineering process must support both system product development and the system user organizations; it must also be compatible with the environments associated with systems development and operation so as to obtain sustainable system developments.

When all of these are accomplished, it will be possible to produce operational systems that are economical, reliable, available, verifiable, interoperable, integratable, portable, adaptable, evolvable, comprehensible, maintainable, manageable, and cost effective, and that lead to a high degree of user satisfaction. These would seem to represent attributes for metrics or to be translatable into attributes for metrics that can measure the quality of an operational systems engineering product. They can be translated into standards with which to measure system performance and systems engineering process effectiveness. Together with cost information, this allows us to obtain cost and operational effectiveness of systems engineering products.

1.8. OTHER RELATED EFFORTS

Much work has been devoted to systems engineering over the past 25 years. The first available text was probably that of Hall [16], who has published a good bit of the early seminal work in systems engineering [17], as well as a recent text [18] that has a philosophical and reflective flavor.

The English school has produced several texts [19, 20] that emphasizes what is called *soft systems methodology*. While conceptually similar to the methodology portion of this book, the soft systems methodology works do not seem to have gained favor in the United States. Possibly this is because there seems to have been little effort to extend these approaches into metrics and systems management areas.

Nadler has been very active in industrial engineering efforts that emphasizes systems engineering, and he has produced a definitive text in this area [21], as well as related papers. Eisner [3] has written a recent text called *Computer Aided Systems Engineering*. Although it is a very respectable work in every way, it is more concerned with the tools and methods for systems engineering and does not focus on systems management concerns to the extent done here.

A recent text by Beam [22] is concerned with pragmatic systems engineering approaches that are directly relevant to detailed design considerations in communication and computer systems. Another recent work on systems architecture and systems design [23] is concerned with the system level details of telecommunication system design. A recent book on systems architecture by Rechtin [24] provides a number of excellent discussions on systems level architecture for large communication and computer systems, including technical direction aspects.

Blanchard [25] focuses on engineering management practices in his recent effort in this area. Design-to-cost is the primary subject in the engineering management text of Michaels and Wood [26], which includes an excellent compendium of recent approaches in this specialized area of systems management. Rouse [27] addresses systems management and systems design issues in his excellent work, but more from the perspective of a systems engineer with interest in human-system interaction issues.

Workers in the field of software engineering have discovered systems engineering and a not inconsiderable number of very useful systems engineering developments are occurring there. This was a major motivating force behind our recent book [11] in this area. There are other related books in the software area which take a systems approach. Some have such systems engineering like names as *Introduction to Systems Analysis and Design* [28], *Managing the Systems Lifecycle* [29], and *A Unified Methodology for Developing Systems* [30], even though the dominant concern in all of these is software.

There are a number of books in the field that are collections of papers. Typical of these are two recent works on systems design, one edited by Rouse and Boff [31] and the other by Newsome, Spillers, and Finger [32]. There are also works in systems design [33], systems management [34], systems engineering processes [2, 6], and related topics.

There is a plethora of company and government reports that deal with this subject [35], especially the more pragmatic aspects of the subject. There are also a number of older books, generally published 20 or more years ago, that have been written from a variety of perspectives [36–40]. Often the perspective taken would now be regarded as the mathematical theory of systems engineering methods, which is not the focus of this text. Warfield, in 1976, published [41] a definitive work relating systems engineering to societal issues. Despite the value of many of these works when first published, many are now out-of-print and available only in selected libraries.

So, we see that we are indeed dealing with a subject that has attracted considerable scholarly attention in the past, at this time [42], and doubtlessly in the future as well.

PROBLEMS

1.1. Consider a systems engineering development issue with which you are familiar or one that is of interest to you. For this systems engineering development effort:

  a. Review the maladies of systems engineering development detailed on page 5. Discuss the extent to which these were problems for the system you are considering. It might be convenient to use some yardstick, such as 0 = no problem at all and 1 = major problem.

  b. Define the phases actually used to develop the system that you are considering. Contrast and compare these phases with the seven-phase methodology for systems engineering described here.

1.2. Reconsider your effort in Problem 1.1. To what extent are the various scales that you have used in this problem subjective? How would you cope with the fact that someone might consider the same system that you have examined and come up with different scores? In what ways would this difference be significant?

1.3. Using Problem 1.1, redefine the effort that actually was undertaken in the system design that you are considering according to the seven-phase effort of the systems engineering life cycle defined in Figure 1.6. If this appears to be a cumbersome set of phases for the effort that you are considering, consider a three-phase effort that involves *definition*, *development*, and *maintenance*. Then decompose any of these phases that involved significant activity into a number of smaller phases. You have just defined your own system life cycle. How does your system life cycle contrast and compare with the one that we have identified in this chapter?

1.4. From a recent newspaper or magazine, select what appears to be a well-written article describing a technological issue of some importance to you and with which you are reasonably familiar. Discuss formulation, analysis, and interpretation of the issue from a systems engineering perspective. Please attempt to identify deficiencies in the article from this perspective.

1.5. The word *system* is used in many contexts. Among these are system of equations, control system, and systems engineering. Provide a definition of the term system that is appropriate to each of these uses.

1.6. Discuss the commonalities and differences among civil engineering, electrical engineering, and systems engineering.

1.7. How does systems engineering differ from philosophy? What are the similarities?

REFERENCES

[1] Sage, A. P., "Knowledge Transfer: An Innovative Role for Information Engineering Education," *IEEE Transactions on Systems, Man and Cybernetics*, Vol. 17, No. 5, 1987, pp. 725–728.

[2] Sage, A. P. (Ed.), *System Design for Human Interaction*, IEEE Press, New York, 1987.

[3] Beniger, J. R., *The Control Revolution: Technological and Economic Origins of the Information Society*, Harvard University Press, Cambridge MA, 1986.

[4] Eisner, H., *Computer Aided Systems Engineering*, Prentice Hall, Englewood Cliffs, NJ, 1987.

[5] Sage, A. P., *Methodology for Large Scale Systems*, McGraw Hill, New York, 1977.

[6] Sage, A. P. (Ed.), *Systems Engineering: Methodology and Applications*, IEEE Press, New York, 1977.

[7] Sage, A. P., "Behavioral and Organizational Considerations in the Design of Information Systems and Processes for Planning and Decision Support," *IEEE Transactions on Systems, Man and Cybernetics*, Vol. 11, No. 9, 1981, pp. 640–678.

[8] Sage, A. P., "A Methodological Framework for Systematic Design and Evaluation of Computer Aids for Planning and Decision Support," *Computers and Electrical Engineering*, Vol. 8, No. 2, 1981, pp. 87–102.

[9] Sage, A. P., "Methodological Considerations in the Design of Large Scale Systems Engineering Processes," in Y. Y. Haimes (Ed.), *Large Scale Systems*, North Holland, New York, 1982, pp. 99–141.

[10] Sage, A. P., Galing, B., and Lagomasino, A., "Methodologies for the Determination of Information Requirements for Decision Support Systems," *Large Scale Systems*, Vol. 5, No. 2, 1983, pp. 131–167.

[11] Sage, A. P. and Palmer, J. D., *Software Systems Engineering*, Wiley, New York, 1990.

[12] Davis, A. M., *Software Requirements Analysis and Specifications*, Prentice Hall, Englewood Cliffs NJ, 1990.

[13] Sage, A. P., *Decision Support Systems Engineering*, Wiley, New York, 1991.

[14] Nadler, G., "Systems Methodology and Design," *IEEE Transactions on Systems, Man and Cybernetics*, Vol. 15, No. 6, 1985, pp. 685–697.

[15] U.S. Department of Defense, *Defense System Software Development*, DOD-STD-2167, June 1985.

[16] Hall, A. D., *A Methodology for Systems Engineering*, Van Nostrand, New York, 1962.

[17] Hall, A. D., "A Three Dimensional Morphology of Systems Engineering," *IEEE Transactions on System Science and Cybernetics*, Vol. 5, No. 2, 1969, pp. 156–160.

[18] Hall, A. D., *Metasystems Methodology: A New Synthesis and Unification*, Pergamon Press, Oxford UK, 1989.

[19] Checkland, P. B., *Systems Thinking, Systems Practice*, Wiley, Chichester, 1981.

[20] Wilson, B., *Systems: Concepts, Methodologies, and Applications*, Wiley, Chichester, 1984.

[21] Nadler, G., *The Planning and Design Approach*, Wiley, New York, 1981.

[22] Beam, W. R., *Systems Engineering: Architecture and Design*, McGraw Hill, New York, 1990.

[23] Chorafas, D. N., *Systems Architecture and Systems Design*, McGraw Hill, New York, 1989.

[24] Rechtin, E. R., *Systems Architecting*, Prentice Hall, Englewood Cliffs NJ, 1991.

[25] Blanchard, B. S., *Systems Engineering Management*, Wiley, New York, 1991.

[26] Michaels, J. V. and Wood, W. P., *Design to Cost*, Wiley, New York, 1989.

[27] Rouse, W. B., *Design for Success: A Human Centered Approach to Designing Successful Products and Systems*, Wiley, New York, 1991.

[28] Hawryszkiewycz, I. T., *Introduction to Systems Analysis and Design*, Prentice Hall, Englewood Cliffs NJ, 1988.

[29] Yourdan, E., *Managing the Systems Life Cycle*, Yourdan Press, New York, 1988.

[30] Wallace, R. H., Stockenberg, J. E., and Charette, R. N., *A Unified Methodology for Developing Systems*, McGraw Hill, New York, 1987.

[31] Rouse, W. B. and Boff, K. R. (Eds.), *System Design: Behavioral Perspectives on Designers, Tools, and Organizations*, North Holland, New York, 1987.

[32] Newsome, S. L., Spillers, W. R., and Finger, S. (Eds.), *Design Theory '88*, Springer Verlag, New York, 1989.

[33] Shakun, M. F. (Ed.), *Evolutionary Systems Design*, Holden Day, San Francisco CA, 1988.

[34] Tinnirello, P. C. (Ed.), *Handbook of Systems Management: Development and Support*, Auerbach Publishers, Boston MA, 1989.

[35] U.S. Department of Defense, Defense Systems Management College, *Systems Engineering Management Guide*, Ft. Belvoir VA, 1986.

[36] Chestnut, H., *Systems Engineering Tools*, Wiley, New York, 1965.

[37] Chestnut, H., *Systems Engineering Methods*, Wiley, New York, 1967.

[38] Good, H. and Machol, R. E., *Systems Engineering*, McGraw Hill, New York, 1957.

[39] Wymore, A. W., *A Mathematical Theory of Systems Engineering*, Wiley, New York, 1967.

[40] Wymore, A. W., *Systems Engineering Methodology for Interdisciplinary Teams*, Wiley, New York, 1976.

[41] Warfield, J. N., *Societal Systems: Planning, Policy, and Complexity*, Wiley, New York, 1976.

[42] Eslaksen, E. and Belcher, R., *Systems Engineering*, Prentice Hall of Australia, Brookvale NSW, 1992.

Chapter **2**

Systems Engineering Processes and Life Cycles

In this chapter, we develop some important notions regarding the process of systems engineering. Basically, a process is a continuous action designed to serve a particular purpose. For a number of reasons, it is desirable to distinguish a number of phases that, together, comprise a systems engineering process. We denote the collection of these phases as the *systems engineering life cycle*. A systems engineering life cycle prescribes a number of phases that should be followed, often in an interactive and iterative manner in order successfully to produce and field a large-scale system that meets user requirements.

There have been a number of process models proposed and used for multiphased life-cycle models in systems engineering. Each of them starts, or should start, by capturing user requirements. These user requirements are then converted to technological system requirements and systems management requirements that will presumably, when satisfied, produce the product or service. Following the requirements phase(s), there is a conceptual, or architectural, design phase, and then a detailed design phase, the result of which is an initial working version of a system. This is evaluated and modified to enable ultimate operational deployment of a functionally useful system, a system that fulfills user requirements. Deployment is followed by a maintenance and modification phase, and potentially other efforts that together describe an extended systems life cycle.

In principal, a complex large-scale system will contain both technological system design and management system design aspects if it is to be realized in an effective and efficient manner, as noted in Chapter 1 and illustrated in Figure 1.2. A system life cycle model may be conceptualized using management or technological system design perspectives or a hybrid of the two. These are not mutually exclusive perspectives, however. In general, we believe that a management system design and development perspective is very much needed, as it leads naturally to an organizational structure for tasks and people. To have only one perspective is to invite, if not to guarantee, failure. Chapter 3 is devoted to a study of the important topic of systems management.

We strongly believe that technology is necessary to ameliorate the problems just delineated. However, there is much evidence [1] that successful application of technology to real-world problem areas must consider the levels of

Symptoms,

Institutions,

Values,

or we will generally be confronted with technological solutions that are looking for problems. Too often, problems are approached only at the level of symptoms: bad housing, inadequate health-care delivery to the poor, pollution, hunger, and so on. Technological fixes are developed, and the resulting hardware creates the illusion that solution of real-world problems requires merely the outpouring of huge quantities of funds. Attacking problems at the level of institutions and organizations would allow the adaptation of existing organizations, as well as the design of new organizations and institutions, to make full and effective use of new technologies. Recent work in technology forecasting and assessment (TF&A), and technology transfer, is directed at this level and we discuss some important concerns related to these in our next chapter. Also of vital importance is the need to deal with problems and issues at the level of values. Serious efforts directed toward resolution of issues of large scale and scope must incorporate the significance of values. Thus, we must be able to identify basic issues in terms of often conflicting human values. These considerations must then be included in system realization efforts to determine fully useful solutions to problems. These realities further justify much emphasis on management system design.

Some necessary ingredients that must exist in order to develop and field large systems, resolve large and complex issues, or manage large systems are the following:

1. A way to deal successfully with issues that involve many considerations and interrelations.
2. A way to deal successfully with issues about which there are far-reaching and controversial value judgments.
3. A way to deal successfully with issues, the solutions to which require knowledge from several disciplines.
4. A way to deal successfully with issues for which future events are difficult to predict.

5. A way to deal successfully with problems in which structural, functional, and purposeful elements of the problem are each given full consideration.

6. A way to deal with issues that are imbedded in uncertain and rapidly changing environments.

7. A way to deal with issues about which the relevant information is very imperfect.

8. A way to deal with issues such that resolution efforts are directed at correction of underlying difficulties and not just relief of symptoms.

Systems engineering is potentially capable of incorporating not only technological perspectives associated with large-scale and large-scope systems, but also institutional and value perspectives. These latter concerns, as well as concerns associated with efficient and effective technical direction, fall into the domain of systems management.

Use of a systems engineering approach for systems development and management is intended to result in a systems engineering development life cycle that addresses the specific system development needs from both technological and management perspectives and that can be tailored to changing requirements and environments. The use of a life cycle, for systems design, development, and implementation,

1. Encourages the identification of what the system is supposed to do.

2. Enhances our ability to establish user, and technological and management system, requirements that are satisfied by the subsequent systems development.

3. Identifies and highlights potentially difficult technological and management system problem areas.

4. Encourages a thorough systematic evaluation of alternative solutions to difficult issues associated with each of the phases in the life cycle.

5. Enables selection of appropriate activities for each phase in the life cycle and enables coordination across these phases.

6. Enables planning for interaction among the various subunits of the system to be fielded, thereby enabling system integration.

7. Encourages development of life-cycle cost and benefit, or operational effectiveness, information.

8. Supports the development of project management strategies throughout the life cycle by enabling project managers to track system acquisition efforts in an accurate manner and to identify the potential for various risks in the design, development, and implementation process.

9. Supports the development of standards and the use of these as disciplines that help insure reliable and trustworthy systems.

10. Supports design, development, and implementation of a reliable, trust-worthy, and high quality systems engineering product that is delivered on time and within budget and one which is sustainable.
11. Supports effective systems management, or management control [2], by enabling the institutions and organizations to be better structured and more manageable.

The use of an appropriate life-cycle model as part of the systems engineering process is essentially equivalent to placing appropriate management controls over the various functions to be performed in producing the desired product or service. As with any formal approach, the use of any specific systems engineering life cycle has advantages and disadvantages compared with another life cycle. However, it is safe to state that the use of an appropriate life-cycle model for design, development, and implementation of a system of large scale and scope generally results in a far better solution than would an approach with no systems management, or management controls, at all. In this chapter, we discuss some systems engineering life cycles, including their advantages and disadvantages.

Thus, this chapter, and the entire text, are concerned with the systems engineering process for major systems and its influences on life-cycle costs and system effectiveness, identification of the structure for an appropriate systems development environment and life cycle, how constituent activities within this life-cycle environment interact to produce an overall product, and ways in which various technological tools and techniques may improve systems development productivity within a, more or less, traditional and specified life cycle. First, we discuss a variety of different systems engineering life cycles. We address also ways in which the phases within the life cycle can be adjusted to take advantage of opportunities to increase the productivity and trustworthiness of systems. Finally, we briefly examine steps within the phases.

2.1. A THREE-PHASE AND A TWENTY TWO-PHASE LIFE CYCLE

In this section, we describe both a very compressed life cycle and a somewhat extended one. In each case, we are concerned with the description of steps and phases in a typical life cycle for acquisition of a major system. Our efforts here are primarily concerned with phases, or macrolevel efforts, in the life cycle. The steps, or microlevel efforts, within each phase are important, but of somewhat secondary concern to the phases themselves. We have more to say about the steps in Section 2.4.

A typical system engineering, or systems acquisition, life cycle might be expected to be comprised of three basic phases as we have discussed and as

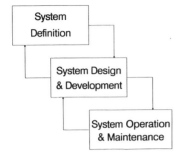

Figure 2.1. A Three-Phase Systems Engineering Life-Cycle Model

indicated in Figure 2.1:

System definition,
System design and development,
System operation and maintenance.

For large systems, these three phases need expansion into a number of more finely defined phases. This will enable the various phases to be better understood, communicated, and controlled in order to support trustworthy systems engineering efforts.

We now describe a life cycle that is comprised of 22 phases. These phases correspond to the phases in the much simpler appearing three-phase model. Here, we identify a set of phases, initially discussed by Beam, Palmer, and Sage [3], that might exist when one client or stakeholder group, such as a government agency or private company, seeks development by another vendor (to be determined) of a large system. The *system definition* effort, or phase, can be expanded into seven phases. The actors in these phases include system developers, clients for the systems study (or the user groups), and champions or funders of system development efforts.

System Definition

1. *Perception of Need:* A deficiency between what a client, stakeholder, or user group would like to have in some situation and what generally they do have results in the identification of a need for some new system or process. This should trigger a set of activities on the part of the user group that results in the identification of a set of requirements for a system or process.

2. *Requirements Definition:* In defining requirements, the user group (or a surrogate) may precondition the system to the use of some specific microenhancement tools or system development methods.

3. *Draft Request for Proposal (RFP):* The user group or a surrogate issues an RFP intended to put as few constraints as possible on the system developer's selection of alternatives.

4. *Comments on the RFP:* The final RFP is typically conditioned by inputs from potential bidders. The benefits to the user, client, or customer of an effective systems acquisition approach should include productivity improvements throughout the systems development life cycle. The major goal, however, is high quality performance of the real delivered system. It is not at all a simple matter to measure the real value of a new system, however, particularly since overall system and organizational performance interact and should never be sacrificed as a result of introduction of a new system. There are many implications to this, especially since system performance is a multifaceted and multiattributed concept.

5. *Final RFP and Statement of Work:* The RFP may not necessarily direct the use of a particular life cycle, macroenhancement approach, or use of particular microenhancement tools. A mature RFP will, however, strongly suggest or even require the demonstration of benefits from the micro- and macroenhancement approaches selected for system development and implementation.

6. *Proposal Development:* Generally, in a competitive manner, proposals are developed by the system development industry. Most proposals will include statements regarding (a) system architecture, (b) selection of approach, and (c) work breakdown structure. This useful construct, discussed in more detail in Chapter 8, more fully enables the determination of the cost of fielding a large system. The proposal itself will certainly display the knowledge and experience of the system developer including, if possible, some specific illustrations of previous developments and description of facilities. The work breakdown structure, both in terms of the scheduling of system development effort, and the nature and timing of the work to be done, is a principle vehicle in which the contractor may display knowledge of system development practices, principles, and perspectives. One major risk in large-scale system development, as perceived by the client or contracting agent, is that, although the special system development approach taken in terms of micro-level tools and components may produce an effective and efficient product, it may not be modifiable without return to the same system development contractor. As desirable as this may appear to the systems engineering contractor, the contractor must be responsible for ensuring that some maintenance/enhancement process, which is acceptable to the customer, can be readily implemented —with or without use of the contractor's own special facilities.

7. *Source Selection:* This step is accomplished by the customer, user group, or some chosen surrogate.

At the end of this phase of effort, a system development contractor is selected. Formally, this completes the system definition phase of the three-phase system development life cycle. It enables the start of a number of system development phases. These phases can be defined more narrowly in terms of eight additional phases comprising system design and development.

System Design and Development

8. *Development of Refined Conceptual Architectures:* After the contract award and before the first critical design review, the selected system development contractor undertakes a substantial evolution of the detailed architecture of the system. This takes place because of new information that has accumulated during the contractor selection process and the associated need for system architecture evolution.

9. *Partitioning of the System into Subsystems:* An orderly systems engineering process should include systems management efforts that result in configuration management of the evolving systems engineering product. This systems management effort includes partitioning of the entire system so integration and interfacing needs are identified, and such that the need for interfacing and integration is minimized to the extent possible.

10. *Subsystem Level Specifications and Test Requirements Development:* A complete systems engineering approach must include means to express and verify subsystem level specifications. It is highly desirable that this include semi-automated or fully automated test generation facilities with test recommendations accompanying all subsystem modules. Such tests should be able to be easily linked to define, or at least to verify, system-level tests.

11. *Development of Components:* The systems engineering contractor or subcontractors develop a variety of hardware and software items including major hardware, interfacing hardware, systems software, and application software.

12. *Integration of Subsystems:* The systems engineering contractor or possibly one or more subcontractors integrate the hardware, software, and associated testing practices. An appropriate systems engineering life cycle must specifically address the systems integration phase of the life-cycle effort. This is generally a far more exhaustive requirement than that of simply linking together a collection of existing subsystems, such as reused software modules. The individual subsystems that comprise a system must be addressed specifically to insure functionality and performance after they are integrated. By inserting integration-oriented hooks into the system development life cycle, it often is possible to use special tools to trace and evaluate operation of a large system. Lacking this, problem diagnosis and detection is far more tedious than it needs to be. Problem correction will be all but impossible.

13. *Integration of the Overall System:* The system engineering contractor integrates the hardware and software with the appropriate integration testing. The considerations here are similar to those in phase 12, with the exception that not all subsystems have been developed at the same technology level or in the same amount of detail. Thus, it is necessary to provide for future systems integration efforts to insure sustainability.

14. *Development of User Training and Aiding Supports:* Computer-based training and aiding support mechanisms are often needed by the systems engineering contractor, and the best tradeoffs between these two approaches to enhancing performance should be determined [4]. Simulation models that emulate portions of the system and its operational environment can be used as training supports. Aiding supports can be provided in cases where a greater degree of operator proficiency is needed. A key concern is that the resulting system be flexible, so that new application requirements can be supported easily and effectively.

Following system development, the system is implemented in an operational environment. The final phase of the three-phase system life cycle now begins. Normal operation and maintenance of the system occurs, which generally leads to system evolution over time. Major proactive modification of existing systems may be required, including retirement of a system and replacement by a new system. At least eight phases of the life cycle that specifically relate to system operation and maintenance can be identified.

System Operation and Maintenance

15. *Operational Implementation or Fielding of the System:* The system engineering contractor, with support from the user or customer group, implements the system.

16. *Final Acceptance Testing:* There are several benefits to an effective approach to final acceptance testing of the implemented system by the user. Acceptance testing often results in minor adjustments to system operation. Ideally, minor changes can be made and documented in the acceptance environment.

17. *Operational Test and Evaluation:* The system user or an independent contractor should be selected for this purpose by the user group.

18. *Final System Acceptance:* By the client or user group.

19. *Identification of System Change Requirements:* As experience with using the system grows, there will be a need for maintenance in order to better adapt the system to its intended use. This maintenance is quite different from "bug fixes" and repair. It includes the need to evolve the system over time in an adaptive and proactive manner to make it responsive to changing needs, and needs that were present

initially but unrecognized in the initial set of requirements specifications. Architectural concerns are of major importance here. Better-defined system structures make it easier for the user to identify those portions of the system that need to be changed to effect a desired alteration in system operation. This could minimize requests that require detailed changes in a large number of subsystem elements. These needs should be addressed as a part of the configuration management effort described in Chapter 4.

20. *Bid on System Changes or Prenegotiated Maintenance Support:* A systems maintenance contractor carries out changes using the same general approach used to develop the initial system. This predisposes the user group toward returning to the same contracting source that developed the initial system, or to another that has the same development capabilities.

21. *System Maintenance Change Development:* The maintenance support contractor can make changes more easily if the history of the original system development has been well-captured and documented and if the system is well-structured and contains an open architecture.

22. *Maintenance Testing by Support Contractor:* The systems engineering process often evolves as a linear sequenced process. Errors in an earlier step of the process may cause errors in a later step. This indicates the need for iteration and feedback from later stages in the process to earlier stages. When potential errors are detected, there should be feedback to earlier activities, and appropriate corrective actions should be taken. Often the only iteration possible is the one of proceeding back to phase 19—system change requirements. If significant defects in the existing system are observed, perhaps brought about by new functional requirements, then iteration back through the entire system acquisition process *may* occur. It appears to be the case in practice, however, that this does not happen often or only happens when there are very significant performance deficiencies. It is this need for iteration back to any of a variety of earlier phases of a system life cycle that leads to the need for a look at the overall objectives of the system life cycle and to critical selection of an appropriate life cycle.

This 22-phase description of the systems engineering life cycle is doubtlessly exhaustive. Roles have been indicated for the client or user group, for the systems engineering group or contractor, and potentially for various subcontractors that possess detailed expertise relative to specific technologies. Their interaction is illustrated in Figure 1.5. The many phases developed here for the systems engineering, or systems acquisition, life cycle are illustrated in Figure 2.2. A recent text [5] provides an excellent discussion of system architecture and design principles that, for the most part, follow from this extensive life cycle.

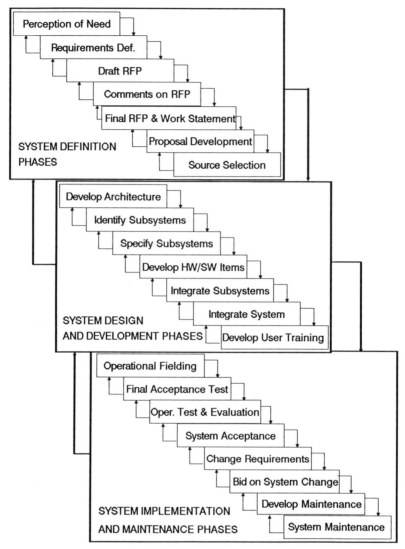

Figure 2.2. A 22-Phase Systems Engineering Life Cycle

Now that we have examined one possible life cycle in some detail, let us examine several others that have been suggested recently. First, some life cycles that have appeared in the systems engineering and management areas are illustrated, then, some software systems engineering life cycles are discussed [6] and their close relationship to general systems engineering life cycles are shown. In the next chapter, we examine additional life cycles from a technical and systems management perspective. The life-cycle concept is ubiquitous throughout much of systems engineering and throughout all of this book on systems engineering.

2.2. MODELS OF THE SYSTEMS ENGINEERING LIFE CYCLE

There have been a number of phase-and-step models of the systems engineering life cycle. Figure 2.3 illustrates a waterfall life-cycle model of three steps and phases. Figure 2.4 illustrates a spiral, or cornucopialike, model of these phases. This cornucopialike life cycle model was initially developed by Hall [7]. The hyperfine structure of this cornucopia indicates interactions among the steps and phases of the systems engineering life cycle. Here, we illustrate the three-step and three-phase model, rather than the more complex seven-step and seven-phase model used by Hall, for convenience. Clearly, the resulting diagram is intricate, especially in terms of the feedback loops of the process. Here, it is possible for any element to follow from, and

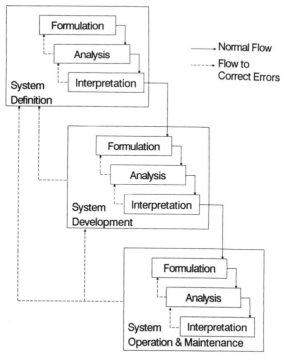

Figure 2.3. An Integrated Waterfall Model of the Steps and Phases of the Systems Engineering Life Cycle

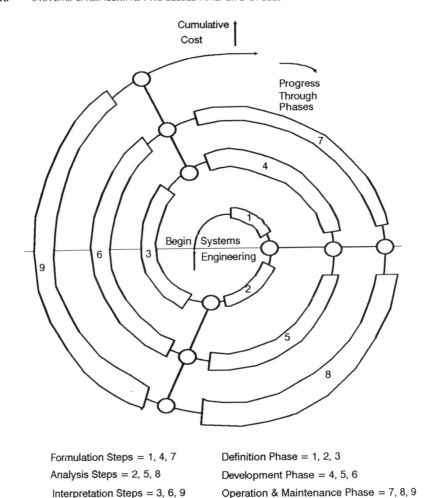

Formulation Steps = 1, 4, 7   Definition Phase = 1, 2, 3

Analysis Steps = 2, 5, 8   Development Phase = 4, 5, 6

Interpretation Steps = 3, 6, 9   Operation & Maintenance Phase = 7, 8, 9

Figure 2.4. Spiral Model of the Activity Structure of Systems Engineering

to lead to any other element, in any possible sequence. Certainly, some of the paths in this complicated diagram are more likely to be traversed than others, depending on both technological reasonableness criteria, systems management desiderata, and standards and legal requirements.

As initially indicated by Hall [7], systems engineering has three major dimensions. The *time dimension* of systems engineering includes the gross sequences or phases that are characteristic of systems work, and extends from the initial conception of an idea through system retirement or phaseout. The *logic dimension* deals with the steps that are carried out at each of the systems engineering phases. The *knowledge dimension* refers to specialized knowledge from various professions and disciplines and to the actors that

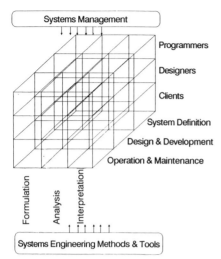

Figure 2.5. The Hall Morphological Box Adapted to a Three-Phase and Three-Step Life Cycle

convey this knowledge. These dimensions of the Hall morphological* box of systems engineering are illustrated in Figure 2.5. We could and should provide support to this extended life-cycle model throughout the systems management process. Also, we could and should support it with and from a great variety of systems engineering methods. Finally, support should be provided not only from below through the provision of effective methods, but from above as well through provision of effective systems management.

In viewing Figures 2.3, 2.4, and 2.5 from this perspective, it is quite clear that it now becomes possible to repeat, or cycle through each of the phases in the life cycle any number of times. In each cycle, or round of the life cycle, the specific efforts at each iterative cycle can be expected to be different. Thus, it becomes possible to recast Figure 2.3 as shown in Figure 2.4 or Figure 2.5. This sort of life-cycle model forms the basis for the spiral software life-cycle model of Barry Boehm, which offers valuable perspectives on managing software system development risk. We describe the software spiral life-cycle model in Section 2.3. Chapter 6 addresses some of the risk-related notions associated with use of this life-cycle model.

Essentially all the life-cycle models for systems engineering incorporate the three-phase notion of system definition, development, and operation. It is of interest to highlight some of these models here. The seven phases of the systems engineering framework of Hall [7, 8] are illustrated in Figure 2.6. The associated life cycle seems to have been initially intended to describe industrial systems. The description of the various efforts are easily modified to describe more general situations. The phases in this life cycle, and a brief descriptive account of the activities at each phase are as follows.

---

*Morphology is the study of structure or form. Morphological studies employ a definite behavioral approach and a specified methodology.

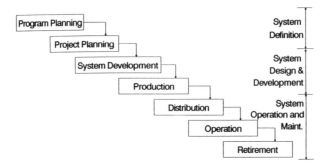

Figure 2.6. The Seven Phases in The Hall Systems Engineering Life Cycle (Time Dimension of the Morphological Box)

System Definition

1. *Program planning,* which is a conscious activity that should result in formulation of the activities and projects supportive of the overall system requirements into more detailed levels of planning, is the first phase in this life cycle. The program planning phase must, if it is to be successful, also include identification of requirements and translation of these into the technological system specifications that are planned for later development.

2. *Project planning,* which is distinguished from program planning by increased interest on the individual specific projects of an overall systems development program, is the second phase. The purpose of this phase is to configure a number of specific projects, which together comprise the program, such that system development can begin.

System Design and Development

3. *System development* results in implementation of the project plans through design of the overall system in detail as a number of subsystems, each of which is described by a project plan. This is the first phase of effort that will ultimately translate the system definition into a product. This phase ends with the preparation of architectures, detailed specifications, drawings, and bills of materials for the system manufacturer or builder.

4. *Production, for manufactured products, or construction, for one-of-a-kind systems* includes all of the many activities needed to give physical reality to the desired system. This could involve, for example, using detailed plans and specifications to construct a new building, manufacture a new product, or produce source code for an emerging software product. A number of related efforts would be needed, such as deter-

mining the sequence, materials flow, required shop floor layouts, and the establishment of quality-control practices. After completion of this phase of effort, we have a systems engineering product that is capable of being fielded or implemented in an operational setting.

System Operation and Maintenance

5. *Distribution, deployment or phase-in* results in delivering systems engineering products or services to users or consumers. This may involve all kinds of distribution facilities, marketing, and sales organizations.
6. *Operations* is the ultimate goal of system development. This phase includes such activities as maintenance.
7. *Retirement, or phase-out* of the system over a period of time and replacement by some modified or new system will generally occur.

From the perspective of this life cycle, the first phased activity is program planning. Working with a large-scale complex system, either from the initial step of creating the system or from an intermediate step of altering the system, is an activity that can be facilitated by effective planning. Creation of an effective plan generally involves a diverse set of resources and a wide spectrum of disciplines. It is necessary to make a conscious effort to project ideas of what is desired into a framework amenable to tests of reality. Many questions can be asked. Can the past and current states of affairs be measured? Does a need for the proposed solution exist? Are there relevant measures of what is to be planned? Is the time frame realistic? Will the plan be accepted and understood by those who will have to carry it out? Clearly, a program plan cannot be established in the absence of knowledge of user requirements for the system. Thus, as we have noted, program planning must necessarily involve information requirements determination and translation of the user requirements for a system into technological specifications. At this phase in the life cycle, these specifications are generally very purposeful initially and are translated to requirements that are functional in nature.

Responsibility for creating and supervising the conduct of a program plan has traditionally rested with systems management. It is generally accepted that the primary tasks of management include planning, organizing, evaluating, and communicating. To conduct these management tasks effectively, one needs a clear statement and understanding of purposeful objectives for the system and the interrelations among the objectives, based upon their relative merits and the costs of implementing them. Thus the need for use of a human value system in the systems planning phase is evident.

For it to be an innovative process, planning necessitates both broad comprehension and goal orientation by those involved. Four often-cited key components of planning are the following:

1. Definition of a goal.

2. Knowledge of current position with respect to the goal.

3. Knowledge of environmental factors influencing the past, present, and future.

4. Determination of best policy given information concerning the above three items.

Definition of the goal is the normative component—a statement of what ought to be. Knowledge of our current position with respect to the goal is the descriptive component—what is. Knowledge of all environmental factors influencing the past, present, and future is the planning horizon—what time, what constraints, and what alterables are allowable. Determination of the best policy from the goal definition, knowledge of the current state, and knowledge of the environment and time frame is the analytic component— what to do and how. These components are clearly applicable to any phase of a systems engineering problem. Formally, they comprise studies in such areas as optimum systems control [9]. Systems control engineering and operations research have contributed much to system development through the microenhancement and analytic efforts of optimization, which are applicable to all phases of a systems engineering effort, even though the primary and intended purpose of the effort may not be optimization.

These steps are applied at each of the phases of the systems engineering life cycle. They may be stated in many ways. As we indicated in Chapter 1, three fundamental steps are involved at each phase: formulation of issues, analysis of the impacts of alternatives that may resolve issues, and interpretation of these impacts in terms of a value system and selection. These three steps can be expanded in a number of ways. We have just indicated a four step expansion. There are many others; for example, Hall identified seven steps. These, and their correspondence to the three-step framework, may be described as follows.

Issue Formulation

1. *Problem Definition:* Isolating, quantifying, and clarifying the need that creates the problem and describing the set of environmental factors that constrains alterables for the system to be developed.

2. *Value System Design:* Selection of the set of objectives or goals that guides the search for alternatives. Value system design enables determination of the multidimensional attributes or decision criteria for selecting the most appropriate system.

3. *Systems Synthesis:* Searching for, or hypothesizing, a set of alternative courses of action or options. Each alternative must be described in sufficient detail to permit analysis of the impacts of implementation and subsequent evaluation and interpretation with respect to the objectives.

Analysis

4. *Systems Analysis:* Determining specific impacts or consequences specified as relevant by the value system. These impacts may relate to such
   important concerns as product quality, market, reliability, cost, and
   effectiveness or benefits.
5. *Refinement of the Alternatives:* Optimizing the system variables to best
   meet system objectives and satisfy system constraints.

Interpretation

6. *Decision Making:* Evaluating the impacts or consequences of the alternatives developed in analysis relative to the value system. This enables
   interpretation of these evaluations such that all alternatives can be
   compared relative to the values. One or more alternatives or courses of
   action can be selected for advancing to the next step.
7. *Planning for Action:* Communicating the results of the effort to this
   point and looking ahead to the next phase. This includes such pragmatic efforts as scheduling subsequent efforts, allocating resources to
   accomplish them, and setting up system management controls. If we
   are conducting a single phase effort, this step would be the final one.
   More generally, it leads to a new phase of effort.

Figure 2.7 indicates the waterfall nature of these seven steps of systems
engineering and their relation to the three-step systems engineering framework. The seven steps of systems engineering are generally carried out in an
iterative fashion at each phase of the systems engineering life cycle. Since the
efforts and purposes of the various phases of the life cycle are quite different,
the specific tools and methods that are appropriate to accomplish each of the

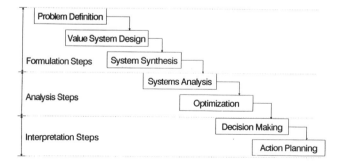

Figure 2.7. Interpretation of the Seven Steps of the Logic Dimension of Hall's Systems
Engineering Life Cycle (Each Step is Repeated at Each Phase of the Systems Life Cycle)

Steps → Phases ↓	Formulation			Analysis		Interpretation	
	Problem Definition	Value System Design	System Synthesis	Systems Analysis	Alternative Refinement	Decision Making	Planning for Action
Definition — Program Planning							
Definition — Project Planning							
Des. & Dev. — System Development							
Des. & Dev. — Production							
Operation — Distribution							
Operation — Operation							
Operation — Retirement							

Figure 2.8. A Matrix of Activities in the Systems Engineering Life Cycle

seven steps can be expected to be quite different from one systems engineering process phase to another. It is possible to go back to refine and improve the results of any earlier step as a consequence of the results of any later. In each of the seven phases of systems engineering, it is necessary to conduct each of the seven steps. Figures 2.4 and 2.5 illustrated this sequencing of the phases and steps of systems engineering and Figure 2.3 illustrated the sequencing for a three-phase and three-step life-cycle model.

When we combine the phases of systems engineering with the steps, we obtain an activity matrix for systems engineering. The phases are the time sequence of activities, and the steps are the logical sequence of activities. A presentation of the plane defined by the phases and steps within each phase is shown in Figure 2.8. This is sometimes called a Hall activity matrix, after its originator. We illustrate it, here, for both the three- and seven-phase and -step systems engineering frameworks.

Many U.S. corporations have advocated use of life-cycle models for the systems engineering process. A recent brochure from the Computer Science Corporation [10] describes two of these systems engineering life-cycle models. One of these is the International Data Corporation system integration model. This is comprised of five phases and is illustrated in Figure 2.9. The Computer Science Corporation itself has proposed a six-phase process, called the *Digital System Development Methodology* (*DSDM*), which is illustrated in Figure 2.10.

American Management Systems [11] has devised a "design first" approach to information systems engineering that is comprised of seven steps as illustrated in Figure 2.11. This approach explicitly rejects postponing design considerations until after subsystems have been specified in detail. It was

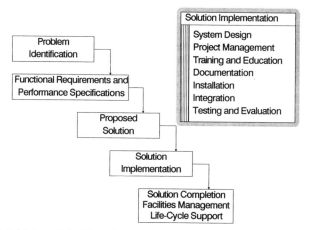

Figure 2.9. Highlights of the Five-Phase Methodology of International Data Corporation

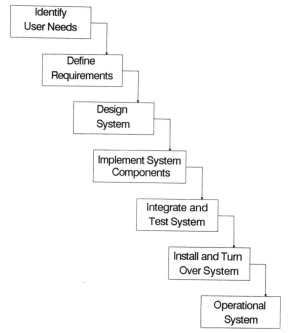

Figure 2.10. Highlights of the Systems Engineering Phases of Computer Science Corporation's Digital System Development Methodology (DSDM)

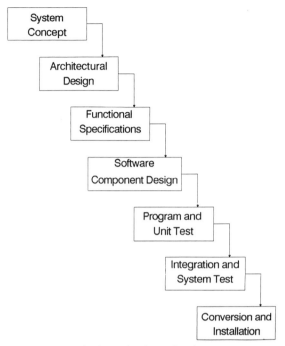

Figure 2.11. The Seven-Phase (Software) Life Cycle of American Management Systems

apparently felt that such a postponement will lead to the development of functional specifications that are biased toward a specific design approach and that, therefore, might exclude consideration of other potentially more appropriate design alternatives. In the AMS life-cycle model, overall technical and functional design approaches are developed during the very first phase of effort, the system concept phase. These approaches are evaluated for important impacts: feasibility, risk, schedule impacts, and other costs and benefits. The system concept is prepared by a small team of senior people with major functional and technical skills and experiences. Everyone works as a single unit under the guidance of a chief architect during this phase of effort, which culminates when the client selects a design that offers the best overall chance of succeeding.

A number of other life-cycle approaches to systems engineering could be cited. For the most part, these would not be dramatically different from the ones that we have examined here. The area of software systems engineering is especially ripe with recent life-cycle models. Let us now examine some of these.

2.3. SOFTWARE SYSTEMS ENGINEERING LIFE-CYCLE MODELS

Credit for the first introduction of a systems engineering based life cycle for use in software development is often given to Royce [12]. The model introduced by Royce was termed the *Waterfall Model*. It embodied existing systems engineering approaches to building large-scale software systems. Figure 2.12 presents this particular waterfall model for the software production life cycle.

Since its initial development and use by Royce, many modifications and iterations have been made to the waterfall model concept for software production. Almost all of these define five to seven phases for systems engineering software development processes. Boehm [13], for example, has been especially concerned with the economic importance of regularizing the development of software and has further developed and popularized the use of the waterfall life cycle for software development. He identified an overall set of seven phases for the waterfall software development life cycle, as shown in Figure 2.13. The similarity to the systems engineering life-cycle modeling construct is distinct and apparent.

The activities to be carried out at each of the seven phases to implement the waterfall software development life cycle model of Boehm, illustrated in Figure 2.13, are as follows.

1. *Systems Requirements Specification:* It is assumed that the software system user group and the systems requirements elicitation group are sufficiently informed about the intended purpose for the new or modified system such that they can identify and develop the system level requirements in sufficiently complete detail that preliminary software requirements can be

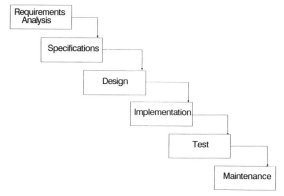

Figure 2.12. Waterfall Life-Cycle Model for Software Development of Royce

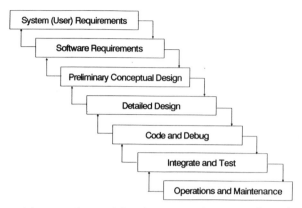

Figure 2.13. Modification of Waterfall Software Development Life-Cycle Developed by Boehm

specified. All of this is to be done before preliminary design may be initiated. This phase of effort is very purposefully oriented.

2. *Software Requirements Identification:* Development in the software requirements phase focuses on the outcomes of the system requirements identification effort that was carried out in phase 1. This phase is concerned with translating the purposeful requirements identified in phase 1 into structural and functional requirements. Explicit concern is with structure and function in terms of:

   a. The nature and style of the software to be developed.

   b. The data and information that will be required and associated structural frameworks.

   c. The required functionality, performance, and various interfaces that may be needed.

   Requirements for both the system and the software are reviewed for consistency and then reviewed by the user to be certain that the software requirements faithfully interpret and meet the user-oriented requirements identified in the previous phase. A software requirements definition document is generally produced in this phase. It becomes a technology and management guideline throughout all subsequent development phases, including validation and testing.

3. *Preliminary Conceptual Design:* The software requirements developed in phase 2 are converted into a preliminary and conceptual software product design in this phase. Thus, phase 3 is primarily aimed at the further interpretation of the software requirements in terms of software system level architecture. The product of this phase of software system development is identification and definition of the data structure, software archi-

tecture, and procedural activities that must be pursued in the next phase. Data items and structures are described in abstract terms as a guide to the detailed design phase to follow. Instructions that describe the input, output, and processing that are to be executed within a particular module are developed here. Preliminary software design involves representing the functions of each software system functions in a way that they may readily be converted to a detailed design in the next phase of effort.

4. *Detailed Design:* The effort in the preliminary conceptual design phase results in much insight as to how the system will ultimately work at structural and functional levels in order to satisfy the software system requirements that were identified in phase 2. Typical activities of the detailed design phase include definition of the program modules and intermodular interfaces, which are necessary precursors to writing source code for the software system under development. Specific attention is directed to data formats in the detailed design phase. Detailed descriptions of algorithms are provided. All of the inputs to and outputs from detailed design modules must be traceable back to the system and software requirements that were generated in phase 1 and phase 2. During this phase, the conceptual software design concept is fine-tuned.

5. *Code and Debug:* In this phase, the detailed design is translated into machine-readable code, generally in a high-level programming language or perhaps a fourth generation language. After the software design requirements are written as a set of program units in the appropriate high-level programming language, the resulting code is compiled and executed. Generally, "bugs" are discovered, and debugging and recoding takes place to insure the integrity of the overall coding operations of this phase. Automated code generators can sometimes perform all, or a major portion, of this task, and produce code almost guaranteed to be bug-free.

6. *Integration, Testing and Preoperation:* In this phase, the individual program units are integrated and tested as a complete system in an effort to ensure that the software requirements specifications obtained in phase 2, and hopefully the system level requirements identified in phase 1, are satisfied. The testing activities and procedures are primarily concerned with software logic functions. All programming statements should be tested at this phase to determine proper functioning, including whether inputs and outputs are operating properly. After system testing, the software is operated under controlled conditions to verify and validate that the entire package satisfies the software technical specifications, and ideally the system level requirements as well.

7. *Operation, Evaluation, and Maintenance:* This phase of the waterfall life cycle is often the longest in time, and often the most costly as well. In phase 7, the system is installed at the user location and evaluated, and then put into actual operation. Maintenance is primarily the process of improving the system to accommodate evolving system level requirements.

These phases normally take place in the sequenced manner described. However, there may be considerable iteration and interaction among the several phases in the life cycle. Each software development unit would generally adapt the activities in the life cycle to accommodate such factors as:

1. Familiarity of the personnel of the firm with the product being developed.
2. Detailed needs of users for the software to be developed.
3. Economic, legal, system integration, and scheduling constraints.

In an ideal situation, the above phases would be carried out in sequence. When all of the phases have been completed, the software product is delivered and performs, as intended by the client, in accordance with the original statement of requirements. Actually this rarely happens either for hardware or software development. What usually happens is that one or more phases need to be repeated, as a result of deficiencies in the process that are discovered after initiation of effort. It is easy to see how the need for this arises. As an example, software system development begins with elicitation of an initial set of user requirements. Development proceeds and an initial detailed design results. Upon viewing the results of this detailed software system design, the user discovers that certain requirements were omitted or misunderstood. The software development process then returns to the requirements phase. Further work is needed with the user until the user is satisfied that all needs have been satisfied.

As the software product evolves, it is often found that some of the initial requirements cannot be supported by the system because of earlier decisions. The user is involved to help identify a solution for the problem. This should call for revisiting the system requirements phase and tracking the implications of the newly resulting requirements through subsequent system development phases. However, such multiple iterations are seldom fully accomplished, generally because of contractual obligations or due to shortage of time or funds.

Another form of the waterfall software development life cycle is that used by the U.S. Department of Defense, and denoted DoD STD 2167-A, which is shown in Figure 1.8. This model splits the system development process into hardware and software development efforts. In this sense, this DoD standard incorporates some notions of concurrent engineering, a subject we consider later.

The spiral model of the software systems engineering development life cycle was introduced by Boehm in 1986 [14] and refined somewhat in 1988 [15]. Basically, it represents several iterations of the waterfall life-cycle model with possibly alternative approaches to software development at each iteration. Its use is intended to result in a risk-management approach to development of software products. Figure 2.14 illustrates some of the central

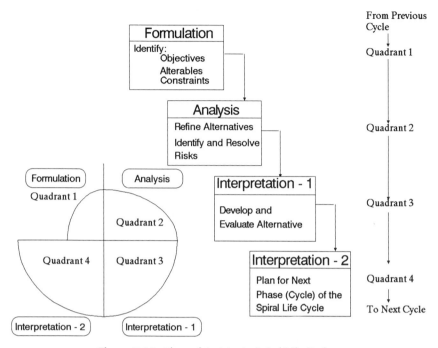

Figure 2.14. Flow of Activity in Spiral Life Cycle

concepts. When used as proposed by Boehm, the spiral life-cycle model requires the use of prototyping, and also calls for an assessment of the *risk*, and translation of this into management control policies, of continuation of software development at each cycle of the development spiral. This extension of the conventional waterfall model explicitly incorporates the notion of feedback and iteration across the phases of the development life cycle. The waterfall model is modified as needed to incorporate prototyping and risk analysis [16].*

One version of the spiral model for a software development life cycle is shown in Figure 2.15. Use of the spiral life-cycle model proceeds by an iterative sequencing of the several phases of development at each specific time that a different type of prototype is developed. This allows for an evaluation and management of risk facets before proceeding to a subsequent set of phases associated with development of a software product. The opportunity to assess such important risk-related facets as cost, schedule, reliability, and performance enables the software systems engineer to deter-

---

*We will consider notions of risk, reliability, and quality assurance in some detail in Chapter 6.

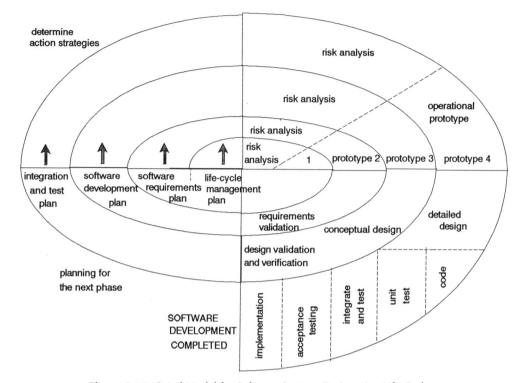

Figure 2.15. Spiral Model for Software Systems Engineering Life Cycle

mine whether to continue on with development, or to seek an alternative path for development, or to abandon the development process as infeasible from a risk management perspective.

In Figure 2.15, the length associated with the radial dimension is used to conceptually reflect the cost parameter as costs accumulate over the several phases of software development. The angular dimension represents the cumulative progress made in software life-cycle development at any particular phase in the life cycle. In the spiral model of Figure 2.15, prototypes that offer insight into the development task at hand are introduced. An iterative cycle of risk assessment, prototype construction, evaluation of the resulting product, and formalization of the various lessons that are thereby learned is produced. This is assumed to continue until it is possible to develop a low-risk system requirements document and associated systems management plan. In this figure, we illustrate a situation where four prototypes are constructed. The actual number of prototypes that will be needed is a function of development uncertainties and risk management policies. Once agreement has been reached relative to the system requirements specifications, Boehm suggests using a standard waterfall model to manage system implementation and integration.

The spiral model activities may be based upon use of the seven steps of the fine structure of systems engineering, as discussed earlier in this chapter and in Chapter 1. These seven steps follow from the three fundamental steps of formulation, analysis, and interpretation, which are implemented at each of the phases of the systems engineering life cycle. It is convenient to view each cycle in the spiral model as corresponding to one, or more than one in some cases, phases of the life cycle. Thus a phase, or a collection of phases, comprises a cycle or round of the spiral life cycle. The steps undertaken in each round may be described as follows.

1. **Issue Formulation**

    1.1. Determine the *objectives* of the software to be developed within the particular cycle being considered in terms of such important factors as performance, functionality, ability to accommodate change, and so on.
    1.2. Identify the *needs*, *constraints*, and *alterables* for the particular cycle at hand, such as budgetary constraints and development time specifications. Determine whether an operational need can be satisfied by a particular software development effort, as generally specified by a specific configuration management plan.
    1.3. Identify *alternative* means of satisfying the stated objectives for the particular phase of the product development under consideration. Such alternative approaches as buying an existing software product, or designing a new software product, or modifying and reusing existing software should be considered.

2. **Analysis**

    2.1. Determine and *evaluate* the alternative impacts in light of the objectives and the constraints. In this step, contingencies that involve significant project risk will be identified.
    2.2. Formulate cost-effective strategies to address the risk situations identified in step 2.1. This may include prototyping, simulation, user interviews, benchmarking, analytic modeling, or a combination of these as well as other strategies for risk reduction and risk management.
    2.3. Determine the risk remaining to complete the project and select a sound development strategy that is based on systems management of the identified remaining risk.

3. **Interpretation**

    3.1. The original hypothesis, that an operational need can be satisfied by the product of a software development effort, is tested. This may involve examination of market factors, the effectiveness of the particular software product, and so on.

3.2. If the hypothesis tested fails, the development spiral is stopped.

3.3. If the hypothesis tested does not fail, another cycle of spiral development may be conducted until software product implementation and successful operational test and evaluation results.

3.4. Alternately if the result of the hypothesis test is satisfactory but considerable risk remains, it may be desirable to disaggregate and partition the development effort through the creation of additional parallel spirals.

3.5. Develop more explicit plans for the next cycle of development, if any. Review this explicit plan and make a commitment to it.

The most interesting part of this spiral development effort is the interpretation step. Each time this step is encountered, for each cycle of the spiral model, an assessment is made as to whether to continue or abort the development process. If the risk assessment indicates the project should be continued, another round of the spiral model commences. Otherwise, the development stops, and an evaluation of the entire project is made. As noted, the plan to continue the development may include partitioning the project into smaller pieces for ease of handling or simply continuing along the path of the spiral model. The decision to proceed may be based on many facets,

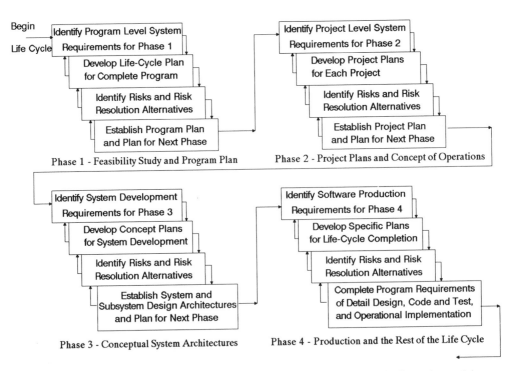

Figure 2.16. Another Systems Engineering Interpretation of the Spiral Life-Cycle Model

such as from an analysis of individual design segments, or from a major review of the overall project requirements. Figure 2.16 illustrates the flow of the activities just described. This is easily seen to be an alternate version of Figure 2.15, which illustrates the cornucopia-like model that Boehm initially used to represent the spiral life cycle. As was the case with Hall's cornucopialike spiral model considered earlier, the spiral model of Boehm is a generalization of other life-cycle models and is particularly well-suited to incorporating submodels, or cycles or rounds, within the complete life-cycle effort.

2.4. STEPS OF THE FINE STRUCTURE OF SYSTEMS ENGINEERING

Systems engineering involves the application of a general set of guidelines and methods useful to assist clients in the resolution of issues and problems that are often of large scale and scope. As we have noted, three fundamental steps may be distinguished at each phase in a system life cycle: problem or issue formulation, problem or issue analysis, and interpretation of the analysis results, including evaluation and selection of alternatives and implementation of the chosen alternatives. As we have also noted, these steps are conducted at each of a number of phases throughout a systems life cycle. This life cycle begins with determination of requirements for a system through system design, development, installation, and maintenance, and ultimate replacement.

The systems engineering paradigm calls for efforts that involve the study of issues in relation to their environment, with due consideration of causal or symptomatic, institutional or organizational, aspects of the problem. The necessity of a systematic, rational, and purposeful course of action is emphasized in formal systems engineering approaches. Systems engineers make use of methods, theory, and data that are based on a variety of disciplines such as behavioral and cognitive psychology, computer science, operations research, economics, and systems and control theory. A serious attempt is made to consider as many relevant aspects of an issue as possible. These aspects typically cut across various fields of knowledge, institutions, and traditional disciplinary boundaries. For example, an issue that initially might appear to be purely economic in nature, might, upon closer inspection, be interwoven with technological, social, political, and environmental problems. A systems engineer will attempt to take all these related fields into consideration when assisting the client in organizing knowledge, such that the client is then better able to formulate, analyze, and interpret the available options through support from the systems engineer. The emphasis in this discussion is clearly the role of the systems engineer as a *support person*. The glue that holds these methods together is a methodology for systems definition, design, development, and operations. This is, of course, the systems engineering life cycle.

There are other ways in which to characterize systems engineering. For example, we may examine the three fundamental steps of formulation, analysis, and interpretation at each phase in the system life cycle. We might view systems methodology as the connecting link between the methods and tools dimension and a systems management dimension as shown in Figure 1.3. The definition of a *methodology* as an open set of organized procedures for problem solving is particularly important. The methodological framework enables the wise selection of appropriate methods with which to accomplish problem resolution. The ingredients of a specific systems engineering methodology are therefore seen to be:

1. A set of *activities* to be accomplished.
2. A set of *methods* and/or *technologies*.
3. A set of *people*.
4. A set of *relations* among the activities, methods, and people.
5. The *environment* in which each of these is imbedded.

The characteristics of the systems engineering approach make it particularly appropriate to ameliorate or resolve many contemporary large-scale problems in a variety of areas. Generally, this is less costly and time consuming in the long run than application of superficial, temporary, and incremental solutions to recurring problems that only relieve symptoms of a problem. Use of an appropriate systems engineering framework, and of the systems engineering methods useful within the framework, leads to efficient and effective use of available methods for dealing with complexity, to a more efficient and effective use of the time allocated to issue resolution efforts; and to more efficient, effective, equitable, and explicable resolution of issues. Adoption of a systems engineering approach results in a better match of problems and methods for problem resolution. In short, an appropriate systems engineering process leads to a product with a lower cost and greater operational effectiveness mix than would generally be obtained by a more random and less systematic approach to development.

We have initially chosen a systems engineering framework involving the three steps of formulation, analysis, and interpretation as the underlying structure ordering the methods we discuss here. Figure 2.17 illustrates another view of the three major steps in this framework and represents a considerable expansion of a portion of Figure 1.4. Supporting Figure 2.17 are various approaches from a variety of areas, including methods from systems science and operations research, programmer microenhancement efforts from computer engineering, and technologies such as very large scale integrated (VLSI) circuit design, and building construction.

Many methods can be used as part of the systems engineering framework of Figure 2.17. The process model of Figure 1.4 indicates how a specified systems management structure influences such portions of the systems engi-

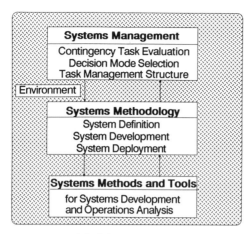

Figure 2.17. Model of the Systems Engineering Process

neering life cycle as requirements specification and decision mode selection. Decision mode selection, which is the decision concerning how to decide, directs the methodology to the use of specific methods for issue formulation, analysis, and interpretation. This meta-level decision influences the process of systems engineering. It enables selection of the particular systems development approach. Meta-level decisions concerning how to decide influence individual judgments, such as that of a human supervisory controller.

A systems effort is generally always conducted in an iterative rather than in only a linear sequenced manner. Typically, after some analysis has been accomplished, certain elements of issue formulation might be reconsidered, and a first preliminary evaluation and selection of alternatives may be made. Only viable alternatives that pass this initial screening are then subjected to more detailed analysis, including exploration of possible implementation plans, before a further evaluation and selection are made.

2.4.1. Issue or Problem Formulation

The first part of a systems effort at problem resolution is typically problem or issue formulation and identification of problem elements and characteristics. The groups affected by the issue and/or proposed solutions are identified, as well as their needs, relevant institutions, fields of knowledge, constraints, alterables, goals for the effort, policy instruments, and actors involved. Subsequently, structuring, or partitioning is often undertaken to facilitate understanding and communication of the problem.

Issue formulation begins with a definition of the problem or issue to be resolved. Problem definition is an *outscoping* activity. That is, it is an activity

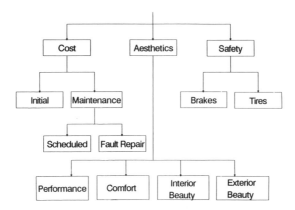

Figure 2.18. Attribute Tree (of Objectives Hierarchy) for Evaluation of Auto Purchase

that enlarges the scope of what was originally thought to be the problem. Problem or issue definition is ordinarily a group activity involving those familiar with or impacted by the issue or the problem, systems engineers, and management specialists. It seeks to determine the needs, constraints, alterables, and societal sectors affecting a particular problem and to determine relationships among these elements.

Of particular importance are the identification and structuring of objectives for the policy or alternative to be chosen, which is referred to as *value system design* in the initial work of Hall. To accomplish value system design effectively, it is necessary to distinguish ends from means or objects from activities. Objectives identified in value system design are quantified as part of the issue evaluation and interpretation step. The analysis step determines the impacts of proposed alternatives upon objectives. Thus, the valuation of an alternative is determined as fairly as possible through the systemic divide-and-conquer process that separates means and ends. It is always necessary to identify objectives for a systems engineering effort and attributes with which to measure the degree of attainment of these objectives by alternatives. Generally, structuring of the attributes in the form of an objectives hierarchy, which often takes the form of an objectives (or attribute) tree, aids the systems engineering process in many ways. Figure 2.18 illustrates one particular attribute tree, which is useful in describing attributes of value for evaluation in an appropriate decision setting. Attribute trees are discussed further in Chapters 7 and 8 when we examine approaches useful for decision analysis, system evaluation, and cost and operational effectiveness determination.

Value is a relative term. We are primarily concerned with the relative values of an alternative policy or decision, which denotes the degree to which we prefer it to other alternative policies. The term *value system* refers to the set of interacting elements that provides an ultimate basis for decision making. Value system design is, therefore, the representation of the properties of an object in a format amenable to metric evaluation. Allocation of resources represents, and results from, a value judgment. If we can characterize the value judgment in a manner that relates to human capabilities and human needs, then the judgment is amenable to reasoned analysis, criticism, and judgment. The conclusion of the process of value judgment is an evaluation of alternatives and the decision to select one or more of them for implementation. *Decision* is the expression of preference for a particular option or alternative that is selected from among a class of options or alternative courses of action.

If a value judgment is an expression of preference, a judgment that has been made with only one option considered must be considered incomplete. Preference implies comparison, which requires two or more class members. The do-nothing-at-all alternative is, of course, one option in most issues. Also, there are cases where a person very familiar with a situation will identify and adopt a single familiar course of action to resolve a given issue. If the person is indeed experientially familiar with the issue at hand, then previous attempts at problem solution have identified a number of alternatives and selected from among these. So, a judgment mode such as *unconflicted change to a new course of action* or *unconflicted adherence to the present course of action* will generally be based upon past successful problem-solving experiences and will not necessarily represent irrational problem-solving behavior. Such judgments could be unsound, however, if the environment has changed and this is unrecognized, as we discuss in Chapter 9.

It is helpful to relate the objectives to needs, alterables, constraints, and societal sectors. Some self- and cross-interaction matrices may be determined, as in Figures 2.19 and 2.20, which show interrelations among the various issue formulation elements. Others can potentially be generated mathematically by Boolean multiplication. This reduces the chances of inconsistencies in the cross-interaction matrices and enables us to view the interactions among problem definition elements.

Essential also is the identification of alternatives, controls, or policies potentially capable of resolving needs. Such identification may range from a simple listing of the actions or policies currently available to influence the system to a comprehensive design effort to conceive systems structures or organizations that might achieve the identified objectives. It is not a simple task to generate meaningful and innovative alternatives. Option generation is a very important and often neglected portion of the problem-solving effort.

This approach to issue formulation is concerned primarily with the answers to four questions.

1. What are the alternative approaches for attaining objectives?
2. How is each alternative approach described?
3. How do we measure attainment of each objective?
4. How do we measure attainment of each alternative approach?

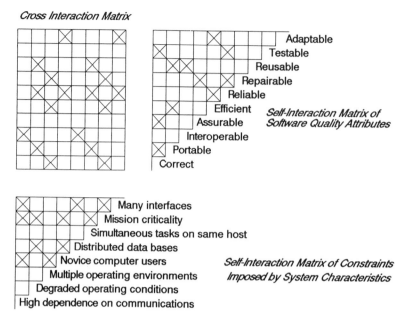

Cross Interaction Matrix

Adaptable
Testable
Reusable
Repairable
Reliable
Efficient
Assurable
Interoperable
Portable
Correct

Self-Interaction Matrix of Software Quality Attributes

Many interfaces
Mission criticality
Simultaneous tasks on same host
Distributed data bases
Novice computer users
Multiple operating environments
Degraded operating conditions
High dependence on communications

Self-Interaction Matrix of Constraints Imposed by System Characteristics

Figure 2.19. Self- and Cross-Interaction Matrices for Software Development Constraints and Needed Quality Attributes

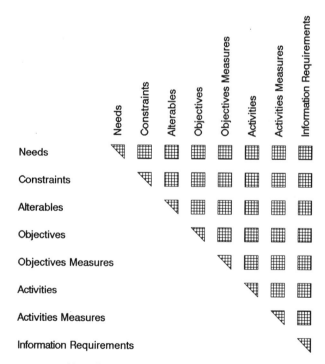

Figure 2.20. Self- and Cross-Interaction Matrices for Issue Formulation

The answers to these three questions lead to a series of objectives and alternative activities, as well as a set of objectives measures and a set of activities measures. Each is needed for broad scope problem definition.

2.4.2. Issue or Problem Analysis

The analysis portion of a systems effort typically consists of two steps. First, the options or alternatives defined in issue formulation are analyzed to assess the expected impacts of their implementation. This is often called impact assessment. Secondly, an alternative refinement or optimization effort is often desirable. This is directed toward refining or fine-tuning viable alternative courses of action through adjustment of the parameters to maximize performance.

Forecasting is an essential ingredient of impact assessment. There are many complications associated with forecasting the impacts of alternative courses of action. Among these are uncertainty and imprecision concerning important future events, uncertainty concerning institutional changes, and uncertainty concerning values and changes in values. A great many approaches have been designed and used for forecasting. These are two general classes of methods: expert opinion methods and modeling or simulation methods. Models are extraordinarily useful in systems engineering. Virtually all of this book is based on models, especially models of the systems engineering process and models that are useful for measurement and management of efforts intended to result in a systems engineering product.

Expert opinions are really mental models formed on the basis of experiential familiarity with a particular issue. A sometimes-occurring difficulty with mental models is that they are difficult to communicate to others. Also, there may be difficulties in accomplishing sensitivity analysis of a mental model to changed conditions and assumptions.

A great deal of intellectual activity has been stimulated by attempts to construct models of systems. A model is an abstract generalization of an object or system. Any set of rules and relationships that describes something is a model of that thing. When we model systems, we enhance our ability to comprehend their behavior and to understand their interrelationships, and our relationship to them. Systems engineering, operations research and related areas have made many contributions toward the improvement of clarity in modeling. A typical result of a systems engineering model is the opportunity to see a system from several viewpoints or to view issues from multiple perspectives.

A system model may be viewed as a physical arrangement, a flow diagram, and/or a set of actions and consequences that can be shown graphically through time as a simplified picture of reality. Improvements in modeling have become more important as systems have become more complex. Usually systems evolve as an aggregate of subsystems that interact with one another to create an interdependent whole. If we understand subsystems and their interactions more clearly, the possibility of their functioning appropriately will be much enhanced. As a prerequisite for such functioning, the complex nature of each subsystem must be understood and the interdependence of subsystems with one another must be explored and comprehended. It would, therefore, seem apparent that a representation of systems and their interactions will enable increased communication between those concerned with systems behavior and systems engineering.

Frequently, all the elements of a mental model are difficult to identify, and the interaction between elements is not clearly indicated. Assumptions are often treated as fixed elements or relations. As we have noted, it is not easy for us to communicate mental models to others. We cannot manipulate mental models effectively. Each of these defects taken separately creates confusions, but as these defects combine and interact, understanding of complex systems is severely hampered. We postulate that systems engineering models transform mental models into forms that are better defined, have clearer assumptions, are easier to communicate, and are more effectively manipulated.

Many formal systems engineering approaches are based on simulation and modeling concepts and methods. Simulation and modeling methods are based on the conceptualization and use of an abstraction, or model, which hopefully behaves in a way similar to the real system. Impacts of policy alternatives can then be studied through use of the model, something which usually cannot be done through experimentation with the real system. Models are dependent on the value system and the purpose behind utilization of a model. We want to be able to determine the correctness of predictions based on usage of a model and thus be able to validate a model. Given the definition of a problem, a value system, and a set of proposed alternative courses of action, we wish to be able to design a model consisting of relevant elements of these three sets, that can help us predict the results of implementing various possible policies.

There are three essential steps in constructing a model.

1. Determine those issue formulation elements that are most relevant to a particular problem.
2. Determine the structural relationships among these elements.
3. Determine parametric coefficients within the structure.

A relatively large number of modeling approaches are described by Gass and Sisson [17] and Atherton and Borne [18].

There are three basic purposeful categorizations of models: descriptive, predictive or forecasting, and policy or planning models. Representation and replication of important features of a given problem is the object of a descriptive model. Good descriptive models are of considerable value in that they reveal much about the structure of a complex issue, and demonstrate how the issue formulation elements impact and interact with one other. An accurate descriptive model must be structurally and parametrically valid. One of the primary purposes behind constructing a descriptive model is to learn about the impacts of various alternative courses of action.

In building a predictive or forecasting model, we must be concerned with determination of proper cause and effect or input/output relationships. If the future is to be predicted accurately, we must have a method with which to determine exogenous or independent "given" variables accurately, the model structure must be valid, and parameters within the structure must be accurately identified. Often, it will not be possible to accurately predict all exogenous variables, and, in that case, conditional predictions can be made from scenarios. Consequently, predictive or forecasting models are often used to generate a variety of future scenarios, each a conditional prediction of the future.

Policy or planning models are useful for much more than predictive or forecasting purposes, although any policy or planning model is also a forecasting model. The policy or planning model must be evaluated ultimately in terms of a value system. Policy or planning efforts must not only predict outcomes of implementing alternative policies, but also present these outcomes in terms of a value system useful for ranking, evaluation, and decision making.

Verification of a model is necessary to ensure that the model behaves in a fashion, and for the purpose, intended by the model builder. If we can confirm that the structure of the model corresponds to the structure of the elements it is representing, then the model is verified structurally. But this offers no assurances that specific predictions made from the model are valid. This is the case since the exact parameters of the model have not been confirmed. Normally, three techniques are used to test the validity of a model. First is the reasonableness test, which we attempt to determine from knowledgeable people that the overall model, as well as model subsystems, respond to inputs in a reasonable way. The model should also be valid according to statistical time series used to determine or estimate parameters and variables within the model. Also, the model should be valid in the sense that the policy interpretations of the various model parameters, structure, and recommendations are consistent with the ethical, professional, and other standards of the group affected by the model. It is very appropriate to verify and validate a model from structural, functional, and purposeful perspectives. We comment on this further in Chapter 4.

2.4.3. Problem or Issue Interpretation

The third step in a systems effort starts with evaluation and comparison of alternatives using the information gained by analysis. Subsequently, one or more alternatives are selected, and a plan for their implementation is designed.

The evaluation of alternative actions must typically be accomplished and implementation decisions made in an atmosphere of risk and uncertainty. The outcome of any proposed policy is seldom known with certainty. One of the purposes of the analysis step is to reduce, to the extent possible, uncertainties associated with the outcomes of proposed policies. Decision making, policy analysis, and planning often involve a large number of decision makers who act according to their varied preferences. These decision makers may have diverse and conflicting data available to them, and, as a result, the decision-making environment may be quite fragmented.

Furthermore, outcomes of decisions are often characterized by a large number of incommensurable attributes. Determining the relative importance of such attributes for purposes of evaluation and choice making is difficult. Also, inadvertent biases, such as those due to a nonconscious ideology or selective perception, are prevalent in many cognitive activities. Unaided evaluations, judgments, and associated decisions are influenced by many heuristic procedures that may lead to selection of poor approaches for human information processing. It is often quite difficult to disaggregate the valuations associated with policy outcomes from the events leading to these outcomes. This confounding of values with facts can lead to extreme difficulties in communication and in efforts at judgment and choice. One of the purposes of a model is to ameliorate these difficulties. For this reason, and others, models are very useful in any analysis process, and in the interpretation process as well.

It is important to note that there is a clear and distinct difference between the refinement of individual alternatives, or the optimization step of analysis, and the evaluation of sets of refined alternatives. In some cases, refinement or optimization of individual alternative policies is not needed in the analysis step. But evaluation of alternatives is always needed in the interpretation step; if there is but a single policy alternative, then there really is no alternative at all. Clearly, the efforts involved in the interpretation step of evaluation and choice making interact strongly with the efforts in the other steps of the systems process.

There are a number of methods for evaluation and choice making that are important.

1. *Decision Analysis:* Involves identification of action alternatives and possible consequences, identification of the probabilities of these consequences, identification of the valuation placed by the decision maker upon these consequences, computation of the expected value of the consequences, and aggregating or summarizing these values for all consequences of each action. In doing this we obtain an evaluation of each alternative course of action in terms of the value system, and the one with the highest value is the most preferred action alternative or option.

2. *Multiple Attribute Utility Theory:* Facilitates comparison and ranking of alternatives with many attributes or characteristics. The relevant at-

tributes are identified and structured, and a weight or relative utility is assigned by the decision maker to each basic attribute. These attribute measurements are then used to compute an overall worth or utility for each alternative. Multiple attribute utility theory allows for various types of structures and for the explicit recognition and incorporation of the decision maker's attitude toward risk in the utility computation.

3. *Policy Capture or Social Judgment Theory:* Assists decision makers in making values explicit and insuring that their decisions are consistent with their values. In policy capture, the decision maker is asked to rank a set of alternatives. Then, alternative attributes and their attribute measures are elicited by elicitation from the decision maker for each alternative. A mathematical procedure involving regression analysis is used to determine the relative weight of each attribute that will produce the ranking specified by the decision maker. The result is fed back to the decision maker who, typically, will express the view that some of their values, in terms of the weights associated with the attributes, are different. In an iterative learning process, preference weights and/or overall rankings are modified until the decision maker is satisfied with both the weights and the overall alternative ranking.

Each of these approaches is very important and we will examine them further in Chapter 7. Economic considerations, including cost-effectiveness notions are of much value also. These will be discussed in Chapter 8.

2.4.4. Considerations in the Choice of Systems Engineering Methods

As we have noted, one function performed by the systems engineer is assisting a client in organizing knowledge to deal with complex issues of large scale and large scope. Systems engineering efforts thereby provide assistance to clients in problem-solving or issue-resolution tasks. There typically are different *experts* in the technologies associated with systems engineering that are potentially available to assist in problem solving. Many of these are experts are familiar with particular technologies, or the application of specific systems engineering methods. They often claim that their particular technology or systems engineering method is "the best." The claim is made that their preferred method will yield results that exceed the quality of, or are more effective than, approaches proposed by others. The person responsible for the decision often is not sufficiently familiar with the various analysis methods to be able to judge which approach would be most appropriate for a particular issue. As a result, the choice of approach may not be determined by the issue needs, but rather by the marketing ability of the competing analysts, and perhaps by prior experiences of the decision maker with analytical assistance.

Because of a possible mismatch between problem and method and because many vendors are likely to overstate the potential power of their technology or approach, thereby raising higher expectations than are justified, many experiences with formal analysis, design, and development are characterized by frustration. Often different, sometimes even conflicting, conclusions have been obtained by those advocating different approaches to the same issue. This has added to the feeling that systems engineering analysis results can be manipulated, and has generated mistrust among managers and decision makers toward the objectivity of analysts and the utility of analytical methods. Among other things, this suggests very careful matching of methods to problems to be resolved by application of various methods.

The problems that may result from less than the most cost-effective use of systems engineering efforts are potentially very expensive. Among the consequences is the expenditure of large sums of money to support development of technologies and systems that do not lead to the best results that could be obtained for the same price, or perhaps even at considerably lower cost. Identification and use of an approach that encourages selection of appropriate methods is very important. A systems engineering approach, similar to that used to resolve actual problems themselves, is appropriate. First, the needs and constraints of the problem situation are assessed. Then, we identify candidate approaches. Finally, we study and compare the candidate approaches, select a subset, and use them in the systems engineering process.

Some general guidelines concerning the identification of candidate methods for resolution of specific issues are appropriate and useful. These guidelines should be used with an awareness of their limitations, as must the systems engineering process itself. The methods most appropriate for a given situation are very much dependent on the characteristics of the task and the operational environment and the experiential familiarly of the problem solver or client with these.

The choice of a systems engineering approach or methodology will primarily be determined by these concerns:

1. The location and number of prospective participants and their experience with the systems engineering task at hand and the operational environment into which it is embedded.
2. The available time and funds for the overall systems engineering effort.
3. The specific life-cycle approaches, in terms of steps and phases, that are considered as candidate approaches.

We suggest a cost- and operational-effectiveness evaluation of potential approaches before making a final choice of a specific systems engineering process, including the methods that are associated with this process. Thus, the approaches suggested here should be useful both in selecting an appropriate system product for development and for selecting the systems engineering process used for development of the systems engineering product.

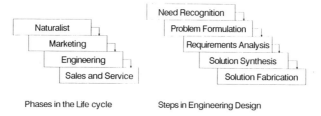

Phases in the Life cycle   Steps in Engineering Design

Figure 2.21. Phases in the "Design for Success" Life Cycle of Rouse and Steps in the Engineering Design Phase

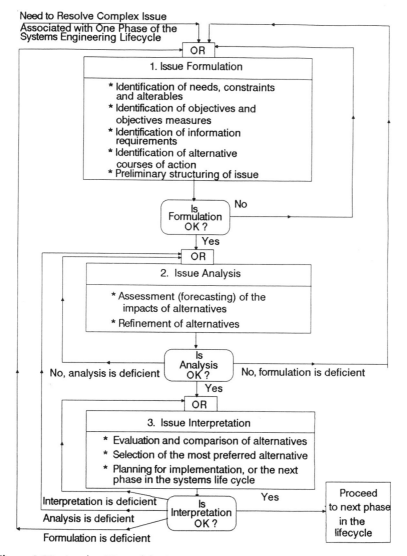

Figure 2.22. Another View of the Iterative Nature of the Systems Engineering Steps

2.5. SUMMARY

In this chapter, we have presented a number of life cycles of systems engineering processes. It is normally very useful to think of a systems engineering process as being described by a number of phases that comprise the time dimension of systems engineering. Within each phase, there are a number of steps. These comprise the logic dimension of systems engineering. Figure 2.21 illustrates the four-phase and five-step model of Rouse [19]. Within each of these phases and steps, there may—and generally will—exist the need for iteration. Figure 2.22 indicates a number of possible iterative loops in the three-step model for systems engineering.

We have indicated that systems engineering is a management technology. In the next chapter, we turn to a discussion of systems management. These rely quite closely on many of our discussions here.

PROBLEMS

2.1. Compare the ways in which management may exercise control over a systems engineering development project with and without use of a systems engineering development life cycle. Which approach do you recommend and why?

2.2. The software development life cycle intended for use by the U.S. Department of Defense is called DoD Standard 2167-A. Compare this life-cycle model with the systems engineering life cycle and the traditional waterfall life cycle and identify the differences and similarities.

2.3. It is claimed that the spiral software development life-cycle model differs from both the systems engineering life-cycle model and the waterfall software development life cycle. Indicate similarities to the systems engineering and waterfall software development life cycle; also indicate potential differences.

2.4. Compare ways in which we could break down a systems engineering development project using both the waterfall and a spiral systems development life cycles. Are there significant differences in the two decomposition approaches?

2.5. The spiral software development model introduces risk management into the software development process and proposes an approach to identifying and treating risk in software development. Show how risk management may be introduced into other systems engineering life-cycle models covered in this chapter.

2.6. The evolutionary software development life-cycle model [20, 21] is based on the approach of building successively more functional prototypes. Is this a significant departure from the systems engineering approach to software development? Indicate how the operational software development life cycle may be recast to incorporate the phases of the systems engineering life cycle.

2.7. Consider a system of your choice. Write a brief description of what the three phases in system development might have included. Also, descriptive appropriate steps for each phase, using the three-step model.

2.8. Recall a contemporary technological problem with which you are reasonably familiar. Analyze this problem in terms of one of the frameworks for systems engineering presented in this chapter. What relevant elements associated with your selected problem immediately suggest themselves? What alternatives exist? What are the critical decisions associated with your problem? What disciplines appear relevant to solution of the problem?

2.9. Recall a contemporary societal or organizational problem with which you are reasonably familiar. Analyze this problem in terms of one of the frameworks for systems engineering presented in this chapter. What relevant elements associated with your selected problem immediately suggest themselves? What alternatives exist? What are the critical decisions associated with your problem? What disciplines appear relevant to solution of the problem?

2.10. From a recent newspaper or magazine, select what appears to be a well-written article of moderate length concerning a controversial societal or technological issue. Dissect the article to the point that the issue is formulated according to the systems engineering framework presented in this chapter. Outline the various problem elements. Does your formulation aid in a better understanding of the controversial issue? How much of the controversy arises from or at the level of symptoms? Institutions? Values? What are the technological perspectives, needs perspectives, and value perspectives associated with the controversy and any proposed solution to it?

2.11. Consider a current policy question being faced by a contemporary systems engineering contractor. Describe the policy question in terms of the framework for systems engineering presented here. Pay particular attention to the steps and phases of systems engineering. What needs, constraints, and alterables can you identify as being part of the problem definition elements for each of the phases in your chosen life-cycle model? What are the appropriate objectives and objective measures that form the value-system design step for these phases? What are the appropriate activity and activity measures that constitute the system synthesis step for each phase? How do these elements fit together such that decisions and policies might be determined at each phase of the life cycle?

REFERENCES

[1] Chen, K., Ghausi, M., and Sage, A. P., "Social Systems Engineering: An Introduction," *Proceedings of the IEEE*, Vol. 63, No. 3, 1975, pp. 340–343.

[2] Anthony, R. N., *The Management Control Function*, Harvard Business School Press, Cambridge MA, 1988.

[3] Beam, W. R., Palmer, J. D., and Sage, A. P., "Systems Engineering for Software Productivity," *IEEE Transactions on Systems, Man, and Cybernetics*, Vol. 17, No. 2, 1987, pp. 163–186.

[4] Rouse, W. B., "Conceptual Design of a Computational Environment for Analyzing Tradeoffs Between Training and Aiding," *Information and Decision Technologies*, Vol. 17, No. 4, 1991, pp. 227–254.

[5] Beam, W. R., *Systems Engineering Architectures and Design*, McGraw Hill, New York, 1990.

[6] Sage, A. P., and Palmer, J. D., *Software Systems Engineering*, Wiley Interscience, New York, 1990.

[7] Hall, A. D., "Three Dimensional Morphology of Systems Engineering," *IEEE Transactions on Systems Science and Cybernetics*, Vol. SSC 5, No. 2, 1969, pp. 156–160.

[8] Hall, A. D., *Metasystems Methodology*, Pergamon Press, Oxford, 1989.

[9] Sage, A. P., and White, C. C., *Optimum Systems Control*, 2nd ed., Prentice Hall, Englewood Cliffs NJ, 1977.

[10] Computer Science Corporation, *Systems Integration—The Process*, Corporate Brochure, 1990.

[11] Forman, F. L., and Hess, M. S., "Strategies for Developing Successful Information Systems," American Management Systems, Special Topics Report, 1989.

[12] Royce, W. W., "Managing the Development of Large Software Systems: Concepts and Techniques," *Proceedings WESCON*, 1970, pp. 1–70.

[13] Boehm, B. W., "Software Engineering," *IEEE Transactions on Computers*, Vol. 25, No. 12, 1976, pp. 1126–1241.

[14] Boehm, B. W., "A Spiral Model of Software Development and Enhancement," *ACM SIGSOFT Software Engineering Notes*, Vol. 11, No. 4, 1986, pp. 14–24.

[15] Boehm, B. W., "A Spiral Model of Software Development and Enhancement," *IEEE Computer*, Vol. 21, No. 5, 1988, pp. 61–72.

[16] Boehm, B. W. (Ed.), *Software Risk Management*, IEEE Computer Society Press, Washington DC, 1989.

[17] Gass, S. I., and Sisson, R. L., *A Guide to Models in Governmental Planning and Operations*, Sauger Books, Potomac MD, 1975.

[18] Atherton, G., and Borne, P. (Eds), *Concise Encyclopedia of Modeling and Simulation*, Pergamon Press, Oxford, 1992.

[19] Rouse, W. B., *Design for Success: A Human-Centered Approach to Designing Successful Products and Systems*, John Wiley, New York, 1991.

[20] McCraken, D. D., and Jackson, M. A., "Life Cycle Concept Considered Harmful," *ACM Software Engineering Notes*, Vol. 7, No. 2, 1982, pp. 29–32.

[21] Gladden, G. R., "Stop the Life Cycle. I Want to Get Off," *ACM Software Engineering Notes*, Vol. 7, No. 2, 1982, pp. 35–39.

Chapter **3**

Systems Management

In this chapter, we discuss some aspects of the management of systems engineering processes—*systems management*. There are two fundamental goals for our efforts here: (1) description of systems management activities directed at the ultimate production of trustworthy systems that are generally of large scale and scope; (2) specification of costs and effectiveness indicators that occur throughout the life cycle of design, development, and implementation of large systems. Generally, system development costs will be a function of the benefits that might be obtained from alternative levels of realization of client needs. Our efforts in this chapter are concerned primarily with systems management perspectives. Chapters 4–9 are concerned, in part, with the metrics and methods needed to establish necessary indicators for systems management.

3.1. INTRODUCTION

There are many definitions that could be provided for the term management and the related term systems management. For our purposes, appropriate definitions are as follows:

Management consists of all the activities undertaken to enable an organization to cope effectively and efficiently within its environment. This will generally involve planning, organizing, staffing, directing, coordinating, reporting, and budgeting activities in order to achieve identified objectives.

Systems Management is the organized and integrated set of procedures, practices, technologies, and processes that contribute to efficient and effective accomplishment of management subgoals. Systems management efforts are designed to lead to achievement of overall plans or objectives of an organization.

The word organization appears several times in these definitions, as does the word plan. It is important to note that a given organization or business will, or at least should, have a plan or a set of plans to achieve the overall objectives of the organization. We use terms like *enterprise, organization* and business in a very general sense to refer to private and public sector groups that have been arranged in a systematic fashion in order to achieve purposeful objectives. Many of these objectives relate to the way in which the organization provides products and services to its customers or clients. Such a plan generally aims to satisfy both general and specific needs. The general needs are those of the organizational units within the enterprise. A framework evolves out of these high level plans and policies that enables identification of plans and subsequent activities to fulfill the needs of a specific client for a systems engineering product or service. Each of these concerns is important. The first relates to the way in which the enterprise organizes itself, and the second relates to the way it serves customers.

More often than not, specific systems engineering programs include identification of a client's needs; development of technical specifications for hardware, software, and systems management processes that will fulfill these needs; and a resulting system design, implementation, and operational test and evaluation. It is important here to provide some notions concerning systems management efforts, as they are crucial to success of the overall systems engineering process and the resulting product or service.

3.2. ORGANIZATIONAL MANAGEMENT

There are a variety of definitions of an organization. Some of those that are relevant to our systems engineering discussions here are the following:

A system of consciously coordinated activities of two or more people [1].

Social units deliberately constructed to seek specific goals [2].

Collectives that have been established on a relatively continuous basis in an environment with relatively fixed boundaries, a normative order, authority ranks, communication systems, and an incentive system designed to enable participants to engage in activities in general pursuit of a common set of goals [3].

A set of individuals, with bounded rationality,* who are engaged in the decision-making process [4].

---

*Roughly speaking *bounded rationality* infers the practice of suboptimization or limited optimization of objectives. It occurs because of limited time, limited information, or limited human ability to optimize. Chapter 9 discusses bounded rationality.

Organizations can be viewed from a *closed-system* perspective. In this view, an organization is an instrument designed to enable pursuit of well-defined specified objectives. From this perspective, an organization will be concerned primarily with four objectives:

Efficiency,
Effectiveness,
Flexibility or adaptability to external environmental influences,
Job satisfaction.

Four organizational means or activities follow from this [5]:

Complexity and specialization,
Centralization or hierarchy of authority,
Formalization or standardization of jobs,
Stratification of employment levels.

In this view, everything is functional and tuned such that all resource inputs are optimized and the associated responses fit into a well-defined master plan.

March and Simon [6], among others, discuss the inherent shortcomings associated with this closed-system model of humans performing machine-like tasks. Not only is the human-as-machine view believed to be inappropriate, but there are pitfalls associated with viewing environmental influences as "noise," as must necessarily be done in the closed system perspective. March and Simon's broadened view of an organization is known as the *open-systems* view. In the open-systems view of an organization, concern is not only with objectives but also with appropriate responses to a number of internal and external influences.

Other authors have expanded upon these views. For example, Weick [7, 8] describes organizational activities of enactment, selection, and retention. These activities assist in the processing of ambiguous information that results from an organization's interactions with changes in the environment. The overall result of this process is the minimization of information equivocality, or ambiguity, such that the organization is able to understand its environment, recognize problems, diagnose causes, identify policies to potentially resolve problems, evaluate the efficacy of these policies, and select a priority order for problem resolution. Minimization of informational ambiguity is a key management task in Weick's view. This allows organizational leadership to develop tactics for achievement that are ideally free of ambiguity.

In this view, the result of the activities of the organization lead to the *enacted environment* of the organization, which contains an external part—for example, the activities of the organization in product markets; and an internal part—the result of organizing people into a structure to achieve organizational goals. Each of these environments is subject to uncontrollable exogenous influences due to economic, social, and other changes. *Selection activities* allow perception, framing, editing, and interpretation of the effects of the organization's actions on the external and internal environments. Identification of a set of relationships believed to be important follows from selection activities. *Retention activities* allow admission, rejection, and modification of selected knowledge in accordance with existing retained knowledge and integration of previously retained organizational knowledge with new knowledge. There are a potentially large number of cycles that may be associated with enactment, selection, and retention. These cycles generally minimize informational equivocality and allow for organizational learning such that the organization is able to cope with very complex and rapidly changing environments.

A very important feature of most realistic contemporary organizational models is that of *organizational learning*. An organization, or individual, has *learned* when it uses the results of experience obtained at some time as the basis for changed behavior at some later time. Much organizational learning is neither beneficial nor appropriate. For example, organizations and individuals may use improperly simplified and distorted models of causal and diagnostic inferences and improperly simplified and distorted models of the contingency structure* of the environment and task in which these realities are embedded. They obtain poor results, do not realize this, and learn poorly.

Group cohesion, conformity, and reinforcing beliefs often lead to what has been called "groupthink" [9, 10] and an information acquisition and analysis structure that enables information processing only in accordance with the belief structure of the group. The result is selective perceptions and neglect of potentially disconfirming information, thereby precluding change of beliefs and adoption of appropriate strategies to achieve the task at hand. There is no evidence whatever that suggests that systems engineering organizations are any less vulnerable to these difficulties than other types of businesses.

Many management studies show that, in practice, plans and decisions are the result of interpretation of standard operating procedures. Improvements are obtained by careful identification of existing standard operating procedures and associated organizational structures. The resulting *organizational process model*, originally due to Cyert and March [11], functions by relying on standard operating procedures that constitute the memory or intelligence bank of the organization. Only if the standard operating procedures fail will the organization attempt to develop new standard operating procedures.

Organizational learning results when members of the organization react to changes in the internal or external environment of the organization by detection and correction of errors. An error is a feature of knowledge that

---

*The *contingency task structure* of a task situation includes the task itself, the environment into which the task situation is embedded, and the experimental familiarity of the person performing the task with the task situation and the environment into which it is embedded.

makes action ineffective (we have more to say about the subject of human error in Chapter 9). Ideally the detection, diagnosis, and correction of error produces learning. Individuals in an organization are agents of organizational action and organizational learning. Argyris [12] cites two information-related factors that inhibit organizational learning:

1. The degree to which information is distorted such that its value in influencing quality decisions is lessened.
2. The resulting lack of receptivity to corrective feedback.

A major claim by systems designers is that errors in using and designing systems are due not simply to probabilistic random events that might be removed through improved system operator or system designer training, or through better system designs [13, 14]. Instead, it is argued that errors are due to two generally more important sources.

1. Errors represent systematic interference and incongruities among models, rules, and procedures.
2. Errors represent some dysfunctionality of the effects of adaptive learning mechanisms.

One of the fundamental notions in studies of human errors is that there is an intimate association between human intent and human error. Realistic efforts to discuss human error and to design systems that can cope with human error, therefore, consider the different types of human intentions and associated errors. In a very insightful work, Reason [15] indicates the importance of knowing:

1. Whether human actions are directed by conscious intent.
2. Whether human actions proceed as planned.
3. Whether human actions achieve the desired result.

Six types of individual actions result from this observation, as indicated in Figure 3.1. Only two, represented by the shadowed boxes, are correct actions. There is no reason to believe that this figure is not also applicable to organizations. Successful system designs create systems that encourage intentional action and successful performance by individuals and organizations. Chapter 9 is concerned with human error, or more generally cognitive ergonomics.

The discussion thus far has focused on concepts. We have presented a description and interpretation of some recent studies in behavioral and organizational theory that have direct relevance to systems management. The primary classical organizing principles supporting organizational behavior

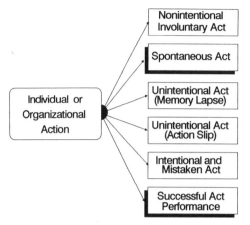

Figure 3.1. Reason's Taxonomy of Human Actions

include the following:

1. Division of labor and task assignment.
2. Identifying standard operating principles.
3. Top down flow of decisions.
4. Formal and informal channels of communication in all directions.
5. The multiple uses of information.
6. Organizational learning.

We must be conscious of these principles if we are to produce systems management and technical direction perspectives that are grounded in the realities of human desires and capabilities for growth and self-actualization. Also, we must be aware of error possibilities, both on the part of the system architects and developers and on the part of system users. These concerns are very realistic and influence the design of information and decision support systems [16, 17], software systems [18], and many other areas within systems engineering that have a strong human interaction component [19]. They have strongly influenced what we will call strategic quality assurance and management principles, a subject we examine in Chapter 5.

Much of our discussion to this point may seem oriented more toward management philosophy than toward systems management practice. Doubtlessly, this is correct. We maintain strongly, however, that successful systems management practice will embody these principles, prescriptions, perspectives and philosophies. Our systems management philosophy supports the pragmatic management of large systems engineering projects. This philosophy should also be incorporated into the processes we use and the systems that we develop, such that they are suitable for human interaction.

There have been many attempts to classify management functions. Among these is the function-type taxonomy of Anthony [20], who describes three planning and control functions:

Strategic Planning Function: Strategic planning is the process of choosing the highest level policies and objectives, and associated resource allocations and strategies for achieving these. According to Anthony, strategic planning is unsystematic in that the need for strategic decisions may arise at any time and that the threats and opportunities that lead to strategic decisions are not discovered systematically or at uniform intervals.

Management Control Function: Management control is the process through which managers influence other organizational members to help them achieve organizational strategies. Management control decisions are those decisions made for the purpose of assuring effectiveness in the acquisition and use of resources to achieve strategic plans.

Task Control Function: The process of task control has as its major objectives the efficient and effective performance of specific tasks. In an earlier work, Anthony described task control in terms of two related functions. *Operational control functions* were accomplished for the purpose of assuring effectiveness in the performance of operations. *Operational performance functions* were associated with the day-to-day decisions made while performing operations.

These three planning and control processes relate to one another as indicated in Figure 3.2. Although there is often considerable variation among the many task control systems that may be found in a given organization, nearly all such systems that may include interaction between one task manager and a team of nonmanagers, or perhaps with an automated system. Task control functions are generally well-structured. Management control involves the interaction of managers in resolving unstructured issues in a manner that

Figure 3.2. Flow of Organizational Information and Associated Planning and Control Decision Flow

supports achievement of the strategic plan of the organization. Management control concentrates on the activities that occur within various *responsibility centers* of the organization. It would appear that the system management function described here, and elsewhere in the systems engineering literature, is fundamentally similar to that of Anthony's management control.

Other discussions of organizational management concentrate on activities of a management team, as contrasted with the decisions that they make. It is not unusual to find seven identified management functions or tasks in the classic enterprise management literature:

1. *Planning:* Identification of alternative courses of action to achieve organizational goals.
2. *Organizing:* Structuring tasks and granting authority and responsibility to achieve organizational plans.
3. *Staffing:* Selecting and training of people to fit various roles in the organization.
4. *Directing:* Creation of an environment and an atmosphere that will motivate and assist people to accomplish assigned tasks.
5. *Coordinating:* Integration and synchronization of performance, including the needed measurements and corrective actions, so as to achieve goals.
6. *Reporting:* Ensuring proper information flow within the organization.
7. *Budgeting:* Appropriate allocation of economic resources needed for goal achievement.

This POSDCORB theory of management is a very common one and is described in almost all classical management texts.

It is quite clear that these functions are not at all independent of one another. Since details of these functions are provided in essentially any introductory management guide, they are not pursued in any further detail here. It is important to note that, collectively, these are the tasks of general enterprise management. They apply to systems management in its *planning activities*, which involve anticipation of potential difficulties and the identification of approaches for detection of problems, diagnosis of causes, and determination of promising corrective actions. They apply also to systems management and its *control activities*, which involve controls exercised in specific situations in order to improve efficiency and effectiveness of task controls in achieving objectives.

Planning is a prominent word in much of the foregoing. We can identify three basic types of, or levels for, plans:

Organizational plans,
Program plans,
Project plans.

One of the major differences in these types of plans is the duration of the plans. Organizational plans are normally strategic in nature and can be expected to persist over a relatively long time. Program plans are intended to achieve specific results, such as a program to design, develop, and install a new information system capability for an airline. For large programs, it is generally desirable to disaggregate the program plans and controls into a number of smaller projects.*

Three fundamental activities are involved in systems management. These are the same activities, or steps, followed in the problem-solving or systems engineering process: (1) formulation of issues, (2) analysis of alternatives, and (3) interpretation of the impacts of the alternatives. These steps are encountered at several levels, or *phases*, of organizational management activities. They begin at the strategy level and result in the preparation of strategic plans, which are then ultimately converted into tactical and operational plans and implemented as management controls or task controls through an effective planning process.

At this highly aggregated level, it may be difficult to envision specific systems management activities. Although there are many more narrowly defined steps into which the aforementioned three steps can be partitioned, three general and nine specific levels of formal activities appear especially important steps in systems management.

1. Issue formulation.
   1.1. Environmental monitoring.
   1.2. Environmental understanding.
   1.3. Identification of information needs.
   1.4. Identification of alternative potential courses of action.
2. Issue analysis.
   2.1. Identification of the impacts of alternatives.
   2.2. Fine tuning the alternatives for efficiency and effectiveness.
3. Issue interpretation.
   3.1. Evaluation of each alternative.
   3.2. Selection of a best alternative.
   3.3. Implementing the selected alternative.

The managers of specific systems engineering functional efforts, as well as managers in general, perform each of the activities that we have just identified. These activities are appropriate at each phase in a systems management effort. There also exists the need for contingency or crisis management plans that can be implemented if the initially intended plans prove unworkable. A severe form of contingency is that of a crisis situation.

---

*There is, again, no standard set of terminology. The software engineering literature, for example, uses projects where we have used programs here. It is a reasonable speculation that this is done to avoid confusion with program(ming) lines of code.

A management crisis exists whenever there is an extensive and consequential difference between the results that an organization hopes to obtain from implementation of a strategic plan and what it actually does obtain. This difference may have a variety of causes. A common element is that the organization has misjudged either the environment in which it is operating or the impacts of its chosen courses of action on the environment. A crisis may occur because of failure to identify a potentially challenging new opportunity or as a result of an existing situation that is threatening to the health or survival of the organization. The preferred solution in either case is crisis avoidance. An acceptable, but somewhat less preferred, solution is extrication from crisis situation. The key to each of these solutions to a crisis is effective management of the internal organizational environment such that it is better able to cope with the external environment.

Some activities may be performed at an intuitive level, based on experiential familiarity with particular task requirements. Some should be performed in a formal analytical manner because they are initially unstructured and unfamiliar. In each case information is of critical importance. Information is needed about the external environment to facilitate understanding of that environment. The first step in issue formulation is identification of a set of information needs relative to systems management objectives. In parallel with this, preliminary identification of potential alternative courses of action is made, and these potential courses of action further act to frame the information needs for proper analysis and evaluation of these alternatives. All of these activities are accomplished as part of issue formulation. The issue analysis and issue interpretation steps are equally rich in terms of their need for information.

Systems management and management control are vitally concerned with the processing (broadly defined to include acquisition, representation, transmission, and use) of information in the organization. Generally, information is now recognized as a vital strategic resource and is so treated here. A simple three-step reasoning process leads to this conclusion.

1. Organizational success depends upon management quality.
2. Management quality depends upon decision quality.
3. Decision quality depends upon information quality.

There are a number of similar ways in which this reasoning process could be stated. In terms of the three planning and control functions of Anthony, for example, we could state the following.

1. Organizational success depends upon strategic planning quality.
2. Strategic planning quality depends upon information quality and management control quality.
3. Management control quality depends upon information quality and task control quality.

These relations, and that of organizational learning, are illustrated in Figure 3.2. Quality is an often used term in this discussion. We will have much more to say about quality in our next three chapters.

Earlier we noted that one of the major tasks of management planning and control is that of minimizing the ambiguity of the information that results from the organization's interaction with its external environment [7, 8] in order to

1. Better enable the organization to understand its environment.
2. Detect or identify problems in need of resolution.
3. Diagnose the causes of these problems.
4. Identify alternative courses of action or policies to correct or resolve problems.
5. Analyze and evaluate the potential efficacy of these policies.
6. Interpret these in accordance with the organizational culture and value system.
7. Select an appropriate priority order for problem resolution.
8. Select appropriate policies for implementation.
9. Augment existing knowledge with the new knowledge obtained in this implementation such that organizational learning occurs.
10. Maintain total quality through strategic planning and systems management.

This task of minimizing informational ambiguity is primarily that of systems management or management control. It is done subject to the constraints imposed by the strategic plan of the organization. In this way, the information presented to those responsible for task control is unequivocal. This suggests that the task control function receives information inputs primarily from those at the management control level. Figure 3.3 represents a partial view of the resulting organizational information flow. It suggests that there are planning and control activities at each of the three functional levels in an organization. The nature of these planning and control activities are different across these levels, however, as is the information that flows into them.

Let us examine some specific implications of management control, or systems control, for the production of trustworthy systems of hardware and software, and associated interfaces that facilitate successful human interaction. Regardless of the hierarchical level at which planning is considered: *A plan is a statement of what ought to be, together with a set of actions or controls that are designed to cause this to occur*. Of course, the interpretation of ought may vary considerably as a function of the level at which planning is accomplished. Also, there may be a number of uncertainties that may act to prevent a normally useful set of activities from achieving the objectives that they should achieve.

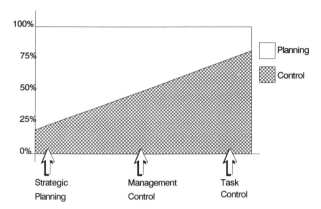

Figure 3.3. Relative Importance of Policy Formulation (Planning) and Policy Execution (Control) in Anthony's Three Functional Planning and Control Processes

In general and as stated, planning involves the following.

1. Identification of goals to be realized, some of which may already be fulfilled to some extent, at the particular level of planning under consideration.
2. Identification of current position relative to the goals such that it becomes possible to specify a set of needs that, when fulfilled, will lead to goal realization.
3. Identification of past, present, and future environments such that it becomes possible to understand effects of the constraints on, and alterables of, realistic courses of action.
4. Identification of suitable alternative courses of action designed to lead to need fulfillment and goal attainment.

These planning elements are associated with an organization's internal environment, including its culture and standard operating policies, and the external environment. We also need to identify measures or metrics such that we can determine success in need satisfaction, goal attainment, and activity accomplishment. These should be linked together such that we identify and understand relationships among the elements of planning.

Implied in the identification of strategic planning options is an analysis of the external and internal environments of an organization.

1. An external environment analysis is needed to identify the present context in which the issue being considered is embedded and to forecast possible future situations.

2. An internal environment analysis, at the level of the organization where planning is being accomplished, is needed in order to determine available resources and to identify the organizational culture.

We can divide this still further. For example, a *general environment* or *management control environment* consists of those elements that affect all organizational activities within a specific domain: cultural, demographic, technological, and so forth. Also, a *task environment* consists of those elements specifically affected by, and affecting, the particular organization and alternative course of action in question.

At this point in a planning effort, we have scoped out the issue considerably and identified a number of possible courses of action. We have accomplished formulation of the issue. The major planning ingredients needed for a complete and useful plan are as follows:

1. *Realistic objectives.*
2. *Identification of a course of action together with suitable and observable activities measures.*

To achieve these, we need to analyze the options that have been generated to determine their impacts on needs. In dealing with a large and complex issue, a variety of systems analysis tools may have to be used. After this analysis, we need to obtain an interpretation, reflecting the value system of the clients. Management control efforts are carried out with respect both to programs that ultimately result in the delivery of operational systems, and to the operational and task-related activities of the systems engineering organization itself. The interpretation activities of management control include evaluation of these impacts and selection of an alternative course of action, with respect either to program deliverables or to adaptation of operational efforts within the corporation to improve performance. It is for this reason that we include iteration and feedback through learning in Figure 3.2.

In any large and complex effort, it is necessary to break a program plan down into several project plans. A successful project plan must identify and detail

What is to be done.
Who will do it.
What resources will be used.
In what time period.

The course of action element, "what is to be done," must meet the needs of the client and must possess sufficient quality and functionality. Sufficiency is a very subjective term that depends on the clients needs, priorities, and available resources to meet these.

There are many causes of systems management failures. Among those which have been identified are:

1. Difficulties in defining work in sufficient detail for the level of skills available.
2. Problems with organizing, building, and managing the systems development team.
3. Systems management team members are reassigned prior to project completion.
4. Failure of clients to adequately review and understand requirements specifications.
5. Insufficiently defined configuration management plan.
6. No firm agreement on program plan, project plans, or configuration management controls.
7. No adequate set of standards.
8. No operational level quality assurance or configuration management plans.
9. No clear role or responsibilities defined for development program personnel.
10. Program perceived as not important to various individuals or organization.
11. No risk management or crisis management provisions.
12. Inability to measure program performance.
13. Poor communications between program management and members of the systems development organization.
14. Poor communications with customer, client, or sponsor.
15. Difficulty in working across functional lines within the system development organization.
16. Improper relations between program performance and the reward system.
17. Poor program and project leadership.
18. Lack of attention to early warning signals and feedback.
19. Poor ability to manage conflict.
20. Difficulties in assessing costs, benefits, and risks.
21. Insensitivity to organizational cultures.
22. Insufficient program and project plans and procedures.
23. Apathy by program or project management.
24. Rush into project initiation before adequate identification of configuration management plans.
25. Unrealistic schedules and budgets.
26. Lack of attention to strategic quality assurance and management.

Most of these will cause systems engineering projects to fail to finish their scheduled activities on time and within costs. And, this list is incomplete!

There is often disagreement within various groups concerning which of these factors are most important and which is likely to be responsible for the failure of systems engineering projects. For example, top level systems managers generally indicate that front-end planning, including identification of system level requirements, is very important. Program and project engineers usually consider this to be less important. On the other hand, program and project engineers are likely to perceive technical complexities of a program as very important and wish to devote significant effort to understanding these. Program management often considers this relatively less important than such factors as identification of requirements and establishment of requirements specifications. Both groups generally identify customer changes in specifications during program completion as being a major factor in time slippage and cost overruns. On the other hand, clients doubtlessly do not perceive their requirements as changing. They perceive that these were very poorly identified initially, and perhaps poorly translated to specifications, which were then implemented in an error prone manner. These differences in perceptions strongly support the use of prototyping techniques and the use of support systems for issue exploration and judgment, so as to enable full understanding of tasks to be undertaken as early as possible in the systems engineering life cycle. Most importantly, they suggest the incorporation of risk management procedures throughout the life cycle, and operational- and strategic-level quality assurance and management of both process and product. Operational-level quality assurance and management will be examined in Chapter 4. Strategic-level quality assurance and management will be the subject of Chapter 5.

3.3. ORGANIZATIONAL STRUCTURES

In general, there are three types of systems engineering organizations. The first and most often found is the *functional organization*, sometimes called a *line organization*. In the functional organization, one particular group is asked to perform one phase of activities associated with developing a systems engineering product or service. This does not mean that one person is necessarily asked to perform all phases of a systems development effort. Rather, it means that a team of people within a fixed structure are asked to do this. Each manager of a section in the organization is given a set of requirements to be met. Each manager is able to exercise more or less complete authority over the activities going on within that particular section.

A representative functional organization structure is depicted in Figure 3.4. Note that this is a typical hierarchical structure in which management authority for a specific systems engineering program is vested in top-level management. The various line supervisors are given tasks, generally associated with life-cycle phases of effort and report to the top management office.

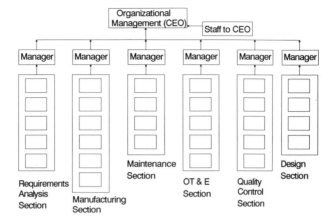

Figure 3.4. Functional Line (Hierarchical) Systems Engineering Organization

There are several advantages to this type of management structure:

1. The organization is already in existence prior to the start of a given systems engineering development program and this enables quick start-up and phase-down of programs.
2. The recruiting, training, and retention of professional people may be easier, as these people generally remain with one, often relatively small, unit throughout much of their career. For this reason, functionally structured organizations are often very people oriented.
3. Standards, metrics, operating procedures, and management authority are already established and have typically become a part of the organizational culture of the systems engineering unit in question.

Of course, there are a number of readily identified limitations also.

1. No one person has complete management responsibility or authority for the program except at a very high management level. That person is often in charge of several development programs.
2. Interface problems within the organization are often difficult to identify and solve.
3. Specific programs may be difficult to monitor and control because there is no single management champion for the effort being undertaken.
4. Functions may tend to perpetuate themselves long after there is any real need for them.

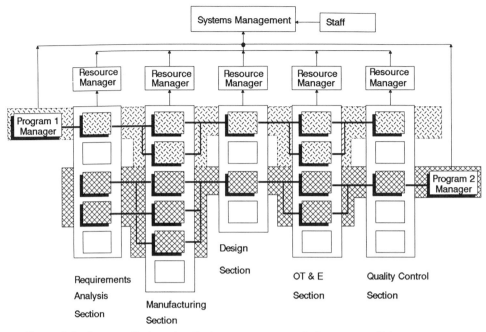

Figure 3.5. Systems Engineering Project Management (Obtaining Staff from Resource Managers) Organization

The *program organization* is one possible remedy for this. In a program organization, or project organizations within a given program, one person is vested with overall responsibility to insure that one specific program is accomplished in a timely and trustworthy manner. In a program management organization, people are usually excised from an existing functional organization and wedded together under the management of the program or project manager.

Figure 3.5 illustrates some of the features of this superposition of program authority on a line organization. Figure 3.6 illustrates the resulting program organization, which may and generally will include a number of projects within a given program that, taken together, comprise the program. There are several advantages to a program or project structure:

1. There is one person in a central position of responsibility and authority for the program—the program manager.
2. That person, or a designee, will generally have authority over all system-level interfaces, especially those that cut across the phases of the development effort and projects within those phases.

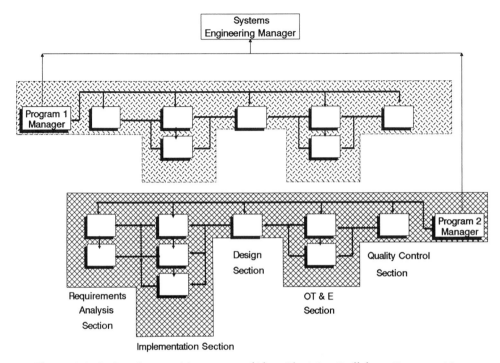

Figure 3.6. Project Systems Management (After Obtaining Staff from Resource Managers)

3. Decisions can generally be made very quickly at the program level because of the new centralized organization, shown in Figure 3.6, which puts the program at a central management level in the organization.

The limitations to program and project organizations are related to their strengths.

1. The project organization must be formed from the existing line organization.
2. Recruiting, training, and retention of people to work in a program office is more difficult than in a functional organization, as the resulting programs and the projects within a program are product-oriented rather than people-oriented.
3. The benefits of economy of scale cannot be achieved for other than very large programs, as there will often be only one or two people in a given technical specialty area associated with a program or project.

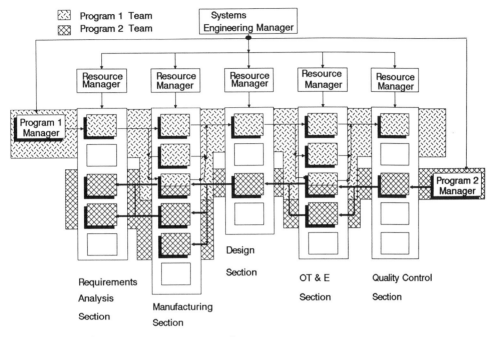

Figure 3.7. Matrix Structure for Systems Management of Programs

4. Programs tend to perpetuate themselves, just as a given functional line will tend to perpetuate itself in a strictly functional organization.
5. Often, it is necessary to develop standards, metrics, techniques, and procedures for each program undertaken. Standards are not the same across all programs, which makes cost–benefit and other comparisons across programs very difficult.

The *matrix program* or *project*, organization has been proposed as a way to, ideally, combine the strengths and minimize the weaknesses of the functional organization and the program organization. Figure 3.7 illustrates how people in a matrix organization are managed both by functional line-type supervisors, often called *resource managers*, and by program or project managers. In this management structure, any given person in the functional line may be working on more than one program or project at a given time.

In this sort of management structure, the program and project managers have responsibility for short-term supervision of the people working for them on various programs. The resource manager would be responsible for longer-term management of these same people. As with the other two management types, there are advantages and limitations to this form of management structure. Some of the advantages are as follows.

1. There is central position of responsibility and authority over the program being undertaken.
2. The interfaces between the various specialty functions can be controlled more easily than in line or functional management programs.
3. It is usually easier to start and terminate a program than in program management organizations.
4. Standard operating policies, technical standards, metrics, and procedures are generally already established, as these are the responsibility of the typically longer term functional line management.
5. Professional staffing, recruiting, education, and training are easier and retention of the best staff members is higher than is typically the case in program management organizations.
6. Because the structure is more flexible, it is potentially possible to obtain more efficient and more effective use of people, than is possible in either functional or program management.

Potential problems with matrix management are as follows:

1. Responsibility and authority for human resources is shared between the line resource manager and one or more program or project managers.
2. It is sometimes too easy to move people from one project to another, especially compared with the project management organization. This may lead to personnel instability and other morale problems.
3. Because of its complexity, greater organizational understanding and cooperation are required than in either program or functional organizations.
4. There is often greater internal competition for resources than in either the program management or functional organization.

The function of systems control or systems management is needed regardless of the program and project structure used. There are many responsibilities of top-level systems management.

1. Coordination of issue identification with the client, and translation of user needs into system specifications.
2. Identification of the resources required for trustworthy development of an organizational system of hardware and software and appropriate interfaces with human operators of systems and with existing systems.
3. Definition and coordination of software and hardware identification, design, integration, and implementation.
4. Interfaces with top-level management of both the client and system developer.

There are a number of front-end problems facing the typical systems engineering program management team. Generally the most difficult of these involve human communications and interactions, system requirements that change over time, perceptions of systems requirement that change over time when the requirements did not change, sociopolitical problems, and the lack of truly useful automated planning and management tools. A program plan requires written documentation including sufficient details to indicate that the plan is thoroughly developed. As a minimum, this must include evidence that the following are present:

1. There exists an understanding of the problem.
2. There exists an understanding of the proposed solution.
3. The program is feasible from all perspectives.
4. Each of the associated projects benefits the program.
5. The program and project risks are tolerable.
6. There exists an understanding of project integration needs.
7. The overall effort is cost effective.
8. The overall effort is conducted from quality assurance and management perspective at both strategic and operational levels in order to insure processes and developed products that are trustworthy and sustainable.

We have discussed most of these needs and how to determine their possible fulfillment. Our efforts in the remainder of the text expand on this discussion.

One of the major needs in systems engineering program (and project) management is management planning and control and associated monitoring. It is necessary to monitor progress of the program and projects associated with the program and their respective controls in order to know how well the effort at any specified instant of time is proceeding according to the established schedule and budget. Change is attempted, if and as needed, to insure that the actual program schedule and performance is as close as possible, or needed, to that planned. These general systems management needs are independent of the structure of the systems engineering organization.

Figure 3.8 illustrates the necessary feedback and iteration in the monitoring and controlling of a program that follows after start-up of the development portion of the systems engineering life cycle. An appropriate sequence of steps is as follows.

1. Monitor progress of the program and projects.
2. Compare actual progress with that contained in the plan.
3. Monitor the quality of the process and the evolving systems engineering product.
4. Revise the plan as needed.

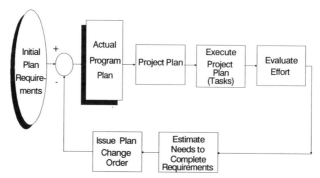

Figure 3.8. Monitoring and Control as a Feedback and Iterative Process to Aid in Systems Management through Strategic Planning

5. Define and utilize operational metrics for evaluation of evolving product quality in terms of
   - *Audits:* Formal examination, generally by an external team, of project management and development.
   - *Reviews:* Formal examination, generally by internal project management, of plan and project documents, people, and products.
   - *Inspections:* Formal examination, generally by an internal peer group, of deliverable parts of the projects.
6. Establish procedures for
   - Understanding of required functions.
   - Identification of critical milestones.
   - Identification of task responsibility.
   - Crisis management.

If monitoring and controlling are not performed effectively, it is very likely that both individual project and overall program progress will suffer, which will not be noticed until it is too late to effectively and efficiently deal with this slippage. We require an appropriate strategic-level approach to process quality assurance and management to do this effectively.

Another important need is organizing and scheduling the technical details to be performed on a program and its constituent projects. The first step in this is to identify the key tasks to be accomplished in terms of:

1. The system itself that is under development.
2. Documentation of the system.
3. Hardware and software tests, reviews, and evaluations.

4. Interface and systems integration considerations.
5. Configuration management functions.
6. System installation and training requirements.

The principal task here is to identify schedules for people and resources such that there will be no unpleasant *surprises* as the systems engineering effort progresses. Essential in this is the need to *communicate* these key tasks and requirements to all concerned.

Staffing is a very important need for systems engineering success at all levels and phases of effort. Organizations are no better than the people who belong to them. This is as true of systems engineering organizations as it is of any other type. One fundamental and often overlooked notion in productive staffing is what we choose to call *the staffing quality principle.* In its simplest form, this states that six people who can each jump one foot does not equal one person who can jump six feet. Few would disagree with this notion as stated. Yet, it is often very hard to apply this principle to programs where a few good people are really needed for program success, and a larger number with lesser talents would not be useful.

3.4. SYSTEMS MANAGEMENT LIFE CYCLES

In this section, we discuss several systems management life cycles that have been proposed for development of new technology products and systems. In large part, our efforts are based on References [21] and [22].

Lack of an appropriate systems engineering life cycle for systems management can make technology developments occur at a much slower rate than might otherwise be possible. Also, the development costs may be much higher than if an efficient and effective process were followed. Such higher costs and slower development rates may impair the ability of organizations, and indeed even of nations, to compete in our increasingly competitive world.

There is much contemporary interest [23] in technology transfer or infusion, innovation, entrepreneurship, and other efforts associated with the effective and appropriate development of technologies for increased competitiveness and satisfaction of marketplace demand. Using a systems engineering approach to technology development enables the focusing of research and development. This should enhance the potential for productivity, including commercialization, of the resulting technological products or services. Here, we describe a framework and architecture for a systems engineering-based approach to these issues.

We envision a multiphase effort not unlike that associated with the general systems engineering process that we have described in Chapter 2. An operational set of critical success factors for systems management will provide:

1. The ability to quickly identify ideas and potential approaches that are worth pursuing and those that are not worth pursuing.
2. The ability to identify a reasonably short, reasonably low-cost sequence of activities that will result in a cost-effective and societally desirable implementation of a systems engineering product, process, or service.
3. The ability to identify a systems engineering program, and specific associated projects, that allow accomplishment of the first two steps and provide for detailed implementation efforts.
4. The ability to identify impediments or barriers to successful program implementations, and either to remove the barrier or provide a mechanism for disengagement from the potentially unsatisfactory systems engineering effort.

Accomplishing this will require a quick-response, action-oriented approach to reviewing the technology under consideration, the systems management of this technology, and the marketplace potential for the technology as well as the products and services that result from its use. To be sure, there are other challenges associated with the overall process of systems management. The U.S. National Research Council [24] sponsored a 1987 study that identified eight critical needs or prescriptions:

1. Integration of technology into the overall strategic objectives of the firm.
2. Being able to get into and out of technologies faster and more efficiently.
3. Accessing and evaluating technologies more effectively.
4. Accomplishing technology transfer in an optimum manner.
5. Reducing new product development time.
6. Managing large, complex, interdisciplinary, and interorganizational projects and systems.
7. Managing the organization's internal use of technology.
8. Optimal leveraging of the effectiveness of technical professionals.

These critical success factors are associated with the entire life cycle of system development. Although they were developed specifically for the United States, these prescriptions appear sufficiently generic that they are

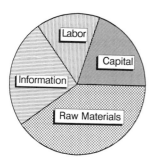

Figure 3.9. Typical Mixture of Four Primary Factors of Production

universally applicable. These can, therefore, be used as some of the factors to evaluate technologies proposed for development. There are, of course, many other factors that affect technologies. These include technology transfer or infusion [25], national and international standardization [26], and issues that affect technology development, planning, and management [27–30].

A critical need in systems management is an approach that is responsive to the rapid shrinking of time between technology conceptualization and product emergence. The major causes of this shrinkage appear to be the increased intensity and significance of international competitiveness and the technological changes made possible by information technology, such as computer-aided design, manufacturing, and production methods. These support many basic industrial developments, such as energy supply, electrical power networks, transportation, manufacturing, and agricultural production. They also support increased information support to production. One result of this shrinkage is a needed shortening of the life-cycle development process.

In our view, information needs to be a separately identified factor that is explicitly included in the usual listing of the three primary factors of production; capital, labor, and raw materials, as noted in Figure 3.9. Only now is this need beginning to be recognized [31, 32]. Among relevant research that also explores this new phenomenon of the increased role of information in the production process are Cohen and Zysman [33], who suggest that American industry has not adapted to flexible manufacturing systems as quickly as it might have; Kaplan [34], who faults U.S. management accounting systems for failing to adapt to new production patterns; and Huber [35], who explores how the modern corporation is adapting its organizational structure to the forces of advanced information technologies.

There are many ways in which the critical ingredients of a systems management approach could be described. As just mentioned, we could describe these as production, capital, raw materials, and knowledge. In a previous discussion, we noted that the steps to be accomplished in each phase of an emerging engineering technology effort involved the interaction

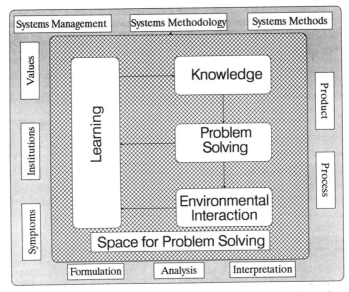

Figure 3.10. Critical Ingredients in Systems Management of Technologies for Development or Transfer

as shown in Figure 3.10 of the following:

1. Problem-solving steps.
2. Knowledge.
3. Learning.
4. Environmental interaction strategies.

We are surely describing a dynamic process that involves the interaction of many variables. An important issue is the use of systems management and integration procedures that effectively and efficiently cope with this process in ways that support continued and sustainable quality, productivity and competitiveness. Thus, we see that an appropriate systems management process must, necessarily, be associated with these four steps and the many interactions shown in Figure 3.10.

There are a number of key strategy elements affecting systems engineering for system and technology developments, including the following.

1. Technological and societal need for a technology.
2. Available technology base.
3. Research and development process management strategies.
4. Market and other external factors.
5. Standards, including technological regulations and legislation.

There are a variety of ways in which we might conceptualize a model to describe the resulting flow of technological development. In Figure 3.11, we envision three primary gateways to control the development and flow of technology from either a push or pull standpoint. As noted in this figure, the *push of technology* is basically scientific in nature in that it includes development of all feasible scientific discoveries. However few, if any, successful products emerge only because of technology push. The *pull of society*, or the marketplace, is basically the pull of the Maslow [36] hierarchy of needs.* It has been postulated [37, 38] that the growing universality of technology now makes successful innovation much more frequently driven by market pull than it is by technological push.

New product development is limited by technological capabilities and systems management capabilities. When the resulting technological systems design and management systems design needs are satisfied, there is really, therefore, only a push from feasible technological innovations. There are, of course, approaches that encourage and stimulate technology push [39]. Usually these involve enhanced communications and attention to scanning and targeting of potential market areas for the new technology. The three gateways—a technological gateway, a management gateway, and a societal gateway—determine what can ultimately flow through to society in the form of a realized system or technology. The major factors affecting (or filtering or monitoring) the flow through these gateways are also shown in Figure 3.11. The nature of each of these gateways varies across different technologies, developing organizations, and nations. There are, of course, other gateways that can be identified.

Any conceptual diagram such as this is necessarily incomplete. What we have shown in Figure 3.11 does not adequately represent the dynamics of the process, the distributed nature of the process, or the many feedback loops involved. It is very clear that technological development alters societal values, certainly the more pragmatic ones, and this in turn acts to change the nature of the societal pull for technological products and services. The process of technological innovation is distributed in space and time, and innovation is

---

*As developed by Abraham Maslow, the basic human needs hierarchy posits five basic human needs:

Physiological needs, such as goods, clothing, and health.

Safety needs, such as stability, security, and order.

Belongingness needs, such as affection and affiliation.

Esteem needs, such as self respect, prestige, and success.

Self-actualization needs, enabling realization of development potentials.

The needs hierarchy provides a useful framework for evaluating societal and marketplace demands for the goods and services that can be provided by systems engineering based approaches. It can, of course, be augmented by other analytical techniques, as we have noted.

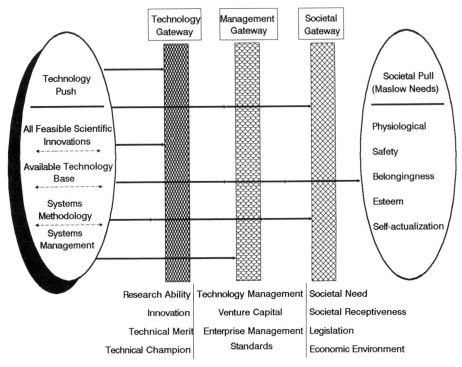

Figure 3.11. Three Gateways for Successful Systems Management of Technology Development

clearly a function of a given organization's technological and market expertise relative to a particular technology innovation or development area.

For there to be appropriate motivation for systems engineering efforts, there must be a perceived need for the system or at least a felt need to accomplish system development.* As a consequence, we need to envision and consider technological needs, systems management needs, and societal needs associated with the process that result in new products, services, or systems. These are the three gateways shown in Figure 3.11. If there is a societal need, then it becomes possible for a technological product to actually and ultimately emerge into the marketplace.

For a new system to be developed, there must be an available technology base that supports development. Existing large investments in production facilities may encourage innovations that capitalize this investment. On the other hand, the existence of a large investment in one form of technology may well impede the propensity to allocate resources to an entirely new

---

*These needs may be established, for example, by a client who funds development. Alternatively, they may be forecast needs, such as occurs prior to a firm undertaking development of a system or product that will subsequently be marketed.

approach that could make the old approach obsolete. At least initially, this might be viewed as a very significant impediment to technology and system development. There may well be, for example, manual methods of production that would initially become obsolete due to introduction of a new technology. Of course, a longer term view of the development situation might show that the initially displaced workers could, upon retraining, enter the workforce more productively at a higher skill level. Furthermore, the very fact that there exists one satisfactory way to do something often provides an intellectual bias against thinking about new methods of approach. Thus, a successful technology or system developer must be motivated and prepared to demonstrate that a new and potentially innovative approach is *better* in some societally acceptable and sustainable ways.

It is possible to characterize existing conditions in an organization along several dimensions relative to development and implementation of a new technology or system. Two questions seem to be of primary importance relative to exploitation of a potential technology development venture. They can be expressed in slightly different form for individuals, groups, and organizations. In generic form, they are as follows.

1. Which new technology or system development markets should a unit enter?
2. How should the unit enter* these markets so as to maximize the likelihood of success and the reward to be obtained from success and at the same time to control the risk of failure and the losses to be suffered in the event of a failure?

A potential new technology or system can be nurtured by one unit through use of one, or through a combination, of the following two basic approaches:

1. Internal development of the technology or system.
2. Venture funding of others and subsequent acquisition, or transfer, of the technology.

There are many ways through which the questions just posed could be resolved. In part, the appropriate development strategy depends upon an analysis of four related questions:

1. How new is the technology to the unit?
2. How new is the market the unit?
3. How familiar is the unit with technological development needs?
4. How familiar is the unit with the market?

The responses to these questions leads to a 16-cell selection matrix, shown in Figure 3.12, that determines the extent to which a specific unit might be able

---

*Entry may occur through internal development of a new technology or through a technology transfer process. There are a number of related infrastructure questions and questions of system integration associated with either approach.

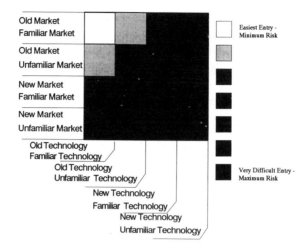

Figure 3.12. Experiential Familiarity with Market and Technology Effect Ease of Entry and Risk

to determine solutions to the many potential problems that may arrive in making a potential new technology operational. The terms *base technology* and *base market* are used to describe technologies and markets with which a unit is presently concerned. Roberts and Berry [40] have described appropriate entry strategies for the nine cells of Figure 3.12 (left three columns and top three rows) that are most supportive of success in development of a new technological product, service, or system.

Newness is the key concern in the matrix entries in Figure 3.12, which indicates a basic 16-cell model of experiential familiarity with technology and market. Certainly, much about many technologies will be new to the unit considering development. But a potential innovation may be a new technology or a new market, in general—or for a specific company or nation. Also, there are questions of existing technologies with which a new technology must be integrated, and experiential familiarity with the system integration process can also be expected to vary. To expand on these concepts, such as to be able to indicate generic costs and effectiveness indices, including success and failure possibilities, is an objective that we will address in Chapters 7 and 8. A dimension that characterizes the client for the technology is also needed.

Another appropriate dimension for consideration is one that represents the type of unit involved in a possible technology or system effort, and the nature of the system itself. Horwitch and Prahalad [41] have identified three ideal organizational modes, and we can easily add a fourth that concerns the individual innovative researcher.

1. The technological innovation process practices by the individual researcher in an academic or industrial research environment.

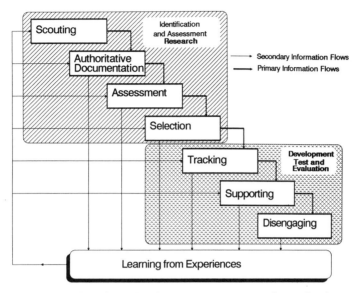

Figure 3.13. Phases in the Life Cycle of Systems Management for Technology Development

2. The technological innovation processes found in small, high-technology-oriented firms.
3. The technological innovation processes that occur in large corporations with numerous products and markets.
4. Those processes found in conglomerates, multiorganizations, and transnational multisector enterprises.

These could easily be expressed in terms of new system developments. The types of technologies and systems most suitable for potential development in each of these four modes of operation will be different, as will the appropriate risk behavior. It would seem reasonable to augment this model to allow consideration of other modes, such as those due to individual entrepreneurs and government development assistance. Also, this taxonomy could be enlarged through consideration of the, potentially very different, roles of the system developer in organizations of different generic sizes and purposes. These might include, for example: individual, small to midsize, large, and multinational.

Also the type of *coordination structures* or patterns of information flow and decision making among the individuals and computer systems concerned with technology development is very important. The study of Malone and Smith [42] illustrates that, both for human organizations and computer systems, these structures* are essential in determining production costs, coordination costs, and system vulnerability to crises of various types. It is particularly interesting to associate these different patterns of information flow and coordination with different approaches to development and/or transfer of technologies so as to obtain the most appropriate relationships between organizational communications and coordination and the development and operational implementation of specific technologies.

A systems life cycle for technology identification and assessment and preliminary implementation is now suggested. The life cycle is comprised of two major phases, which can be broken into seven more specific phases.

1. Technology Identification and Assessment
   1.1. *Scouting* and identification of requirements specifications for candidate engineering technologies.
   1.2. *Authoritative information documentation* concerning technological, economic, and societal need for, and feasibility of, the technologies.
   1.3. *Assessment* and evaluation of the technologies.
   1.4. *Selection* of appropriate technology for initial development and implementation.
2. Preliminary Development, Test, and Evaluation
   2.1. *Tracking* of the progress of development and implementation concerning all aspects of the candidate engineering technology.
   2.2. *Supporting* the operational implementation of the technology in ways that are meaningful to the technology itself and the results obtained in the earlier phases of the process.
   2.3. *Disengaging* from projects that prove to be productive and that have been successfully transferred, or that indicate productivity or risk potentials beyond critical thresholds.

These seven phases are also formally the seven phases that can, with slight modifications in the activities for each phase, be used to identify and nurture emerging technologies to the point where they might lead to new products and services.† The critical attributes of potentially successful technologies should be identified in the initial phases of the systems engineering life-cycle

---

*Decentralized markets, functional hierarchies, product hierarchies, and centralized markets are the fundamental four structures with functional hierarchies and centralized markets being further characterized as small scale or large scale. The historical evolution over time of these is in the order listed. Since market pull is generally the dominant force in the long-term success of technological innovations, it is appropriate to devote abundant attention to establishing coordination structures and associated perspectives that will enable successful development of a selected technology.

†These seven life cycle phases are used in literature of the National Science Foundation *Emerging Engineering Technology* Research Initiative, and were first suggested to this author by Dr. Nicholas DeClaris of the National Science Foundation and the University of Maryland. They have been investigated by this author under NSF Grant EE-8820124.

process. Among these attributes are innovativeness, timeliness, cost-effectiveness, and profitability of the products, concepts, or services of the technology under consideration. Identification of productive environments for potential technology development and transfer candidates is also a need. It is not difficult to characterize the appropriate environment as one in which a highly motivated group of people are free to pursue potentially unusual ideas, as well as not so unusual ideas. The environment should be one that recognizes and rewards success and recognizes that there will be some failures. In addition, must be a sense of urgency based on the awareness that the utility of any need, idea, or actual product can be quickly outdated.

Figure 3.13 illustrates the iterative nature and sequencing of these phases. In particular, it shows the learning that *should* occur as the process is repeated. It is potentially important to note that these phases do not form a complete systems engineering life cycle. They can be embedded into such a life cycle, however.

The *scouting and requirements specification* phase has as its goal the identification of societal and market needs, potential technology development and/or transfer activities, and implementation of the resulting technology as a product, process, or system. The effort in this phase should result in the identification and description of preliminary conceptual technology development and/or transfer characteristics that are appropriate for the next phase of the process. It is important to note that it is necessary to translate operational deployment needs into requirements specifications so that these needs can be addressed by technology development and transfer efforts. Thus we see that information requirements specifications are affected by, and effect, each of the other phases of the systemic framework developed here.

A result of the scouting and requirements specifications phase is a clear definition of alternative technology development and transfer issues, which permits deciding whether to undertake the authoritative information documentation and assessment phases. If the scouting and requirements specifications effort indicates that marketplace and societal needs can be satisfied in an efficacious manner, then documentation should be prepared concerning specifications that have resulted from this preliminary conceptual first phase of effort. Initial specifications for the following phases of effort should typically also be prepared, as they will be needed to do an assessment of candidate technologies.

At the end of this phase, preliminary documentation objectives and plans for the assessment effort are also determined. In general, each phase of effort should always include plans to implement the next phase of the effort.

Authoritative information documentation includes preparation of definitive position papers on each of the identified candidate technologies. This should be accomplished in considerable detail and should encompass all of the important areas of concern that we have just discussed. The authoritative preliminary conceptual design typically results in specification of the content, associated architecture, general design constructs, and research agendas for the product, process, or service that should result. The primary goal of this phase is to conceptualize a prototype that is responsive to the requirements identified in the previous phase. Preliminary concept papers, prepared according to the requirements specifications, should be obtained from this effort. Rapid prototyping is desirable in most cases. Feasibility and the potential payoffs from it should be investigated as a part of this phase.

The desired product of the *assessment* phase is a set of detailed evaluations of the proposed alternative technology initiatives. There should exist a high degree of confidence that a useful product will result from the detailed technology development and transfer effort, or the entire effort should be redone from phase one or possibly abandoned. Another result of this phase should be a refined set of specifications for the supporting and disengaging phases of the technology development effort.

After technology development and/or transfer projects are selected and funded, there is a need to *track* the efforts of the team engaged in the specific technology development. A preliminary development product, design, process, or system is produced at the end of the supporting phase. This will generally not be a final product or service design, but rather the result of implementation of prototype designs and development products. User guides for the product should be produced such that realistic test and evaluation of the results of the effort can be conducted.

Preliminary evaluation criteria should be obtained as a part of scouting and authoritative information documentation phases and possibly modified during the following two phases of the effort. A set of specific evaluation test requirements and tests are evolved from the objectives and needs determined in the earlier phases, such that each objective measure and critical evaluation issue can be measured by at least one test instrument.

If it is determined that the systems engineering product under development cannot meet technology and marketplace needs, the process reverts iteratively back to an earlier phase, and the effort continues with potentially modified objectives and redirected efforts. An important by-product of evaluation is determination of ultimate performance limitations for an operationally realizable technology and identification of those protocols and procedures for use with the technology that will be most effective in achieving overall goals.

The last phase of a new technology development effort is *disengaging*. This occurs when the technology is ready for full-scale production and introduction into the marketplace. Alternately, disengaging should occur when consensus is reached that further continued development of the technology yields little benefit compared to the costs involved.

This systems engineering life cycle for the technology development process is doubtlessly extensive and exhaustive. It needs to be applied with wisdom and maturity. This is particularly the case since exhaustive, time-consuming continual changes in systems requirements may prohibit the life-cycle process from ever being completed. An important advantage of the process just described is that it allows for formal consideration of the interactions among the phases of the effort, and attempts to view the whole process within a contextually realistic setting.

The need has been sensed to identify potentially innovative technologies at the earliest possible point in the initial research phase such that the subsequent development, evaluation, and test of these emerging technologies can be enhanced and nourished, especially with respect to the potential for their ultimate development into new products and services. Many studies of enhanced productivity suggest that a much closer working relationship between academia and industry assists in the transfer of knowledge [43] from academia to industry and hence on to the consumer in the form of new products or services. The phased life cycle of identification and assessment, research, development, test and evaluation suggested in Figure 3.13 potentially supports this relationship.

Three of the major ingredients in successful systems development are the following:

1. Identification of a systems development process model and an associated life cycle that enables development from the new-technology concept to ultimate production and marketing of cost-effective and sustainable system products and services.
2. Appropriate mix of people and disciplinary specialties to ensure success.
3. Incentives to encourage the foregoing.

The development of an appropriate new-product development life cycles and associated systems management process should include strategic quality assurance and management efforts at each of the life-cycle phases in order to ensure reliability, applicability, and maintainability, quality, and trustworthiness of the evolving systems product. This suggests that we may consider several systems management life cycles. One of these may be appropriate for preliminary research, development, test, and evaluation (RDT&E) of a new technology. A second may be useful for system acquisition based on this technology.

There are a number of other systems management life cycles. We now examine some of them. We have used the term *phase* throughout to describe portions of the systems engineering life cycle. This is a standard term used in much, but by no means all, of the systems engineering literature. The terms stage, milestone, and phase are all generally used in different sources to indicate the same thing. First, we review the life cycle for new-product development suggested by the U.S. National Institute of Standards and Technology (NIST, formerly the National Bureau of Standards) [44]. The NIST documentation indicates that

New product development is the process whereby a new technology or technical product is developed, from the point of conception to the point where the

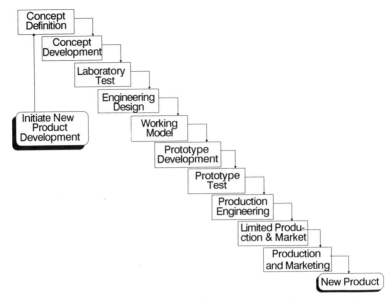

Figure 3.14. NIST Systems Development Life-Cycle Methodology for New Products

product is ready for production and marketing. ...The rapid loss of competitiveness of U.S. industry in international markets is an extremely serious problem with wide-ranging consequences for our material well-being, our security, and our political influence. Its causes are many, but among them certainly are the slow rate at which new technology is embodied in commercial products and processes, and the lack of attention paid to manufacturing. We need to compete in world markets with high-value-added products, incorporating the latest innovations, manufactured in short runs with flexible manufacturing methods. We need research, management, and manufacturing methods that support change and innovation.

The NIST Life-Cycle Methodology. NIST, through their Energy Related Inventions Program (ERIP), uses a ten-phase model for new-product development as indicated in Figure 3.1.4.

The Defense Systems Acquisition Life Cycle. The life-cycle approach to systems acquisition of the U.S. Department of Defense is illustrated in Figure 3.15 and described in DoD Instruction 5000.2, dated September 1, 1987. This system acquisition life cycle is comprised of five phases (or milestones).

1. *Milestone 0—Program Initiation/Mission-Need Decision:* Mission need is determined and program initiation is approved, including authority to

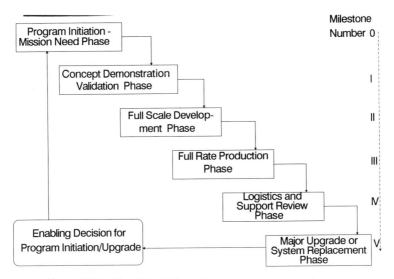

Figure 3.15. The U.S. Defense Systems Acquisition Life Cycle

budget the program. A concept-definition analysis is performed. Primary consideration during this initial acquisition phase is given to the following four steps:

1.1. Mission area analysis.
1.2. Affordability and life-cycle costs.
1.3. Feasibility of a modification to an existing system to provide the needed capability.
1.4. Operational utility assessment.

2. *Milestone I—Concept Demonstration/Validation Decision:* A concept demonstration/validation effort is performed. If successful, the development continues. Primary attention is paid to seven areas of effort:

2.1. Program alternative trade-offs.
2.2. Performance/cost and schedule trade-offs, including the need for a new-development program versus buying or adapting existing military or commercial systems.
2.3. Appropriateness of the acquisition strategy.
2.4. Prototyping of the system or selected system components.
2.5. Affordability and life-cycle costs.
2.6. Potential common-use solutions.
2.7. Cooperative development opportunities.

The efforts in this phase establish broad program cost, schedule, and operational effectiveness and suitability goals and thresholds, allowing the program manager maximum flexibility to develop innovative and cost effective solutions.

3. *Milestone II—Full-Scale Development Decision:* Approval for full-scale development (FSD) is made at this life-cycle phase. As appropriate, low-rate initial production of selected components and quantities may be approved to verify production capability and to provide test resources needed to conduct interoperability, live fire, or operational testing. Decisions in this phase precede the release of a final request for proposals (RFP) for the FSD contract. Primary considerations in this phase of activity are described by 13 steps:

   3.1. Affordability in terms of program cost versus the value of the new or improved system and its operational suitability and effectiveness.
   3.2. Program risk versus benefit of the added capability.
   3.3. Planning for the transition from development to production, which will include independent producibility assessments of hardware, software, and databases.
   3.4. Realistic industry surge and mobilization capacity.
   3.5. Factors that affect program stability.
   3.6. Potential common-use solutions.
   3.7. Results from prototyping and demonstration/validation.
   3.8. Milestone authorization.
   3.9. Personnel, training and safety needs assessments.
   3.10. Procurement strategy appropriate to program cost and risk assessments.
   3.11. Plans for integrated logistics support.
   3.12. Affordability and life cycle costs.
   3.13. Associated command, control, communications, and intelligence (C^3I) requirements.

   Decisions at this phase result in the establishment of more specific cost, schedule, operational effectiveness and suitability goals, and thresholds than were possible in earlier phases. Particular emphasis is placed on the requirements for transitioning from development to production.

4. *Milestone III—Full Rate Production Decision:* The decision made at this phase regards activities associated with the full-rate production/deployment phase. Primary considerations at this phase involve 12 steps, many of which are similar to steps in Milestone II.

   4.1. Results of completed operational test and evaluation.
   4.2. Threat validation.
   4.3. Production cost verification.
   4.4. Affordability and life-cycle costs.
   4.5. Production and deployment schedule.
   4.6. Reliability, maintainability, and plans for integrated logistics support.
   4.7. Producibility as verified by an independent assessment.
   4.8. Realistic industry surge and mobilization capacity.
   4.9. Procurement or milestone authorization.

4.10. Identification of personnel, training, and safety requirements.

4.11. Cost effectiveness or plans for competition or dual sourcing.

4.12. Associated C^3I requirements.

5. *Milestone IV—Logistics and Support Review Decision:* The decision process at this phase identifies actions and resources needed to ensure that operations readiness and support objectives are achieved and maintained for the first several years of the operational support phase of the life cycle. Primary considerations at this phase of the life cycle are the following:

5.1. Logistics readiness and sustainability.

5.2. Weapon support objectives.

5.3. Implementation of integrated logistics support plans.

5.4. Capability of logistics activities, facilities, and training and manpower to provide support efficiently and cost-effectively.

5.5. Disposition of displaced equipment.

5.6. Affordability and life-cycle costs.

6. *Milestone V—Major Upgrade or System Replacement Decision:* The decision process at this phase encompasses a review of a system's current state or operational effectiveness, suitability, and readiness to determine whether major upgrades are necessary or deficiencies warrant consideration of replacement.* Primary considerations at this phase of the life cycle are

6.1. Capability of the system to continue to meet its original or evolved mission requirements relative to the current situation.

6.2. Potential necessity of modifications to ensure mission support.

6.3. Changes in technology that present the opportunity for a significant breakthrough in system worth.

There has been much present experience, including some criticism as well, of the Defense Systems Acquisition Life Cycle. This acquisition cycle is related to, but different from, the various categories of research and development funded by the DoD. There are six categories identified; each are prefixed by the number 6 in the DoD funding breakdown structure.

6.1. *Research* includes all efforts of study and experimentation that are directed toward increasing knowledge in the physical, engineering, environmental, and life sciences that support long-term national security needs.

6.2. *Exploratory development* includes efforts that are directed toward evaluating feasibility of solutions that are proposed for specific

---

*This phase actually addresses a system many years, normally five to ten, after deployment if the acquisition is *successful*. If there are major flaws, the initially deployed system may not be satisfactory and maintenance may be required from the day of operational implementation. As is now well known, this has been a particular problem with software acquisition.

military issues that need applied research and prototype hardware and software.

6.3A. *Advanced technology development* is concerned with programs that explore alternatives and concepts prior to the development of specific weapons systems.

6.3B. *Advanced development* is involved with proof of design concepts, as contrasted with development of operational hardware and software. The Milestone I decision, just discussed, occurs during advanced development.

6.4. *Engineering development* results in hardware and software for actual use that is developed according to contract specifications. A program moves from advanced development to engineering development coincident with a Milestone II decision by the Defense Acquisition Board.

6.5. *Management and support* involves the support of specific installations or those operations and facilities that are required for general research and development use.

Analytical New-Product Decision Model. Developed by Pessimier [45], this model emphasizes administrative and interface issues. This life-cycle model is comprised of six phases that lead to development of a new product.

1. The *search phase* emphasizes studies designed to locate potentially profitable additions to a unit's product line or capabilities. Market studies, research and development work, and acquisition studies all qualify as part of a well-organized search program.

2. The *preliminary economic analysis phase* involves quick, low-cost studies to eliminate weak proposals. Preliminary weights for relative desirability are assigned to the remaining promising proposals.

3. A *formal economic analysis phase* is then conducted. This phase entails careful, detailed studies to clarify, improve, and appraise the proposals that survive preliminary economic analyses. The end product of this phase is a recommendation to abandon a technology development proposal, to defer action pending additional studies, or to proceed to develop a product, system, or capability.

4. The *development phase* involves transformation of the development plan into a tangible product, through software coding, manufacturing, or other development activity.

5. The *product testing phase* involves operational product or system test and evaluation and market tests to measure the reactions of consumers, resellers,* or final buyers to the product, system, or service.

---

*Generally, this term denotes firms who buy the products of others for incorporation into larger systems. Often, they are called system integrators.

6. The *commercialization phase*, which is the final phase of development, involves full-scale production and marketing operations to establish the product in its desired place in the organization's product line.

Because timeliness in the development and introduction of a new product is so important to a firm's sales and profits, good administration and management of new-product development are essential to a firm's growth and health. This aspect of commercialization is emphasized by Pessimer, who disaggregates the commercialization phase into five related stages.

1. *Introduction:* The initial stage during which the product is being introduced to both product resellers and consumers.
2. *Growth:* The stage during which the product gains acceptance and finds its place in the market.
3. *Competition:* Competitive offerings appear and find their places in the market.
4. *Obsolescence:* The product's competitive disadvantages indicate that the product will soon need modification or replacement.
5. *Termination:* The product is phased out in favor of an improved product or is dropped from the product line.

Maintenance, which includes product transitioning through retrofit, is another activity which may well follow the obsolescence phase and is included as a part of the termination portion of commercialization.

Other Systems Management Life-Cycle Models. There are a number of other systems management life-cycle models. For example, the National Society of Professional Engineers has suggested a six phase model [46].

1. The *conceptual phase* demonstrates that the new-development concept is valid, by applying a test-of-principles model.
2. The *technical feasibility phase* shows that engineering development is feasible within the technological state of the art.
3. The *development phase* makes needed improvements in the technology, and in actually producing and testing a single operational version of the product or system.
4. The *commercial validation and production preparation phase* develops manufacturing techniques and establishes a test market for the new product.
5. The *full-scale production phase* builds manufacturing facilities and puts the new product into production.
6. Finally, the *product support phase* continues consideration of a variety of engineering improvements, such as maintenance, to insure that the product remains competitive.

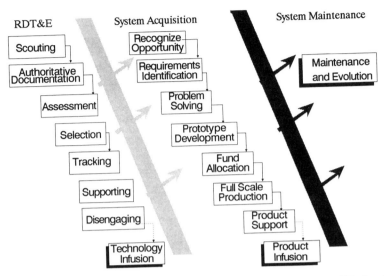

Figure 3.16. Technology Development / Transfer Support to The General Technological Innovation Process Involving Emerging Technology RDT & E, System Acquisition, and Maintenance

Generally these phases are preceded by an emerging technology development, or RDT and E, effort. In Figure 3.16, we illustrate how the phased life cycle for emerging technology development, which we discussed earlier in this chapter, may be embedded into a typical product development life cycle. There are opportunities for technology and product transfer, or infusion, within this expanded life cycle.

3.5. EVALUATION OF SYSTEMS MANAGEMENT LIFE CYCLES FOR PRODUCT DEVELOPMENT

At several points in the phased development of a system or technology, it is desirable to evaluate whether a new-product development strategy should be adopted. There are many factors that need to be considered in doing this. Large-scale system and technology research and development consumes financial and other resources, often for a significant time period. It is invariably necessary to recognize that the benefits of developing one particular new-product concept must be weighed against the costs of foregoing other opportunities (see Chapters 7 and 8 for a more detailed analysis of these issues).

There are a number of issues to be resolved through these evaluation efforts.

1. Determination of an appropriate specific process to use for the identification and evaluation of potential new products to be developed.
2. Identification of the groups that should be involved in the evaluation process.
3. Identification of the criteria used to determine the length and type of support for the development effort.
4. Establishment of appropriate criteria for determining potentially major alterations to the development effort, perhaps due to the emergence of a crisis situation, possible continued development according to a crisis management, or termination of the new product development effort.

Each of these issues includes a number of multiattributed criteria for program evaluation. Dutton and Crowe [47] provide the following excellent summary of important criteria.

1. Technological merit.
   1.1. Technological objectives and significance.
   1.2. Breadth of interest of strategy.
   1.3. Potential for new discoveries and understandings.
   1.4. Uniqueness of proposed development strategy.
2. Social benefits.
   2.1. Contribution to improvement of the human condition.
   2.2. Contribution to national pride and prestige.
   2.3. Contribution to international understanding.
3. Programmatic (management) issues.
   3.1. Feasibility and readiness for development.
   3.2. Technological logistics and infrastructure.
   3.3. Technological community commitment and readiness.
   3.4. Institutional infrastructure and implications.
   3.5. International involvement.
   3.6. Cost of the proposed strategy.

These attributes may be viewed in several ways. For example, they can be seen as the various gateways for the development of new products, such as shown in Figure 3.11.

Many authors have described gateway concepts, similar to this one. For example, Karger and Murdick [48] view the life-cycles phases as screens that allow successful products to flow into the marketplace. Figure 3.17 illustrates another extension and conceptualization of this concept, which can be extended further back into the research life cycle. The screens could also be modified to accommodate the notion of technology push and market pull.

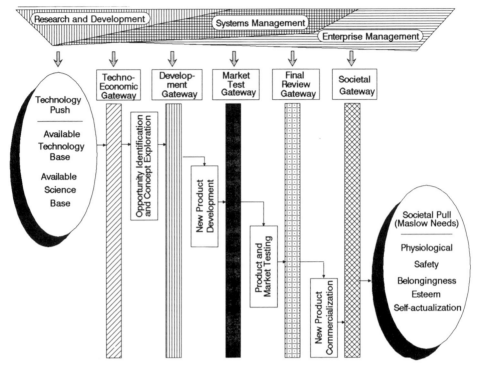

Figure 3.17. Expanded Set of R & D, Systems and Enterprise Management Gateways for Product Development

Quite clearly, the *go* and *no-go* decisions of Gruenwald [49] may be viewed as the result of using evaluatory screens such as those shown here.

In addition to obtaining an evaluation of proposed new-product development strategies, an appropriate approach to evaluation should also allow for full exploration of the functions needed to ensure satisfactory development of the new product. As indicated by Roberts in his definitively edited work on technological innovation [23], people are needed who fill these roles:

1. *Idea Generators.* Contribute ideas from technology push or market pull considerations to extend the ultimate potential of the new product under development.

2. *Idea Exploiters.* Also referred to as innovators or technology development champions or research enterpreneurs, they take new product development ideas and attempt to get them fully explored, supported, and adopted.

3. *Management Leadership*. Also business leaders, they see to it that the various planning, scheduling, monitoring, and control functions are carried out effectively.

4. *Information (and Knowledge) Gatekeepers*. Provide informed realities technology, capital, manufacturing, standards, and market potential.

5. *Sponsors*. Although not directly involved with the development strategy (in order to ensure objectivity), they provide leadership and resources from the highest levels to enable development of the technology or to restrict it when it proves cost ineffective.

Each of these actors have appropriate roles to play at various phases in a systems management life cycle to insure sustainable developments.

To focus the resources and activities of an organization on the needs and desires of the consumer for a potential new product, one must use market research. Kinnear and Taylor [50] state that the purpose of marketing information input is to narrow decision-making error and to broaden perspectives within which decisions are made. It may be the case that, even before a new-product development effort is undertaken, one should have a good idea as to the conditions in the marketplace with respect to the proposed new product or service. Market research can provide this information. There are a number of risk-management implications involved in this. Some of these are examined in Chapters 6, 7, and 8.

It has been often stated that there are three basic components in any marketing research undertaking:

1. Making certain that the right marketing research questions are asked.
2. Using appropriate marketing research techniques and controls.
3. Presenting marketing research findings in a clear, comprehensible format that leads to management action.

Of course, these questions can be restated such that they apply to any phase of the systems life cycle. These motivate many discussions of large-scale design and development, such as the work of Rouse [51] discussed in Chapter 2.

3.6. SUMMARY

In this chapter, we have discussed a number of important issues relative to systems management. First, we provided some philosophical perspectives on systems organizations and then indicated a few pragmatic ways in which an organization could be structured. A considerable number of our discussions related to some conceptual ways that might be used to evaluate various system development proposals. We discussed six important systems engineering-based evaluation strategies. It is desirable to summarize them here.

The *gateway concept* suggests that a technology, to reach a mature stage in which it yields useful products or services, must pass through three gateways: the technology gateway, the management gateway, and the societal gateway. Passing through the technology gateway requires research ability, innovation, technical merit, and a technical champion [52, 53]. All feasible scientific innovations, the available technology base, and systems design methodology are employed, and help provide what is popularly referred to as *technology push*.

The multiphase life cycles of systems management track a technology or system through a number of distinct phases of a life cycle, or what is often called the *technology pipeline*. Feedback loops and iteration are an important part of this methodology. For example, the identification of market needs in one phase of the life cycle might be substantially modified as a result of assessments conducted in a later phase. The importance of these feedback loops in the technology innovation process is increasingly becoming recognized [54], especially with the current focus on sustainable development.

The *newness matrix*, illustrated in Figure 3.12, suggests an approach for analyzing the risks, hazards, and uncertainties associated with introducing a new technology or a new system. Such uncertainties, both in the market place and in the technology itself, constitute, in the terminology of business, the risk factor. Risk, of course, is of fundamental importance in all decision making strategies.

The newness matrix is particularly relevant in the early stages of a technology's development, where there are numerous uncertainties. As persuasively argued by Florida and Kenney [55], American industry has tended to rely on major technological breakthroughs, rather than incremental improvements in technology, as the major mechanism for technological progress, with substantial competitive disadvantages as a result. The newness matrix approach attempts to focus attention on just the types of problems that this breakthrough mentality may skip over. Newness, or uncertainty, of the market may be due to any of the following.

1. New uses for a product or system always bring about uncertainties.

2. User skepticism about improved performance characteristics, especially when technologies are developed with the notion that they substitute for existing technologies by providing higher performance at an acceptable increase in price, raises the important concern that the consumer may not be particularly impressed with the performance improvement.

3. Requirement for behavior adjustment by the user may result in the most imaginative and potentially useful new technology failing, because

users cannot, or will not, adjust their behavior to meet the needs of the new technology or system.

4. Competitive technologies are often highly dynamic, adding enormous uncertainty to markets.
5. Unpredictable technological developments can add enormous uncertainties to markets.
6. Legal barriers in the form of regulatory requirements can add considerable uncertainty to the technology or system adoption process.

The other axis of the newness matrix is technology uncertainty. This may be due to any of the following factors.

1. Innovativeness of new technology present many uncertainties in development. Almost by definition, the more innovative a technology, the more uncertainty it presents. The difficulty is identifying what is genuinely a technological innovation and what is simply an extension of existing technology.
2. The number of constituent technologies that make up a product or system may provide for a geometric, rather than an arithmetic, increase in the number of uncertainties involved in an innovation.
3. Manufacturing difficulties frustrating new technology development are potentially numerous.
4. Institutional changes required to introduce the new technology may be very difficult to predict.

This leads to a final observation that each of the phases in the new product development life cycle are inexorably linked and that there does need to be a strong coupling, particularly at the strategy development level, among all phases in the development life cycle. It further suggests an interrelated common set of evaluation metrics for each phase in the new-product development process. The most important development questions that suggest these metrics include:

1. The overall goal of each phase.
2. The key business disciplines involved.
3. The research and development role for product development.
4. The marketing role for planning.
5. The marketing role for market development and product management.
6. The ultimate manufacturing role.

7. Finance and economic systems analysis concerns.
8. The functional outputs of each phase.
9. Reliability, adaptability, and maintainability (RAM) concerns.
10. The possible reasons for failure at each phase, and associated risk management.
11. Strategies for excellence in development through strategic quality assurance and management.

Conversion of these to metrics and use of them to select development strategies for new product evolution, and to assist in the systems management and technical direction of the actual process of new product evolution, is a major step in the achievement of an appropriate methodology for strategic systems management of all facets of technology development.

PROBLEMS

3.1. Marvin [56] identifies ten key factors to success, which are very similar to the phases for systems management discussed in this chapter. The ten factors, stated in the form of attributes, are as follows.
1. Take time for development by concentrating on the development of new products, rather than doing this as an afterthought.
2. Know the phases of the process to ensure successful product development program.
3. Join knowledge with ability, because all of the required talents and abilities should be brought together and incorporated in the group responsible for new product development.
4. Draw on experience, because experience will support efficiency and enable developments to move swiftly and surely toward the desired objectives.
5. Develop planning techniques, so as to provide tested systems of organized attack on problems and make the efficient application of experience and ability in problem solving possible.
6. View problems in perspective, so as to be able to see oneself and the product or service under consideration as others see them.
7. Know the competition, because familiarity with the state-of-the-art technology and trade customs provides the background for evaluating the soundness of specific system and product development ventures.
8. Break away from the past, to enable system and product development to be undertaken in an atmosphere of complete independence

and with the view that the old ways of doing a job should be cast aside.

9. Protect product ideas as assets, by protecting basic ideas from being revealed and corporate interests from being disclosed indiscriminately.

10. Provide adequate facilities, as fully maintained and utilized facilities are essential to the success of any product development undertaking.

Marvin identifies ten questions, the answers to which are useful in analyzing a new-product development program.

1. Do we have time to do the job?

2. Do we understand all the problems involved?

3. Do we have the ability (technical knowledge and skills) to tackle product development programming?

4. Do we have the experience necessary?

5. Do we know how to plan a successful product development program?

6. Will we be able to put development programs in their proper perspective?

7. Are we familiar with the practices of our competitors?

8. Can we break away from past practices, concepts, and viewpoints?

9. Are proprietary product ideas protected?

10. Do we have the plant and facilities for product development?

The desired answer to all of these questions is of course "yes." The more "no's" one has, the less the probability of success. Thus, we see that the suggested life-cycle development approach can be converted into a set of metrics to be used for evaluation of proposed new-product development strategies.

Contrast and compare the implicit life cycle identified here with some of what we have obtained in this chapter. How can the ten attributes be used to evaluate a proposed technology development? How do these compare with the approaches suggested in this chapter?

3.2. Kuczmarski [57] suggests that the expanded process suggested better enables people to cope with the two distinct phases for new-product development that emphasize the preliminary effort of goal or direction setting.

1. Direction setting involves identifying:
    1.1. Corporate objectives and strategies.
    1.2. New product blueprints.
    1.3. New product diagnostic audits.
    1.4. New product strategies.

     1.5. Categories of application for the new product.

     1.6. Category analysis and screening.

  2. New-product development involves

     2.1. Category selection, through analyzing and ranking potentially attractive categories by studying the role of new products in the company and choose categories that provide the most attractive possibilities for idea generation.

     2.2. Idea generation, by generating ideas in selected categories through a variety of problem-solving and creative approaches.

     2.3. Concept development, by developing concepts, conducting initial screens, and setting priorities by taking ideas that pass the initial screens and developing descriptions of the product.

     2.4. Business analysis, by conducting business analysis of selected concepts through formulating a potential market and conducting competitive assessments.

     2.5. Screening, or filtering concepts to determine prototype candidates while keeping in mind financial forecasts developed in the business analysis, and filtering the remaining concepts through all performance criteria.

     2.6. Prototype development, or developing an operating prototype of the product and run product-performance tests.

     2.7. Market testing, or determining customer acceptance and running marketing trials in order to determine consumer purchase intent, and testing the product in either a simulated market or actual market trials.

     2.8. Plant scale-up and manufacturing testing, or initiating plant scale-up and production to determine roll-out equipment needs and manufacturing the product in large enough quantities to identify bugs and problems, and run product-performance tests.

     2.9. Commercialization, or developing plans to introduce the new product to the trade and consumers.

     2.10. Post-launch checkup, or performance monitoring of the new product.

Illustrate a life-cycle systems management model this incorporates this. Contrast and compare it with the NSPE life cycle and the life cycle of Rouse.

3.3. We have already noted that systems management is very important to new-product development. Kuczmarski [57] provides an example of a company president's perception of an ideal new-product environment:

  1. Top-level endorsement and high visibility for new products.

  2. New products tied to long-range corporate objectives and financial plans.

3. Agreed-upon new-product charter covering objectives, category areas, and screening criteria.

4. Small-company flexibility and expediency.

5. Clear identification of responsibilities.

6. High level of communications and interdepartmental cooperation.

7. Collegial, teamwork involvement.

8. Front-loaded process.

9. Portfolio of product improvements, line extensions, and "new" products to balance risk.

10. Atmosphere where failures are accepted along with the successes.

11. Resources and consistent commitment to do the job, in terms of money and people.

Attempt to evaluate the structural models for systems management discussed in this chapter in terms of this environmental characterization.

3.4. In the Directed Research for Product Development Model, Gruenwald [49] presented a systems management life cycle to address new-product development objectives. There are seven phases to this life cycle and they may be described as follows.

1. *Phase 1. The Search for Opportunity—Compilation of Available Data:* Information on the industry, sales data, technology, consumer interests, and the competition are required. The effort is begun by performing an industry analysis, which should include sales volume and trends, basic technology, competition, customer definition, and other pertinent factors such as foreign trade, regulatory restrictions, and so on. The next step is to identify opportunities by defining targets, forecasting rough volume and share, performing a risk-ratio analysis, performing a feasibility study, performing a war-gamelike assessment of the competitive reactions, examining technical hurdles, and considering legal and policy issues. On the basis of this, a decision is made as to whether to proceed, that is *go* or *no go*.

2. *Phase 2. Conception:* In this phase, we translate market facts into product concepts and customer positioning communications before making commitments to exhaustive product oriented and directed research and development. There are six steps in this phase.

3. *Phase 3. Modeling (Prototypes):* This involves bringing proposed new products closer to reality in the form of prototype products and prototype communications. The steps in this phase involve developing descriptors, developing prototypes, and making a *go* or *no-go* decision on the basis of the information obtained from using these.

4. *Phase 4. Research and Development:* This phase covers a number of different activities including checking outside scientific resources,

pilot plant production, analyzing the factors necessary to scale up from pilot plant production to full-scale commercialization, controlled tests, and feasibility studies. On the basis of the information that results from these steps, a decision is made whether to proceed to the next phase.

5. *Phase 5. Marketing Plan:* This involves the development of a marketing plan and a *go* or *no-go* decision to proceed to the next phase is made on the basis of the information obtained at this phase.

6. *Phase 6. Market Testing*

7. *Phase 7. Major Introduction:* Expands upon the market testing phase.

Gruenwald has attempted, through use of this relatively exhaustive life-cycle methodology, to minimize risks and maximize success probability. This is clearly a systems engineering-based approach to address new-products goals in an organized, phase-by-phase iterative method. The important decision considerations that are addressed after a new product has been defined as follows:

1. Is there a latent demand for the product?

2. Can a product be made that will satisfy the market?

3. Can the company be competitive with the product and within the field?

4. Will the entry be profitable and satisfy the corporate charter, as well as other company objectives?

Contrast and compare this life cycle with others that have been discussed in this chapter.

3.5. Identify a set of three phased life cycles that accomplish emerging technology development, new system acquisition, and system operation and maintenance. Contrast and compare this systems management model with those in this chapter. How may sustainable development concerns be addressed through these life cycles?

REFERENCES

[1] Bass, B. M. (Ed.), *Stogdill's Handbook of Leadership*, Free Press, New York, 1981.

[2] Etzioni, A., *Modern Organizations*, Prentice, Hall, Englewood Cliffs NJ, 1964.

[3] Hall, R. H., *Organizations, Structure, and Process*, Prentice Hall, Englewood Cliffs NJ, 1977.

[4] Mintzberg, H., *The Nature of Managerial Work*, Harper and Row, New York, 1973.

[5] Hayre, J., "An Axiomatic Theory of Organizations," *Administrative Science Quarterly*, Vol. 10, No. 3, 1965, pp. 289–320.

[6] March, J. G., and Simon, H. A., *Organizations*, Wiley, New York, 1958.

[7] Weick, K. E., *The Social Psychology of Organizing*, Addison Wesley, Reading MA, 1979.

[8] Weick, K. E., "Cosmos Versus Chaos: Sense and Nonsense in Electronic Contexts," *Organizational Dynamics*, Vol. 14, No. 3, 1986, pp. 50–64.

[9] Janis, I. L., and Mann, L., *Decision Making: A Psychological Analysis of Conflict, Choice, and Commitment*, Free Press, New York, 1977.

[10] Janis, I. L., *Crucial Decisions: Leadership in Policymaking and Crisis Management*, Free Press, New York, 1989.

[11] Cyert, R. M., and March, J. G., *A Behavioral Theory of the Firm*, Prentice Hall, Englewood Cliffs NJ, 1963.

[12] Argyris, C., *Reasoning, Learning and Action: Individual and Organizational*, Jossey-Bass, San Francisco, 1982.

[13] Rasmussen, J., Duncan, K., and Leplat, J. (Eds.), *New Technology and Human Error*, Wiley, Chichester, 1987.

[14] Rasmussen, J., and Vicente, K. J., "Coping with Human Error Through System Design: Implications for Ecological Interface Design," *International Journal of Man-Machine Studies*, Vol. 31, 1989, pp. 517–534.

[15] Reason, J., *Human Error*, Cambridge University Press, Cambridge, 1990.

[16] Sage, A. P. (Ed), *Concise Encyclopedia of Information Processing in Systems and Organizations*, Pergamon Press, Oxford, 1990.

[17] Sage, A. P., *Decision Support Systems Engineering*, Wiley, New York, 1991.

[18] Sage, A. P., and Palmer, J. D., *Software Systems Engineering*, Wiley, New York, 1990.

[19] Sage, A. P. (Ed.), *System Design for Human Interaction*, IEEE Press, New York, 1987.

[20] Anthony, R. N., *The Management Control Function*, Harvard Business School Press, Boston MA, 1988.

[21] Sage, A. P., "Systems Management of Emerging Technologies," *Information and Decision Technologies*, Vol. 15, No. 4, 1989, pp. 307–326.

[22] Smith, C. L., and Sage, A. P., "Systems Management of Technology Infusion and New Product Development," *Information and Decision Technologies*, Vol. 15, No. 4, 1989, pp. 343–358.

[23] Roberts, E. B. (Ed.), *Generating Technological Innovation*, Oxford University Press, New York, 1987.

[24] National Research Council, *Management of Technology: The Hidden Competitive Advantage*, National Academy Press, Washington DC, 1987.

[25] Robinson, R. D., *The International Transfer of Technology*, Ballinger, Cambridge MA, 1988.

[26] Cargill, C. F., *Information Technology Standardization: Theory, Processes and Organizations*, Digital Press, Bedford MA, 1989.

[27] Edosomwan, J. A., *Integrating Innovation and Technology Management*, Wiley, New York, 1989.

[28] Arthur Young Information Technology Group, *The Arthur Young Practical Guide to Information Engineering*, Wiley, New York, 1987.

[29] Gallo, T. E., *Strategic Information Management Planning*, Prentice Hall, Englewood Cliffs NJ, 1988.

[30] Tinnitello, P. C. (Ed.), *Handbook of Systems Management, Development and Support*, Auerbach, Boston MA, 1989.

[31] Parker, M. M., Benson, R. J., and Trainor, H. E., *Information Economics: Linking Business Performance to Information Technology*, Prentice Hall, Englewood Cliffs NJ, 1988.

[32] Repo, A. J., "The Value of Information: Approaches in Economics, Accounting, and Management Science," *Journal of the American Society for Information Science*, Vol. 40, No. 2, 1989, pp. 68–85.

[33] Cohen, S. S., and Zysman, J., "Manufacturing Innovation and American Industrial Competitiveness," *Science*, Vol. 239, No. 4844, 1988, pp. 1110–1115.

[34] Kaplan, R. S., "Management Accounting for Advanced Technological Environments," *Science*, Vol. 244, No. 4920, August 25, 1989, pp. 819–823.

[35] Huber, G. P., "A Theory of the Effects of Advanced Information Technologies on Organizational Design, Intelligence, and Decision Making," *Academy of Management Review*, Vol. 15, No. 1, 1990, pp. 47–71.

[36] Maslow, Abraham H., *Motivation and Personality*, Harper & Row, New York, 1970.

[37] Frey, D N., "R & D to the Marketplace: A New Paradigm?" *The Bridge*, Vol. 19, No. 1, 1989, pp. 16–20.

[38] Roberts, E. B., "Managing Invention and Innovation," *Research and Technology Management*, January 1988, pp. 11–29.

[39] Souder, W. E., "Improving Productivity Through Technology Push," *Research and Technology Management*, Vol. 32, No. 2, 1989, pp. 19–24.

[40] Roberts. E. B., and Berry, C. A., "Entering New Businesses: Selecting Strategies for Success," *Sloan Management Review*, Vol. 26, No. 3, 1985.

[41] Horowitch, M., and Prahalad, C. K., "Managing Technological Innovation: Three Idea Modes," *Sloan Management Review*, Vol. 17, No. 2, 1976.

[42] Malone, T. W., and Smith, S. A., "Modeling the Performance of Organizational Structures," *Operations Research*, Vol. 36, No. 3, 1988, pp. 421–436.

[43] Sage, A. P., "Knowledge Transfer: An Innovative Role for Information Engineering Education," *IEEE Transactions of Systems, Man, and Cybernetics*, Vol. SMC-17, No. 5, 1987, pp. 725–728.

[44] National Bureau of Standards (NBS)/Industrial Technology Services (ITS), "New Programs and Directions at the National Bureau of Standards," March 1988.

[45] Pessemier, E. A., *New-Product Decisions, An Analytical Approach*, McGraw-Hill, New York, 1966.

[46] National Society of Professional Engineers, *Engineering Stages of New Product Development*, NSPE, 1420 King Street, Alexandria VA 22314, 1990.

[47] Dutton, J. A., and Crowe, L., "Setting Priorities among Scientific Initiatives," *American Scientist*, Vol. 76, 1988, pp. 599–603.

[48] Karger, D., and Murdick, R. G., *New Product Venture Management*, Gordon and Breach, New York, 1972.

[49] Gruenwald, G., *New Product Development: What Really Works*, NTC Business Books, Lincolnwood IL, 1985.

[50] Kinnear, T. C., and Taylor, J. R., *Marketing Research*, McGraw-Hill, New York, 1983.

[51] Rouse, W. B., *Design for Success: A Human-Centered Approach to Designing Successful Products and Systems*, Wiley, New York, 1991.

[52] Rubenstein, A. H., Chakrabarti, A. K., O'Keefe, R. D., Souder, W. E., and Young, H. C., "Factors Influencing Innovation Success at the Project Level," *Research Management*, Vol. 19, No. 3, 1976, pp. 15–20.

[53] Lehr, L. W., "Stimulating Technological Innovation: The Role of Top Management," *Research Management*, Vol. 22, No. 6, 1979.

[54] Kline, S. J., "Innovation Is Not a Linear Process," *Research Management*, Vol. 28, No. 4, 1985, pp. 36–45.

[55] Florida, R., and Kenney, M., *The Break-Through Illusion: Corporate America's Failure to Move from Innovation to Mass Production*, Basic Books, New York, 1990.

[56] Marvin, P., *Product Planning Simplified*, American Management Association, New York, 1972.

[57] Kuczmarski, T. D., *Managing New Products: Competing Through Excellence*, Prentice Hall, Englewood Cliffs NJ, 1988.

Chapter 4

Operational and Task Level System Quality Assurance Through Configuration Management, Audits and Reviews, Standards, and Systems Integration

The systems engineering life cycle is intended to enable evolution of a trustworthy system that has appropriate structure and function to support identified purposeful objectives. We have described the essential nature of the systems engineering process in our previous chapters. This chapter continues with our discussions of methodological and management issues in systems engineering. The focus is, however, shifted slightly from concerns that more directly affect the life cycle of design and development, and associated systems management, to concerns that relate to *quality assurance* (QA) or *system quality assurance* (SQA), as a part of systems management or management control. Figure 4.1 illustrates the relationship among such quality assurance issues as configuration management and the acquisition of trustworthy systems. In particular, it illustrates how the notion of quality assurance through configuration management, audits and reviews, standards, and systems integration provides essential support to systems management. We examine these issues here through a discussion of the following:

Configuration management
Concurrent engineering
Audits and reviews
Standards
Systems integration

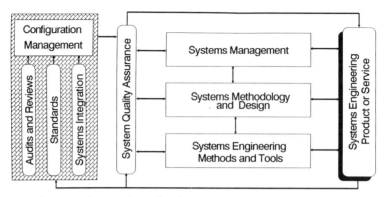

Figure 4.1. Three Fundamental Levels of Systems Engineering and Configuration Management for Quality Assurance of Systems Engineering Products or Services

There are a number of other related issues, such as strategic level quality assurance and management, or total quality management (TQM) and total productive maintenance (TPM) that are examined in succeeding chapters when we consider metrics for quality assurance, verification and validation, maintenance, and test and evaluation.

4.1. INTRODUCTION

In the previous chapters, we examined a number of systems engineering life cycles. The simplest life cycle is comprised of three phases

System definition
System design and development
System implementation and maintenance

as we indicated in Figure 2.1. There are a number of critical events and activities, often called *milestones*, that should be planned to occur at designated points in the systems life cycle. This suggests that the simple life cycle model of Figure 2.1 might be expanded, but still within the three phases, to include evaluation and reviews, and associated feedback, to insure production and delivery of an operational system that is of high quality and that satisfies user needs or requirements. These quality assurance efforts should be ubiquitous throughout all phases of the systems engineering life cycle. Figure 4.2 illustrates this expansion for the three-phase life cycle model. It shows a number of opportunities for efforts such as reviews and audits to verify and validate system baselines for configuration management and to

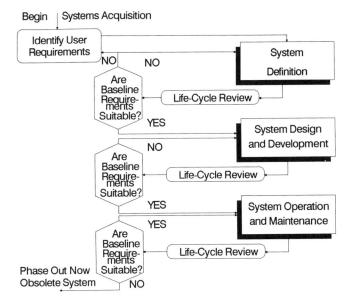

Figure 4.2. A Three-Phase Systems Engineering Life-Cycle Model Illustrating Baselines and Reviews for Quality Assurance

assure system quality. In general, baselines are designated points in the systems life cycle at which some important features of the system configuration are defined in detail and which serve as reference points for further efforts in the systems life cycle. A baseline is much like a benchmark and seems to establish important configuration information. We discuss baselines in greater detail later in this chapter.

There is general and universal agreement that quality should be the primary driver of the entire systems engineering process and of virtually everything else as well. However quality is difficult to define. Many of us might say that we know it when we see it, even though we cannot define it. There are many problems with this sort of attitude, especially when we cannot act entirely as individuals but must function as part of a group or team. If we cannot define quality, then we will doubtlessly have difficulties in communicating it, or in assisting others to recognize it. We might view quality assurance as a proactive effort during system acquisition and operation. It would enable us to detect the presence, or potential presence, of faults or errors in either system design or operation. The diagnosis of the causal factors leading to real or potential flaws would also be supported through quality assurance procedures, in terms of the reviews and audits undertaken. Maintenance efforts are initiated based on these review and audit results, and lead ultimately to correction of the errors or faults in system design or

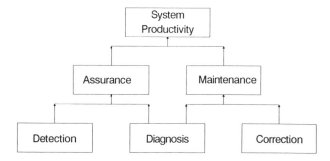

Figure 4.3. System Productivity Roles for Quality Assurance and Maintenance

operation, or to establishment of better procedures for system operation. Figure 4.3 illustrates these notions.

Quality is a subjective term and a multiattributed one as well. Simply stated: *system quality is the degree to which the attributes of the operational system enable it to perform its specified end-item functions so as to achieve client purposes in acquiring the system.* The IEEE Standard for Software Quality Assurance Plans [1] uses a similar definition for quality assurance: *"Quality assurance is a planned and systematic pattern of all actions necessary to provide adequate confidence that the item or product conforms to established technical requirements."* Each is a reasonable definition of operational level product quality. Each contains the notion of a metric to indicate the degree of quality or degree of conformance of a product or service to the requirements of the user or client. The first definition does place somewhat greater focus upon the need for the system to achieve the purpose of the human stakeholders who serve as clients or customers for the system.

Three steps support overall quality assurance efforts:

1. Identification of quality assurance plans.
2. Identification of quality assurance tasks in terms of schedules, responsibilities, and an organizational structure to enable implementation of the QA plan.
3. Description of the tools and approaches that will be used to accomplish these tasks.

We need appropriate metrics or indicators of system quality to obtain an early warning indicator of potential difficulties and make appropriate design changes early in the systems engineering life cycle. Conformity with standards of good practice, legality, and so on, are intrinsic to these definitions. Such standards become a set of guidelines against which the results of a systems engineering effort may be reviewed and audited and (potentially) approved.

There are two kinds of standards: (1) for the practice of systems engineering as a professional activity and (2) through which to judge the worth of the product of a systems engineering effort. The first may be called systems management standards. The second is a system performance standard. Each is needed. There are other standards, of course. These two generic types are the most important for our efforts here.

There are also important notions of systems integration that are needed in quality assurance, as it will be rare that a system does not need to be compatible with some existing system. We wish to examine both types of standards, and systems integration concerns as well, in this chapter.

One inevitable aspect of system evolution, regardless of the life cycle approach taken, is that *changes* in the initial system concept will occur. If these changes are not carefully controlled and documented, several forms of chaos may occur. Among these are systems that may initially function but that are so poorly documented and understood that they cannot be easily maintained or changed over time as user needs change. A way to manage change for the productive benefit of the systems engineering process is much needed.

The purpose of *configuration management* (CM) is to manage change throughout a systems development process. Some of the many causes of change in the life cycle process of systems engineering are as follows:

1. During system development, it is discovered that the system specifications have not been properly identified and that a system designed to satisfy the initially identified requirements will not actually satisfy user requirements.
2. At some phase in the systems acquisition life cycle, it is determined that the user requirements have changed and that system specifications need to be changed to reflect this.
3. During system evaluation, or perhaps at other phases in the life cycle, a fault in the evolving system is detected and the subsequent correction of this fault requires a change in the system design concept.

In each of these cases, there is a need to know that a change is necessary, what kind of change is needed, that all stakeholders to the systems acquisition process agree with the change and are informed of pending changes, and that careful records are kept of changes to be made. These records should include

1. The rationale supporting the changes.
2. The people who made the changes.
3. What the changes were.
4. When the changes were made.

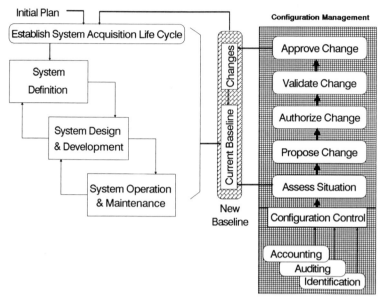

Figure 4.4. A Three-Phase Systems Engineering Life-Cycle Model Illustrating Configuration Management Process

All of these are configuration management functions and support the primary quality assurance purpose of configuration management. This purpose is provision of system acquisition stability in producing a trustworthy system. In the first part of this chapter, we discuss the concept of configuration management (CM). Figure 4.4 illustrates some CM activities that occur throughout the systems engineering life cycle. Configuration management generally is a necessary ingredient to assure operational or product-level quality.

Also inherent in quality assurance is the notion of *testing* as the primary tool of system quality assurance. This includes the concept of system design for testability, as there will be no useful way to determine functionality of a system if it is not testable. To do this requires appropriate system metrics. It is not good enough to identify attributes of system functionality; we must have attribute measures of functionality as well. Although an attribute need not necessarily be quantifiable, an attribute measure should be. Thus the term *attribute measure* is synonymous with *metric*. The latter part of our efforts in this text will be concerned with metrics and with the use of metrics for such important purposes as quality assurance and maintenance.

We envision a three-stage process in system quality assurance and possible subsequent maintenance efforts. System quality assurance is primarily determined with detection of the existence of faults. Second are some efforts at determination of the location and type of flaw, that is to say, diagnosis of the fault. Finally, system maintenance is concerned with proactive and reactive fault correction. Thus maintenance is given a very general interpretation, as discussed in more detail later in this chapter. Figure 4.2 illustrates how these relate together to enhance system productivity, as we have noted.

System quality assurance and associated testing can be conducted from a perspective of structure, function, or purpose. From a structural perspective, systems would be tested in terms of micro-level details that involve such hardware issues as thermal suitability of specific VLSI chips, or such software issues as programming language style, control, and coding particulars. From a functional perspective, operational level system quality assurance and testing involves treating the system as a blackbox and determining whether the input–output performance of the system conforms to the system technical requirements specifications. This is generally known as *validation* testing in the software systems engineering literature. From a purposeful perspective, a system must be tested to determine whether it does what the client really wishes it to do.

Notions of system quality assurance indicators are closely related to notions of systems management. It is the systems management, or management control, process that produces product-level quality assurance requirements. Generally, a client or user group is initially concerned only with the quality of the system (product or service) acquired and delivered to it in terms of usability of this system for an assumed set of purposes, which are often difficult to specify in advance. But, the quality of the end product is a direct function of the quality of the process that produced the end product. We are necessarily, therefore, concerned with trustworthy systems engineering processes and trustworthy systems engineering products. Notions of strategic level quality assurance and management, which lead primarily to process level improvements, and from process improvements to product improvements, are examined in Chapter 6.

Quality assurance is the name often associated with efforts that lead to the equivalent of a guarantee for an acquired system. Generally, this guarantee is with respect to the quality attrribute measures exceeding minimum performance standards on all quality attributes. But there are other approaches that allow compensatory trade-offs. *Quality control* is the act of inspecting an established product, that is to say the result of a specific system development process, to make sure that it meets some minimum defined set of standards. Here we are much more concerned, as systems engineers, with the more general term quality assurance, which also implies design for quality, and not just inspections to eliminate the unworthy. Quality assurance involves those systems management processes, systems design methodologies, and system acquisition techniques and tools that act to ensure that the resulting system meets or exceeds a set of multiattributed standards of excellence.

System *verification* is the activity of comparing the product produced at the output of each phase of the life cycle with the product produced at the output of the preceding phase. It is this latter output that serves both as the input to the next phase and as a specification for it. System *validation* compares the output product at each stage of the systems acquisition life cycle, or occasionally at only the final product phase, to the initial system specifications. Often these activities are performed by people outside of the systems engineering development organization, and the term *independent verification and validation* is sometimes used in these cases. Figure 4.5 illustrates some of the differences in these important concepts and their pertinence and pervasiveness throughout a typical systems engineering life cycle.

As initially defined, verification and validation generally do not address the appropriateness of the system requirements and may not, as a consequence, determine whether a system really satisfies user needs. Whereas verification seeks to determine whether the system and associated subsystems of hardware and software are being built correctly, validation seeks to determine whether the right product has been produced from an assumed set of correct specifications. So, verification and validation are quality control techniques and a part of overall system quality assurance.

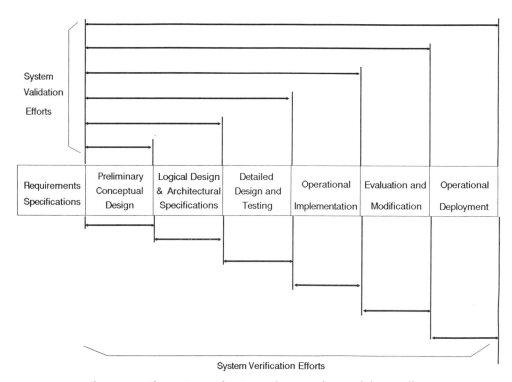

Figure 4.5. Phasewise Verification and Across-Phase Validation Efforts

The formal process of determining that a system product is suitable for an intended application is often termed *system certification*. The certification of system is a warranty or certification that the system will perform in accordance with agreed-on requirements.

Obtaining a high degree of system productivity requires a number of related system product assurance approaches. Managing change requires appropriate measures for control and communication at the management level and at the task level. Status accounting and auditing are important activities that assist in this. A number of cross-referencing efforts are needed to properly document configuration management efforts. We will discuss some of these issues later in this chapter, in Sections 4.3 and 4.4.

In the remainder of the chapter, we will discuss operational and task level quality assurance through the use of standards that help to assure trustworthiness and systems integration to insure that the new system is operationally functional and compatible with existing, or embedded, systems.

4.2. CONFIGURATION MANAGEMENT

Configuration Management (CM) is the systems management process that identifies needed functional characteristics of systems early in the life cycle, controls changes to those characteristics in a planned manner, and documents system changes and implementation status. Determination and documentation of who made what changes, why the changes were made, and when the changes were made, are the functional products of configuration management. Configuration management is a subject of considerable interest in systems engineering management [2], particularly in software systems engineering management [3]–[6]. There are four essential functions or baseline efforts that comprise configuration management in the initial DoD configuration management standards [7] that have served to foster most subsequent developments in this area.

1. *Configuration Identification:* Identification of specifications that characterize the system and the various subsystems, or *configuration items** (CI) throughout the life cycle. This specification becomes more precise and explicit as activities move to later phases of the life cycle. Partitioning of a system into subsystems is a systems management and technical direction decision, and frequently involves major judgmental concerns.

---

*Of course, you expected CI to be an acronym for configuration identification. That there are two CIs here is illustrative of one of the difficulties of excessive use of acronyms. Sorry, but the idea to use them did not originate here! It is important to note some of the more common acronyms and not also get lost in a sea of alphabet soup. Hopefully, we have struck a reasonable balance between these two extremes.

2. *Configuration Control:* Characterization and direction of proposed changes. These may be important changes, called class-I changes in the DoD literature, that affect structure and function and associated costs and schedules. Or they may be relatively minor changes that are more in the nature of parametric changes, such as editorial changes in documentation or substitution of functionally equivalent hardware items, that are not associated with structural modifications, Class-II changes in the DoD literature. Thus, change control involves evaluation, approval, and coordination of changes to any functional baselines. All changes to a specified configuration baseline are closely evaluated and controlled, and are generally under the review and approval authority of a *Change Control Board* (CCB), or *Configuration Control Board* (CCB). Changes may be prioritized according to their criticality as emergency, urgent, or routine. It is the duty of the CCB to deal with each of these.
3. *Configuration Status Accounting:* Generally use of a management information system (MIS) to ensure traceability of the configuration baseline as it evolves over time. This MIS records changes and provides reports that document these changes.* Configuration status accounting should encompass:
   3.1. The time at which each baseline came into existence.
   3.2. Descriptive information about each configuration item (CI), including the time of each change.
   3.3. Such configuration change decision information as disapprovals, approvals, delay decision, and so on.
   3.4. Detailed information about each engineering change proposal (ECP) made by the systems engineering contractor, including such information about each change as decision status.
   3.5. Status of documentation of the baseline.
   3.6. Deficiencies to a proposed baseline that are uncovered by a configuration audit. Configuration status accounting is primarily a reporting function.
4. *Configuration Audits and Reviews:* These are needed to validate achievement of overall system or product requirements, which ensures integrity of the various baselines, each of which are interpretations of the requirements at more and more narrowly defined levels. Most importantly, these make the systems engineering product as visible and understandable as possible to stakeholders, including both system developers and clients, and enhances the traceability of the system as it evolves throughout the acquisition life cycle. There are three basic kinds of audits in configuration

---

*In the software engineering area, the term *program support library* (PSL) is often used to denote this MIS. A PSL generally supports not only configuration control, but also program development and management [2]. There are four primary ingredients in a PSL: internal machine readable libraries, external hardcopy libraries, and standard computer and office procedures.

management, functional configuration audits (FCA), system configuration audits (SCA), and physical configuration audits (PCA). These operational product level audits are performed prior to any proposed major baseline change. There are a number of reviews as well, such as requirements, specifications, and in-progress reviews. We provide an expanded description of audits and reviews in the next section.

The essential tasks in configuration management involve identifying the configuration of a system at discrete points in time. Changes to this configuration are controlled and documented such that integrity and traceability of the resulting configuration is made clear throughout the systems life cycle. Configuration management is the means through which a systems engineering team assures trustworthiness and integrity of the design, development, and cost trade-off decisions that determine performance, producibility, operability, and supportability of the fielded system. CM efforts begin at the first phase of the life cycle and extend throughout as the system is made operational. Configuration changes occur throughout the life cycle. Changes are controlled to ensure cost effectiveness, and are documented so that all system users are made aware of the current state of the system.

The concept of baseline management is central to configuration management. By definition, *baselines* are designated points in the systems life cycle where some important features of the system configuration are defined in detail. Configuration control is the primary driver in configuration management; the identification, accounting, and auditing efforts serve to provide formal evaluations and records of control efforts. Figure 4.4 has illustrated these important notions. It provides an overview of the configuration management process and illustrates our discussions up to this point.

There are three formally defined DoD baselines. These, and their definitions, are as follows.

1. *Functional Baseline:* Established at the end of the phase in the life cycle that deals with concept exploration (CE). The functional baseline contains the initially determined system specifications that are contained in a request for proposal (RFP) document, as described in Chapter 2. The system level requirements specifications define the technical specifications for the program.

2. *Allocated Baseline:* Allocated functional specifications for subsystems, or projects. Each configuration item is generally associated with an allocated baseline and it is the allocated baseline that provides the specified basis for the detailed design and development activities in the systems engineering life cycle.

3. *Product Baseline:* Established for each project or subsystem, and the associated configuration item, the product baseline generally contains specifications for each project or subsystem. The functional configura-

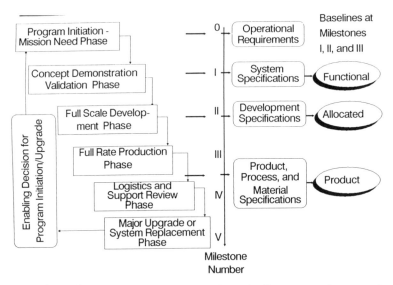

Figure 4.6. The Defense Systems Acquisition Life Cycle Illustrating Milestones, Specifications, and Baselines

tion audits (FCA) and physical configuration audits (PCA) serve to verify and validate the subsystems in each configuration item.

Figure 4.6 illustrates the point, or phase, or milestone, in the system life cycle at which these baselines are established. The DoD system acquisition life cycle is shown in this figure, although we can clearly associate these baselines with any of the systems engineering life cycles that we have discussed. It is important that the many interfaces in the system life cycle, which represent interfaces with a great variety of people and products, be properly established. It is especially important that baselines be established at appropriate times. If they are established too early, design and development creativity and the ability to make trade-offs is reduced. If they are established too late, unnecessary and expensive change to partially developed systems will, more often than not, result. This is one of the crises that result from lack of appropriate configuration management.

In general, baseline selection should be responsible to specific system development needs. It might be reasonable to associate one baseline with each of the phases in the systems engineering life cycle. There are many alternate baseline possibilities. For example, it may be appropriate to disaggregate the functional baseline into two baselines.

1. *User Requirements Baseline:* Established after user requirements for the system to be developed have been established. The initial user require-

ments baseline would include the original set of user system requirements and associated documentation. Succeeding requirements are
incorporated into a modified baseline. This baseline will naturally tend
to be oriented toward purpose and function.

2. *System Specifications Baseline:* Established after user requirements have
been translated into an initial set of technical system specifications.
This baseline first includes the original technical system specifications,
for hardware and for software, and is updated as technical system
specifications change, or evolve, over time.

In a similar way, alternate names for the baselines may be somewhat more
appropriate in specific systems acquisition efforts. The allocated baseline is,
in essence, a conceptual design baseline. It includes critical design architectures and how these are related to technical specifications. In a similar way, it
may be more appropriate to use the term *subsystem baseline* to describe
those portions of the allocated baseline that are identified later in time. This
would describe task performance requirements for each subunit of hardware
and software that comprise the projects in a systems engineering management plan (SEMP), and the configuration management plan (CMP).

Further disaggregation may be desirable. The product baseline is generally
established to control the latter portions of the life cycle. It may be desirable
to associate additional baselines with these phases. The operational implementation and maintenance portions of the systems engineering life cycle are
often very critical in the production of trustworthy systems. Thus, it may be
worthwhile to separate baselines for operational implementation and/or
maintenance. Definitions for such baselines follow.

1. *Systems Integration Baseline:* Used to assist in management of the
operational implementation needs for interoperability and architectural
compatibility across new and existing systems.

2. *Maintenance Baseline:* Established at the time of operational deployment of a system to better manage maintenance and other enhancement and evolution efforts.

Figure 4.7 illustrates the relationships among these two baselines sets.*
Figure 4.8 illustrates the DoD standard 2167A systems engineering life cycle
[8], established primarily for the control of software project development, and
both the three and six baseline sets that we have discussed here. It illustrates
many of the audits and reviews that are conducted, as part of quality
assurance efforts, throughout the life cycle (see Section 4.4 for a discussion of
reviews and audits, including acronym definitions for Figure 4.8). Figure 4.8

---

*There are other baseline possibilities. For example, we might suggest an audit baseline.
However, this sort of baseline would extend across each of the life-cycle phases and would not be
concentrated upon a single phase of effort.

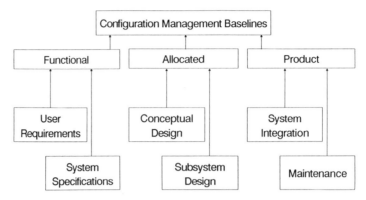

Figure 4.7. Relations Among Baseline Sets

also illustrates the notion of concurrent development of hardware and software, or *concurrent engineering*. This is a subject of much current interest and relevance, and we examine it briefly in our next section.

We now discuss some pragmatic management issues associated with configuration management. The Configuration Control Board (CCB) is generally headed by, or responsible to, a configuration manager. This board has a number of configuration management responsibilities, including

1. Developing CM guidelines and practices for a specific program.

2. Documenting and distributing these guidelines and practices.

3. Identifying the initial systems baseline and tracking the evolution of this baseline over time.

4. Assuring that unauthorized changes are not allowed to be made to the baseline.

5. Insuring that all proposed changes are evaluated before a new baseline is established.

6. Assuring that changes made are properly documented.

7. Adjudicating competing claims and disputes, and approving requests for exceptions to CM policy.

8. Managing system configuration libraries that represent baseline control (maintain a master copy of the currently approved baselines), document control (maintain, distribute, and store documents relating to the system being acquired), deliverable control (to enable forwarding and distribution of all documents), and release control (to effectively and

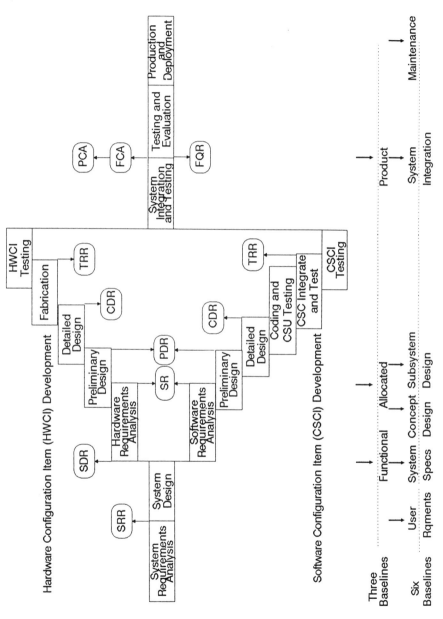

Figure 4.8. System Development Baselines, Reviews, and Audits

efficiently modify master copies of system development documents and distribute appropriate change notices).

For small programs, one person might perform configuration management functions. For very large programs, a number of people are needed. The configuration control board is a part of this team, and the other persons in the configuration management office are, in effect, staff members for the CCB. It is generally necessary for the client or customer for a systems acquisition effort to be represented on a CCB. Configuration control management efforts may, therefore, be a significant part of an overall systems acquisition effort. Configuration management efforts need proper configuration, also.

Baseline control is a very important function of configuration management. There are necessary trade-offs that must be made in baseline control. The principle challenge is to ensure the maximum flexibility possible to the extent that this aids in the acquisition of trustworthy functional systems. On the other hand, the baseline must be safeguarded against unauthorized change.

A major task is that of ensuring system integrity. Every proposed baseline change must be thoroughly tested in order to verify both that the intended change produces desired results and that the change does not simultaneously degrade some other functional capability. This is not at all an easy task. If for example, there are n subsystems, we need to do much more than insure that a change made in one of them does not influence operations of the other $n - 1$ subsystems. There is a problem of second-order changes as well. It may turn out that a change in one subsystem does not formally produce a change in behavior of any other subsystem. It may also turn out that a change in a second subsystem also does not produce a change in the behavior of the other $n - 1$ subsystems. However, when both of the suggested changes are made, there may be changes in the behavior of a number of other subsystems.

This suggests that there will always need to be some testing to be sure that all changes made up to a point do not cause problems. The systematic application of this sort of testing is called *regression testing* in much of the software engineering literature. For this reason, configuration management is defined to be a systems management function at the program, rather than at the project, level.

Of course, not all proposed changes to a system affect formal baselines. Changes to a system that do not affect functional baselines are generally under the purview of a discrepancy review board (DRB). This board, where one exists, takes user identified performance discrepancies or defects, and determines which system feature may require correcting. Although proposed changes that might result from the efforts of a CCB require customer or client approval, those that result from the efforts of a discrepancy review board do not.

In DoD literature, problems detected in a system or its documentation are classified into five priority levels.

1. *Priority 1* problems prevent the achievement of mission essential capability as specified by a baseline requirement, prevent an operator's achievement of a mission essential capability, or pose threats to personnel safety.
2. *Priority 2* problems adversely affect the achievement of a mission essential capability, or adversely affect an operator's achievement of such a mission. System performance is adversely affected, and no corrective solution is known.
3. *Priority 3* problems are essentially priority 2 problems, except that a work-around corrective solution is known.
4. *Priority 4* problems create operator inconvenience or annoyance, but do not affect mission essential capabilities.
5. *Priority 5* problems are of all other types.

In Chapter 9, we will provide an expanded discussion of human errors, including their interrelations with system errors.

There are many responsibilities of top-level systems engineering management, including the following.

1. Assisting in definition of the issue at hand, identification of client needs and requirements, and translation of the identified user needs and requirements into an initial set of technical system specifications.
2. Identification of the resources required for trustworthy development of a system that is of high quality and that, therefore, will satisfy user needs and requirements.
3. Interactive iteration of initially identified requirements and specifications with the user, so as to obtain a set of final requirements and specifications that are cost effective and within budget constraints;
4. Definition and coordination of software and hardware program and project plans, and other needs of the life-cycle phases for design, development, evaluation and test, system integration, implementation, and maintenance.
5. Maintaining communication and interfaces with all stakeholders to the effort including top-level management of both the client, user, and system development groups.
6. Developing, through use of a total quality assurance and management process, a trustworthy system product that is sustainable over time.

It may appear by now that the critical success factors affecting a systems acquisition effort, and its management and technical direction, need to be

TABLE 4.1   Table of Contents for Typical Systems Engineering Management Plan for Configuration Management

Title Page
Preface
Table of Contents
List of Figures
List of Tables
1. Executive Summary
2. Introduction
   2.1   Program Objectives
   2.2   Program Description
   2.3   Program Scope
   2.4   Program Overview
   2.5   Program Organization and Responsibilities
   2.6   Program Deliverables
   2.7   The Systems Engineering Management Plan Components
   2.8   Program Milestones
3. Program Management Approach
   3.1   Systems Management Assumptions, Needs, Constraints
   3.2   Systems Management Objectives
   3.3   Program Risk Management
   3.4   Program Staffing
   3.5   Program Monitoring and Control
   3.6   Program Integration Approach
4. Program Technology
   4.1   Technical Description of System Development Projects
   4.2   Project Methods, Tools, and Techniques
   4.3   Project Procedures and Support Functions
5. Program Quality Assurance
   5.1   Quality Assurance Plan
   5.2   Maintenance Plan
   5.3   Documentation Plan
   5.4   Operational Deployment Plan
   5.5   Configuration Management Plan
      5.5.1   CM Objectives and Overview
      5.5.2   CM Organization, Charter, Members, Duties
      5.5.3   CM Methods
         Baselines
         Configuration Identification
         Configuration Control
         Configuration Auditing
         Configuration Status Accounting
      5.5.4   CM Implementation Plan
         Procedures
         Personnel
         Budget
         Implementation Milestones
   5.6   System Security Plan

TABLE 4.1   (*Continued*)

6. Budget and Resource Management Plan
    6.1   Work Breakdown Structure
    6.2   Cost Breakdown Structure
    6.3   Resource Requirements
    6.4   Budget and Resource Allocation
    6.5   Program Performance Schedule
    Additional Contents
    Reference Materials
    Definitions and Acronyms
    Index
    Appendices of Supporting Material

identified and resolved at the initial phases of system development. This is, indeed, correct. Effective systems management not only demands the generation of plans but also requires communicating them. There are many reasons for preparation of written plans, which should be accomplished at both the program and project levels. Generally, a good way to do this is through the systems engineering management plan and a configuration management plan, which may be a part of the SEMP. Table 4.1 presents the generic components of a typical project plan.

Specific goals for a systems engineering management plan and the written documentation concerning it include sufficient details to indicate that the plan is satisfactory in the sense of producing a cost-effective and trustworthy system. At a minimum, this must include evidence that the following are present.

1. There exists an understanding of the problem.
2. There exists an understanding of the proposed solution.
3. The system acquisition program is feasible from all perspectives including costs, effectiveness, and timeliness.
4. The projects within the program each benefit the program.
5. The system acquisition risks are tolerable and explicable.
6. There exists an understanding of system integration and maintenance needs.
7. The fielded system is a quality system.

If quality assurance cannot be provided, system development should probably not proceed.

One of the major needs in systems management is monitoring and control of programs and projects. It is necessary to monitor progress of the program and projects associated with the program to know whether the effort, at any specified instant of time, is proceeding according to schedule. If not, either because of a change in the requirements or for other reasons, we can take steps to get the effort back on schedule. Thus, systems management and change management are each inseparable parts of a systems engineering management plan. Figure 3.8 illustrated the necessary feedback and iteration in monitoring and controlling. An appropriate sequence of steps and key tasks to be accomplished were discussed in the concluding paragraphs of Section 3.3.

Preparation of a systems engineering management plan, including a configuration management plan, as we have already noted, is a fundamental task in system quality assurance. The duties of a system quality assurance team, which is intimately associated with the SEMP and which may often both write and implement the plan, include the following.

1. Development and implementation of quality assurance standards for a specific system development effort, including the practices, procedures, and policies that comprise the system quality assurance plans.
2. Development and implementation of metrics, testing tools, and other quality assurance techniques.
3. Implementation of the resulting quality assurance plan, including documentation of a final quality assurance report.

The essential components of the system quality assurance plan (SQAP) include identification and implementation of these items.

1. The scope and purpose of the plan.
2. The organizational structure for implementing the plan, including specific tasks to be performed by members of the group.
3. Identification of documents that need to be prepared and methods to determine quality and adequacy of this documentation.
4. Metrics, standards, procedures, and practices, including reviews and audits, that will be used in implementing the plan.
5. Methods that will be used in collecting, maintaining, and recording quality assurance information.

If we desire to implement the approaches suggested here as a part of systems management or management control to provide system quality assurance, identification of associated methods and metrics for system measurement and test are required. In these efforts, it is critical to develop approaches that increase the possibility of early detection of the potential for errors. We must also be able to apply appropriate diagnosis and correction at these early phases of the life cycle, when such changes consume much less time, effort, and money than later on. We turn our attention to these topics in later chapters, especially in Chapters 6 and 9.

There are many perspectives that can be taken relative to system quality indicators. It is possible to speak of the metrics that should be used at the different phases in the systems engineering life cycle. This would represent an internal assessment of quality. An external assessment of a developed and implemented system would generally need to be made also. This would ensure that concerns of revision and transition to evolving and new requirements are addressed, as well as concerns of operational functionality and trustworthiness. This reflects the fact that, throughout its life cycle, a system being acquired will typically undergo

- Identification of need and specification of requirements
- Initial design and development
- Controlled introduction to a customer market
- Release to customers
- Modification to meet evolving needs
- Transition to a new environment.

Attributes, and attribute measures or metrics, are needed to serve project management needs during initial design and introduction of the system to the

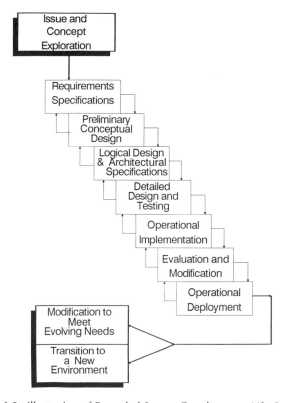

Figure 4.9. Illustration of Extended System Development Life Cycle (SDLC)

client group. There are also needs to modify, or a transition to a new environment and perhaps a new customer base as well. Figure 4.9 illustrates this proactive view of these extended phases in the evolution of a new system that will satisfy client requirements. From this figure and the discussion here, we see that systems quality assurance efforts should serve the need of systems management in planning and implementing an efficient and effective systems acquisition venture, as well as obtaining rapid solutions to problems that may occur at any phase in the systems engineering life cycle, systems acquisition life cycle, or systems development life cycle (SDLC).

4.3. CONFIGURATION MANAGEMENT AND CONCURRENT ENGINEERING

Often, it is desired to produce and field a system in a relatively rapid manner. The life-cycle process could indeed be potentially accelerated in time if it were possible to accomplish phases of the life cycle in a more or less concurrent, or simultaneous, manner. *Concurrent engineering* is a systems engineering process approach to the integrated coincident design and development of products and systems [9]. Concurrent engineering is intended to cause systems acquisition, broadly considered to include all phases of the life cycle, to be explicitly planned to:

1. Better integrate user requirements so as to result in high-quality cost-effective systems.
2. Reduce system development time through this better integration of life-cycle activities.

The basic tasks in concurrent engineering are much the same as the basic tasks in systems engineering. The first step is that of determining what the customer wants. After the customer requirements are determined, they are translated into a set of technical specifications. The next phase involves program planning for development of a product that satisfies the customer. Often, especially in current engineering, this involves examining the current process, especially at the systems management or management controls level. This process is usually refined to best deliver a superior quality product as desired by the customer, within cost and schedule constraints.

The effort needed to establish user or customer requirements for a system or product is not to be undertaken in a cavalier manner. There are a number of approaches that can be used to determine requirements (some of these are examined in Chapter 6). In concurrent engineering, the very early and very effective configuration of the systems life-cycle process takes on special significance. This is so because the simultaneous development efforts need to be very carefully coordinated and managed or the opportunities abound for significant cost increases, significant product time increases, and significant deterioration in product quality. The use of coordinated product design teams, improved design approaches, and careful and critical use of standards are among the aids that can be brought to bear on concurrent engineering needs. There are particularly critical in concurrent engineering because of the group nature and simultaneous nature of the system development effort.

There is much need for a controlled environment for concurrent engineering and for system integration needs as well. This requires several integration and management undertakings.

1. *Information Integration and Management:* It must be possible to access information of all types easily. It must be possible to share design information across the levels of concurrent design in an effective and controlled manner. It must be possible to track design information, dependencies, and alterations in an effective manner. It must be possible to effectively monitor and manage the entire configuration associated with the concurrent life-cycle process.

2. *Data and Tool Integration and Management:* It must be possible to integrate and manage tools and data such that there is interoperability of hardware and software across the several layers of concurrency.

3. *Total Systems Engineering:* It must be possible to insure that the process is directed at evolution of a high-quality product and that this product is directed at resolution of the needs of the customer or user in a trustworthy manner that is warmly endorsed by the customer.

Concurrent engineering clearly requires much upfront planning·so that simultaneous development of subsystems may occur in a trustworthy manner. What this suggests is that a potentially larger number of subsystems be identified such that it is possible to design and develop them in parallel. Compression of the phases of the life cycle and at least partial parallel accomplishment of some of them is somewhat more problematic. The macroenhancement approaches to system development, especially software system development [10], appear particularly useful in this regard. These include prototyping as a means of system development, use of reusable (sub)systems, and expert system and automated program generation approaches. Each of these allows, at least in principle, compression of the acquisition life cycle in a manner that is compatible with acquisition of a trustworthy system.

Through use of approaches such as these, it is hoped to obtain systems that are of high quality, have a short deployment time, and low life-cycle costs.

1. High quality, in terms of system performance, suitability, and reliability in a large variety of operational environments over a sustained time period.

2. Short deployment time, for new product and service designs, and for delivery and maintenance of existing product designs.
3. Low life-cycle costs, for system design, development, maintenance, and retrofit or phaseout.

In a very insightful work, Winner et al. [11] identified three critical activities and a number of technical capabilities that support these.

1. Obtain early, complete, and continuing understanding of customer requirements and priorities. This requires capabilities for obtaining information concerning comparable products, processes, and support. It requires identification of complete and unambiguous information on this new system and product, including support needs. Finally, it requires synthesis or translation of user requirements into design specifications for the system product and the validation of design specifications.
2. Translate the system requirements and specifications into optimal products and manufacturing and support processes that can be performed concurrently and in an integrated fashion. There are many required capabilities. These include managing information and data concerning the system product and development process and dissemination of product, process, and support data to the concurrent teams.
3. Continuously review and improve the system product and support characteristics. This includes intelligent oversight to enable impact assessment of changes, and proactive, concurrent availability of current design.

These provide a very useful set of critical success factors for the concurrent engineering process.

It appears reasonably clear that there is no single unique best way to approach concurrent engineering. It would appear to be more a philosophy of and approach to strategic management than anything else. Of course this strategic management needs to be translated into management controls or systems management, and thence to task and operational level effort that results in a system. How a particular organization approaches concurrent engineering is a function of their organizational traditions and culture. In Chapter 6, we examine some strategic level quality assurance and management approaches that are particularly relevant to concurrent engineering.*

---

*The *total quality management*, or strategic quality assurance and management approach that we describe in Chapter 6, involves a process orientation and a commitment to continued improvement; a major focus on customer satisfaction; a system engineering approach to issue resolution; a major focus on worker empowerment through motivation, self-direction, teamwork, education, and training; and strong management leadership and passion for excellence through quality. This would seem essential, or highly desirable at the very minimum, for successful adoption of a concurrent engineering approach.

The major emphasis on team performance in concurrent engineering suggests that efforts that involve computer shared group work [12] and/or group decision support systems [13] may prove to be of considerable value in enhancing team performance in concurrent engineering. Formally, there is little that is very new in the subject of concurrent engineering. But, we should be very careful to not dismiss the strengthened needs for the strategic planning and systems management necessary to ensure success in concurrent systems engineering.

4.4. DEFINITIONS FOR QUALITY ASSURANCE AND ASSOCIATED REVIEWS AND AUDITS

A rather large number of acronyms are used today to describe various systems engineering quality assurance indicators. Many of these have originated in software systems engineering [14]–[16], often after having been transferred to it from more generic systems engineering efforts. Among these, we list the following alphabetically by acronym.

CIL—Configuration Item Log: This document is comprised of a table that describes the time schedule and associated initiation and completion dates for all events associated with configuration items. A *configuration item index* (CII) is generally associated with the CIL.

CMP—Configuration Management Plan: See system configuration management (SCM).

CDR—Critical Design Review: The objective of a CDR is to examine the system design description (SDD) to determine the extent to which it satisfies the requirements of the system requirements specifications (SRS). Special concern is associated with determining that all design specification change requests have been accommodated.

DB—Design Baseline: The DB is a formal milestone in many system development life cycles (SDLC) that begins at the end of the initial system development phase. This baseline is authorized by the preliminary design review (PDR). Associated with the DB is a design baseline configuration item (DBCI).

FB—Functional Baseline: The FB is a formal milestone in the SDLC. Generally, this baseline is operationally established by the results of the initial design concept review (IDCR). The functional baseline configuration item (FBCI) is a document that describes the FB. It is included in the functional description (FD) of a system.

FCA—Functional Configuration Audit: An FCA, sometimes called a *functional audit* or *FA*, is an audit held just prior to final system delivery in order to verify that all of the requirements of the system requirements specifications (SRSs) have been met by each of the subsystems and associated documentation. The FCA is sometimes replaced or augmented by a final development review (FDR).

FD—Functional Description: This document provides a description of the technical system specifications. The purpose of the FD is to provide a clear description of the operational or functional capabilities of the system to be developed and fielded.

Feasibility Study: The FD provides a detailed discussion of the practicality of the system. Generally, a FS results from the System Design Request (SDR).

FQR—Formal Qualification Review: The FQR, or formal qualifications test (FQT), is an acceptance testing procedure designed to determine whether a final system or subsystem, one that is ready for integration and implementation, conforms with the final system technical specifications and requirements.

IDCR—Initial Design Concept Review: The IDCR is usually the first formal review in the SDLC and occurs after the preliminary conceptual design phase of effort. The functional baseline (FB) is usually established as a result of the IDCR.

IPA—In Process Audit: At various times in the system development process, in-process audits (IPAs) may be conducted to evaluate consistency of the system design at that phase of the life cycle, including the following.

1. System performance versus conceptual design requirements.

2. Hardware and system interface specifications.

3. Functional requirements and design implementations.

4. Functional requirements and review-audit prescriptions.

It should be noted that an IPA may include both a functional configuration audit (FCA) as well as a physical configuration audit (PCA).

MP—Maintenance Plan: This document discusses management controls and task level details associated with the after operational deployment maintenance of the system being fielded.

MR—Managerial Review: Managerial reviews are held periodically to determine the extent of execution of the SQAP. The team engaged in this activity is often called the management review team (MRT).

OB—Operational Baseline: This milestone is established at the end of the last phase of the SDLC, and before the extended SDLC begins. An operational baseline configuration item (OBCI) is generally included in the OB.

OTE—Operational Test and Evaluation: Operational test and evaluation (OTE) generally involves determination by an outside independent organization of the extent to which an operational system product meets the requirements and needs of the client or user.

PB—Product Baseline: The PB is a formal baseline established at the end of the functional configuration audit (FCA).

PCA—Physical Configuration Audit: The purpose of a physical configuration audit (PCA), sometimes called a physical audit (PA), is to evaluate whether the system and associated documentation are both internally consistent and also suitable for delivery to the client. This audit examines actual detailed system structure, such as code in software and VLSI chips in hardware. When a complex system is under development, it may be beneficial to conduct the FCA and PCA in incremental fashion. Certain configurations items may be available for audit prior to others assuming all interdependencies with other CIs have been validated in accordance with the formal test plans and procedures. The case where subsystems of large complex systems may be under development at different locations, by different contractors, requires special attention because of the concurrent and distributed nature of the life-cycle effort that make it infeasible to have full-system physical configuration audit in one central location. These practices are usually implemented by contract language or configuration management plans (CMP) for a systems engineering project sponsored by the DoD.

PDR—Preliminary Design Review: The purpose of a preliminary design review (PDR) is to enable an evaluation of the extent of acceptability of the preliminary system design, as specified by a preliminary version of the system design description (SDD). It represents a technical review of the basic design approach for each major system subsystem. System development and verification tools are identified. If changes are recommended, particular care is used to determine their propagation throughout the system development life cycle in a consistent manner.

SCM—System Configuration Management: System configuration management (SCM) is the process of identification of system configuration at specific points in time along the life cycle so as to enable maintenance of the traceability and integrity of the system configuration throughout the development life cycle. Although this is not formally a part of quality assurance, the specific steps to determine quality assurance activities are directly related to SCM results. A systems configuration management plan (SCMP) or configuration management plan (CMP) specifies activities to be included in SCM.

SR—Specifications Review: After hardware and software specifications have been obtained, there is a need to separately review each of these. Sometimes, these are denoted software specifications review (SSR) and hardware specifications review (HSR).

SDD—System Design Document: The system design document (SDD) is prepared to indicate the design specifications for the system that is to be developed. A SDD defines the system architecture, modules, and interfaces for a system that (presumably) satisfies the specified system requirements. It also contains the computer code that describes or specifies the system capabilities.

SDP—System Development Plans: The processes, procedures, activities, and standards that are used for system management, including system quality assurance and management, are often denoted by the collective term *system development plans* (SDPs).

SDR—System Design Review: The system design review (SDR) occurs after configuration management has been accomplished and the SDD has been written. The SDR should describe each of the major subsystems that comprise the overall design specifications, such as databases and internal interfaces. Tools required for verification are also identified as a by-product of this effort.

SDT—System Development Test: The SDT is the first major test of a newly produced system. It involves testing of both the individual subsystem units, and functional testing of the entire system to determine the extent to which it satisfies the system specifications.

SQA—System Quality Assurance: Formally, system quality assurance (SQA), or just quality assurance (QA), is the extent or degree to which a system product is in conformity with established (technical) requirements.

SQAP—System Quality Assurance Plan: The plan or systematic effort contemplated to determine, or measure, the extent of system quality assurance is known as the system quality assurance plan (SQAP).

SQPP—System Quality Program Plan: There is generally a meta-level planning activity that leads to the specific SQAP. The system quality program plan is the guiding force behind this effort. It is this program plan that leads to the projects that comprise the SQAP.

SRR—System Requirements Review: A review to evaluate, and determine the adequacy of, the requirements that are stated in the system requirements specifications (SRS) is called a system requirements review (SRR). Planning for system testing is accomplished here.

SRS—System Requirements Specifications: The system requirements specification (SRS) should clearly and precisely describe each of the essential requirements for the system as well as the external interface. Each of these specifications should be defined so that it is possible to develop an objective metric that can be used to verify and validate achievement level by a prescribed method.

STR—System Test Report: The systems engineering contractor, or a surrogate subcontractor, is obligated to perform evaluations of each hardware configuration item (HWCI) and computer software configuration item (CSCI). These evaluations, together with the results of the formal qualification review (FQR) or formal qualifications test (FQT), are reported in the system test report (STR).

SVVP—System Verification and Validation Plan: The system verification and validation plan (SVVP) describes the inspection, analysis, demonstration, or test methods that are used to:

1. *verify* that the specifications in the SRS are implemented in the design expressed in the system design document or SDD, and that the design expressed by the SDD is indeed implemented by the resulting system of hardware and software; and
2. *validate* that the operational specifications are in compliance with the requirements that are contained in the SRS.

SVVR—System Verification and Validation Report: The results of the execution of the SVVP and the results of all reviews, audits, and tests of the SQA plan are contained in the system verification and validation report (SVVR).

TRR—Test Readiness Review: The system development contractor performs evaluations of hardware and software development products using understandability, consistency, document traceability, design appropriateness, and adequacy of test coverage as evaluation attributes. The results are presented at a test readiness review (TRR).

Figure 4.10 illustrates how many of these reviews and audits flow from one to another, and how they are matched to the various phases in the systems engineering life cycle. The matrix in Figure 4.11, when completed, will contain the many activities involved in system quality assurance and systems management. A major objective in development of a system quality assurance plan (SQAP) is to complete the entries in this matrix and to accomplish them. Figure 4.8 also indicated many of these reviews and audits in the DoD 2167A life cycle.

This concludes a brief discussion of the various plans, reviews, and audits associated with configuration management and associated system quality and system product quality assurance. Some of these may be implemented as metrics on a continuous scale, and some as metrics on a binary (acceptable or not acceptable) scale. Each has uses, and a continuous scale may be more or less appropriate than a binary scale, depending on the intended use. Some of these are performed in a *static* fashion, through the examination of code, system hardware, documentation, and a prototype constructed to emulate customer requirements. Others may be performed through functional operation of subsystems or execution of the programming code or of the system hardware product itself.

It is possible to develop or identify a number of attributes of system quality and related operational functionality concerns. It would be possible to provide a brief definition of these to enable us to measure these for a given systems engineering product or service. An alternate and potentially more appropriate method of approach is to structure these attributes so that we obtain an attribute tree that enables the meaning of each of these to become

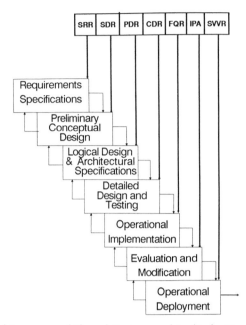

Figure 4.10. Typical Sequence of Phased Reviews and Audits for Configuration Management and Quality Assurance

	QUALITY ASSURANCE			PROJECT MANAGEMENT		
SDLC - PHASE 1	SRR			MR		
SDLC - PHASE 2	SDR			MR		
SDLC - PHASE 3	PDR			MR		
SDLC - PHASE 4	CDR			MR		
SDLC - PHASE 5	FQR			MR		
SDLC - PHASE 6	IPA			MR		
SDLC - PHASE 7	SVVR			MR		

Figure 4.11. Activity Matrix to be Completed as Part of Quality Assurance and Configuration Management

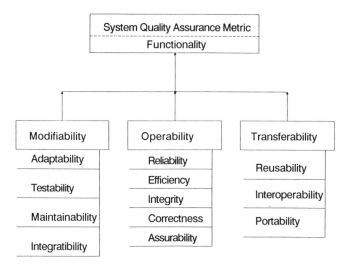

Figure 4.12. One Possible Systems Engineering Quality Attribute Tree (Functionality)

apparent through the hierarchical structure. Generally, a structure that relates important performance attributes should be developed for each specific system product, as attributes of importance will vary from one system under development to another. Figure 4.12 illustrates some system quality assurance attributes of system maintenance from the perspective of maintenance types. Here, system evaluation would be scored according to present operability, future modifiability, and future transferability (or transitioning) to a new environment. As we repeat often in this book, we mean much more than fixing faults and bugs by the term maintenance.

Such an attribute tree,* of course, needs to be used with prudent and mature judgment. If, for example, one particular system scored high on operability and modifiability but low on transferability, it should not be considered seriously as a candidate for transitioning. However, it would be acceptable to design such a system if, for some reason, it is known that the system will never be a candidate for this transitioning.

It would also be possible to structure the quality assurance attribute tree such that the first-level attributes are those of structure, function, and purpose. In doing this, we tend to evaluate the system along dimensions that correspond to specific testing instruments. Figure 4.13 represents this sort of structure from the viewpoint of the user or client who is especially concerned with satisfaction of purpose, from the viewpoint of the system developer who

---

*Much of our discussions in Chapters 7 and 8 are specifically relevant to the use of attribute trees for evaluation, decision making, and cost-effectiveness analysis of systems.

Figure 4.13. Perspectives on Operational Level System Quality Assurance

is generally concerned with structure, and from the joint perspective of functionality as the linkage between structure and purpose.

Sometimes an initially identified set of attributes turns out to be difficult to understand for purposes of identifying quantifiable attribute measures and scoring. In this case, it is usually beneficial to disaggregate those at the lowest level that was initially defined into lower level attributes. Figure 4.14 shows this for each of the attributes defined in Figure 4.12.

Operational level product or system quality assurance can be enhanced through walkthroughs and inspections, formal verification and validation efforts, and other approaches that provide confidence that a systems engi-

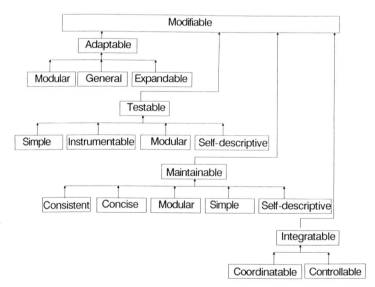

Figure 4.14a. Disaggregation of Quality Assurance Attributes Affecting Modifiability

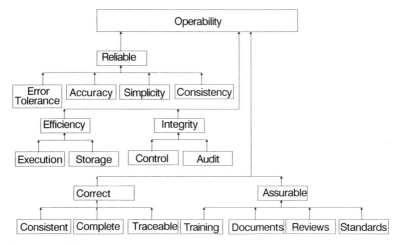

Figure 4.14b. Disaggregation of the Operability Quality Assurance Attribute

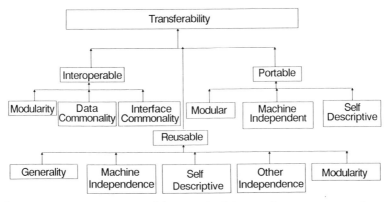

Figure 4.14c. Disaggregation of the Transferability Quality Assurance Attribute

neering product conforms to established technical specifications, and to the user requirements. These are expensive and time-consuming activities. They must provide benefits that justify their cost, or their use may be counterproductive. This is especially the case when they result in system development delays without a corresponding increase in system functionality. It is very important that we keep the normative component of quality assurance in mind. Although it is important that we detect and ultimately correct errors, it is much better to establish systems engineering life-cycle design and development procedures such that error occurrence is minimized and the operational system is trustworthy and produced in a cost-effective manner.

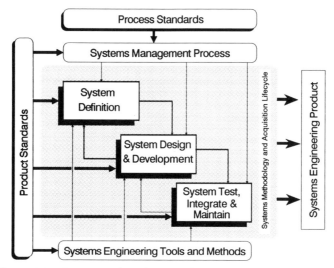

Figure 4.15. Process and Product Standards in the Systems Life Cycle

4.5. THE ROLE OF STANDARDS IN SYSTEMS ENGINEERING

It is easy to set up a dialectic concerning the benefits of standards in systems engineering. It can be easily argued that any attempt at standardization brings about unwise restrictions on management and technological innovation. On the other hand, it is virtually impossible to integrate system products if they are not built according to some reasonably common and agreed-on official specifications. Not even cables connect together unless the plugs are compatible. Very few of the modern engineering systems that we enjoy in our home or use in the office would be anywhere near as useful if their development were not influenced by standards. The parts of the system simply would not fit together.

There are many types of standards. In the most general sense, standards embody the behavioral norms and mores of a particular social group. Standards are metrics, often quantifiable, that a group agrees to adhere to in order to influence behavior.* In our previous discussions, we noted that standards could be of a process or product nature. Figure 4.15 illustrates one possible expansion of this definition to include the three primary phases of the systems engineering life cycle and the systems management approach being used. We can be concerned with product or process standards at any of

---

*The ten commandments represent a mandate or standard to which many people subscribe. The various units of distance measurement and the monetary units in use in a given country also constitute standards. The Occupational Health and Safety Act (OHSA) is another set of standards. It has been enacted into law in order to insure safety in the workplace.

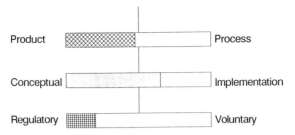

Figure 4.16. Illustration of Mix of Standards Types in a Specific Standard

these design and development methodology phases, or at the systems management level. In a similar way, we can develop product standards for the methods and tools used in the various phases of the systems engineering life cycle.

Also, standards may be of an implementation or a conceptual nature. A conceptual standard would influence all products of a generic class. A programming language standard that absolute GOTO statements not be used in any higher level programming language would represent a conceptual standard. A standard ADA language that did not have GOTO statements would represent an implementation standard.

Standards may be voluntary on the part of the group espousing the standards, or they may be *regulatory*, which are generally imposed. With voluntary standards, there may be no effective enforcement vehicle to ensure compliance. Regulatory standards that require adherence to a rigid standard may reduce vigor in dealing with changing conditions. There are, of course, trade-offs in choosing standards, and, indeed, in whether standards should be used at all.

We may also have personal or organizational internal standards, that are unique to a given individual or organization, and external standards, which are promulgated by industry, or national or international organizations and are intended to be adhered to by many, ideally all, organizations. In general, internal standards and voluntary standards are equivalent. Regulatory standards are normally equivalent to external standards.

Figure 4.16 illustrates some potential scales for this three-attribute view of standards. To be most effective, a standard should be agreed upon by all stakeholders involved.

An enlightened standard should have certain essential properties.

1. A standard should be approved by relevant stakeholder groups and should express a consensus view of these stakeholders.
2. A standard should be documented and widely available to relevant stakeholder groups.

3. Standards should relate to processes as well as to product, thereby providing regulations for human action and intent, as well as providing minimum acceptable specifications for product or system attributes.

4. There should be reasons supporting every standard, or component of it, and these should be explicated so as to be understood and appreciated by those who are obligated to follow, or conform to, the standard.

5. Illustrations should be provided of the benefits associated with following a standard and the problems associated with not following it.

6. Standards should be kept up to date and abreast of contemporary societal needs for them.

7. Standards should be introduced to increase the common good and to encourage competition in the marketplace and should never bar potential competition from entering a market with an otherwise useful product or service.

8. Standards should be applicable, complete, consistent, enforceable, modifiable, traceable, usable, unambiguous, and verifiable.

9. Product standards should support operations and maintenance.

10. Process standards should support the efficient and effective production of trustworthy and high-quality systems that are sustainable.

11. Standards that are unenforceable, and in particular those that represent an imposition of a solution, should not be established.

A standard expresses a minimum acceptable requirement. Although most standards express minimum acceptable scores for a process or product on each of a number of attributes, standards may also allow for trade-offs among attribute scores. Closely related to the concept of a standard is that of a guideline. While standards are requirements, guidelines are simply suggestions for products and processes. The major difference between a guideline and a standard is that a guideline offers considerably greater flexibility and room for judgment.

There are a large number of systems engineering-related standards and guidelines that have been promulgated by the many groups having interests in this area. An excellent discussion of standards, from both a general technology and information technology perspective, is available in the book by Cargill [17]. Much of our discussion is based upon this excellent source.

There are two primary types of standards organizations, formal and informal. The *formal* standardization bodies are responsible for the definition and promulgation of defined public standards. They are comprised of national and international standardization groups, professional organizations, and trade associations. These public standards are generally called formal or accredited standards.

Informal, or industry standards are created by the informal standards bodies. The purpose of these is to assist in the implementation of formal standards. As initially formulated, these informal standards are perhaps better called specifications. An informal standard or specification is generally intended for submission to a formal standards group for subsequent approval, perhaps after modification, as a recognized formal standard.

In the United States, standards, including systems engineering standards, are established by the American National Standards Institute, the Institute of Electrical and Electronics Engineers, the American Society of Mechanical Engineers, the National Institute for Measurements and Standards, the Electric Industries Association, the Human Factors Society, the Department of Defense and others. The National Institute of Standards and Technology (NIST), formerly known as the National Bureau of Standards (NBS), is the major government organization involved in standardization. The NIST provides a national standards laboratory and is also specifically charged to assist university, government, and private sector industry in technology transfer and in maintaining technological competitiveness.

The American National Standards Institute (ANSI) is a coordinating body for standards activities in the United States. Five U.S. engineering societies have joined to create the American Engineering Standards Committee (AESC) and to support ANSI. ANSI does not create standards. Rather, it manages and coordinates private sector standards activities. Under the leadership of ANSI, three types of organization can create a standard. These are the accredited sponsor, the accredited organization, and the accredited standards committee. Each attempt, in somewhat different ways, to obtain openness, equity, and effectiveness in the standards they promote and the way they promote those standards.

Under the accredited sponsor (AS) approach, an approved or accredited organization may invite comments on a proposed standard. When deliberations have been concluded with due process, the resulting standard is submitted to ANSI for publication as a standard. In this approach, a group of stakeholders reach consensus on a standard to which they will adhere. The programming language standard ADA was developed in this manner. When there is substantial stakeholder disagreement, this approach often has difficulties.

Under the Accredited Organization (AO) approach, a standard is established by a group of experts in the subject area. Usually, professional societies and trade groups take this approach to the creation of a standard. They notify ANSI of their intent and agree to abide by the consensus decision. The resulting standards are known as American National Standards (ANS). Challenges to this approach usually concern whether the group possesses the necessary expertise or whether the group is truly representative of the various stakeholder groups.

Under the American Standards Committee (ASC) approach, an ad-hoc committee is formed to develop a standard. A secretariat is formed to provide legal, management, and economic support for the undertaking and to maintain administrative contact with ANSI. The ASC controls its own agenda, and usually operates through a number of subcommittees, publishing iterative versions of a standard until a consensus is finally accepted. This approach is most often used when the issues are highly contentious.

The ANSI encourages these groups to formulate standards and to submit them to ANSI for approval as American National Standards (ANS). While other than ANS may appear, the lack of endorsement by the ANSI will generally preclude widespread acceptance.

In order to have the highest stature, a standard should be agreed to and accepted by international standards bodies. There are a number of international standards groups. These include the International Organization for Standardization (ISO), the International Electrotechnical Commission (IEC), and the International Telecommunication Union (ITU). The ISO is a federation of standards groups of participating nations and is concerned with standards in all technological areas except electrical and electronics engineering and telecommunication systems. ANSI and other national standards groups have membership in the ISO. The membership of international standards bodies is generally comprised of countries, professional organizations, and trade associations.

There are a plethora of regional standards groups such as the European Computer Manufacturers Association (ECMA). Under the CCITT (Comite Consultatif International de Telegraphique et Telponique—International Telegraph and Telephone Consultative Committee) approach, the CCITT functions as a committee of the ITU. The CCITT is typically considered a peer with ISO. The CCITT, as a committee of the U.N. chartered ITU with treaty-status government representatives, typified perhaps by the United States when represented by the State Department, is the formal peer to the ISO, which is not governmental at all. It is a strictly voluntary and nongovernmental membership effort.

Most developed nations have national standards groups that function in a similar manner to those in the United States. These include the Association Francaise de Normalization (AFNOR), which is a French national standards body. The secretariat of many of the ISO committees is, by design, from the AFNOR. The British Standards Institute (BSI) has similar responsibilities for the United Kingdom. Deutsches Institut für Normung (DIN) is the German national standards body. The Japanese Industrial Standards Committee (JISC) develops and promulgates Japanese standards, and also represents Japan on the ISO and IEC.

Standards should be set within constraints imposed by the systems engineering life cycle. For example, standards must reflect efforts taken in the early phases of the life cycle. Also, standards must reflect the priorities and goals of the systems engineering effort. It must be possible to enforce any standards that are set in an efficient and effective manner. In general, a standard should not be introduced when a more flexible guideline would be sufficient.

Configuration management, for its part, can contribute to effective standards. The various tests, reviews, and audits that are accomplished throughout the systems engineering life cycle are intended, in part, to insure that the systems engineering product adheres to relevant standards.

We strongly support the use of standards to insure systems engineering quality and trustworthiness, and believe it is possible to develop standards that enhance the efficiency, effectiveness, and other usability attributes of systems, such as sustainability. The simple reason for this is that standards provide the common base of understanding needed for management controls over system development for trustworthiness and use or customer satisfaction.

4.6. SYSTEMS INTEGRATION

New systems are not to be sold or deployed in a vacuum. Usually they are improved versions of, or are additions to, an existing system. The new system will normally evolve from an existing system in the fashion shown in Figure 4.17. The new system may be delivered as a result of some contracted effort with an external systems engineering contractor, or it may be developed in-house. Systems integration is the process through which a number of products and services, both hardware and software, are specified and assembled into a complete system that will achieve the intended functionality. In many ways, this definition is sufficiently broad that there is reason to question whether systems integration and systems engineering are not equivalent concepts. We consider systems integration to be a part of systems engineering, although it is not always defined this way. The major difference is that the systems integrator is an assembler of products and services that have

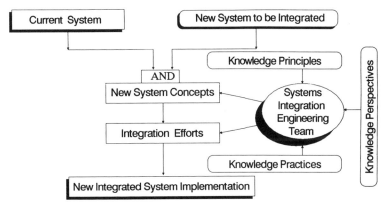

Figure 4.17. The Systems Integration Engineering Process

been originated by others. Systems engineering also includes the possibilities for producing systems of hardware and software. Clearly, a system integrator uses, or should use, the systems engineering process in their system integration activities. Clearly, the systems integrator also must be knowledgeable about all facets of the systems acquisition process.

There is an inherent relationship between systems integration engineering and standards. This exists because both system users and system developers, and purveyors or marketers, have a common need for standards that are system independent and specific developer independent. The term *open systems architecture* is now used to describe any of several generic approaches the intent of which is to produce open systems that are inherently interoperable and connectable without the need for retrofit and redesign. The term *open systems interconnection* (OSI) is often used to describe a generic framework or infrastructure, perhaps stated in the form of a model, that provides a basis to ameliorate incompatibility conditions. The term, which originated in Europe in the early 1980s, was initially intended to set forth a seven-step strategy to enable telecommunications networks to function together. While the OSI framework has led to a number of important telecommunication standards, no operational OSI standard has been produced to date.

An appropriate open systems architecture standard must be explicitly defined such that anyone desirous of using it can use it for implementation purposes and must satisfy the other desirable attributes of standards that we discussed in Section 4.5.* Systems integration, which is fundamentally concerned with the technological and management issues needed to bring about functional operability of systems, is very concerned with these issues also, although perhaps from very slightly different perspectives. The overall tasks of a systems integrator include the following.

1. *System Definition:* Identifying user requirements and technological specifications for a system, including needs for systems integration to insure compatibility with existing and possible future systems.

2. *System Design and Development:* Identification of an appropriate architecture or preliminary conceptual design for the system, including appropriate interfaces to existing systems, evaluating the performance of the system, potentially modifying the system architecture for better performance and enhanced interoperability, and thereby establishing an effective open architecture for the system to be developed.

3. *System Test, Integration, and Maintenance:* Insures that the operational system is cost effective and of high quality.

---

*It is interesting to conjecture on the relations between open system architecture standards and proprietary system architecture standards. If there is major consumer interest in and demand for a particular system, other system developers may be essentially forced to make their product conform to it. MS-DOS and Windows are, in effect, proprietary system architecture standards.

These are just the phases of the systems engineering life cycle, modified slightly to explicitly recognize the role of the systems integrator and the concomitant need for an open systems architecture. The confluence of systems integration, open systems architectures and standards for these may be expected to lead to an open systems environment that would

1. Reduce the system acquisition costs.
2. Reduce system integration costs.
3. Protect current investments in hardware and software.
4. Allow increased independence in acquiring new systems and in modifying and maintaining existing systems, and through this process.
5. Maximize the quality and effectiveness of integrated system products in resolving user and customer issues and problems.

Current developments in this area include the development of the Portable Operating System Interface (POSIX) Open Systems Environment. There exists, as of January 1992, a first version of a POSIX Standard. This is a standard operating system interface that is based on the UNIX operating system and supports application portability at the source code level. A standard of this sort attempts to cope with application program interfaces and external environment interfaces to enable applications software to run on a platform that provides interfaces to the external environment. This mammoth undertaking illustrates the major coordination needed among the many participants in a large systems effort [18], in this case in the area of information systems engineering, and is portentous of many future efforts needed to insure continued advances in information technology [19] through enhanced systems engineering.

The result of a contracted systems integration effort is normally intended for use in a larger system that is under development. Both technical and functional integration are generally achieved through systems integration. The first deliverable in this effort is generally a *system operational implementation and integration plan* that will identify, analyze, and prioritize technical and functional integration problems, and then propose solutions. To do this efficiently and effectively requires the following.

1. Outstanding capabilities in systems management.
2. Thorough understanding of contemporary technology for the development of trustworthy implementation and system integration plans.
3. Superior capabilities with respect to assembly, installation, and test of systems, including such hardware and software skills as are needed to design and implement special purpose utilities and subsystems needed in the assembly of larger systems.

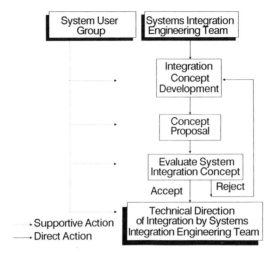

Figure 4.18. Interaction of Systems Integration Engineering Team and User Group

4. Noteworthy proficiency in logistic and maintenance support for systems.

5. Technical direction and cost control abilities so as to be able to deliver an effective and trustworthy system within budget and on time.

These systems engineering capabilities must be present throughout the complete life cycle of development of the system, as it is highly unlikely that they will suddenly appear at the latter phases of the life cycle that involve integration and test of an operational system. Generally, it will be necessary for a systems implementation and integration contractor to provide integration concepts and designs. Often, the user group will have the contractual fight to approve or disapprove of these concepts as they are developed. In many instances, the systems integration contractor will be obliged to assist the user group in the management and coordination of other contractors, primarily through review of proposed subsystem design specifications. Figure 4.18 depicts this process. It indicates the primary role of the integration contractor and the systems management and technical direction role that is required as part of the effort.

Both top-down and bottom-up approaches generally are needed and used for systems integration. The top-down approach primarily is concerned with long-term issues that concern structure and architecture of the overall system. The bottom-up approach is concerned with making parts of the existing system more efficient and effective so they can be incorporated into the newly overall system. The overall integration design must take into consideration existing hardware and software that can be changed and any

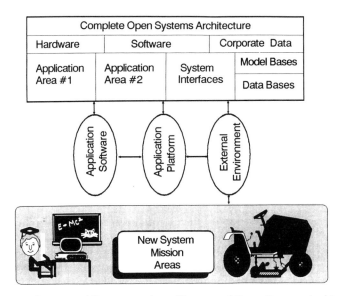

Figure 4.19. Conceptual Representation of Integrated Open Systems Architecture

other constraints on the proposed system, including existing and evolving system architectures.

Of central concern in a systems integration effort is the system-level information architecture. Conceptually, this might appear as shown in Figure 4.19. The mission areas for a system normally vary from case to case. The primary tasks of the systems integration team are design of the overall system architecture and integration of subsystems into this architecture. The first of these tasks calls for top-down systems engineering, whereas the second requires management and technical direction of contract work and bottom-up approaches to achieving interfacing and interoperability of existing systems. These phases of effort are illustrated in Figure 4.20.

Systems integration engineering requires attention to both technology and management problems on the part of both the implementation and integration teams. Technical tasks generally include assessing the impact of architectural changes on both the system under development and its stakeholders. The system integration team should also provide systems management support relative to technical systems and scheduling matters. These generally involve cost studies of possible acquisitions, and configuration management studies.

When a positive "go-ahead" for an integration decision is made, a systems integration contractor generally is asked to prepare design specifications and a recommendation on how to accomplish system implementation and integration. The contractor assists in systems management and technical direction of

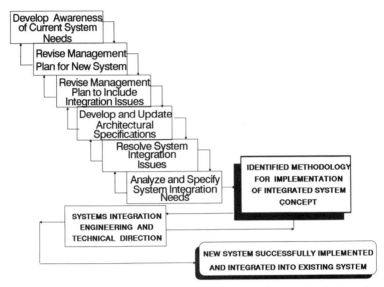

Figure 4.20. Phased Life-Cycle Activities in Systems Integration Engineering

the system integration and test effort. This may include performance of a systems integration task order, in which case the system user group independently prepares the task order and determines the acceptance criteria. The contractor would then establish a management control system to ensure quality assurance during the latter portions of the systems life cycle. This would include operational testing and evaluation accordance with contractor-prepared and user-approved criteria.

The first efforts in systems integration should generally be to obtain, from a variety of sources, identification of

Where the system user or stakeholder group is.

Where the user group needs to be.

How it should get there relative to systems integration.

This situation assessment defines the needs, constraints, and alterables of the system. Of course, a preliminary version of this assessment is required to bring the effort to this stage.

Once the situation assessment is complete, potential additions and modifications to the existing system must be identified. The impacts of these alternatives on the resulting system are then analyzed. This step should allow for some adjustment of parameters for each alternative implementation to permit optimization of performance. The systems under consideration should be either immediately interoperable with existing systems, or at least able to be integrated with some degree of effort. There is a major need for an evaluation methodology to validate the software, the hardware, the human interfaces, and the trustworthiness of the resulting system. These topics are presented in the configuration management and review and audit portions of this chapter.

Cost and effectiveness indices are determined for each alternative. These are included in planning documents for systems augmentations. These documents identify potential integration opportunities within the existing environment of computer hardware, software, communications, and physical plant. Also included with each evaluation should be an analysis of risk factors affecting each alternative. Risk has numerous attributes and should be fully explored. We explore these concepts in much of what follows.

The system integration documents we have been discussing should be prepared in a preliminary version during the initial evaluation of user requirements and system specifications. And, they should generally be revised prior to undertaking systems integration efforts. These documents should also include management plans for all phases of the system integration life cycle.

Invariably, the goals of those working on a systems acquisition project include the following.

1. To identify new technology approaches that enhance functionality of the new system.
2. To identify significant "cost-drivers" that represent a high percentage of total costs of the system.
3. To identify methods that reduce costs while simultaneously retaining benefits and on-time delivery of the operational system.
4. To field a quality system, within the constraints set by schedule and price, that is of high quality, and trustworthy in terms of satisfaction of customer needs.

These objectives apply to systems acquisition and to systems integration.

Operational deployment of a system, and related system integration concerns, is an iterative and evolving process. Systems that once fit well into a complete system may not do so at some future point. This evolving design and development concept is shown in Figure 4.21. The iteration and feedback are essential to ensuring continuing functionality of the system.

Firms that provide systems integration services generally do not also produce hardware or software, except, as already noted, in minor ways through the writing of utility software routines and the design of minor hardware interface items. These invariably are intended to support the utilization of in-production and in-use hardware and software. They are (almost) always totally professional services firms. A principle advantage to

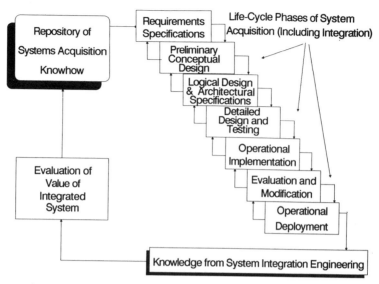

Figure 4.21. Evolving Nature of Systems Integration Engineering

this is that a system integrator may be more disposed to unbiased selection of the best systems for integration.

Such firms normally operate through one of the following five types of contracts [20].

1. The prime contract to a single source.
2. The consultant agreement.
3. The joint venture contract of shared ownership.
4. Hybrid agreements.
5. Fixed-price *vs.* time-and-materials contract.

The most usual form of contract that is accepted by a system integration firm, occasionally called a *systemhouse*, is a fixed-price contract for the purchase of hardware and software* and a time-and-materials contract for the professional services portion of the effort.

Systems integration has four fundamental dimensions [21].

1. *Integration Technology:* Supports transfer of data across different subsystems. This process includes file transfer protocols, document protocols, and remote procedure calls. Automatic data transfer, common database structures for different applications, and process-to-process communications through well-defined functional interfaces and interaction protocols are examples of how integration technology is accom-

---

*Hardware and software are generally in-production and in-use products with established prices.

plished. Some form of integration technology is generally necessary for overall systems integration, but is never sufficient to insure it.

2. *Integration Architecture:* Structures subsystem design to insure easy and secure data sharing across subsystems. Storage of common data in data bases requires functional interoperability if the data is to be shared. Accomplishing distributed data storage through use of an integration architecture that has direct access to data or functional access by activating other systems is a need.

3. *Semantic Integration:* Insures either that the same concepts mean the same thing in different portions of the system or that there exists a translation mechanism that will resolve semantic inconsistencies so as to allow information exchange across systems. Semantic inconsistencies invariably exist when different subsystems are procured from different vendors.

4. *User Integration:* Enables a system user to concentrate on the tasks to be accomplished and not the specific details of the technological system being integrated. This generally requires easy access to different applications and systems, uniform user interfaces, consistent data, and consistent use of semantic concepts.

The three perspectives on knowledge we discussed in Chapter 1—practices, principles, and perspectives—interact with these four integration aspects, as indicated in Figure 4.22. One major objective of any overall system acquisition effort should be to reduce implementation risks and enhance

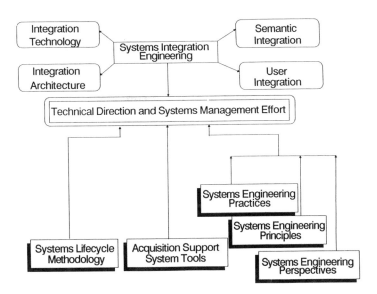

Figure 4.22. Principal Knowledge Ingredients in Systems Integration Engineering

trustworthiness of the resulting system. Whether this should be accomplished through the efforts of a system integration contractor exclusively or whether a more general engineering effort that might include production of new hardware and software is clearly a matter for judgment and choice based on the particular issues at hand.

Human concerns are especially important in systems integration. Six objectives are particularly important.

1. To convert human performance requirements into systems engineering specifications.
2. To accommodate systems design to human cognitive requirements.
3. To fully maximize physiological and cognitive workload potentials.
4. To identify and accommodate human error potentials in preliminary system design.
5. To predict human-system performance capabilities and limitations early in the life cycle and adjust the design accordingly.
6. To quantify human performance factors so as to obtain system designs that are very appropriate for successful human interaction.

The U.S. Army Research Institute's *Hardware Manpower* (HARDMAN) tools [22] are designed to assist in this. These tools enable us to project manpower, personnel, and training requirements at an early phase of alternative design strategies. Costs and other inputs can be analyzed to predict performance. Although system integration needs occur at later portions of the life cycle, attention to the primary system integration needs at an early phase enables optimal trade-offs between hardware and manpower while developments costs are still relatively low. It results in technological and management systems (including decision support systems) that enable machines to do what machines do well and humans to do what humans do well.

Systems integration should be capable of efficiently and effectively coping with future user needs for hardware and software acquisition. Conceptual architectures and frameworks for open systems architectures and integration [23, 24] are particularly important in this regard. These are needed to accommodate the identification of requirements for, and the subsequent development and operationalization of, an integrated system that is responsive to contemporary needs, especially for systems that can function in contemporary high-velocity environments.

This requires an approach that recognizes that a systems integrator and a systems user will have different perspectives on development of the system. It is only natural that a system user will be almost exclusively interested in the ultimate product and its trustworthiness. On the other hand, a systems engineer or systems integrator can be expected to have very strong interests in the process that is undertaken to insure delivery of the product. Through this interest in process, there will naturally occur an interest in product, as

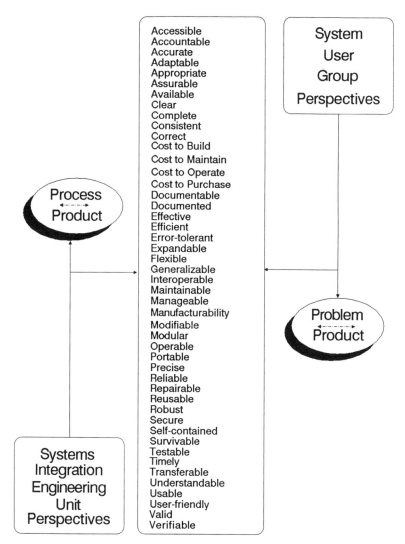

Figure 4.23. Multiple Perspectives on Process and Product of Systems Integration

well. Ultimately, products or systems are valued in terms of their ability to resolve issues or problems, and this is what the ultimate customers for a system desire. Figure 4.23 indicates some aspects of this multiple perspectives view of systems integration. To accommodate each of these perspectives, we need a strategic level approach to quality assurance and management—one that produces total quality management, more fully discussed in Chapter 5.

A fundamental goal of strategic planning and associated systems management is to balance these perspectives. Figure 4.24 illustrates some of the

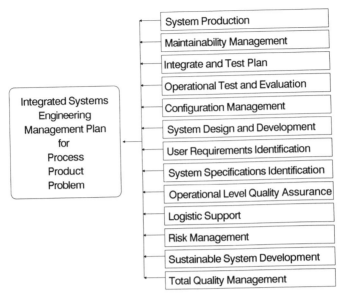

Figure 4.24. The Need for Process Integration at the Level of the Systems Engineering Management Plan

concerns of systems management as expressed in the components of the systems engineering management plan (SEMP). Thirteen elements of the SEMP are identified in this figure. It is a task of systems management to develop each of these elements and to accomplish it in such a fashion that the system customer is satisfied with the system product in terms of its ability to resolve problems. We see, then, the need for integration of problem, product, and process as essential to systems integration.

4.7. SUMMARY

We have covered quite a bit of territory in this chapter. We began with a discussion of configuration management for system quality assurance, and then showed that the major mechanisms to support CM and QA are appropriately planned reviews and audits.

Next, we argued that standards are essential to CM and QA. Finally, we discussed system integration concerns associated with the implementation of trustworthy systems. In the next chapter, we discuss the important subject of quality assurance and management, from a strategic perspective. This leads to process quality which is absolutely necessary, although not sufficient for the development and fielding of high-quality, cost-effective system products that are sustainable over time.

PROBLEMS

4.1. Develop a checklist of activities to be accomplished when operational implementation of a office automation system is to be achieved. Discuss the relationships among the activities in your checklist. When in the development life cycle should these activities be identified?

4.2. Discuss the relationship between system integrate and test as a part of the implementation phase of the systems life-cycle and maintenance efforts.

4.3. Describe the cost influencers for systems integration. Generally, development changes that change costs at one phase of the systems life cycle will have offsetting effects at other phases. Prepare a report indicating the trade-off concerns associated with changing the resources allocated to the implementation phase of the life cycle.

4.4. Please discuss the statement, *A system acquisition life cycle with lower integration costs is always to be preferred to one with higher integration costs.*

4.5. What trade-offs would you expect among system integration and implementation and system maintenance? Is it possible that lack of attention to integration will increase maintenance costs.

4.6. Identify and describe the major influencers of a make or buy decision. The make decision will involve using a complete systems engineering process, whereas the buy decision will involve use of a systems integrator.

4.7. Identify how design practices, design principles, and design perspectives influence systems integration considerations.

4.8. A system, such as the one for office automation considered in Problem 4.1, could be developed for a large corporation by purchasing hardware and then engaging a software development team to design appropriate software. Alternately, one could be developed by enforcing a contract requirement that no new software products be developed, except for small utility programs, and that all software satisfy an "in-production in-use" requirement that the software exist in the commercial marketplace a month or so before contract proposals are due. The purpose of this requirement would be to avoid vaporware. Please contrast and compare the two approaches. In the process of doing this, please define vaporware.

4.9. Is the assumption in Problem 4.8 that the hardware be purchased before the software is developed, or procured, a good one? Please redo the problem using the requirement that hardware and software systems be developed or procured together.

4.10. In Figure 4.23, we identified a rather long laundry list of functionality attributes of acquired systems. Please categorize these on a one-dimensional scale with product at one end of the scale and process at the other end. Please categorize these according to a proper location on a line connecting these two characteristics. One of the potential development problems is that we may not specify these attributes, either at all or perhaps not with sufficient specificity. Consider a systems acquisition life cycle. At which phase should each of these attributes be specified and measured? At what phase in the life cycle will the impact of poor quality be realized and noticed. Would your answer differ depending upon whether we are going to use a systems acquisition or a systems integration process? How?

4.11. A decade old Rome Air Development Center (RADC) study resulted in identification of a number of *quality criteria* for software:

Access audit

Access control

Accuracy

Communication commonality

Communicativeness

Completeness

Conciseness

Consistency

Data commonality

Error tolerance

Execution efficiency

Expandability

Generality

Instrumentation

Machine independence

Modularity

Operability

Self-descriptiveness

Simplicity

Software system independence

Storage efficiency

Traceability

Training

Propose an assessment of software quality based on these attributes. Please write a brief paper outlining how the configuration management process should incorporate these attributes. What would be the role of these attributes in a standard for software quality. This particular RADC study suggested two scoring systems: one based on linear weighted sums and the other on a binary scoring procedure. Contrast and compare these two approaches for use as standards.

4.12. The attributes noted in Problems 4.10 and 4.11 could easily be used in a checklist-like fashion to accomplish reviews and audits. Please write a formal plan to do this at each of the phases in a typical systems development life cycle.

4.13. What is software configuration management (SCM)? Four processes involved in SCM are software configuration identification (SCI), software configuration control (SCC), software configuration status accounting (SCSA) and software configuration auditing (SCA). How do these relate to the software development life cycle? How do they relate to verification and validation? Software quality assurance? Please discuss the relations between software configuration management and configuration management.

4.14. There are a number of standards that deal, at least in part, with systems and software quality and its assurance. Please prepare a brief review of four of these. Develop the suggested taxonomy of quality developed in each of these and contrast and compare your results.

4.15. Is it possible for a system to be correct and to be neither reliable nor of high quality? Please explain.

4.16. *Requirements volatility* is a term first used by Barry Boehm to mean the tendency of the system level and software level requirements to change during the software development process. How does requirements volatility effect systems and software quality, reliability, and maintainability?

4.17. Write a project paper in which you associate the three CM baselines with at least two of the life cycles that we have discussed thus far in this text. Please be sure to discuss possible difficulties associated with this, and possible needs for additional baselines. Be sure to explicitly consider appropriate points in each life cycle for establishing relevant baselines.

4.18. Write a brief paper contrasting and comparing the roles for standards and guidelines in configuration management, systems integration, and quality assurance. Is the statement *standards are more important than guidelines* relevant, true, or what?

4.19. Write a brief paper discussing the need for integration of problem, product, and process concerns in systems integration. Relate this to the subjects of configuration management, concurrent engineering, and quality assurance reviews and audits.

4.20. In looking around in various places, one can easily find discussions that suggest improvement in systems engineering productivity through better:

a. Methods and tools

b. System design methodologies

c. Systems management

d. Process and product improvement technologies

e. Technology transfer

f. Use of reusable software, prototyping, automated SW generation

g. Education and awareness

h. Risk management

i. Environmental understanding for enhanced sustainability

k. Understanding of cost and operational effectiveness issues

Please provide a brief commentary on prospects possible through each of these. In particular, how are these factors related in terms of enhancement of systems engineering productivity and sustainable development [25].

4.21. We have discussed an elementary three-phase model for system (product) evolution that is comprised of system definition, system design and development, and system operation and maintenance. Furthermore, we have indicated a number of decompositions of this three-phase model into other more descriptive models that may be used to evolve an operational systems product. The guiding hand that supports the systems engineering life cycle is the systems management process. In keeping with our notion of three essential stages, we might describe the systems management process as one of:

a. Identification of an appropriate strategic plan for system development in terms of such variables as needs, environments, risks, costs, and so on.

b. Identification of system product project plans and schedules to ensure effective product delivery.

c. Monitoring and review of efforts, including tracking actual progress against plans, reviewing and monitoring the actual product development effort, and updating the risk management and product development plans as needed.

Please provide some elaboration of these notions. In particular, please relate your discussion to Barry Boehm's *Spiral Life-Cycle Model* and to configuration management.

REFERENCES

[1] IEEE Standards Board, *Software Engineering Standards*, IEEE Press, New York, 1987.

[2] Blanchard, B. S., *System Engineering Management*, Wiley, New York, 1991.

[3] Bersoff, Edward H., "Elements of Software Configuration Management," *IEEE Transactions on Software Engineering*, Vol. SE-10, No. 1, 1984, pp. 79–87.

[4] Babich, Wayne A., *Software Configuration Management: Coordination for Team Productivity*, Addison Wesley, Reading MA, 1986.

[5] Humphrey, Watts S., *Managing the Software Process*, Addison Wesley, Reading MA, 1989.

[6] Bersoff, E. H., and Davis, A. M., "Impacts of Life Cycle Models on Software Configuration Management," *Communications of the ACM*, Vol. 34, No. 8, 1991, pp. 105–118.

[7] U.S. Department of Defense Military Standard MIL-STD-483A, *Configuration Management Practices for Systems, Equipments, Munitions, and Computer Programs*, Government Printing Office, Washington DC, 1970.

[8] U.S. Department of Defense, *Military Standard: Defense System Software Development*, DoD-STD 2167A, 1988.

[9] Rosenblatt, A., and Watson, G. F., "Concurrent Engineering," *IEEE Spectrum*, July 1991, pp. 22–37.

[10] Sage, A. P., and Palmer, J. D., *Software Systems Engineering*, Wiley, New York, 1990.

[11] Winner, R. I., Pennell, J. P., Bertrand, J. P., and Slusarczuk, M. M. G., "The Role of Concurrent Engineering in Weapons System Acquisition," Institute for Defence Analyses Technical Report R-338, December 1988.

[12] Grief, I. (Ed.), *Computer Supported Group Work: A Book of Readings*, Morgan Kaufman, San Mateo CA, 1988.

[13] Sage, A. P., *Decision Support Systems Engineering*, Wiley, New York, 1991.

[14] Card, D. N., *Measuring Software Design Quality*, Prentice Hall, Englewood Cliffs NJ, 1990.

[15] Evans, M. W., and Marciniak, J. J., *Software Quality Assurance and Management*, Wiley, New York, 1987.

[16] Vincent, J., Waters, A., and Sinclair, J., *Software Quality Assurance: Practice and Implementation*, Prentice Hall, Englewood Cliffs, NJ, 1988.

[17] Cargill, C. F., *Information Technology Standardization: Theory, Process, and Organizations*, Digital Press, Bedford MA, 1989.

[18] Institute of Electrical and Electronics Engineers, *Draft Guide to the POSIX Open System Environment*, STD1003.0/D11, March 1991.

[19] Sage, A. P., "Information Technology," in S. Parker (Ed.), *McGraw Hill Yearbook of Science and Technology*, McGraw Hill, New York, 1992.

[20] Beutel, Richard A., *Contracting for Computer Systems Integration*, The Michie Company, Charlottesville VA, 1991.

[21] Nilsson, E. G., Nordhagen, E. K., and Oftedal, G., "Aspects of Systems Integration," *Proceedings of the First International Conference on Systems Integration*, IEEE Computer Society Press, April 1990, pp. 434–443.

[22] Booher, H. (Ed.), *People, Machines, and Organizations: A MANPRINT Approach to System Integration*, Van Nostrand Reinhold, New York, 1990.

[23] Rossak, W., and Prasad, S. M., "Integration Architectures: A Framework for System Integration Decisions," *Proceedings 1991 IEEE Systems, Man and Cybernetics Conference*, Charlottesville VA, 1991, pp. 545–550.

[24] Rossak, W., and Ng, P. A., "Some Thoughts on Systems Integration: A Conceptual Framework," *Journal of Systems Integration*, Vol. 1, No. 1, 1991, pp. 97–114.

[25] Hatch, H. J. "Accepting the Challenge of Sustainable Development," *The Bridge*, Vol. 22, No. 1, 1992, pp. 19–23.

Chapter **5**

Strategic Quality Assurance and Management

In the previous chapter, we examined a number of issues relative to quality assurance and management. Discussions were primarily aimed at assurance of system or product quality. Most of the approaches advocated are at the levels of operational and task management. Although necessary, these approaches often are not sufficient. Efforts at the level of strategic management are also required. In this chapter, we examine issues of quality at the strategic level. This generally results in a set of strategic quality assurance and management plans, and these, in turn, enhance process quality at the level of systems management or management control. Appropriately this chapter is titled Strategic Quality Assurance and Management. A somewhat equivalent title would be Total Quality Management.

First, we give a very brief historical overview of quality. Then, we examine several approaches to quality assurance through strategic management.

5.1. INTRODUCTION TO AND HISTORY OF QUALITY

Notions of quality have been around for a long time. Everyone wants it,* at least when one is speaking of quality at a sufficiently high axiological level. Historically, the notions of quality were reactive and inspection oriented. This approach evolved into the use of statistical quality control techniques. More recently, notions of quality assurance and management for quality have

---

*At more places than the General Electric Company it is said that "quality is our most important product." Perhaps the central point in this chapter, and surely in most of the strategic quality assurance and management literature, is that much more is needed than just saying it! It must be put into practice.

emerged. A recent book by Garvin [1] provides a very readable historical account of the emergence of notions of quality.

Formal product inspections became necessary only with the rise of mass production technology some two centuries ago. When products were individually crafted by artisans, the name of the artisan was the equivalent of a quality warrant. Once products became mass produced and interchangeable, especially when parts are intended to be portions of a larger and more complex product, product inspections were felt to be necessary. Notions that the quality control inspector is responsible for work quality were pervasive in the early 20th century writings of the "father" of "scientific management" in America, Frederick W. Taylor [2]. His efforts are responsible for transferring the machine shop into the then-modern factory production system. There are six principles generally associated with this approach:

1. Segmentation of work planning and the associated task scheduling from the actual execution of the work effort.
2. Determination of the most efficient physical movements necessary to accomplish a task, determination of the optimum allocation of rest periods such as to maximize worker efficiency, and very careful measurement of the time required to perform tasks.
3. Thorough design of the workplace layout and identification of the best way to utilize workplace tools.
4. Utilization of productivity incentives through a system of piece rates and bonuses, to encourage high worker productivity.
5. Establishment of authority and cooperative relationships between workers and management that allowed workers to be considered as individuals while conforming to task requirements at the same time.
6. Vigilant selection and training of workers, reward of the most productive workers, and discharge of unproductive or otherwise unsuitable workers.

Quality control through inspections was viewed as a major management responsibility in the 1920s writings of many, such as Radford [3]. In these early efforts, quality control was to be achieved primarily through such inspection activities as counting and scoring, and then repairing faulty products where possible.

It became evident fairly quickly, however, that not every assembly-line product could be adequately or inexpensively inspected; and the concept naturally arose that statistical approaches to quality inspections would be more cost effective [4]. The doctrine of total quality control emerged in the 1950s with the publication of a definitive handbook on the topic by Juran [5], who was to become revered in Japan for introducing quality control technologies. The notion that "quality control is everybody's job" became popular. Feigenbaum [6], one of the early, seminal, and prolific professionals in this area, disaggregated quality control into three parts in the early 1960s:

1. *New Design Quality Control:* To insure that the design of the product permits proper manufacturing.

2. *Incoming Materials Quality Control:* To insure that the materials of protection meet quality standards.

3. *Product Quality Control on the Shop Floor:* To insure that the manufacturing process does not introduce faults.

Cost analysis, product specifications, and the use of careful measurements to determine product reliability and quality were the major tools of these efforts at securing quality. Management, planning, and control were from this point forward integrated into the process of statistical quality assurance, at first at the operational level and later at the strategic management level.

Soon thereafter emerged professionals who were called quality control and/or reliability engineers [7]. Their efforts went beyond simple statistical sampling and included planning for and managing product quality. Nevertheless, the focus was still at the operational level. Although reduction in product defects occurred, the lack of attention at the strategic level that could lead to systems management approaches to process quality often left product quality as an elusive concept.

The Taylor approach to management and associated quality control was later dubbed *Theory X Management* by McGregor [8]. Its basic premise is that management should monitor worker performance through frequent time and motion studies. With appropriate control, workers can become as efficient and predictable as machines. Current characterizations of Theory X are as follows, although it should not be assumed that Taylor would present the precepts of his scientific management principles in this manner.

1. The management of an organization is responsible for profit. The elements of profit are capital, resources, and people.

2. It is necessary to direct and control people, who are fundamentally lazy.

3. The major responsibilities of management are the direction of workers to insure that they fulfill their roles in the organization.

McGregor's belief was that humans are not as inherently slothful as these Theory X precepts would indicate and would instead respond to management based on the Maslow Hierarchy of human needs (see Chapter 3 for a discussion of these needs). In other words, humans would respond better to a management effort based on interpersonal relations than they would one based on optimization of task performance. He proposed a *Theory Y*, that would replace external control of human behavior by internal control: self-control, self-motivation, and self-direction. There are five precepts of Theory Y, which could be restated with a smaller or larger number of precepts.

1. The management of an organization is responsible for profit. The elements of profit are capital, resources, and people.

2. People are basically motivated for success and are capable of assuming accountability and acting in a responsible manner. Management does not have to instill these attitudes in workers.

3. Whether work is a source of satisfaction and will be willingly performed or is a source of punishment to be avoided is a controlled variable for the organization.

4. Most people will learn not only to accept responsibility, but also to seek it if the organization's structure and reward system encourages these virtues.

5. The major goal of management is to empower people, to enable them to become committed toward organizational objectives, and to recognize and fully develop their inherent potentials consistent with support of the organization.

Other schools emphasized these points during the 1960s. Maslow himself [9] evolved a Theory Y-type management approach. In another seminal study, Herzberg [10] identified 16 factors that affect job attitudes. These are achievement, recognition, the work itself, responsibility, advancement, growth opportunities, company policies and administration, supervision, relationships with supervisor, work conditions, salary, relationships with peers, personal life, relationships with subordinates, status, and security. Six of these could be categorized as *motivators* that promote organizational success—achievement, recognition, the work itself, responsibility, advancement, and growth opportunities. These reflect the top two elements in the Maslow hierarchy of needs: esteem and self-fulfillment. Herzberg's other 10 factors were called *hygiene* factors. These factors, when they are unsatisfactory, lead to job dissatisfaction. They are company policies and administration, supervision, relationships with supervisor, work conditions, salary, relationship with peers, personal life, relationships with subordinates, status, and security. The hygiene factors reflect the lower three elements in the Maslow hierarchy of needs: physiological needs, safety and security needs, and love and relationship needs.

On the basic of this and related research, Herzberg suggested seven management principles, which he called *vertical job loading*, that would enrich jobs through improving motivation.

1. Remove controls while retaining accountability, to increase the sense of responsibility and personal achievement.
2. Increase the accountability of individuals for their own work, to increase the sense of responsibility and recognition.
3. Give workers complete "natural" units of work to accomplish, to increase the sense of responsibility, achievement, and recognition.
4. Increase the authority of workers for the activities they perform, to increase the sense of responsibility, achievement, and recognition.
5. Make periodic reports, such as performance evaluations, directly available to workers, rather than only to supervisors, to increase the sense of internal recognition.
6. Introduce and encourage workers to perform new and more difficult tasks that they have not previously performed, to increase growth and learning.
7. Assign workers to specific and specialized tasks and thereby enabling them to become experts, and to increase their responsibility, growth and advancement.

In addition, Herzberg cautions against what he called *horizontal job loading* efforts. He viewed these as challenges to increase numerical production amounts, adding one meaningless task on top of another, and rotating assignments "horizontally" through a number of routine jobs, each of which are in need of enrichment.* A recently published book [11] reprints many of the earlier seminal works in this area, including the work of Herzberg. The works edited by Hax [12] and Schein [13] also contain a number of excellent discussions on these points.

One of the potential difficulties with Theory Y is that there is no explicit provision for dealing with conflict. This is less a problem in a Theory X organization, since organizational leadership can eliminate potential conflict quickly by "getting rid of the disturbers of the peace." A Theory Y organization, by contrast, is adaptive and creative, and tries to increase individual initiative through interpersonal relationships. The resulting multitude of individual innovators tend to compete for power and resources, and the result is often a confrontational environment that is difficult to coordinate.

---

*We might question the use of the words "meaningless" and "in need of enrichment," in this summary of Herzberg's writings. Horizontal rotation of workers (or management, for that matter) might be a perfectly legitimate way both to teach employees new skills and to give an organization greater resiliency in case specialized workers get sick, retire, or move on. Of course, such horizontal rotation is not a substitute for more formal training programs or other methods of enhancing employees' work experiences.

Theory Z [14] was proposed to cope with this difficulty. Theory Z management stresses arriving at major decisions through a consensus that is based on shared values. Many ascribe the Japanese quality achievements to adoption of Theory Z principles [15]. Ouchi identified 13 steps necessary to implement the Theory Z organizational philosophy.

1. Know and appreciate the Theory Z philosophy.
2. Audit the organization's existing management philosophy.
3. Define an appropriate management philosophy for the organization.
4. Implement the new philosophy.
5. Develop the needed interpersonal skills.
6. Evaluate progress in adopting Theory Z.
7. Involve the union, if any.
8. Stabilize employment.
9. Identify and implement a system for slow evaluation and promotion.
10. Broaden the path of career development.
11. Implement Theory Z programs at the lowest organizational levels first.
12. Seek areas of continued improvement.
13. Continuously develop wholistic relationships.

The *quality circle* and *company-wide quality control* efforts in Japan appear to be roughly equivalent to Theory Z management. Actually, the company-wide quality control (CWQC) [16, 17] concept in Japan involves four principles:

1. Inclusion of quality-related efforts at all phases of the systems engineering life cycle, not just the phase specifically concerned with system production.
2. The participation by all people in the organization in quality-related efforts.
3. Setting the objective of continuous improvement in quality.
4. Careful consideration of the customer's definition of quality.

The quality circle (QC) is one aspect of consensus-building toward quality in Japan, although there is some doubt, as indicated by Garvin [1], of the impact of QC on quality in Japan. There are seven principle tools of the quality circle: Pareto charts, causal diagrams, stratification, check sheets, histograms, scatter diagrams, and control charts. We discuss some of these in Chapter 6.

Today, it is recognized that the operational task level quality assurance programs we discussed in Chapter 4 are necessary but not sufficient to assure appropriate quality of products. The earlier approach of statistical inspections to discover product defects has evolved to systems management controls at the strategic level to assure quality throughout the life cycle of a product.

Generally, Deming is credited with much of the early work leading to the current focus on quality [18] in Japan. Many others were involved in this effort, including Crosby [19, 20], Juran [5] and Feigenbaum [6]. Crosby identified 14 facets of and for strategic quality assurance and management: management, quality improvement team, quality measurement, cost of quality evaluation, awareness, corrective action, zero defects planning, quality education, zero defects day, goal setting, error caused removal, recognition, quality councils, and do it over again. As we will see, many of the identified approaches for quality assurance through strategic management have a great deal in common.

5.2. FROM INSPECTIONS TO TOTAL QUALITY MANAGEMENT

In the previous section, we can see the emergence of quality consciousness at four levels:

1. Inspections to identify, and cull out, defective products.
2. Statistical quality control, to provide cost-effective metrics on the performance of manufacturing technology.
3. Operational task level quality assurance and management through quality measurements and planning and design to provide for operational controls that will insure quality.
4. Strategic quality assurance and management to insure market competitiveness through the involvement of all organizational elements involved in the process and product under consideration.

Clearly these concerns are not mutually exclusive. Inspections will, for example, surely be a part of strategic quality assurance and management. However, strategic management for quality assurance will not *depend* upon inspections. In brief, we might state that these four levels correspond to

Inspected quality.
Controlled quality.
Built-in or manufactured quality.
Proactively managed quality.

Although each is important, it is the latter two that will result in maximum competitiveness of processes and products. Figure 5.1 illustrates these four quality levels conceptually. Together, these comprise what has been called total quality management (TQM). Stuelpnagel suggests [21] that TQM involves customer satisfaction, continuous improvement, robust design, variability reduction, statistical thinking, management responsibility, supplier integration, quality control, education and worker training, teamwork, cul-

Figure 5.1. Four Approaches to Quality

tural change, and stakeholder interfaces. These ingredients involve each of the four levels of quality noted in Figure 5.1.

5.3. QUALITY AND QUALITY ASSURANCE DEFINITIONS AND PERSPECTIVES

There are many perspectives and definitions of quality, reflecting a range of views of the philosophers, economists, operations researchers, managers, artists, and engineers who have addressed the topic. Philosophers tend to be concerned with axiological and definitional issues. Those in product marketing are more likely to be concerned with customer appeal and satisfaction, and those in engineering with design and manufacturing for quality. It is often not clear whether quality is conceived objectively or subjectively, and whether it may be measured absolutely or only relatively.

As systems engineers, we believe that there are metrics that can be used to measure software quality. In Chapter 4, we identified a number of attributes

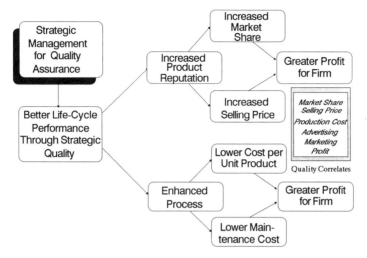

Figure 5.2. Effects of Increased Quality through Strategic Quality Assurance

of quality systems, including such dimensions as

Purposes served or performance objectives.
Operational functions performed.
Structural features.
Reliability, availability, and maintainability.

Doubtlessly, there are other dimensions, such as beauty or aesthetic appeal and we can disaggregate each of these dimensions. Figures 4.12 through 4.14 present a number of attributes associated with operational level quality assurance. Quality is also influenced by such factors as cost to produce, selling price, demand or market share, advertising strategy, marketing strategy, maintenance strategy and profit [1], as shown in Figure 5.2.

Our discussions in Chapter 4 were concerned primarily with operational level quality considerations; in this chapter we examine strategic quality assurance and management. This is a subject of much contemporary interest, especially regarding strategies for creating successful new products, systems, and organizations [22]. Basically, strategic quality assurance involves systems management of processes, aiming to secure product quality. Quality assurance practices below the level of systems management generally produce operational level quality control and inspections for product quality only. This relationship is illustrated conceptually in Figure 5.3.

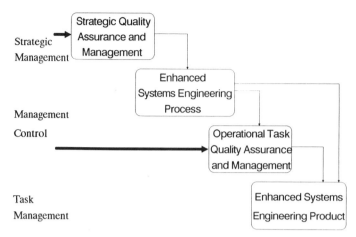

Figure 5.3. Illustration of Potentially Increased Benefit from Strategic Quality Assurance and Management

Garvin [23] has identified five perspectives on quality.

1. *Transcendental Perspective:* From this perspective, quality is synonymous with *inherent excellence*, and is recognizable by all. In this view, quality cannot be precisely defined, but "you know it when you see it." A popular book appeared two decades ago that took this perspective [24], which we might call a wholistic view. Wholistic means completely seeing things as a whole or unified pattern of experiences, without the necessity to view the whole through examination of each of the parts that comprise the whole, or gestalt. A wholistic perspective is necessarily an experienced perspective.

2. *Product Based Perspective:* From this perspective, quality is a descriptive, measurable attribute, made up of microlevel attributes. This is the approach we adopted in Chapter 4, and it provides a holistic definition, where holistic means seeing the whole through examination of the parts.

3. *User or Customer Perspective:* Each user has their own view of quality, so this perspective is necessarily subjective. It is the approach taken, at least in principle, by advocates of the subjective utility theory of decision analysis that we examine in Chapter 6. The challenge it presents product planners is combining these individual preferences to gain meaningful insights into aggregated market preferences. The user perspective on quality reflects the demand side of market equilibrium, a subject we will discuss in Chapter 8.

4. *Manufacturing Perspective:* The manufacturing, or production, perspective, on quality focuses on standardization in system design and production. From this perspective, a quality product is designed, produced, and fielded correctly to specifications, *the first time*. Product reliability, availability, and maintainability are critical attributes, and statistical quality control inspection is used to detect and correct defects in the production system. Cost reduction and on-time fielding of products are important aspects of this perspective. In summary, quality means conformance to requirements and specifications, including those relating to cost and schedule, from the manufacturing perspective.

5. *Value Perspective:* From this perspective, a quality product is one that provides sufficient performance at an acceptable price. This approach blends quality as a measure of goodness with quality as a measure of utility. Quality, then, can be maximized only for certain customers, unless the product has very universal appeal.*

Each of these perspectives has its uses, but in the final analysis, it is the end user or "customer" who sets the standards for quality. Of course, there are numerous steps involved in moving an emerging technology to an acquired system, and those attributes that account for end user quality perceptions need attention at all these steps. But all the fundamental determinants of quality relate to final use.

Garvin [1, 23] has suggested eight dimensions, or attributes, of quality:

1. Performance of the end use system or product.
2. Features of the end use system or product.
3. Reliability of the end use system or product.
4. Conformance of the end use system or product to internal and external, quality and performance standards.
5. Durability, or maintainability of the end use system or product.
6. Character of the human aspects of serviceability of the end use system or product.
7. Aesthetic aspects of the end use system or product.
8. Perceived overall quality of the end use system or product.

*A reasonably good example of the potential problem here is that of considering an expensive Rolex watch. While it is doubtlessly a watch of very high quality, it may have very little utility for a specific purpose that I may have in mind, keeping time accurately, that is not also possessed by a watch of much lower price. From a value perspective, then, what is the value of a Rolex? This becomes even more complicated when we bring in the obvious reality that if I do not want the Rolex, I can (perhaps) sell it at auction at a rather high price, buy another quality watch of much lower price, and have much money left over. So, is not the Rolex watch of high value from a value perspective? Doubtlessly it is.

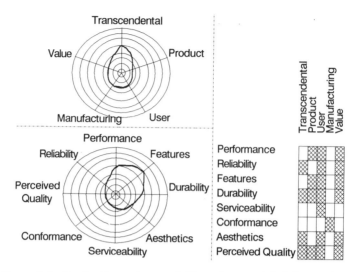

Figure 5.4. Quality Perceptions and Dimensions, and their Interactions

Figure 5.4 illustrates how these quality measures, and the perspectives discussed earlier, interact. We have, of course, identified many other attributes of quality in Chapter 4. Let us now turn our attention to an examination of several different approaches to achieving quality and assuring quality through strategic management or total quality management at the levels of process and product.

5.4. APPROACHES TO QUALITY ASSURANCE THROUGH STRATEGIC MANAGEMENT

There have been a number of approaches to quality assurance through strategic management. We discuss some of the most important ones here.

5.4.1. Managerial Psychology

In a seminal work first published in 1958, Leavitt [25] describes management as the *interaction* of three activities:

Pathfinding.
Decision making.
Implementation.

These are interconnected and strongly influenced by the organization's history, current problems, and environment. To emphasize one management activity at the expense of another can have very negative impacts on the organization.

Three types of individuals correspond to these activities:

1. Pathfinders or artisans who are effective at getting things done.
2. Decision makers, who may often have the right solutions to the wrong problems.
3. Implementers, or compromising salespeople, who may often have no vision.

Leavitt is very concerned with the role of humans in organizations, and provides a considerable amount of thinking and philosophy concerning human nature and the management of people.

1. We are self-centered and long for praise.
2. We are not as good as we would like to think; but pointing out this fact usually does not accomplish anything.
3. The wholistic is at least as important as the holistic.
4. As information processors, humans are both flawed and excellent.
5. We are creatures of our environment, responsive to external rewards and punishments but are also strongly driven from within.
6. Actions speak louder than words—people will distrust us when we speak words that mismatch our deeds.
7. We need meaning for our lives and will do a great deal for an institution that will provide this meaning for us. We simultaneously need independence and dependence.
8. Management controls should let people guide themselves.
9. Organizational systems that are calculated to tear down self-images are not useful.
10. Positive reinforcement will work wonders. It will encourage people to place useful items onto their agendas. Label a person a loser and he will start acting like one.

Positive reinforcement is one of the major thrusts of Leavitt's thinking, which calls for activities that are

Specific.
Immediate.
Achievable.
Provide intangible feedback.

Another key tenet of Leavitt is that people must believe in their tasks and their organization to perform optimally. Thus, a sense of purpose should be a major goal of management. This can be accomplished through values, stories, myths, and legends of the organization.

To summarize, Leavitt's theory of management rests on four principles:

1. Acceptance of limits to human rationality.
2. A realization of the need of all employees for
   2.1 Meaning,
   2.2 A modicum of control,
   2.3 Positive reinforcement.
3. A realization that actions and behaviors shape attitudes and beliefs, and vice versa.
4. The realization that values and distinctive cultures are important ingredients in organizational and human efforts.

We suggest that these are features of organizational success and that they relate strongly to strategic quality assurance and management of process and product.

A somewhat more recent view of organizational excellence is provided by the best selling, and authoritative, book *In Search of Excellence* [26]. Let us now examine some of the prescriptions for quality presented in this excellent work and related and subsequent books.

5.4.2. Strategies for Excellence

One of the major studies of American corporate success stories was published in 1982 by Peters and Waterman [26], who researched 36 noteworthy American companies over the period 1961 through 1980. In this seminal work, the critical success factors of a number of American companies were identified, and from these elements of excellence were distinguished. These are presented as lessons from America's best run companies. Given this study approach and the relatively low velocity environments at the time of the study, it would be expected that these might be descriptive characteristics of organizations that have achieved excellence over a period of relatively little turbulence.

A major lesson is that a central role for management is harnessing the social forces in the organization to shape and guide values. Stated somewhat differently, structural problems cannot easily be considered apart from people problems. The authors identified seven key factors in an organization:

Structure.
Strategy.
People (staff).

Management (style).

Systems and procedures.

Guiding concepts and shared values (culture).

Present and hoped-for corporate strengths and skills.

Two of the "S" factors, strategy and structure, are considered organizational "hardware". The other five are human factors related.

In promoting innovation, Peters and Waterman are particularly concerned that tools do not substitute for thinking, that intellect does not overpower wisdom, and that analysis does not impede actions. To this end, they identify the following eight attributes for excellence:

1. *Bias for Action:* While formal analytical approaches are needed, one must avoid the *"paralysis of analysis."* Successful companies know who they are and where they are going. With this vision, a bias for action can thrive.

2. *Close to the Customer:* A successful organization is customer-oriented, understanding its customers, their desires, and needs.

3. *Autonomy and Entrepreneurship:* A successful enterprise allows its employees to assume responsibility and act in an enterprising manner, even though they "generate a reasonable number of mistakes" in the process.

4. *Productivity Through People:* A successful organization realizes that people are its most important attribute and helps its employees to enhance their abilities and overcome their limitations.

5. *Hands-on and Value Driven:* Successful organizations excel at clarifying and communicating their values and objectives, thereby avoiding vacuous plans and slogans that generate cynicism. Organizational words and deeds are congruent.

6. *Stick to the Knitting:* A successful organization defines its "knitting," or activities that support objectives and remains steadfast in pursuit of these activities. It does not subordinate strategic planning and objectives to momentarily attractive new market niches that are not part of its knitting.

7. *Simple Form—Lean Staff:* The successful organization is a dynamic organism that adjusts its structure to serve evolving needs, ideally employing simple form and lean support staff.* Structure is necessarily subordinate to purpose.

---

*These observations and others also suggest that new developments in information technology, office automation and broadband telecommunications, actually support not only a lean staff concept but also a minimum depth to the management hierarchy.

8. *Balancing Controls with Autonomy:* The successful organization promotes autonomy and entrepreneurship while remaining hands-on and
value driven. Vision and values are clearly guided by management,
while employees are given a relatively free hand to pursue this vision.

These guidelines have important implications for strategic quality assurance and management. Most notably is the realization that

1. Self-generated controls are far more effective than inspector-imposed
   quality controls.
2. Overly complex, rigid, or precise analytic tools are usually counterproductive. Often they solve the wrong problem.
3. Those who implement plans should make the plans or at least come to
   internalize them.

The following approaches to management, stated in the form of slogans, are
generally to be avoided:

1. Bigger is better, because it produces economies of scale.
2. Consolidation always eliminates overlap, duplication, and waste.
3. Make sure that everything is carefully and formally coordinated and
   controlled.
4. Low-cost producers are sure fire winners; survivors always make
   products cheaper.
5. Remove disturbers of the peace.
6. The manager's job is decision making.
7. Control everything; people are just one of the factors of production.*
8. Get the incentives right and productivity will automatically follow.
9. Inspect scrupulously in order to control quality.
10. If you can read financial statements, you can manage anything. A
    business is a business is a business.
11. Top executives are smarter than the market and smarter than their
    subordinates.
12. It's all over if we stop growing.
13. Cost reduction and efficiency are top priority. Quality and effective-
    ness take a back seat to these.
14. Never be overly concerned about the fact that most of the important
    issues are external to the analysis.

---

*See our discussions in the microeconomic theory of the firm in Chapter 8 on just this point, but
from a rather different perspective.

Peters and Waterman note that it is inherently easier to develop a negative argument than to advance a constructive one. Although they advocate formal analytical methods that involve analysis, planning, specifying, and evaluating; they also encourage informal ways to learn and adapt, including to interact, test, try, fail, stay in touch, learn, shift direction, adapt, modify, and perhaps most important, KISS (*keep it simple, stupid*).

The authors discourage management by edict and strongly encourage management by wandering around. While noting that strong companies usually have excellent analytical capabilities, they recommend that major decisions be shaped more by organizational values than numerical analysis. They argue strongly that application of their identified principles will produce the following.

1. A broadly shared culture that motivates employees to search for appropriate solutions to organizational goals.
2. A sense of purpose emanating from the following.
   a. Love of product.
   b. Provision of top quality service.
   c. Acknowledging the innovations and contributions of fellow employees.
   d. The realization that it is peer pressure, rather than orders from the boss, that is the main motivator.

   This sense of shared purpose enables the sought after simultaneous loose–tight properties. It simultaneously permits a balance of firm management guidance and considerable operating autonomy.

Peters has published two more recent studies that, in part, update the central message in *In Search of Excellence*. The first of these [27] reaffirms the merit of leadership by walking around (LBWA). It advocates trust and respect for the creative potential of each person in the organization, and cultivation of this potential through

Coaching.
Confronting issues and problems realistically.
Counseling.
Educating.
Facilitating.
Listening.
Sponsoring.
Teaching.

Peters stresses that LBWA should maintain the established chain of command and not seek to abrogate management authority and responsibility. Quality is deemed to result from this: quality in terms of care of customers and quality in terms of constant organizational innovation.

In a more recent work [28], Peters argues forcefully that there are four major contemporary realities that can create crisis situations:

1. Generic uncertainty throughout the world, due to a variety of factors, for example, dramatic oil price escalation, major exchanges in currency exchange rates, worsening of lifestyles in third-world countries, and less than honest American business and political practices.
2. Technology revolution, due to a number of newly emerging technologies, many of which are associated with the information technology revolution of control, communication, and computation.
3. New competitors for established organizations. There are new domestic competitors due to the entrepreneurial explosion, spin off companies, and firms that have come into being to exploit newly emerged technologies. There are new international competitors to the rapid emergence of Japan and Germany as world powers, and the emergence of some third-world nations as economic and industrial powers. If the book had been written in mid-1991, the enormous changes in the former U.S.S.R. would surely have been listed as a major change in this category.
4. Changing tastes and values, symptoms of which include the growth of a larger affluent class, the growth of a larger impoverished class, and changing attitudes toward work and family.

Figure 5.5 illustrates the changes produced by these contemporary conditions and suggests strategy for coping with these factors. One of the purposes of the new study was to determine the extent to which the initial eight principles for excellence, identified on pages 199 and 200, can predict and are portentous or causally influential of excellence.

It is claimed that a successful organization learns to adjust to the following aspects of our world's current fast-changing, high-velocity environments:

Uncertainties.
End of isolation.
Demise of mass markets and production.
More choices.
Market fragmentation.
Product and service explosions.
Demands for quality and rapid responsiveness.
Greater complexity.
Midsize firms.
Cleaned-up portfolios and more competitive enterprises.

These have the potential to produce a crisis condition in organizations that are not capable of coping with the new challenges. These changes affect

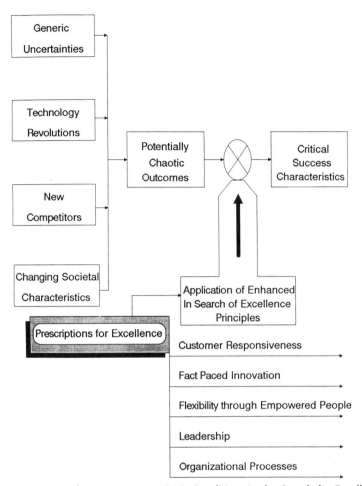

Figure 5.5. Coping with Contemporary Crisis Conditions in the Search for Excellence

virtually all structural aspects, functions, and goals of contemporary organizations. To cope with these contemporary realities, a number of suggested changes must occur.

1. The organization must shift from a current focus on mass markets to niche markets. The productions characteristics most relevant to these new markets are short production runs and continual differentiation of both new and mature products.

2. The current images of domestic and export markets must be reoriented to an image of a single global international market. Product and market development should be internationally oriented from the start.

3. Manufacturing practices must shift from the current emphasis on volume to frequent product design changes, short runs, flexibility, and automation.

4. Sales and service components of organizations must be elevated from second-tier operations to essential sources of value and new ideas.

5. Innovation must no longer be exclusively driven by central research and development. Innovation must now be shaped by autonomous and decentralized units, and driven by customer satisfaction through product improvements.

6. People must no longer be seen as requiring microlevel control and direction. They must be regarded as the prime source of value added in the organization.

7. Organizational structure should be kept nearly flat. Traditional hierarchical and functional layers should be broken up, such that self-managed teams are encouraged.

8. Leadership must cease being central strategy oriented only. It also must become change and shared value oriented as well, with staff management functions supporting line functions.

9. Management information systems must become decentralized and distributed such that they truly serve information use as a strategic organizational weapon.

10. Financial management and control must become decentralized, with the financial staff becoming organizational team members rather than only centralized police.

Peters sees these principles as implying five prescriptive areas:

1. *Customer Responsiveness Prescriptions:* Encourage organizational listening and quick responsiveness.

2. *Fast-Paced Innovation Prescriptions:* Encourage organizational flexibility and responsiveness in new competitive environments.

3. *Flexibility Through Empowered People Prescriptions:* Suggest a relatively small number of hierarchical layers in an organization and increased rewards to individuals for quality and productivity.

4. *Leadership Prescriptions:* Promote adaptability among employees.

5. *Organizational Processes Prescriptions:* Address the need for appropriate systems management and metrics.

These top-level prescriptions, illustrated in Figure 5.5, can help achieve the following five value-added strategies:

1. *Providing a Top-Quality Product:* Mount a quality-improvement revolution in order to ensure that quality is always defined in terms of customer perceptions.

2. *Providing Superior, Even Fanatical, Service:* Emphasize customer-related intangible elements, and measurement of the resulting customer satisfaction.

3. *Achieving Extraordinary Responsiveness:* [Also called total customer responsiveness (TCR)]. Establish bold new partnerships with customers and aggressively create new markets.

4. *Becoming International:* Sell, and perhaps produce, throughout the world.

5. *Creating Uniqueness:* Achieve an understanding of this uniqueness of organization and purpose, both inside and outside the organization.

These critical value-added strategies are made possible through the incorporation of four organizational capabilities [28].

1. The organization should become obsessed with listening for facts, values, and perceptions, through use of what is called a Customer Information System (CIS).
2. The organization should turn manufacturing and operations into marketing weapons, in addition to these remaining as traditional cost centers to be optimized. The organization should use new technology to increase flexibility, being aware that structure and attitude change must accompany new technology introduction. Through this, the organization should develop a new model of the role of manufacturing.
3. The organization should make sales and service forces into heroes by paying them well, recognizing their accomplishments, and providing them with outstanding tools and the opportunity to participate in the structuring of their jobs and support system. Give a lot to them, expect a lot from them, and prune if this does not occur.
4. The organization should achieve fast-paced innovation by investing in application-oriented ventures within the organization. This may be accomplished through the following four strategies:
   4.1. Pursue a team concept for product and service development, and involve suppliers, distributors, and customers, while keeping in mind dangers of matrix management and use of only partially committed people.
   4.2. Encourage pilots and prototypes of everything.
   4.3. Practice creative swiping, by borrowing ideas aggressively, and adapting and enhancing them to the effort at hand.
   4.4. Adopt systematic word-of-mouth marketing.
   Four tactics are proposed to help implement these strategies:
   4.5. Support committed champions of innovation.
   4.6. Practice purposeful impatience and reward innovation.

4.7. Support, and identify potentially fast failures—by rewarding well thought out mistakes that are soon corrected,* being sure to learn from these mistakes.

4.8. Set precise goals that enable the definition and measurement of innovation.

Taken together, it is suggested these four courses of action should create an environment conducive to innovation.

Turning to the issue of quality, Peters sees 12 primary attributes of a quality revolution. These are the critical success characteristics associated with Figure 5.4.

1. Management is obsessed with quality.

2. There is a guiding system, or ideology, that provides both a passion for and a systematic process to support quality.

3. Quality is measured by the work team or department, in order to avoid bureaucratization, and not by quality control inspectors. Quality measurement begins at the start of a project, extends throughout the production process, and must be visible and reportable.

4. Quality is rewarded.

5. Everyone is trained in techniques for quality assessment.

6. The quality of every major function in the organization is measured in a cooperative and nonadversarial manner.

7. There is no such thing as an insignificant improvement.

8. There is constant stimulation of quality enhancement efforts.

9. There is a parallel organizational structure devoted to quality improvement. There is also the recognition that this parallel structure must be carefully controlled such that it does not deteriorate into a new, apathy-creating, layer of bureaucracy.

10. Everyone is a player in the organization's quality assessment and management process.

11. Quality improvement is the primary source of cost reduction, through simplification of the production process.

12. Quality improvement is a never-ending objective.

The customer must be viewed as the final judge of quality. It is noted that the customer for public goods is sometimes difficult to definitively identify (see Chapter 8 for a discussion of public goods and public bads). This will create

---

*It is important to differentiate between errors that are intentional and mistaken acts which do not accomplish intents, and unintentional acts or slips and lapses, that result from improper implementation of action plans. We will discuss human errors in Chapter 9 and provide fuller explication of why this is quite important.

Create Total Customer	Create New Specialized Niche Markets
	Provide Top Quality Products
	Provide Superior Service
	Achieve Extraordinary Responsiveness
	Become International and Global
Responsiveness & Launch a Customer Revolution	Create Uniqueness of Product /Service
	Create Obsession with Listening
	Make Manufacturing a Marketing Weapon
	Make Heros of Sales and Service Forces
	Pursue Fast Paced Innovation

Figure 5.6. Prescriptions for Customer Responsiveness

additional difficulties in defining quality assurance and management strategies in public sector efforts.

In our discussions thus far, we have noted the importance of creating total customer responsiveness, through product quality and responsiveness, and the role of innovation in developing innovative products. In this management revolution handbook, Peters has identified the following as necessary to achieve excellence.

Flexibility achievement through empowering people.

Developing leadership for change at all organizational levels.

Building responsive organizational systems.

Taken together, there are a total of 50 prescriptions. Figures 5.5 through 5.10 illustrate a hierarchical structure for these.

It is of interest to contrast and compare these 50 prescriptions with the eight principles identified in *In Search of Excellence*. Examination of these eight, illustrated in Figure 5.11, indicates that the 50 prescriptions just discussed appear both to be more finely grained and detailed versions of the eight principles of excellence and also to incorporate activities necessary for organizational change to accommodate such realities as generic uncertainties, technology revolutions, new competitors, and changing societal conditions. It is these 50 prescriptions that should be applied in the face of potentially chaotic outcomes to yield critical success characteristics of organizations that can cope well with crises and chaotic conditions. In a steady-state world,

Pursue Fast Paced Innovation & Create an Organizational Capacity for Innovation	Invest in Application Oriented Small Starts
	Pursue Team Product & Service Delivery
	Encourage Prototypes of Everything
	Practice Creative Swiping
	Encourage Systematic Verbal Marketing
	Support Committed Champions
	Be Innovative in Daily Affairs
	Support Fast Failures
	Set Quantitative Innovation Goals

Figure 5.7. Fast-Paced Innovation Prescriptions

these prescriptions presumably become equivalent to the eight basic principles that are identified in *In Search of Excellence*.

These numerous prescriptions for excellence are not at all mutually exclusive, and there are many interactions among them. Figure 5.12 illustrates interactions among the 10 prescriptions for empowering people and the 10 prescriptions for developing a new view of leadership. Figure 5.13

Prescriptions for Empowering People in the Organization	Involve Everyone in Everything
	Use Self-Managing Teams
	Listen, Reward, Recognize Performance
	Invest Time & Significant Effort in Recruiting
	Train and Retrain Present People
	Provide Incentive Pay for Everyone
	Provide Term Employment Guarantees
	Simplify/Reduce Organizational Layers
	Reconceive Role of Middle Management
	Eliminate Bureaucratic Rules/Conditions

Figure 5.8. Peters' Prescriptions for Empowering People

	Cope with Paradoxes and Conventional Wisdom
	Develop an Inspiring and Enabling Vision
	Manage and Lead by Example
	Manage Visibly - Reduce Information Distortion
Prescriptions for Creating	Pay Attention - Be a Compulsive Listener
Love of Change and New	Emphasize Front Line People as Heros
Leadership View	Set the Context for and Delegate
	Pursue Horizontal Management - Bash Bureaus
	Evaluate Everyone on Love of Change
	Create a Sense of Urgency

Figure 5.9. The Ten Leadership Prescriptions

illustrates hypothetical relations between five prescriptions for excellence and eight key elements of systems engineering. In each and every individual case, these prescriptions for excellence need to be mapped into the specific strategic plans of an organization. From this perspective, the 50 prescriptions are really meta-level prescriptions. Cross-interaction matrices could potentially be used to link them to specific strategic plans and management controls, as shown conceptually in Figure 5.13.

	Develop Important Metrics and Measures
	Identify and Use Effecive People Evaluations
Prescriptions for Systems and	Decentralize Information, Authority, Planning
Organizational Processes	Set Conservative Goals and Growth Targets
	Demand Total Integrity in Everything

Figure 5.10. The Five Prescriptions for Excellence of Systems and Organizational Processes

Top Level Principles from	Organization Has a Bias for Action
	Organization Stays Close to the Customer
	Autonomy and Entrepreneurship
	Organizational Productivity through People
In Search of Excellence	Hands-on and Value-Driven Organization
	Organization "Sticks to the Knitting"
	Organization has Simple Form and Lean Staff
	Organization is Simultaneously "Loose-Tight"

Figure 5.11. Peters' Top Eight Principles Found in the Initial *In Search of Excellence* Study

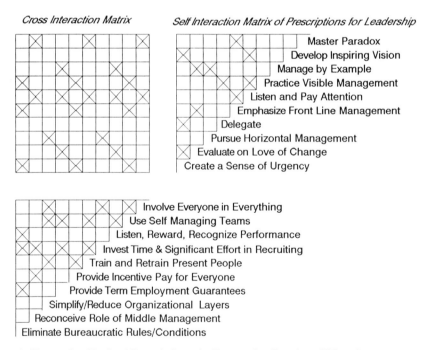

Self Interaction Matrix of Prescriptions for Empowering People at All Levels

Figure 5.12. Hypothetical Self- and Cross-Interaction Matrices for Prescriptions for Leadership and Empowering People at All Levels

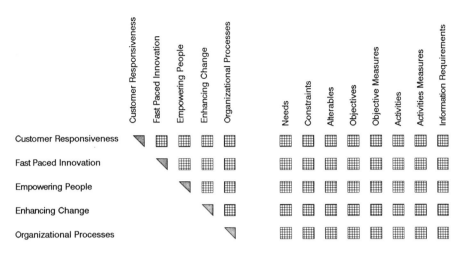

Figure 5.13. Needed Mapping of General Prescriptions for Excellence to Specific Strategic Plan and Management Controls for an Organization

5.4.3. Total Quality Management (TQM)

Over the past 10 years, Deming has become an American motto for quality. He, together with other quality control professionals such as Juran and Feigenbaum, have been national heroes in Japan for 40 years. The Japanese have ascribed much of their success in improving quality and productivity to their adoption and extension of Deming's principles. The aforementioned book by Garvin [1] provides a historical account of much of this success in Japan.

The term *total quality management* is generally understood to refer to the continuous improvement of processes and products through use of the following.

Objectives and associated measures with which to determine objective satisfaction.

An integrated systems management approach.

A life cycle process of product development through organizational teamwork.

The methods of statistical quality control.

Achievement of customer satisfaction.

An essential feature of this is the notion of the *totality* of the approach. It involves efforts at the level of strategic planning, management control, and operational and task control. It involves approaches to process and product and focuses on problems and issues addressed by the product.

In his books, Deming [18, 29] describes a number of obstacles to quality management, specifically in the United States, including the following.

1. *Lack of Constancy of Purpose:* Organizations may not stick close to the knitting in a continuous manner over time, to borrow some of the words of Peters.

2. *Focus on Profit Over the Short Term Only:* Deming espouses the view that current American management is often obsessed with short-term profit, to the exclusion of developing processes, products, and services that support success over the longer term. This leads to a lack of interest in research, education, training, and other efforts that reflect a constancy of purpose.

3. *Emphasis on Performance Ratings and Annual Reviews:* Deming sees this as "management by fear." It promotes a short-run, reactive inspection and evaluation of what has occurred, rather than a long-term proactive effort to promote achievement orientation. Besides, these views tend to be more focused on the shortcomings of people rather than the management system, or process, which is more often than not, in Deming's view, the real culprit.*

4. *Mobility of Management:* Rapid management turnover encourages a short-term view, rather than long-term commitments to quality and productivity.

5. *Management by Current Statistics:* There are a number of important factors, such as customer satisfaction, quality, and productivity improvements, that may not easily be measurable over the short run, but have important long-run consequences for profitability. It can be devastating to ignore these.

Unfortunately, these obstacles to success are neither mutually exclusive nor collectively exhaustive.

Deming thinks American management is in a crisis situation, and he has proposed strategies to lead us out of the crisis. In this section, we discuss some highlights of Deming's approach, generally called total quality management (TQM), although the term seems not to have originated with Deming. This approach has its roots in the quality control theories of Shewhart [4] and others, even though statistical quality control is not explicitly mentioned. In particular, Shewhart was concerned with

1. Special or peculiar causes that relate to the performance of specific individuals and called *assignable causes*;

---

*One of the culprits here is the improper use of data concerning errors to evaluate performance. We will provide a number of additional commentaries on this point in our discussions of risk management later in this chapter, and in our discussion of evaluation metrics in Chapters 7 and 8 and in our discussions of cognitive ergonomics in Chapter 9.

2. Systems management difficulties, called *system causes* and which are only resolvable by top management.
3. Cooperation and communication among all elements of the enterprise (marketing, design, engineering, production, sales, accounting, etc.).
4. Statistical quality control, including a number of new charting approaches.

There are 14 aspects of TQM as first published by Deming [29], and restated in the 1986 revision of this book [18].

1. Create constancy of purpose for improvement of product and service.
2. Adopt the new philosophy.
3. Cease dependence on inspection to achieve quality.
4. End the practice of awarding business on the basis of price alone. Instead, minimize total cost by working with a single supplier.
5. Continually improve planning, production, and service.
6. Institute training on the job.
7. Adopt and institute leadership.
8. Drive out fear.
9. Break down barriers between staff areas.
10. Eliminate slogans, exhortations, and targets for the work force.
11. Eliminate numerical production quotas for the work force and numerical financial goals for management.
12. Remove barriers that rob people of pride of workmanship. Eliminate the annual rating or merit system.
13. Institute a vigorous program of education and self-improvement for everyone.
14. Involve everybody in the company in this transformation.

Each are important and deserve additional commentary. To this end, we provide a brief commentary and interpretation of the 14 prescriptions in the TQM approach of Demming.

1. *Create Constancy of Purpose:* The organization's objectives should be very clear, and should address the organization both as it exists today, and as it is meant to exist in the future. Innovation and faith in the future are both called for. Continual improvements, even of a very small nature, will eventually produce improvements. These result from the elimination of such *quality-destroying characteristics* as delays and defects in purchased products, and adverse human behavior and attitudes. They also result from long-range planning, continuous edu-

cation and training, elimination of unsafe and degrading working conditions, and improved communications up and down the organization.

2. *Adopt the New Philosophy:* The philosophy embodied in the other 13 prescriptions should be explicitly adopted. Adoption requires more than endorsement; it requires action. Endorsing and espousing anything without manifestation of change will almost always lead to cynicism regarding the organization's sincerity.

3. *Cease Dependence on Inspection:* One of Deming's major arguments is that lasting quality is not created by screening out defective products through massive inspection at the output stages of production. Instead, quality comes by improving the production process, requiring major efforts at the strategic management level. Deming proposes methods of statistical process control to achieve such improvements. These methods are explained in his book, *Out of the Crisis* [18], its predecessor [29], and much of the statistical quality control literature. (Some of these methods are discussed in Chapter 6.)

4. *End the Practice of Choosing Suppliers Based Solely on Price:* The lowest-price components often are not the least expensive ones, especially over the long term. If the lowest-price components are not of good quality and if the vendor does not provide appropriate maintenance services, it is likely that these components will create quality problems, and long run costs will be increased. Deming advocates working with a single supplier when its quality and service meet the needs of the organization. Above all, we should buy for quality and not for price alone. Suppliers are an integral part of strategic management for total quality.

5. *Continually Improve Processes:* Improving productivity should be a never-ending task. The objective should be not to fix problems once and forever but to commit to continued improvement through process improvement. This approach should be implemented across all aspects of an organization's activities.

6. *Institute Training Programs for Quality Improvement:* While education and training costs money, lack of education and training may cost more over the long run. Productivity improvements are largely achieved by people, so these people should be educated or trained for the tasks they need to perform. Unfortunately, investments in education and training often do not appear on the financial records of an organization as assets. Therefore, they are often not looked at as investments or benefits, but merely as costs which lower profits.

7. *Adopt and Institute Leadership:* The most fundamental role of management is to provide leadership, and this leadership should focus on continual quality improvement. Management must take a leadership role in implementing the TQM philosophy, particularly when introduc-

ing contemporary approaches to systems management and management control.

8. *Drive Out Fear:* Employees may avoid expressing ideas and admitting mistakes for fear of losing status, position, or even their jobs. However, the process of organizational improvement is enhanced if every lesson that is learned is also communicated. Thus, management must assure that people feel sufficiently secure that they will proactively communicate lessons learned from their mistakes.

9. *Break Down Barriers Between Organizational Units:* Quality is best achieved by having people in each functional area of the organization understand the other functional areas, and communicate regularly with people in these areas. For example, people in design and manufacturing need to work with those in marketing to introduce new high-quality products and systems that are market oriented.

10. *Eliminate Slogans:* Slogans such as "quality counts" are useless. Indeed, they are counterproductive without appropriate methods, controls, and management commitment. Continual identification and correction of problems with processes and products will lead to quality improvements; slogans alone will not.

11. *Eliminate Numerical Quotas and Goals:* The mandate to pursue specific quantitative targets holds back the best people and frustrates below-average people who cannot achieve them. Although it is important to be able to predict production rates and sales, these metrics themselves should not become the primary objectives of either the organization or its management. Operational quality assurance and control metrics have definite uses, as discussed in some detail in Chapter 4, and throughout this book.*

12. *Remove Barriers to Pride of Workmanship:* Job performance should be evaluated and rewarded in terms of the work quality and productivity of the individual, not through an organizationally mandated, inspection-oriented, rating form. People will take pride in and be motivated by individual judgments of their workmanship. Similar results do not occur through use of rating forms. Deming thinks that performance appraisal, particularly when accompanied by inflexibility and inappropriate metrics, is the number one problem facing American management today.

---

*One of the major problems with numerical quotas and goals is that these are often objectives measures, and not fundamentally objectives. Sadly, we often substitute objectives measures for objectives, and then pursue the measures as if they were the objectives. One good illustration of this is the Scholastic Achievement Test (SAT). This is supposed to be a *measure* of educational aptitude. Yet, the use of this instrument has resulted in many students setting the objective of obtaining a sufficiently high score on the SAT. We can even buy books and take short courses designed to help achieve this sublimated objectives measure!

13. *Institute Education and Self-Improvement:* Broadly based, not just specifically job related, education and other self-improvement efforts are important for everyone in the organization. The organization can encourage education with reimbursement incentives. Much more important, however, is top management's personal commitment and establishment of educational objectives.

14. *Put Everybody to Work:* The quest to achieve each of the 14 TQM prescriptions is likely to require a major transformation of the organization. This effort can be substantially facilitated by obtaining a consensus throughout the organization and involving everyone in the effort.

Deming's 14 points seem quite simple and straightforward, but they are not necessarily easy to implement. It would be easy to create chaos through the inept application of these principles. Thus, the approach to implementation of TQM, or any related approach to strategic quality assurance and management, is a critical issue.

It is interesting to note that only prescriptions 3, 4, and 5 relate specifically to quality improvement and explicitly include statistical quality control notions. Others relate more to a philosophy of management (1, 2, 8, 10), organization structure and functions (7, 8, 9, 11, and 14), and to developing human resources (6, 8, 12, 13).

The 14 points can be grouped into five categories, as illustrated in Figure 5.14. This suggests that a function such as Chief Quality Officer (CQO) might well become an important position in organizations that adopt the new management philosophy. This would only be so if the CQO works in such a

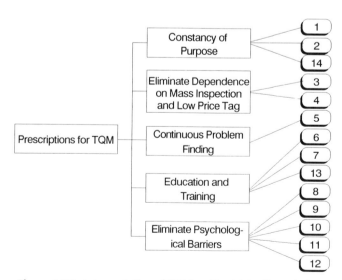

Figure 5.14. Interpretation of TQM as Five Major Prescriptions

Figure 5.15. Organizational Implementation of Quality Management

manner that quality remains everyone's objective. Such a CQO would necessarily need a perspective that encompassed much more than classical mathematical statistics, however. The CQO would also be involved in strategic management and systems management efforts that include *detection* of problems with process and/or products, in *diagnosis* of the root causes of the difficulties, and in generating alternative courses of control effort to effect a *correction* of the difficulties through improved quality and productivity. With the notion of quality remaining everyone's objective, each member of the organization is a quality-responsible person. In that sense, the office of the CQO would be a support office for organizational members.

5.4.4. Other Related TQM Studies

There have been a considerable number of examinations of these and similar concepts and we examine some of them here. Benson, Saraph, and Schroeder [30] propose a three-stage model of quality management. The organizational quality context is first determined. This leads to a determination of organizational change needed to accommodate quality deficiencies, which in turn leads to an organizational response. Feedback about quality performance then impacts the organizational quality context. This model is shown in Figure 5.15.

In a previous study [31], these authors identify 78 lowest-level metrics of quality management, which may be structured as follows.

1. Acceptance of quality responsibility by top organizational leadership
   1.1. Extent to which the top organizational leadership assumes the responsibility for quality performance.

    1.2.   Acceptance of responsibility for quality by major department heads or those who report to the top leadership.

    1.3.   Degree to which top executives and major department heads are evaluated for quality performance.

    1.4.   Extent to which the top management supports long-term quality improvement processes.

    1.5.   Degree of participation by major department heads in the quality improvement process.

    1.6.   Extent to which top management objectives include quality performance.

    1.7.   Specificity of quality goals within the organization.

    1.8.   Comprehensiveness of the goal-setting process for quality.

    1.9.   Extent to which quality goals and policy are understood within the organization.

    1.10. Importance attached to quality by top management compared with cost and schedule objectives.

    1.11. Amount of review of quality issues in top management meetings.

    1.12. Degree to which top management considers quality improvement as a way to increase profits.

    1.13. Degree of comprehensiveness for the strategic quality plan.

2. Visibility and autonomy roles for the quality department

    2.1.   Visibility of the strategic quality department.

    2.2.   Quality department's access to top management.

    2.3.   Autonomy of the quality department.

    2.4.   Utilization of quality department's staff as a resource by the rest of the organization.

    2.5.   Amount of coordination between the quality department and other departments.

    2.6.   Effectiveness of the quality department in improving quality.

3. Extent of quality related training for all employees

    3.1.   Specific work-skills training (technical and vocational) available for hourly employees throughout the organization.

    3.2.   Team building and group dynamics training for employees.

    3.3.   Quality-related training given to hourly employees.

    3.4.   Quality-related training given to managers and supervisors.

    3.5.   Training in total quality concepts, including the philosophy of company-wide responsibility for quality, throughout the organization.

    3.6.   Training of employees to implement quality circle programs.*

    3.7.   Training in basic quality assurance and control statistical techniques.

    3.8.   Training in such advanced statistical quality control techniques as design of experiments and regression analysis.

---

*We discuss quality circles later in this section. Some of the analysis tools associated with quality circles will be discussed in Chapter 6.

    3.9. Commitment of top management to employee training.

    3.10. Availability of resources for employee training.

4. Understanding of customer requirements for product/service design

    4.1. Thoroughness of new product/service design reviews before the product/service is produced and marketed.

    4.2. Coordination among affected departments in the product/service development process.

    4.3. Emphasis on quality of new products/services emphasized as compared to cost or schedule objectives.

    4.4. Extent of analysis of customer requirements in product/service development process.

    4.5. Clarity of product/service specifications and procedures.

    4.6. Extent to which implementation/producibility is considered in the product/service design process.

    4.7. Extent to which sales and marketing people consider quality a saleable attribute.

    4.8. Quality emphasis by sales, customer service, marketing, and public relations personnel.

5. Quality control and management of suppliers

    5.1. Extent to which suppliers are selected based on quality rather than price or schedule.

    5.2. Thoroughness of the supplier rating system.

    5.3. Reliance on reasonably few dependable suppliers.

    5.4. Amount of education of supplier by organization.

    5.5. Technical assistance provided to the suppliers.

    5.6. Involvement of the supplier in the product development process.

    5.7. Extent to which longer term relationships are offered to suppliers.

    5.8. Clarity of specifications provided to suppliers.

    5.9. Responsibility assumed by purchasing department for the quality of incoming products/services.

    5.10. Extent to which suppliers have programs to assure quality of their products/services.

6. Clarity of quality process management at all levels

    6.1. Use of acceptance sampling to accept/reject lots or batches of work.

    6.2. Use of statistical control charts to control processes.

    6.3. Amount of preventive equipment maintenance.

    6.4. Extent to which inspection, review, or checking of work is automated.

    6.5. Amount of incoming inspection, review, or checking.

    6.6. Amount of in-process inspection, review, or checking.

    6.7. Amount of final inspection, review, or checking.

    6.8. Importance of inspection, review, or checking of work.

    6.9. Self-inspection of work by workers.

    6.10. Stability of production schedule/work distribution.

    6.11. Degree of automation of the process.

6.12. Extent to which suppliers have programs to assure quality of their products/services.
6.13. Clarity of work or process instructions given to employees.
7. Extent of appropriate use of quality data and reporting mechanisms
    7.1. Availability of cost quality in the organization.
    7.2. Availability of quality data (error rates, defect rates, scrap, defects, etc.).
    7.3. Timeliness of the quality data.
    7.4. Extent of quality data collected by the service/support areas of organization.
    7.5. Extent to which quality data (cost of quality, defects, errors, scrap, etc.) are used as tools to manage quality.
    7.6. Extent to which quality data are available to hourly employees.
    7.7. Extent to which quality data are available to managers and supervisors.
    7.8. Extent to which quality data are used to evaluate supervisor and managerial performance.
    7.9. Extent to which quality data, control charts, and so on are displayed at employee work stations.
8. Quality awareness by employees
    8.1. Extent to which quality circle or employee involvement type programs are implemented in the organization.
    8.2. Effectiveness of quality circle or employee involvement type programs in the organization.
    8.3. Extent to which employees are held responsible for error-free output.
    8.4. Amount of feedback provided to employees on their quality performance.
    8.5. Degree of participation in quality decisions by hourly/nonsupervisory employees.
    8.6. Extent to which quality awareness building among employees is ongoing.
    8.7. Extent to which employees are recognized for superior quality performance.
    8.8. Impact of labor union on quality improvement.
    8.9. Effectiveness of supervisors in solving quality problem.

The results of this study do indicate that the management control strategy used to influence the organizational quality context is indeed very influential of the quality management approaches adopted. If this were not the case, and indeed if beliefs concerning ideal quality management strategies were free of organizational context, then there would be little reason to have top-level managerial influences of the type advocated here. In their papers, Benson, Saraph, and Schroeder [30, 31] show that these techniques have been widely used and do, indeed, have a major influence on insuring quality.

As we have seen, a primary means of pursuing strategic quality assurance and management is through total quality management, which we may define as total and integrated management control process that specifically addresses all aspects of system and product quality during all phases of the life cycle. Thus, TQM is focused on the process of both technological and management system design. It is concerned with the life cycle of system acquisition, the product or service under development, and the capability and trustworthiness of this product in meeting customer needs. Total customer satisfaction is a major objective of TQM, as compared with the often-used approach of fielding the least expensive design that conforms to minimum customer requirements.

In an earlier discussion, we noted the Theory Z approach to management and quality. We briefly re-examine this, and a Theory W approach, within the context of TQM. One implementation of the consensus approach advocated by Theory Z is the quality circle, which normally incorporates the elements we now describe. Although the actual characteristics of a quality circle do vary, perhaps considerably from implementation, most involve a number of relatively common notions [32].

1. *Membership:* Comprised of employees, generally volunteers, from a particular work unit. Often, not all volunteers can be accommodated, and some form of rotation of membership is needed.

2. *Decision Making and Spending Authority:* Generally this is not given to quality circles, and quality circles generally lack budgets and other organizational resources to conduct studies. They are volunteer efforts.

3. *Meeting Agendas:* Always focus on approaches for improving product quality, often through productivity and cost reduction techniques. Agendas are limited to quality and productivity discussions, and do not include broader issues that deal with organizational issues or quality of work life.

4. *Compensation:* People are generally paid for attendance at QC meetings, which are usually held on company time, although in Japan QC meetings are commonly held on employee time. There may be intangible rewards for successful groups such as banquets and medals, but usually there are no direct pecuniary benefits.

5. *Information Sharing:* The more general aspects of organizational performance are not normally shared.

6. *Meeting Frequency:* Averages twice a month; meetings generally last one or two hours.

7. *Meeting Leadership:* Generally carried out by a facilitator who is not part of management. The facilitator's primary objective is to assist the QC workers to organize their meetings, identify solutions to quality related issues, and prepare presentations for management.

8. *Top-Down Implementation:* Often, they are mandated by top management, and implemented at lower organizational levels.

While the quality circle may, at first glance, identify a number of potential disadvantages to this approach, most disadvantages flow from the reality that information and knowledge convey power. Thus, there must be some shift of corporate power that occurs through implementation of a QC as influence can indeed occur without authority [33]. Pragmatically, there will almost always be resistance to management change if management does not participate in the QC efforts that identified the potential need for change. Often also, QC ideas have been developed without sufficient understanding of the constraints on action, and the resulting admissible alternatives that are available to an organization. The sad reality of this is that hopes are often raised and then necessarily dashed as solutions are proposed that cannot, or will not, be accepted. A possible resolution to this dilemma is management participation, at least in an advisory capacity, in quality circles.

Perhaps the most widespread implementation of TQM is the practice of Quality Function Deployment (QFD). QDF was first used in 1972 by Mitsubishi, at its Kobe shipyard, and then adopted by Toyota, who developed it in a number of ways [34]. The purpose of QFD is to promote integration of organizational functions and to facilitate responsiveness to customer requirements. As described by Clausing [35], QFD is comprised of structured relationships and multifunctional teams. The members of the multifunctional teams attempt to insure that all information regarding customer requirements and how best to satisfy these requriements is identified and used, that there exists a common understanding of decisions, and that there is a consensual commitment to carry out all decisions. Interaction matrices, as discussed in our previous chapter, may be of considerable assistance in this effort. We provide additional commentary on QFD and several other related approaches in Chapter 6.

We referred earlier to the consensus-based difficulties inherent in Theory Y and Z management efforts, and Boehm and Ross have [36] have recently proposed a *Theory W* management theory to ameliorate these. The W in Theory W refers to "win–win" situations that enable the many participants in a systems engineering development effort each to win. These include customers, users, bosses, subordinates, project managers, and maintainers in the original reference. We might add marketeers, advertisers, and others. The theory aims to be simultaneously simple—by being easy to understand and apply; general—in addressing most technical, management, and human situations likely to be encountered; and specific—in providing useful advice to the situation at hand. The primary technique of Theory W is negotiations among stakeholders to convert lose–lose, or win–lose situations into win–win situations.

Boehm and Ross recommend approaches identified by Fisher and Ury [37]:

1. Separate people from the problem at hand.
2. Focus on the interest of the parties involved and not on advocacy positions.
3. Identify options that allow for the mutual gain of all stakeholders.
4. Insist on using identified, objective, and explicit criteria for judgment and choice.

The authors identify a three-phase management process comprised of a number of steps within each phase:

1. Establish a set of win–win prerequisites for the development effort
   1.1. Understand the way in which people wish to *win*. Identify the key stakeholders, in particular the customer, and develop scenarios that indicate what each stakeholder would consider to be a winning outcome.
   1.2. Establish reasonable expectations concerning what can be achieved with the technology at hand and the humans involved. A major difficulty in this regard is the tendency for the systems management team to underestimate the costs and time required to field a system. Developing a reasonable set of trade-offs between performance, including quality and cost, and schedule is essential.
   1.3. Match the tasks of the systems engineering team to various win–win options.
   1.4. Provide a supportive environment, especially in the form of education, training, and support for the systems development program.
2. Structure a win–win systems acquisition process
   2.1. Establish a reasonable strategic system acquisition process and associated program plan. We have identified typical program planning approaches in Chapters 2, 3, and 4. Table 4.1 presents a typical outline for a systems acquisition effort that is applicable here.
   2.2. Use this plan to control the project.
   2.3. Identify and manage all risks in a win–win fashion.
   2.4. Maintain constant involvement of participants.
3. Structure a win–win system product
   3.1. Design the product for the user, developing a quality product that is service oriented, easy to learn and use, easy to modify, cost effective, and designed for human interaction.

The Theory W win–win approach is gaining considerable recognition, as evidenced by recent developmental efforts in the support technologies [38–41] that might better enable it. One of the major purposes of Theory W management is risk management, to which we turn our attention in Chapter 6.

Barry [42] suggests a set of fourteen activities of a process to support implementation of a TQM strategy.

1. Identify the present management and measurement system, including the extent to which it is customer and supplier driven.
2. Assess the structural and cultural facets of the organization, including its quality culture.
3. Identify the top management commitment to TQM.
4. Create a strategic organizational vision and philosophy that is supportive of TQM.
5. Identify an appropriate TQM strategy for the organization under consideration.
6. Identify a management control structure to implement to TQM strategy.
7. Identify education and training needs to accomplish the implementation.
8. Identify resource needs for implementation of the TQM strategy.
9. Select suppliers of this education and training program.
10. Identify quality standards and metrics to insure continuous quality improvement, efficiency, and effectiveness of the process-driven adaptation to TQM.
11. Institutionalize TQM such that it becomes a part of the organization's culture and normal mode of doing business.
12. Monitor and evaluate the results of the TQM implementation.
13. Proactively adapt the specific TQM process to the organization in question on the basis of actual operating experience with TQM.
14. Continue the improvement through TQM.

These might represent specific activity steps to be incorporated into the appropriate portions of Figure 5.15 to provide an overall picture of the implementation of a total quality management concept. There are, of course, a number of variations on these specific steps that might be described [43, 44] and a number of case stories of its implementation [45].

Here, we discuss only one of these variations. Juran [46] identified a detection, diagnosis, and detection-like set of activities that will lead to total quality management, explicitly through statistical quality control approaches. These are as follows:

1. Assign priority to production projects.

2. Conduct a Pareto analysis of symptoms of quality control problems.
3. Identify hypotheses concerning cause of symptoms.
4. Test hypotheses through collection and analysis of data.
5. Select appropriate hypotheses.
6. Design experiments for particular quality diagnostic efforts.
7. Obtain management approval for experimental designs.
8. Conduct specific experiments to determine product defects and diagnose their cause.
9. Propose corrective strategies or remedies.
10. Test corrective strategies or remedies.
11. Apply corrective action to effect a remedy.
12. Proceed to establish quality control at a higher standard.

The first five of these efforts involve planning for statistical quality control. Activity steps 6 through 8 involve diagnosis and detection of specific product quality deficiencies. Steps 9 through 12 involve providing corrective action to processes, such as to remove, or ameliorate, defects in the resulting product.

Juran [47] identifies these three major efforts as a trilogy and compares them with analogous steps in financial control processes as follows.

Quality Control	Financial Control
Quality Planning	Budgeting
Quality Control	Cost Control
Quality Improvement	Cost Reduction

In the quality planning effort, internal and external customers, and their needs, are identified. Generic product features that are responsive to customer needs are identified, as well as associated quality goals. A process is then designed to meet or exceed product specifications, including quality specifications. The capability of the process to produce the intended result is then established.

Quality control involves determining what to control, determination of appropriate metrics and product performance standards, measuring actual performance, and identifying appropriate action to ameliorate product performance deficiencies. Quality improvement is concerned with identifying specific improvement areas, pragmatic organization to effect the improvement, implementation of corrective action, operational demonstration that the correction is effective, and applying appropriate controls to remain in this higher quality state.

There are at least three topics we need to discuss in order to bring an approach such as this to fruition. We need to be able to identify requirements. We need to be able to identify the possible risks that might materialize and appropriate methods to use to cope with, or manage, these risks. We need to be able to develop sufficient analysis tools to enable communication of these to appropriate people in the organization. This is the subject of Chapter 6.

We see a number of common features among many of the strategic quality assurances and management approaches. The Deming approach to TQM is perhaps an exemplar of all of them. Among the many efforts that also relate to total quality management, often in the more generic sense that we have referred to as strategic quality assurance and management, are works by Rosander [48] on applications of Deming's 14 points to the service industry, DoD documents in this area [49, 50], and so on.

5.5. SUMMARY

In this chapter, we have discussed a number of strategic quality assurance and management issues. We provided a brief overview of the development of system quality efforts. These began with inspection of output products to eliminate the unworthy, but this approach was soon replaced by statistical quality control at the operational level. This, in turn, evolved into a strategic management approach to quality improvement. Top management support is essential for strategic quality assurance efforts that include: employee empowerment, teamwork, education, and training; a major focus on the customer, and metrics to enhance product quality through process improvement.

We discussed a number of contemporary approaches to strategic quality assurance and management, many of which involve total quality management, denoting the use of a strategic management process. Many of our discussions suggested risk management for quality as a very important area for effort. We consider this, and related topics, in much of the remainder of this book.

PROBLEMS

5.1. Identify some quality relevant difficulties with systems engineering products and services that are not easily solved through the operational and task level quality assurance and management approaches of Chapter 4. How are these quality assurance issues addressed at the level of strategic management?

5.2. What are essential differences between product and process quality? How do these affect strategic and operational level quality notions?

5.3. Prepare a case study of one American company that has attempted to implement strategic level quality assurance and management approaches. Discuss and illustrate how their approach follows and differs from the efforts discussed here.

5.4. Suppose that you worked one year in a pure Theory X organization, one year in a pure Theory Y organization, and one year in a pure Theory Z organization. Describe the attributes that you would need to have to make maximum progression as a worker in each of these three. How do your descriptions differ, and why?

5.5. The claim is often made that until roughly the 1960s, U.S. industrial efforts were the most productive in the world, but that there has been a decline since then, and a major decline beginning in the mid to late 1980s. What has caused this, and why?

5.6. The major categories for the Malcolm Baldridge National Quality Award and their point values are:

Leadership (100)

Information and Analysis (60)

Strategic Quality Planning (90)

Human Resource Utilization (150)

Quality Assurance of Products and Services (150)

Quality Results (150)

Customer Satisfaction (300)

Please contrast and compare these with the Deming 14 points, and the strategies and prescriptions for excellence of Peters. How do the 1000 points assigned to the various categories for this U.S. national award separate out among these approaches for quality assurance and management.*

5.7. Prepare a brief paper identifying barriers to implementation of total quality management.

5.8. Contrast and compare what you feel would be the major differences involved in implementing total quality management in a large state university in the United States as contrasted with implementation of TQM in a high technology systems engineering firm with 500 employees.

5.9. A private company's critical quality elements might include people requirements, business strategies, process strategies, supplier requirements, and customer requirements. How are each of these addressed by TQM?

5.10. We have noted three focus areas for systems engineering: process, product, and problem. How would you expect the focus on each of these to differ across Theories X, Y, and Z, and the Deming, Juran, and Peters quality procedures?

---

*Further information concerning this award may be obtained from the National Institute of Standards and Technology, Gaithersburg MD 20899.

5.11. From [19] and [20], identify Crosby's 14 steps to quality improvement. Please contrast these with the steps advocated by Deming and Juran.

REFERENCES

[1] Garvin, D. A., *Managing Quality: The Strategic and Competitive Edge*, Free Press, New York, 1988.

[2] Taylor, F. W., *The Principles of Scientific Management*, Harper, New York, 1911.

[3] Radford, G. S., *The Control of Quality in Manufacturing*, Ronald Press, New York, 1922.

[4] Shewhart, W. A., *Economic Control of Quality of Manufactured Products*, Van Nostrand, New York, 1931.

[5] Juran, J., Ed., *Quality Control Handbook*, McGraw Hill, New York, 1951.

[6] Feigenbaum, A. V., *Total Quality Control*, 3d ed., McGraw Hill, New York, 1983.

[7] Budne, T. A., "Reliability Engineering," in C. Heyel (Ed.), *Encyclopedia of Management*, Van Nostrand, New York, 1982.

[8] McGregor, D. M., *The Human Side of Enterprise*, McGraw Hill, New York, 1960.

[9] Maslow, A., *Eupsychian Management*, Irvin, Homewood IL, 1965.

[10] Herzberg, F., "One More Time: How Do You Motivate People," *Harvard Business Review*, Vol. 46, No. 1, January 1968, pp. 53–62 (Reprinted in *HBR* in September 1987).

[11] Harvard Business Review Books, *Manage People, Not Personnel*, Harvard Business School Press, 1990.

[12] Hax, A. C., *Planning Strategies that Work*, Oxford University Press, New York, 1987.

[13] Schein, E. H., *The Art of Managing Human Resources*, Oxford University Press, New York, 1987.

[14] Gellerman, S. W., *Motivation and Productivity*, American Books, New York, 1978.

[15] Ouchi, W., *Theory Z: The Art of Japanese Management*, Bantam Books, New York, 1981.

[16] Ishiwkawa, K., *What is Total Quality Control: The Japanese Way*, Prentice Hall, Englewood Cliffs NJ, 1985.

[17] Kogure, M., and Akao, Y., "Quality Function Deployment and CWQC in Japan," *Quality Progress*, October 1983, pp. 25–29.

[18] Deming, W. E., *Out of the Crisis*, MIT Press, Cambridge MA, 1986.

[19] Crosby, P. B., *Quality is Free: The Art of Making Quality Certain*, McGraw Hill, New York, 1979.

[20] Crosby, P. B., *The Eternally Successful Organization*, McGraw Hill, New York, 1988.

[21] Stuelpnagel, T. R., "Total Quality Management," *National Defense*, American Defense Preparedness Associations, 1988, pp. 57–62.

[22] Rouse, W. B., *Strategies for Innovation: Creating Successful Products, Systems, and Organizations*, John Wiley, New York, 1992.

[23] Garvin, D. A., "What Does 'Product Quality' Really Mean?," *Sloan Management Review*, Vol. 26, No. 1, Fall 1984. Reprinted in Hax, A. C., *Planning Strategies that Work*, Oxford University Press, New York, 1987, pp. 152–174.

[24] Pirsig, R. M., *Zen and the Art of Motorcycle Maintenance*, Bantam Books, New York, 1974.

[25] Leavitt, H. J., *Managerial Psychology*, 4th ed., University of Chicago Press, Chicago, 1978.

[26] Peters, T. J., and Waterman, R. H. Jr., *In Search of Excellence*, Harper and Row, New York, 1982.

[27] Peters, T. J., and Austin, N. K., *A Passion for Excellence*, Random House, New York, 1985.

[28] Peters, T. J., *Thriving on Chaos: Handbook for a Management Revolution*, Alfred A. Knopf, New York, 1987.

[29] Deming, W. E., *Quality, Productivity, and Competitive Position*, Center for Advanced Engineering Study, MIT, Cambridge MA, 1982.

[30] Benson, F. G., Saraph, J. V., and Schroeder, R. G., "The Effects of Organizational Context on Quality Management: An Empirical Investigation," *Management Science*, Vol. 37, No. 9, September 1991, pp. 1107–1124.

[31] Saraph, J. V., Benson, P. G., and Schroeder, R. G., "An Instrument for Measuring the Critical Factors of Quality Management," *Decision Sciences*, Vol. 20, No. 4, Fall 1989, pp. 810–829.

[32] Lawler, E. L. III, *High Involvement Management*, Jossey Bass Inc., San Francisco, 1986.

[33] Cohen, A. R., and Bradford, D. L., *Influence without Authority*, John Wiley, New York, 1990.

[34] Hauser, J. R., and Clausing, D., "The House of Quality," *Harvard Business Review*, May-June 1988, pp. 63–73.

[35] Clausing, D., "Quality Function Deployment: Applied Systems Engineering," *Proceedings 1989 Quality and Productivity Research Conference*, University of Waterloo, June 1989.

[36] Boehm, B. W,. and Ross, R., "Theory-W Software Project Management: Principles and Examples," *IEEE Transactions on Software Engineering*, Vol. 15, No. 7, July 1991, pp. 902–916.

[37] Fisher, R., and Ury, W., *Getting to Yes*, Penguin Books, Baltimore, 1983.

[38] Raiffa, H., *The Art and Science of Negotiation*, Harvard University Press, Cambridge MA, 1982.

[39] Fraser, N. M., and Hipel, K. W., *Conflict Analysis: Models and Resolutions*, North Holland, New York, 1984.

[40] Hipel, K. W. (Ed.), "Special Issue on Conflict Analysis: Parts 1 and 2," *Information and Decision Technologies*, Vol. 16, Nos. 3 and 4, 1990, pp. 183–371.

[41] Neale, M. A., and Bazerman, M. H., *Cognition and Rationality in Negotiation*, Free Press, New York, 1991.

[42] Barry, T. J., *Management Excellence through Quality*, ASQC Quality Press, Milwaukee, 1991.

[43] Weaver, C. N., *TQM: A Step-by-Step Guide to Implementation*, ASQC Quality Press, Milwaukee, 1991.

[44] Talley, D. J., *Total Quality Management Performance and Cost Measures: The Strategy for Economic Survival*, ASQC Quality Press, Milwaukee, 1991.

[45] Walton, M., *Deming Management at Work*, G. P. Putnam, New York, 1990.

[46] Juran, J. M., *Leadership for Quality*, Free Press, New York, 1989.

[47] Juran, J. M., "The Quality Trilogy: A Universal Approach to Managing for Quality, *Quality Progress*, August 1988, pp. 19–24.

[48] Rosander, A. C., *Deming's 14 Points Applied to Services*, Marcel Dekker, New York, 1991.

[49] DoD 5000.51G, "Total Quality Management—A Guide for Implementation," February 1990.

[50] SOAR-7, "A Guide for Implementing Total Quality Management," Reliability Analysis Center, Rome Air Development Center, Griffis AFB NY 13441, 1990.

Chapter 6

Information Requirements, Risk Management, and Associated Systems Engineering Methods

6.1. INTRODUCTION

Systems engineering concepts are directly applicable to the development and integration of management and technological processes that support all of the major life-cycle functions needed to produce high-quality and trustworthy systems. Information is the glue that holds together such functions as research and development, design, production, distribution, maintenance, and marketing in a total quality approach. A major purpose of the systems management function is to implement the strategic plan of the organization to provide this total quality approach, including associated risk management.

Some of the major elements of this effort are

System requirements determination strategies.

Information requirements determination strategies.

Risk management strategies.

In this chapter, we wish to provide a number of perspectives on these important facets of systems engineering, and to introduce tools and approaches that aid in fielding large systems. Such tools are generally most needed at the front end of a systems engineering life cycle. They also support efforts at the latter phases of the life cycle that influence system evolution over time. Through appropriate attention to these efforts and also aided by effective approaches to operational and strategic quality assurance and management, systems integration, and standards, it should be possible to achieve the many functional quality objectives for systems engineering illustrated in Figure 6.1.

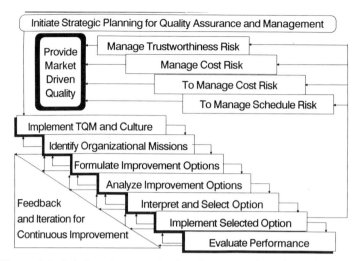

Figure 6.1. Relations Between Quality Planning and Quality Achievement

Much of systems engineering is concerned with the resolution of issues or problems. A *problem* [1, 2] is an undesirable situation or unresolved matter that is significant to some individual or group and that the individual or group is desirous of resolving. Problems have four basic characteristics.

1. There is a detectable gap between a present state and a desired state, and this creates a concern.
2. It may be difficult to bring about concordance between these two states.
3. This situation is important to some individual or group.
4. The situation is regarded as resolvable by an individual or group, either directly or indirectly. Solving a problem would constitute a *direct* resolution. Ameliorating or dissolving a problem, by making it go away, is an *indirect* resolution of a problem.

To say that there is a problem is to imply that there is a situation that deserves attention, thought, and action. If these are not characteristics of a presumed problem, then what exists does not really constitute a problem. There are three essential demands in issue or problem formulation.

1. We need to be able to detect the existence of a potential problem.
2. We need to be able to diagnose the potential cause(s) of the difficulty.
3. We need to be able to correct the situation through determination of a course of action, or correction, that will ameliorate the problem, and implement this such as to respond to the particular situation at hand.

Detection, or *problem identification*, involves the perception of a difference between what we would like to see in a situation and what we do see. *Problem diagnosis* involves the mental, visual, and verbal specification of the problem. However, the two actions of detection and diagnosis overlap, since fully identifying the problem involves constructing an account of the problem's cause. Problem identification should contribute to eventual determination of the causes of the problem.

Failures in problem detection and diagnosis may occur through many different forms of human error. The following are among these.

1. *Setting Improper Thresholds:* Too-high problem recognition thresholds can delay recognition of the problem and too-low thresholds can cause unwarranted attention to insignificant issues.

2. *Failure to Generalize:* The treating of each separate dysfunctional condition or event as a distinct problem, rather than as a symptom of a broader dysfunction, will also create difficulties. This may result in undue attention being paid to the relief of symptoms of difficulties rather than addressing the underlying causes of problems.

3. *Failure to Anticipate:* The reactive attempt to solve problems as they occur, rather than proactively anticipating problems and taking actions that will preclude their occurrence, will cause difficulties. Anticipatory problem identification may involve extrapolating existing conditions, which currently pose no problem, to a future time, when there will be a problem, if appropriate action is not taken.

4. *Failure to Search for and Process Potentially Available Information:* There are a number of information processing biases that result through the use of flawed heuristics. If employed, these will likely lead to improper judgment and choice. Included among these is the failure to search for potentially disconfirming information that would alert us to the presence of a problem by denying support for a hypothesis about one.

We examine some of these in Chapter 9 when we explicitly consider the role of cognitive ergonomics in systems engineering.

There exists a number of characteristics of effective systems efforts that can be identified. These form the basis for determining the attributes of systems and systemic processes themselves. Some of these attributes are more important for a given environment than others. Effective system fielding must typically include an operational evaluation component, which considers the strong interaction between the system and the situational issues that led to the system design requirement. This operational environment evaluation is needed to determine whether a systems engineering process

1. Is logically sound.
2. Is matched to the operational and organizational situation and environment extant.

3. Supports a variety of cognitive skills, styles, and knowledge of the humans who must use the system.
4. Assists users of the system to develop and use their own cognitive skills, styles, and knowledge.
5. Is sufficiently flexible to allow use and adaptation by users with differing experiential knowledge.
6. Encourages more effective solution of unstructured and unfamiliar issues allowing the application of job specific experiences in a way compatible with various acceptability constraints.
7. Promotes effective long-term management.

This listing contains a strong focus on the ability of the system product to meet a customer need or resolve a customer problem. It also supports development of systems that minimize the presence of the four error mechanisms just identified. In our previous efforts, we have already mentioned the innate coupling between problem, product, and process. All three of these are clearly present here. Information and risk management are major ingredients as well.

Figure 6.2 illustrates this transitioning of the requirements through purposeful, functional, and structural stages into technical specifications. The figure also shows, somewhat generically, the analysis and evaluation of these information-based requirements to enable verification and validation of an appropriate set of system technical specifications. The effort involved here is not at all a simple one, and much recent effort has been devoted to the

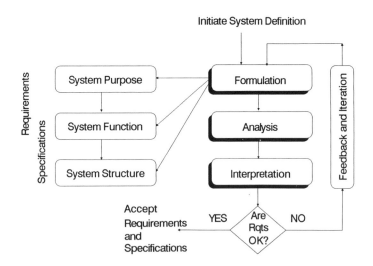

Figure 6.2. Identification of User Requirements and System Specifications

development of appropriate approaches to requirements determination as evidenced by two large edited volumes on this subject [3, 4].

These information relevancy observations are not new. Simon [5] is among those who express the view that a central problem in contemporary organizations is not how to organize to produce efficiently, but how to organize to make decisions, that is, how to effectively determine what information to acquire and how to process and appropriately use this information.

In this chapter, we provide some perspectives on information and systems requirements determination, risk management, and some methods and tools that are useful in identifying requirements for information and systems, and in managing risk. We will begin our effort with a brief discussion of issue formulation. Then we discuss requirements determination, risk management, and tools to aid in these efforts.

6.2. INFORMATION REQUIREMENTS DETERMINATION

Judgments, at least prudent judgments affecting important actions, should never be made without appropriate information. For it is only through information that one becomes aware of the need for judgment and choice activities, and the result of these in the form of a decision. Information is often defined as data of value in decision making. The activities of data acquisition, representation, storage, distribution, and use are generally involved in information processing.

The task of information requirements determination is necessarily involved with all of these, although formally it is concerned with determination of what information is to be acquired. This requires an appraisal of how the information will be converted to useful knowledge. Human and organizational concerns are critical in this appraisal, which is often called *situation assessment*.

It is possible to define information at several levels, as noted by Shannon many years ago [6]. In Shannon's model of communication and information processing, the transmitter, channel, and receiver each treat messages simply as data. At the technical, or syntactic, level, information is transmitted in some natural language over a channel. The system encodes a message into a sequence of data signals, sends the resulting signals over the channel, and decodes the signals in order to reproduce the message. The idea is to have the same syntax that is input to the electronic communication system also be received at the output.

There are generally no possibilities for natural language translation, or for translation into another message with the same semantic meaning in this interpretation of the technical or syntactic level. Electronic communication is deemed successful if the message syntax that is produced by the source is exactly recreated at the destination, regardless of semantic or functional

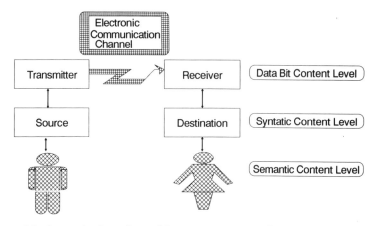

Figure 6.3. Semantic, Syntatic, and Data Bit Content in Electronic Communications

ambiguity contained in the message which may make it imprecise, imperfect, or otherwise flawed.

Shannon's information theory model, as illustrated conceptually in Figure 6.3, consists of five components: the *source* of the information or producer of the message; a *transmitter*, which operates on the message to produce a signal suitable for transmission; a *channel*, which is the medium used to transmit the signal from the transmitter to the receiver; the *receiver* itself, which performs the inverse operation of the transmitter thereby reconstructing the message from the signal; and the *destination*, which interprets and acts upon the message.

The technological or syntactic level approach to information, fundamentally concerned with error-free transmission of data bits representing syntax, is best explained by Shannon's *information theory*. Although these concepts are appropriate for maximizing channel capacity or otherwise optimizing technology components so as to best transmit most effectively the data that is in a given message, they do not directly concern whether the information that is transmitted is appropriate. That is, of course, our major concern in information requirements determination. Systems engineers need to be aware in any given situation whether it is more important to correctly process syntactic information, or semantic information, since some forms of syntactic representation are more appropriate that others for semantic understanding of the underlying knowledge. Subjects such as information presentation and visualization have become of great importance, for these reasons. The design of a communication system that allows a human and an information technology-based system to exchange knowledge requires that differences between the human and the computer be recognized, and that some sort of human-computer communication architecture be developed.

In his truly seminal work, Shannon was concerned exclusively with the technical, or syntactic, level of communication. He makes the rather strong statement that "semantic aspects of communication are irrelevant to the engineering problem." Today, we recognize that higher level concerns in information processing are of great importance to systems engineering, and many other efforts. At the *semantic level*, concern is with the meaning and efficiency of various messages.

Messages meant to support decision making should have the following three basic characteristics:

1. They should be clear, permit rapid comprehension of all relevant aspects of the situation.
2. They should readily explain recommended actions and expected results of these actions.
3. They should help guide the process of judgment and choice and, therefore, should be based on high-quality and relevant information.

These concerns are much more relevant at the semantic (and higher) level than the syntactic (and lower) levels. Lack of attention to these concerns when setting information requirements is often the cause of poor designs.

The classic *value of information* concept, described in most texts on systems engineering and decision making and which we discuss in Chapter 7, results in the evaluation of possible information-gathering activities in terms of the expected economic benefit. If the benefits in obtaining a particular item of information exceed its cost, then the information should be purchased. Although this is a useful concept, there are a number of limitations to this approach.

1. The requirements for a detailed model of the decision situation often are not available.
2. Sensitivity analysis to determine the effects of imprecision in the parameters is not at all easy to perform, especially when there are two or more parameters that need to be tested jointly.
3. The major usefulness of "value of information" lies in efficiency of information use, as contrasted with overall decision effectiveness.
4. It is difficult to compare the value of information sets that are structurally or parametrically inconsistent.

Notwithstanding these limitations, "value of information" is an important concept and is of much potential value in systems engineering.

There are a number of available approaches for the determination of information needs. Taggart [7] has developed an approach called *management information requirements analysis*, that is based on human communication needs and information complexity. This approach examines pertinent reports and management responsibilities and then develops a set of information requirements. The approach carefully defines what a person does and determines a set of elements from this. In part the effort consists of automated examination of these for verb and noun indicators of information needs.

Some programming languages assist in information requirements determination [8, 9]. An example is *problem statement language/problem statement analyzer*. The identified requirements are translated into system inputs, which are stored in a data base for later recall and possible updating. Software tools produce a report directly from the identified data base. A top-down approach is used, which allows development of criteria and definition of system boundaries through examination of the internal and external problem environment. After this is accomplished, objects are utilized to define the conceptual qualities of the proposed system. Relations between each of the identified objects are identified next. These are stored in a centralized database for later use in responding to queries. The efforts of Yadav and Chand [10] are concerned with these same notions and an expert modeling DSS is proposed to study information requirements.

Efforts of this sort may be very useful for identification of technical information and requirements specifications from an existing set of user requirements. Additional efforts are needed to identify an initial set of information requirements, or requirements for a system, from users.

The information requirements determination taxonomy of Davis [11–13] represents a very useful synthesis of research on this subject. According to Davis, two levels of information requirements are necessary. *Organizational level requirements* specify the system structure, portfolios of applications, and boundaries within which individual decisions are to be made. *Application level requirements* determine information processing needs for specific applications. Although the uses of information at these two levels, process and product, are different; the generic procedures for determining information requirements are the same for each level. The specific procedures most useful at a given level will depend upon several factors.

Davis has identified four strategies for determining information requirements. Taken together, these yield approaches that may be designed to ameliorate the effect of three human limitations: limited information processing ability, bias in the selection and use of information, and limited knowledge of appropriate problem solving behavior.

The first strategy is to simply ask people for their information requirements. The usefulness of this approach will depend on the extent to which the interviewers can define and structure issues and compensate for cognitive biases in issue formulation. There are a variety of methods that can be used to assist in this [1].

The second strategy is to elicit information requirements from existing systems that are similar in nature and purpose to the one in question. Examination of existing plans and reports represent one approach of identifying information requirements from an existing, or conceptualized, system.

Davis' third strategy consists of synthesizing information requirements from characteristics of the utilizing system. This permits one to obtain a model or structure for the problem to be defined, from which information requirements can be determined. This strategy would be appropriate when the system in question is in a state of change and thus cannot be compared to an existing system.

The fourth strategy consists of discovering needed information requirements by experimentation. Additional information can be requested as the system is employed in an actual or simulated operational setting and problem areas are encountered. The initial set of requirements for the system provides a base point for the experimentation. This represents an expensive approach, but is often the only alternative when the experience base does not exist. This approach is equivalent to use of a prototype, either an evolutionary prototype or throwaway prototype [14].

Each of these four strategies has advantages and disadvantages, and it is desirable to be able to select the best mix of strategies. One's choice will depend on the amount of risk or uncertainty in information requirements that results from each strategy. Here, *uncertainty* is used in a very general sense to indicate information imperfection. Five steps are useful in identifying information uncertainties and then selecting appropriate strategies [11].

1. Identify characteristics of the utilizing system, technology system, users, and system development personnel as they affect information uncertainty.
2. Evaluate the effect of these characteristics on three types of information requirements determination uncertainties
   2.1. Availability of a set of requirements.
   2.2. Ability of users to specify requirements.
   2.3. Ability of system designers to elicit and specify requirements.
3. Evaluate the combined effect of the information requirements determination process uncertainties on overall requirements volatility.
4. Select a primary information requirements determination strategy.
5. Select a set of specific methods to implement the primary requirements determination strategy.

Figure 6.4 illustrates the use of these steps to identify an appropriate mix of information requirements identification strategies. The uncertainty associated with information requirements determination, that is to say the amount of information imperfection that exists in the environment for the particular task, influences the selection from among the four basic strategies as indicated in Figure 6.5. The factors that influence such information imperfection include

1. Stability of the environment.
2. Stability of organizational management.
3. The previous experience of system users with planning, design, and use of the type of system under consideration.

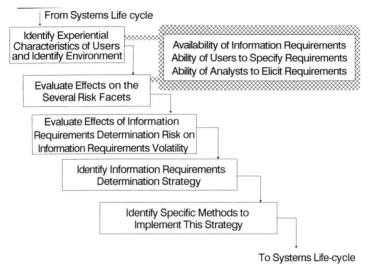

Figure 6.4. Identification of Information Requirements Determination Strategy

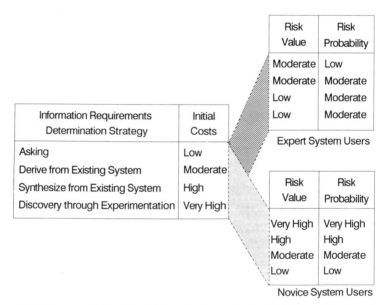

Figure 6.5. Influence of Experiential Familiarity on Desirable Information Requirements Strategy

4. The extent to which the present system is appropriate.
5. The extent to which a change in information requirements will change the usage of present-system resources and thereby degrade present operationality functionality.

Taken together, these enable selection of an approach to determining information requirements that copes with three essential contingency dependent variables due to risk elements in requirements determination.

Information requirements determination must lead to specifications for design and fielding of the system. A purpose of the system definition phase of the life cycle is to determine possibilities of insufficient and/or inappropriate information, that is, information that is sufficiently imperfect so as to make the risk of an unacceptable design too high to be tolerated. The requirements elicitation team should be able to determine the nature of the missing or otherwise imperfect information and suggest steps to remedy this deficiency.

This suggests that an appropriate approach to information requirements determination should enable detection, diagnosis, and correction of faults in both the process of obtaining the requirements and the actual requirements themselves. The type of information that the user of a particular system will wish to, or should, use is dependent on the contingency task structure. Information requirements determination should also aim to assist the systems engineering team in the ultimate fielding of the system. Satisfactorily coping with these needs should result in truly innovative and useful approaches to determining information requirements.

Rockart [15] has developed an approach useful for information requirements determination that is based on "critical success factors," that is, factors important to the success of a mission to be undertaken or to the organization. A number of contemporary applications of management science make use of these critical success factors or success attributes [16].

There have been several related investigations. Pinto and Prescott [17] asked 408 project managers to rate the relative importance of ten critical success factors in each of four phases of a system life cycle—conceptualization, planning, execution, and termination. The ten critical success factors [18, 19] are as follows:

1. Initial clarity of objectives and general guidelines that determine project mission.
2. Top management support in terms of willingness to provide both the necessary resources and the necessary authority.
3. Client consultation and communication, including listening.
4. Project schedule, including detail specification of the work breakdown structure necessary to accomplish identified missions.
5. Personnel recruitment, selection, and training for the project.
6. Access to required technological expertise.

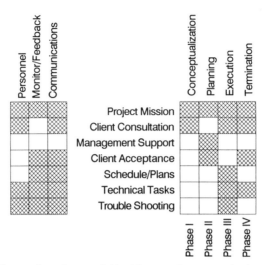

Figure 6.6. Interactions Among Critical Success Factors and Life-Cycle Phases

7. Marketing to insure client acceptance of and enthusiasm for the fielded system.

8. Monitoring and feedback to insure appropriate configuration management throughout the project.

9. Communications and networking to insure that all stakeholders associated with the fielded system are provided with necessary information.

10. Trouble shooting and crisis management to enable various risks to be controlled appropriately.

Figure 6.6 illustrates our interpretation of the interaction results determined in this study. It is particularly interesting to note that project mission is a major critical success factor across each of the phases. Personnel turned out not to be a significant success factor at any life cycle phase.* The two most critical efforts in the conceptual phase of the life cycle are project mission and client consultation. In the planning portion of the life cycle, project mission, top management support, and client acceptance are the most important factors. Factors 8 and 9, monitoring, feedback and communications, are not included because each is closely correlated with the other eight factors and therefore could not be measured independently in this study.

*There seems to be some difficulty here with respect to confounding of factors. It is relatively clear that if all of the nine critical success factors are present except the personnel factor, the overall project would be a success. But, it is difficult to believe that all of these other factors could be present in any large amount without having a suitable and capable set of project personnel. This is the same argument that was apparently used to justify elimination of the monitoring and feedback and communications factors from consideration in this study.

Figure 6.7. Possible Hierarchical Structure Among Critical Success Factors

The sublimation of client consultation to client acceptance that occurs between phases 1 and 2 is probably reflective of the fact that the client is proactively in search of a remedy to a perceived problem in the conceptual or requirements determination phase of the effort. However, some of this initial need diminishes at the time of conversion of the requirements into architectural specifications and the development of project plans in phase 2 of this life cycle. In phase 1, it is absolutely essential for the vendor to consult with the client about what is expected from the system. Once agreement has been reached, the technical staff of the vendor takes control of the project, and highly interactive consultation becomes less important. On the other hand, the vendor needs to keep the customer closely informed as to how the unfolding system will meet the customer's needs.

Shown in Figure 6.7 is a hypothetical tree structure of the ten critical success factors leading to mission success. Although this model does not provide an approach that will directly enable identification of information requirements, it does offer an excellent analysis of project management needs, and therefore includes indirect use in determining information needs.

Information requirements determination has been a subject of much recent interest in software systems engineering [20–22], as well as in other areas [23–26]. Researchers in this area have come to much the same conclusions as obtained in other systems studies. *Humans often have great difficulty in properly specifying requirements in terms of verbal discourse or unstructured paragraphs of natural language.*

This may lead to a significant ambiguity between information and subsequent system specifications [27]. Daft and Lengel [28] make an important related observation when they state that uncertainty and equivocality are two major forces that influence information processing in organizations, and that the major problem is not the lack of data but the lack of clarity in the data that is available. A major difficulty here is that identified requirements only become imbedded into a fielded system after several intervening phases of the systems life cycle. When a fielded system exhibits dysfunctional performance due to *inadequate* requirements, the burden of upgrading the system into an acceptable one is often quite substantial. Therefore, the need for high-quality effort in the initial phases of the systems life cycle cannot be overemphasized.

Many identification requirements identification objectives can be stated. Some of the most important are as follows:

1. Define the problem, or range of problems to be solved, including identification of needs, constraints, alterables, and stakeholder groups associated with operational deployment of the operational system or product that will be fielded.

2. Determine objectives for the operational system or the operational product that will be fielded.

3. Obtain commitment for system design and development from the user group and management.

4. Search the literature and seek other expert opinions concerning the approach that is most appropriate for the particular design and development situation extant.

5. Determine the estimated frequency and extent of need for the system or product to be fielded and the requirements for humans to be able to successfully interact with it.

6. Determine the possible need to modify the initial system design concept and the resulting operational system in order to meet changed user or customer requirements.

7. Determine the degree and type of performance and quality expected from the system.

8. Estimate expected operational effectiveness improvement, or benefits, due to the use of the system.

9. Estimate the expected costs or work breakdown structure of fielding the system, including design and development costs, operational costs, and maintenance costs.

10. Determine typical planning horizons and periods to which system to be fielded must be responsive.

11. Determine the extent of tolerable operational environment alteration due to use of the system to be fielded.

12. Determine which particular systems engineering life-cycle process appears best and the formulation, analysis, and interpretation algorithms that best support all phases of this life cycle.

13. Determine the most appropriate roles for the system or product to perform within the context of the operational environment under consideration.

14. Estimate potential leadership, education, training, and aiding requirements for use of the final system.

15. Provide for these education, training, and aiding requirements for use of the operational system.

16. Estimate the qualifications required of the systems engineering team that will field the system.

17. Identify operational cost and operational effectiveness evaluation plans and criteria for the system to be fielded.

18. Determine political acceptability and institutional constraints affecting fielding of the system.

19. Document analytical and behavioral specifications to be satisfied by the systems engineering process life cycle and the operational system.

20. Determine the extent to which the user or customer group can and will require changes during and after system fielding.

21. Determine potential requirements for availability of the systems engineering contractor after completion of development, operational tests, and complete fielding of the system.

22. Identify a specific set of systems management or management controls that are responsive to total quality requirements.

23. Develop the system level requirements and specifications for process and product that ensure adherence to the resulting total quality management standards and that ensure sustainability.

Many of these objectives relate to user level requirements specifications. Others relate to the system or technical level specifications. The primary products are a set of user level requirements specifications and a set of system level specifications. These are intended to lead to high-quality management prescriptions for the specific effort to be undertaken.

Here, we have discussed a number of approaches relative to information requirements determination. Our present discussion emphasizes the importance of both information requirements and risk management as inherently necessary facets of system engineering efforts. Next, we examine some risk management concerns. Following this, we will discuss several approaches to support information requirements determination, risk management, and total quality systems efforts.

6.3. RISK MANAGEMENT

In this section, we examine a number of relevant aspects of risk, including its effect on system quality and information requirements.

6.3.1. Risk

Certainty is rarely encountered in choice-making situations and *risk* may be thought of as lack of full control over the outcomes of a particular course of action. Stated somewhat differently, risk reflects lack of full information about such an outcome. Risk management should make explicit the often hidden trade-offs inherent in much decision-making activity. In the case of the public sector, such trade-offs often involve regulation, standard setting, legislation, environmental protection, economic efficiency, technological choice, development sustainability and individual liberty. Risk management should also seek improved methods for evaluating and judging these trade-offs.

The traditional approach used to model risky situations employs probability distributions of the outcomes of alternative choices and the associated utilities of these outcomes. Risk and gamble are, therefore, essentially equivalent expressions. Slovic and Lichtenstein [29] have proposed that a gamble can be described by its location on four basic risk dimensions:

1. Probability of winning.
2. Amount to win.
3. Probability of losing.
4. Amount to lose.

These basic dimensions are assumed to be integrated into a contingency structure for decision making. Decision making, in this case and from this perspective, is a form of information processing. There are a number of other frameworks for the description risk-oriented choice making, including mean-variance models, in which risk preference is described in terms of mean risk and risk variance [30].

A *risk unit* could be defined as the probability per unit time of the occurrence of a unit cost burden. In this sense, risk represents the statistical likelihood of being adversely affected by some potentially hazardous event. Thus, risk involves measures of the probability and severity of adverse impacts. Safety may be said to represent the level of risk that is deemed acceptable.

By risk to an individual, we refer to the possibility that an individual will be seriously impaired. This may result from either the hazards of normal living, such as being struck by lightning, or accidents, such as being shot by a hunter. There are voluntary risks that individuals elect to assume, such as those due to one's own smoking, and involuntary risks that individuals do not elect to assume, such as those due to a nearby power plant or being forced to inhale the smoke of others.

Risk issues fall into several categories, which are not at all mutually exclusive [31].

1. Technically complex risks that are easily comprehensible only to those very educated in the specific technology.

2. Risks that can be significantly reduced by applying an appropriate technology.

3. Risks that constitute public problems and whose technical components need to be distinguished explicitly from their social and political components, so that responsibilities may be assigned properly.

4. Risks whose possible consequences appear so grave or irreversible that prudence dictates urging of extreme caution, even before the risks are known precisely.

5. Risks that result from technological intrusions on personal freedom that are made in pursuit of safety.

This list has a clear focus on public sector risk issues. Public sector decision making involving low-probability but high-hazard possible consequences have a number of particularly interesting and difficult aspects [32]. These include aggregation of preferences, limited knowledge, uncertainty, irreversibility, intergenerational effects, distribution of benefits versus risks, counter risks, and second-order intended and unintended consequences.

Risk factors affecting the private as well as public sectors include the following.

1. Market structure of a particular industrial sector.

2. Impact of financial institutions and their linkages to industrial organizations.

3. Prestige and status identification with "advanced" technology development.

4. The role of public sector management in nurturing emerging technologies in the private sector.

5. The risk management strategies and tactics of decision makers.

These factors vary across the nations of the world, and across industrial groups within a given nation. Thus, we note the complex nature of the trade-offs involved in issues involving risk and hazard.

There appears little doubt that risk-associated decision making is primarily influenced by responses to two questions:

1. What are the possible event outcomes and their valuation?

2. How likely is the occurrence of the various event outcomes?

An economically or technologically rational theory of choice typically involves giving numerical measures to these probabilities and values and aggregating these numbers into a single index or merit, which a rational actor is supposed to maximize (this approach is examined in Chapter 7). Therefore, an economic or technologically rational approach to choice making can conveniently be divided into two parts:

1. The proper way to measure probabilities of outcomes and the value measures to be associated with these outcomes.
2. The proper way to combine these measures.

In classical gambles, outcomes are simply different amounts of money. These gambles may be modeled in terms of urn models or random number generators with easily determined objective probabilities. Very early theories ignored the value measure part of choice making. Monetary value and objective probabilities were accepted as appropriate measures, and effort was concentrated on combining these to create statistically probable monetary returns. This model is of great importance in many applications but is inappropriate in many other cases. Expected monetary value is the basis of many models of economic rationality.

The expected utility model, which is of central importance in modern decision-making theories in systems engineering, differs from the expected value model in two respects. First, it substitutes subjective for objective probabilities. Second, it replaces monetary values with the decision maker's subjective utility for outcomes and risk bearing attitude. The rule of combination, however, is still the same. Subjective utility models form the basis for contemporary technological rationality, a subject discussed in Chapter 9.

Risk perceptions depend upon the type of risk, the scope of the risk, which may vary from local to global and from a single individual to an entire society, and the effect of the risk. Risk effects may be completely reversible, perhaps at a very large cost, or totally irreversible, and still perhaps with very large costs. Risk measurement and risk management are complex tasks. It is highly desirable to develop a methodology for risk assessment, and associated choice making, that allows and encourages explicit risk management and such desirable outcomes from risk management as sustainable development.

6.3.2. Risk Management Methodology

As we define it, *risk management** is an approach to managing that is based on identification and control of those areas and events in the systems engineering life cycle that have the potential for causing unwanted change in either the process or product. Risk management requires developing optional plans for managing risk situations that might arise, and then implementation

---

*It is important to define risk. For our purposes, *risk* is the probability or likelihood of injury, damage, or loss in some specific environment and over some stated period of time. Thus, risk involves two elements, probability and loss amount. A related concept is reliability. *Reliability* is the probability that a product or system will perform some specified end user function under specified operating conditions for a stated period of time.

of the appropriate corrective action plan when a risk has eventuated and this is detected and diagnosed as to the cause of the risk.

In developing plans for risk management, we

1. Identify generic risk situations which might arise.
2. Analyze these generic risk situations.
3. Interpret the results of the analysis to enable risk handling through selection of an appropriate risk abatement or risk management tactics.

The second of these efforts involves risk handling at the operational level, or operational risk management of any evolving systems product. In our risk management efforts, we

1. Identify the appropriate risk management operational plan.
2. Execute this risk management plan to detect any developing risks, and diagnose risks according to the plan.
3. Monitor and control the results by taking corrective actions, and adjust the system acquisition plan accordingly.

Figure 6.8 illustrates these risk management steps and Figure 6.9 illustrates risk management within the overall systems engineering life cycle.

We can easily identify several different types of risk in fielding a new system.

1. *Performance risk* results when the fielded system operates in such a manner as to create hazards or poor operating properties. It would be realistic to call this *technical performance risk*. We will do this, and do

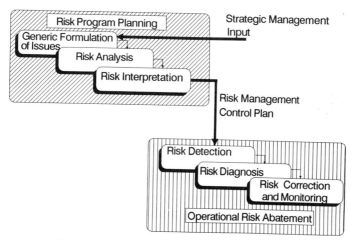

Figure 6.8. The Two Phases of Risk Management

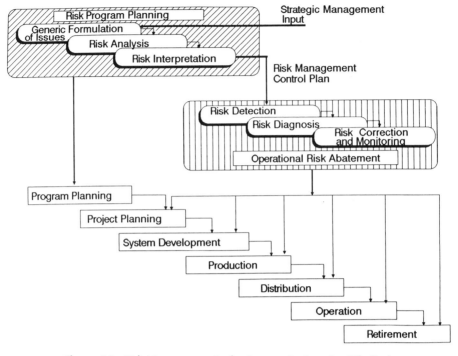

Figure 6.9. Risk Management in the Systems Engineering Life Cycle

not intend to obscure the reality that a technical risk may result in societal, or other, harm that occurs because of poor initial forecasting of technological impacts.

2. *Acquisition schedule risk* results when the intended time for delivery of a fielded system needs to be migrated to some later time.

3. *Acquisition cost risk* results when the anticipated cost of fielding the system increases beyond that forecasted.

4. *Fielded system supportability risk* results when the operational system is unsupportable by planned maintenance efforts.

5. *Programmatic foundation risks* are those created by events outside the formal control of the systems management process.

These risk elements pertain to the product, but are significantly influenced by the life-cycle process used in fielding the system. This is especially the case with programmatic foundation risk, which often results from changes in the external environment. At first glance these facets may seem to ignore some of the many risk considerations indicated earlier. However, this is really not the case, as performance risk, programmatic risk, or fielded system supportability risk generally are caused by one of the aforementioned risk considerations.

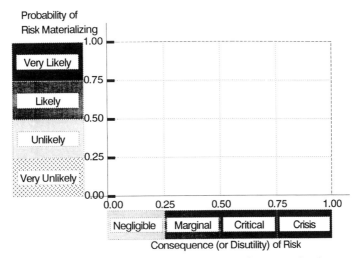

Figure 6.10. Fundamental Two-Dimensional Nature of Risk

It is important to note that technical risk, programmatic risk, and support risk are generally the cause of schedule or cost risks. It is these latter two risks that are most often detected in practice in managing system product development. It is important to be able to diagnose the cause of the detected risk, in order to implement corrective risk management efforts. We also note that these risks may result from causes internal to a given project or external to it. Also, the risks may be long term or short term in nature, and the magnitude of their impacts may be quite varied.

In many instances, it is possible to associate a probability of occurrence with a risk. The probability of a given risk may vary from 0, in which case the risk cannot eventuate, to 1, in which case the risk is certain to materialize. In addition, a given risk can have virtually no impact or be catastrophic. The importance of a given risk element clearly depends upon a blend of the probability of it being realized and the impact that it can have if it is realized. Figure 6.10 illustrates a matrix of risk importance in terms of the probability of the risk materializing and the criticality of the risk element itself.

This leads us to the definition that the *risk associated with an event* is a combination of the probability of the event occurring and the significance of the impact of the occurrences of this event. A risky event is generally thought of as being undesired. We may speak, for example, of the risk of having an automobile accident. But, we can also speak correctly of the risk of winning a lottery, even though this would not be commonly done. If this seems inappropriate, we can always talk about the risk of not winning the lottery. The term *risk of an automobile accident* involves both a risk probability and a risk impact (no pun intended). The risk of a such a negligible consequence accident as scraping a fender against a curb is moderately high. The risk of such a catastrophic accident as a head on collision with both parties traveling at 65 miles per hour is much less.

The term risk is commonly used in several, not fully compatible, ways. Our statement that "risk represents the statistical likelihood of being adversely affected by some potentially hazardous event" may suggest that a risky undertaking is *always* undesired. This is not correct, however. A decision that involves risk is not necessarily an undesirable decision. The risk outcome is bad, but not necessarily the decision itself. We need to carefully distinguish between good and bad decisions and good and bad outcomes. Much more will be said on this in this chapter and in the rest of this text.

Risk events or risk outcomes generally deal with events that are hazardous or otherwise undesirable. Some such events are generally understood to be undesirable in and of themselves. Examples of this would include an outbreak of typhoid fever, an automobile crash, a hurricane, a bolt of lightning, or dying from cancer. In these cases, one generally refers to the risk *of a particular event occurring*. It would be uncommon, but not impossible, to imagine an individual making a conscious decision to have typhoid fever, or to get hit by lightning, or to die from cancer. Such a person would then need to undertake activities that would, with reasonably high probability, result in the desired (although generally undesirable) outcome.

Alternatively, risk may be associated with a decision that is not in and of itself undesirable. This might include, for example, a delicate brain operation, a bold amphibious military operation, or introduction of an incompletely tested new product into the marketplace. In this case, we generally refer to the risk of a decision or *decision under risk*. Figure 6.11 provides a conceptual illustration of how we might speak of this combination of risk event probability and risk event magnitude. Low risk, moderate risk, high risk, and crisis are the terms used here. There are others, and the assumption and use of four risk categories is somewhat arbitrary.

The risk event concept necessarily involves probability of harm and amount of harm. This makes it difficult, but not at all impossible, to determine a single number that can be used to characterize or measure risk. If we were to use a single number to represent risk, we would be unable to distinguish whether we were observing a situation in which there were a very small (nearly zero) probability of a very large loss, or a very large (nearly 1.0) probability of a very small loss. There are also other important considerations as well. For example, we may be much more reluctant to accept a 0.0001 probability of loss of life tomorrow than a 0.0001 probability of loss of life some time over the next 20 years. This just says that consequence must be associated with its evolution over time. Although such single number measures are possible, and we illustrate how to obtain and use them in Chapter 7, the other important considerations cannot appropriately be neglected.

We note that there are at least two opportunities for imperfect information that are associated with risk. We may lack precise knowledge of the

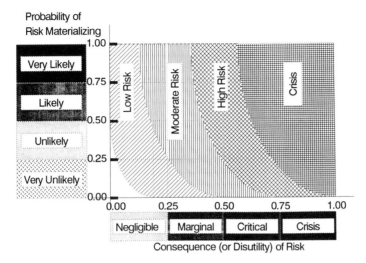

Figure 6.11. Risk Management Categories

probabilities of the various risky events materializing. In addition, we may not have precise knowledge of how individuals value the impacts of these risks. As a result, it is often difficult to identify the precise contours of the curves in Figure 6.11 illustrating the various risk categories. Clearly, we know that we are in trouble if there is a very high probability of occurrences with very significant damage. Similarly, we know that there are few difficulties associated with events with a low probability of occurrence and negligible damage. But what about events for which there is a low probability of truly catastrophic damage, or events with moderate probabilities and moderate damage? How much management effort is warranted to alter these risks? These are significant questions, and ones that we wish to examine here.

As we have noted, systems engineering risks are generally reflected in the failure to deliver a product that is trustworthy, the failure to deliver the product on time, or for that matter at all, or the failure to deliver the fielded system within specified costs. Failures can occur with respect to any of the three of these. The product may not be trustworthy.* It may not be delivered on time. It may not be delivered within allocated costs. Obviously, these are not mutually exclusive possibilities. It does seem that they are collectively exhaustive. Figure 6.12 illustrates the five facets of risk management identified here and the risk management activities that are to be used to control them.

---

*We use trustworthy in a general sense to indicate a product that pleases the customer. A product may fail to be trustworthy because it does not perform as expected, or is not reliable, or is not maintainable, or sustainable over time or for a large number of other reasons.

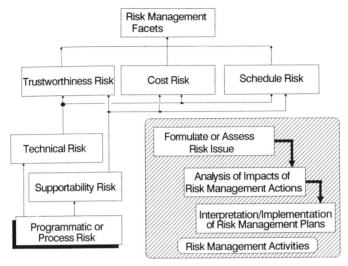

Figure 6.12. Risk Management Facets and Activities at Process and Product Levels

In the context of systems engineering, there are five fundamental factors of risk.

1. *Technical Risk:* Risk associated with fielding a new technology, or a new design. Usually a new technology is used in the hope of providing enhanced performance in the same environment as a present technology or design, or provision of the same system performance level in a new and more challenging environment. Fundamentally, technical risk results from the demand for "greater" performance. Generally, technical risk results from immature technologies or lack of understanding of the impacts of using a given technology that is presumed to be mature. Ordinarily, a product based on an emerging technology involves more risk than one based on an established technology. Technical risk is associated with all elements of product trustworthiness, including functionality, reliability, availability, and maintainability. It is also associated with the fielded system causing harm and becoming unsustainable. This is, of course, a functional deficiency. Technical risk exists at all phases of the systems life cycle and should be addressed by appropriate systems management efforts throughout the systems engineering process.

2. *Acquisition Schedule Risk:* Results from poor forecasting of the time necessary to field a system or produce a product. Of course, this risk can materialize because of influences outside the control of the systems development organization or causes external to it.

3. *Acquisition Cost Risk:* Results from poor cost forecasting. Again, this can be influenced by factors that are external or internal to the system

development organization. Also, technology risks can contribute to cost risks.

4. *Fielded System Supportability Risk:* Results when the system proves to be not fully supportable over time. This might result from the lack of an open systems architecture that makes it difficult or impossible to evolve the system over time. It could result from reliability or operational maintainability concerns. We discussed some of these issues in Chapter 4, and Figure 4.9 illustrates a life cycle including supportability and evolvability needs.

5. *Programmatic Support Risk:* Risk that is associated with activities and events formally outside of the sphere of influence of the systems process management but that exert an influence on the outcome of a given life-cycle process and the resulting systems engineering product. These deficiencies may result from a planned life cycle that is deficient for the development environment or from such necessary alterations in the life cycle as those due to labor strikes, requirements volatility, systems engineering contractor stability, or other causes. There are a number of possible programmatic risks. Among these are the following:

    5.1. Judgments and decisions made by others, often at higher authority levels, that directly affect the system development program.

    5.2. Environmental events that affect the system development program, even though they are not specifically directed at it.

    5.3. Unanticipated changes brought about by imperfect systems management capabilities, such as poor forecasting of design needs or poor transitioning from customer requirements to technical specifications for the system to be produced.

It would easily be possible to further expand on this listing of programmatic support risks. For example, an inexperienced systems management team may fail to anticipate problems in a system under development. Even though this is an environmental risk, it is brought about because of a deficiency in systems management. This leads to technical risk as well. Programmatic support risk, then, includes those risk occurrences caused by events external to the systems engineering organization, events within the organization but external to that part directly responsible for the system under development, and unforeseen problems with the process or the system under development.

As indicated, these five facets of management risk are not at all mutually exclusive, and the occurrence of one risk facet can lead to other risks. In many cases, cost and schedule risks are created when other risk eventuate. Often, the realization of programmatic or environmental risks result in technical and/or supportability risks which, in turn, lead to cost and schedule risks. The realization of a technical risk, such as poor- or low-quality system performance, often leads to programmatic changes that affect system production. Thus, there are causal feedback loops among these risk facets. Figure 6.13 illustrates some typical causal linkages among these risk facets. It is

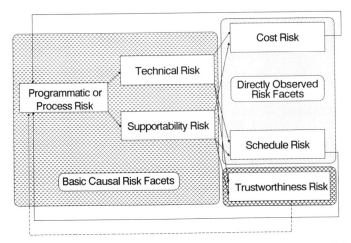

Figure 6.13. Typical Causal Linkages Among Systems Management Risk Facets

reasonable to ask whether a second-order problem such as a cost overrun, that arises from a primary risk facet, say a technical risk, actually a risk in itself? It might be better to call these "associated risks." The practical point is that it is not feasible to calculate the probability or cost of such associated events without reference to the probability of other more primary and causative events. This creates a need for a structuring tool that will support influence relations among probabilistic events.

If systems management controls are each very well conceived and executed, there should be no risks that are directly due to cost, schedule, supportability, or technology risk facets. That is to say, there should be no internal causes of risk. But, external or environmental risks may still exist and these will potentially result in programmatic support risks. These, if they materialize, may well result in realization of other risk facets. In particular, it may lead to very difficult to observe trustworthiness risk issues. This suggests that a major task for proactive systems management is that of determining the root causes of risk, and developing appropriate risk management efforts.

Of course, we never know all we need to know in order to manage risk completely—that is to say eliminate it. And even what we do know loses its value over time, as circumstances invariably change. While changed circumstances may represent an improvement, prudent vision generally suggests that we be prepared for the worst and develop appropriate contingency plans should it occur! We must simply accept the fact that judgments regarding risk must be based on imperfect knowledge. Risk is inherently a present and future oriented concept. There are no past risks. The past has occurred and thus is factual, even though we may not know all of the facts about past history. Thus, lack of historical knowledge may well introduce present and future risks.

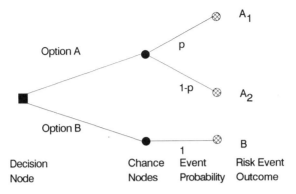

Figure 6.14. A Fundamental Model of Risk Events

One way to approach risk is through the decision tree illustrated in Figure 6.14. A simple example illustrates how such a tree can be used to model a real situation. Suppose that I am becoming a bit unhappy with the speed with which I am able to complete word processing documents on my computer. There is a chance that purchase of a new computer with a faster microprocessor will resolve the speed difficulties and enable more efficient and effective preparation of this book. On the other hand, the problem may be due to my software or my printer, and my personal limitations. Thus, if I buy a new computer, the improvement in speed may be worth the cost, but there is a distinct possibility that the new system will not improve my efficiency or effectiveness in any meaningful manner.

The decision tree in Figure 6.14 presents a risk-free option, that of keeping my present product. There is a risky option, that of purchasing a new product, which may produce strong acceptance by me, the customer, because of increased productivity, or may only achieve weak personal (customer) satisfaction, because it does not increase my efficiency and/or effectiveness. In this case, I have suffered the loss of $3,000, the purchase price of the new machine. Even if I purchase the new machine and it is useful in the short run, there is a chance that it will turn out to be of low quality and will fail early in its lifetime. So, there are some uncertainties due to the nature of the product itself. Thus, we see the need for both current and future information on a variety of topics to make an informed and most appropriate decision.

We see that there are indeed a large number of risk management ingredients and that Figure 6.12 has indicated only the top level risk management facets. There have been many studies of how project and process management relate to product success, and they may enlarge our perspective on risk possibilities.

In one such study, Deutsch [33] identifies a number of factors associated with technical performance, business project adversity, intrinsic management power, and operational level management. His hypothesis is that the project adversity factors and management control influence operational level management, and this in turn influences both business performance and technological system performance. This conclusion seems self-evident. The major research issue investigated, however, is the extent to which it is possible for quality management control to overcome intrinsic development program adversity factors. The factors identified by Deutsch, with project adversity being a primary risk factor, are as follows

1. Technical Performance Factors
   1.1. *User Satisfaction:* The degree to which the users of the system are satisfied by system performance.
   1.2. *Requirements Achieved:* The degree to which the requirements for functional, external interface, operational, and quality performance are satisfied.
2. Business Performance Factors
   2.1. *Cost Performance:* The variance between projected and actual system costs.
   2.2. *Schedule Performance:* The number of months slippage in key schedule milestones.
3. Project Adversity Factors
   3.1. *Project Size and Character:* Complexity of system.
   3.2. *External Interface Adversity:* Complexity of interactions with the surrounding external environments.
   3.3. *Business Constraints:* Difficulty of setting realistic schedule projections.
   3.4. *Technology Development:* Degree and extent of new technology development that is required in order to field a satisfactory system.
   3.5. *Number of Users:* Number of ultimate users, and diversity of their objectives for the system.
4. Intrinsic Management Control Factors
   4.1. *Three-Party Dialogue:* Thoroughness of coordination among user, customer, and contractor.
   4.2. *Requirements Priorities:* Degree of consensus among user, customer, and contractor regarding priority of technical requirements.
   4.3. *Technical Scope Definition:* Clarity, scope, and stability of technical requirements.
   4.4. *Risk Mitigation:* Risk mitigation measures taken before a commitment to full-scale development.
   4.5. *Risk Monitoring:* Thoroughness of risk management processes.
   4.6. *External Interface Control:* Measures to facilitate interaction with elements external to the system.
   4.7. *Multiple Users Reconciliation:* Reconciliation of potentially conflicting technical specifications among user organizations.
   4.8. *Personnel Resources:* Availability and quality of project personnel.

4.9. *Technological Resources:* Quality and amount of technological resources assigned to the project.

4.10. *Strategic Planning:* Scope of strategic planning efforts throughout the systems engineering life cycle.

4.11. *Management Control:* Scope of the operational level systems management measures.

5. Operational or Task Management Factors

5.1. *Business Risk Management:* Degree to which management neutralizes the effects of unrealistic cost and schedule projections.

5.2. *Technical Risk Management:* Degree to which management neutralizes the risk of new and emerging technology developments.

5.3. *External Interface Management:* Degree to which management neutralizes the effect of uncertain system interfaces with external elements.

5.4. *Multiple User Needs:* Degree to which management neutralizes the effect of potentially conflicting requirements from multiple user organizations.

5.5. *Problem Scope Management:* Degree to which management neutralizes the effects of project size, scope, and complexity.

5.6. *Planning, Feedback, and Control:* Degree to which management integrates strategic planning with operations.

The results of this interesting study provide a relatively strong indication that the system acquisition technical and business performance factors are significantly influenced by the identified systems management factors. It turned out that business risk management and problem scope management were the two factors most heavily correlated to performance, but not by any large amount, for the specific software development projects considered here. Two major risk parameters were determined to be the degree of technology development required for a successful system acquisition effort, and unrealistically optimistic cost and schedule projections. In a more generic sense, many of the factors identified in this study are one of the five fundamental risk facets noted earlier: technical risk, acquisition schedule risk, acquisition cost risk, fielded system supportability risk, and programmatic support risk. Clearly, identification of trustworthiness factors is desirable. Clearly also, the strategic management approach used for quality assurance strongly influences these.

6.3.3. Assessment of Risk Management Factors

We can speak of risk in two ways: the risk associated with a specific course of action and the risk associated with a specific outcome that may follow from selection of an alternative course of action. This relates to the need to discuss, and distinguish between, decisions under risk, and risk events or outcomes. Risk assessment may be defined as the procedure in which risk considerations are formulated, analyzed, and interpreted in order to incorporate risk management into decision making. Risk assessments may be performed at an individual, group, organizational, or societal level. A risk assessment process may vary from highly intuitive to very formal and analytical in nature.

The overall risk assessment process may be partitioned into two levels, as we have discussed. The first supports the development of risk program, or risk management, plans. This is a systems management or management controls function. The second supports the actual implementation of these plans for purposes of risk abatement. This is an operational control, or task level, function. Each of these levels involves the formulation, analysis, and interpretation of risk issues.

Risk management, in turn, can be disaggregated into components involving risk avoidance, in which the desired degrees of risk reduction and risk avoidance are determined, and risk acceptance, in which an effort is made to ameliorate the effect of a given risk. It is at the systems management level that these plans are developed, and at the risk abatement level that generic plans for risk alleviation are implemented.

Our suggested approach to the assessment of risk management factors is based upon the three primary steps of systems engineering. As we know, these three steps involve the formulation, analysis, and interpretation of the impacts of action alternatives upon the needs and the institutional and value perspectives of stakeholders. Each of these three steps is important in obtaining an understanding of the effects of risky or potentially hazardous activities and event outcomes.

In the risk impact formulation step, a number of elements representing risk factors of a technological innovation are identified, including needs, constraints, alterables, objectives and potential measures of these, and activities and activity measures. In the analysis step, we forecast the hazards and other consequences of the risk event under consideration. This includes the estimation of the probabilities of occurrence of hazardous outcomes, the impact magnitudes associated with these outcomes, such that subjective interpretation of these in terms of utility becomes possible. Many methods are available, including cross-impact analysis, interpretive structural modeling, economic modeling, cost–benefit analysis, mathematical programming, optimum systems control, and system estimation and identification theory. A large variety of systems engineering tools are available and a complete description would require an encyclopedia [34], and more! This is especially so when we recognize the interaction that exists between risk assessment [35, 36] and the diverse areas of technology forecasting and assessment [37] and business and public policy. An excellent overview of these areas, together with a number of systems analysis methods and procedures for application to the forecasting and management of technology was presented in a recent text by Porter et al. [38].

After formulation and analysis of the risk impacts, we attempt to interpret these impacts. The risk interpretation step specifies individual and group utility for the impacts of a potential system or technological innovation. Finally, choice-making and plans for action are formulated, based upon these interpretations. While estimating risk probabilities, the risk analyst must be careful not to "change" the respondent's perspectives, so as to destroy existing perspectives, create new perspectives, or deepen existing perspectives. Other problems in expressing risk occur when converting risk to cost of risk, assessing the risks of environmental catastrophe, and eliciting estimates of probabilities. A structural model, generally in the form of a hierarchy or tree structure, or in an influence form convertible to a tree structure, provides the framework within which risk probabilities are estimated. The risk impact assessment model is no better than this structural model and aids are very useful to assist in determining and expressing this structure.

There are many risk management factors that lead to information imperfections. These include human error of various types, improper forecasting of technological impacts, overconfidence relative to technological progress and organizational capabilities, and failure to anticipate environmental and/or cultural changes. Each of these affect the interpretation of risk at either the management control or task control level. At the strategic level, effort is directed at a risk program plan that will ameliorate, to the extent possible, both the likelihood of risks materializing and the impact of those that do eventuate.

Among the characteristic ingredients that influence the interpretation of risk events are the following.

1. *Type of Consequence:* Economic, social, technological, organizational, ethical, political, environmental, etc.
2. *Scope of the Risk:* Voluntary or involuntary, time horizon, spatial or geographic distribution, equity distribution over individuals and groups, controllability, observability, etc.
3. *Probability of the Event:* Including both the likelihood that a given event will materialize and the confidence that this estimate of probability is sound.
4. *Magnitude of Impact of the Event:* Including an assessment of the likely influence that is felt, including the distribution of this impact over time and space.
5. *Propensity for risk taking:* Individual, organizational, or societal.

Embedded in these characteristic ingredients are the need for specific risk formulation, analysis, and interpretation efforts, if we assume that risk planning has been accomplished. One of the potential advantages of a systems engineering approach to risk assessment and risk management is the ability to disaggregate these factors into several strategic steps outlined in Figure 6.13 and below. These steps involve:

1. Generic risk issue formulation, including some attempt at structuring the various risk management issues and the identification of alternative risk management plans.
2. Analysis of generic risk management plans, including the determination of probabilities of outcomes for alternative courses of action and the associated impact magnitude of hazards.
3. Interpretation of these alternative risk management plans, so as to enable selection of a specific risk management plan for implementation.

When risks are assessed, attention is often focused on some particular dimension of risk impact believed to be of preeminent importance for the risk abatement decision. For example, people with very little money and a great fear of losing it in a particular decision situation may focus attention on the amount of money that could be lost and base their decision almost exclusively on this aspect, largely disregarding other information provided. They may choose an option that results in a sure outcome of $1,000, rather than accept a risky alternative which might result in an outcome of $10,000 with a probability of 0.95 and an outcome of $0 that is associated with a probability of 0.05. We examine some risk taking behaviors in Chapter 7 after we have provided a background in decision analysis. Beliefs about the relative importance of a few risk aspects may derive from previous experience with a similar risk or from a logical and rational analysis of the decision task, or even from quite irrational fears or prejudices. Whatever may be a person's beliefs about the relative importance of various risk aspects, our argument is that appropriate risk management techniques involving planning and action should be employed.

When individuals or organizations "voluntarily" take risk, regardless of whether they are for personal pleasure or organizational and competitive advantage, they sometimes appear willing to accept rather high risk levels for rather modest benefits return. Under other circumstances, just the opposite is true, and people avoid even risks that involve very low probabilities of small losses. There are two parameters affecting choices that help explain this phenomenon. The first is perception of the ability to personally manage risk-creating situations. In involuntary situations, when individuals or organizations no longer believe they can personally control or influence their risk exposure, the associated risk exposure is perceived to be more significant than if it were a voluntary risk [39].

A second major aspect is the way in which the risk and associated decision situation is framed. This depends greatly upon the model of the risk situation used for judgment and choice. We examine some framing models in Chapter 7. Here we provide some overall perspective for these models. One of the typical results of a formal risk impact analysis is a model that indicates possible outcomes of the events under consideration. The outputs from this step serve as inputs to the interpretation step. There are three principal tasks in the interpretation step of risk assessment, regardless of whether we are at the level of risk management planning or risk management abatement.

1. Determine utilities of possible risk management activity outcomes.
2. Evaluate the effectiveness of proposed risk management activities in terms of these utilities and the probability of their being realized.
3. Select appropriate risk management action options that maximize the overall benefit, or subjective expected utility, and plan for implementation of these options (discussed in more detail in Chapter 7).

6.3.4. Need for a Theory of Risk Assessment

Our discussions in previous sections have indicated the necessity of informed choice making in situations involving risk outcomes. We have indicated that there are many different characteristics of risk, including the following.

1. Individual risks versus organizational or societal risks.
2. Voluntary risks versus involuntary risks.
3. High-probability, low-impact risks versus low-probability, high-impact risks.
4. Delayed impact risks versus immediate consequence risks.
5. Risks with low-value impacts for many versus risks with disastrous ramifications for a few.

The very different characteristics of risk complicate the process of determining suitable risk management plans and risk abatement efforts. Adding to these complications are issues concerning organizational goals, profit and survival, social equity and justice, development sustainability and decision-maker responsibility and ethics. Finally there are analysis concerns with problem structuring and obtaining sufficient and accurate information for decision making. This complicates considerably the notion of acceptable risk, which have been examined [40, 41] in considerable detail but which remain as major challenges.

Decisions by management related to risk often face difficulty in acceptance and implementation because of problems in communications. Covello, von Winterfeldt, and Slovic [42], based on their review of the risk communication literature, indicate four important problem areas that affect risk communications.

1. *Message Problems:* Due to the high level of technological complexity and associated data uncertainties.

2. *Source Problems:* Result from a lack of institutional trust and credibility, expert disagreements, and the use of technical bureaucratic language, which appears designed to obfuscate.

3. *Channel Problems:* Result from selective and biased reporting that focuses on sensational and dramatic aspects and contains inaccuracies and distortions.

4. *Receiver Problems:* Result from inaccurate and improper risk perceptions, overconfidence in the ability to avoid harm, unrealistic demands for certainty, and potential inability or reluctance to make trade-off decisions.

Four generic efforts are proposed to cope with these difficulties. These are information and educational programs, behavior change to encourage personal risk reduction behavior, disaster warnings and other emergency information to provide direction and behavioral guidance, and joint problem solving and conflict resolution so as to involve potentially affected stakeholders in risk management decisions.

Many questions exist pertaining to the proper role of systems engineers with respect to risk management efforts, as well as to the proper role of the scientist, the domain expert, the enterprise manager, and the public. Although risk management efforts cannot eliminate all risk, they can provide advice to clients on how best to allocate resources to maximize utility in the presence of risks.

6.4. SYSTEMS ENGINEERING METHODS FOR REQUIREMENTS, RISK, AND QUALITY MANAGEMENT

There are a relatively large number of systems engineering tools and methods that are appropriate for requirements identification and risk and quality management efforts. In this section, we briefly describe a few of these. Our emphasis is on those approaches that are most suitable for identification and communication of requirements, risk, and quality management efforts.

6.4.1. Network-Based Systems Planning and Management Methods

Network-based systems planning and management methods were developed in the late 1950s in response to the need for a systematic tool for planning and management of large-scale projects. For the most part, these methods enable project planning, scheduling, and controlling. Project completion schedules generally incorporate the order of activities, time constraints, and resource availability constraints. The most widely known of these approaches are the Program Evaluation and Review Technique (PERT) and the Critical Path Method (CPM). Details concerning these approaches are discussed in a rather large number of texts including some initial treatments of the subject [43–45] and some more modern works [46–49]. We describe salient features of these approaches here. These, and other methods that have been developed since, are based on three assumptions.

1. A large project can be disaggregated into a number of separate activities, also called tasks or jobs.
2. There is a particular sequence in which the tasks must be accomplished.
3. Time for completion or duration can be estimated for each activity.

Figure 6.15 indicates a typical listing of activities, precedence or sequence relations, and durations for a hypothetical project involving concurrent development of hardware and software. The three assumptions are fulfilled for this hypothetical project.

In a network management model, the identified precedence relations between various activities are displayed graphically in the form of a network. Often, this is called arrow diagraming. An arrow diagram displays precedence relations among the activities that comprise the network. The nodes in the network may represent various events, such as completion of an activity. The activities themselves would then be illustrated by the branches between nodes. Sometimes, it is necessary to insert dummy activities, or dummy branches, in order to avoid possible ambiguities concerning precedence. This is known as the *activity-on-arrow* representation.

Activity	Duration	Predecessor	Successor
a. Requirements	3	Begin	b
b. Specifications	3	a	h
c. Hardware Design	2	b	d
d. Breadboard Construction	3	c	e
e. Test Breadboard	2	d	f
f. Design Prototype Hardware	2	e	g
g. Test Prototype Hardware	3	f	k
h. Design Software	4	b	i
i. Code Software	5	h	j
j. Test Software	5	i	k
k. Integrate Sware & Hware	3	g, j	l
l. Operational Evaluation	3	k	m
m. Deploy System	3	l	n
n. Document System	2	m	o
o. Train System Users	3	n	p
p. Maintain System	6	o	End

Activity Duration and Precedence for System Fielding

Figure 6.15. Phased Activities for System Fielding Effort

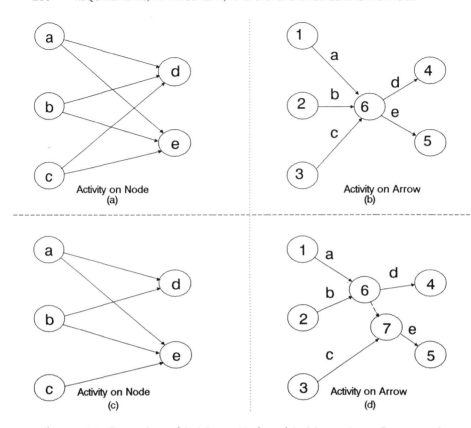

Figure 6.16. Comparison of Activity-on-Node and Activity-on-Arrow Representations

An alternate approach to the activity-on-arrow convention is known as the *activity-on-node* convention and is generally preferred. In it, the nodes represent activities and the arrows represent events or the time points that follow activity completion. Figure 6.16 illustrates simple sequences of activities and events for these two conventions. Although the diagrams of Figure 6.16a and b are basically the same, the activity-on-arrow convention will require the insertion of a dummy node in order that no more than a single arrow is needed to represent a single activity. A dummy node is also needed in the representation of Figure 6.16d to properly represent precedence orderings. In Figure 6.16c, completion of activities *a* and *b* is sufficient to enable the start of activity *d*, whereas completion of activities *a*, *b*, and *c* is needed to enable the initiation of activity *e*. This is the activity-on-node representation and does not require the insertion of a dummy node. Such a dummy node is needed in the equivalent activity-on-arrow representation. Dummy nodes are not needed in the activity-on-node representation, but

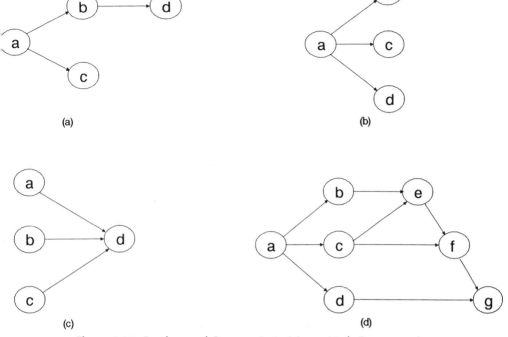

Figure 6.17. Fundamental Concepts in Activity-on-Node Representation

may be used for convenience or clarity in display of a complicated sequence of activities. We use the activity-on-node representation in our efforts to follow.

Nodes are used to represent activities or jobs. Directed lines connecting two nodes are used to represent the time events that occur upon completion of an activity. Several definitions are of interest for the activity on node representation. Figures 6.16 and 6.17 illustrate some of the following terms.

1. An *activity* represents some defined sequence of jobs or tasks that is necessary to complete a project. For convenience, we generally use lower case italic letters to represent activities.
2. A *predecessor activity* is an activity that must *immediately* precede an activity being considered. In Figure 6.17a, activity *a* is a predecessor to *b*, and activity *b* is a predecessor to *d*. Although activity *a* surely must come before activity *d* in this figure, activity *a* is not called a predecessor of activity *d*.*

---

*In directed graph terminology, we are asking for a description of a minimum edge adjacency matrix among the activities. Sometimes this may be difficult to initially establish. A reachability matrix may be established first and the minimum edge adjacency matrix obtained from this.

3. A *successor activity* is an activity that immediately follows the activity that is being considered. Activities *b* and *c* are successor activities to activity *a* in Figure 6.17a and activity *d* is a successor to activity *b*.

4. An *antecedent activity* is any activity that must precede another activity under consideration. Activities *a* and *b* are antecedent activities of *d* in Figure 6.17a. Activity *c* is not an antecedent of activity *d*, and activity *a* has no antecedents.

5. A *descendent activity* is one that must follow an activity being considered. In Figure 6.17a, activities *b*, *c*, and *d* are all descendants of activity *a*.

6. A *burst point*, or burst, is a node with two or more successor activities. Figure 6.17b shows a burst point at node *a*.

7. A *merge point*, or merge node, is a node where two or more activities are predecessors to a single activity. Figure 6.17c illustrates a merge node.

We may use these node and arrow symbols, together with the information contained in the activity analysis, to enumerate paths for the network. In general, approaches to construction of a network planning model involve three efforts:

Activity analysis.
Arrow diagraming.
Path enumeration.

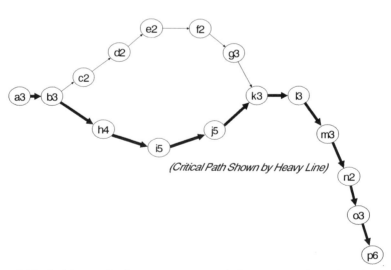

Figure 6.18. Activity on Node Representation of the System Life-Cycle Schedule of Figure 6.15

These comprise the network planning phase. Following this, a scheduling phase of effort is initiated. In this, the particular sequence of activities that takes the longest time to complete is identified. This is called *the critical path*. It is important to note that "critical" is used here in a very special and restricted sense. It refers to time only. Figure 6.18 illustrates the diagram that results from use of the activity analysis results indicated in Figure 6.15 and the arrow diagraming conventions described by Figure 6.17. Generally this is called a PERT diagram.

The time length of the critical path determines the project duration. The critical path is indicated by the thicker branches in Figure 6.18. Any change in an activity on the critical path will affect project duration. Thus, this method focuses attention on critical activities. Such an approach can also be used to estimate project time to completion and can be used to schedule jobs under limited money and human resource availability and to investigate time–cost trade-offs. This approach is very useful for operational level management and task control of a project. These network planning methods have been used in the preparation and management of large projects in many application areas.*

CPM and PERT may be viewed as variations of one another. CPM analysis typically includes *forward pass* and *backward pass* computations. The forward pass computation uses a given starting point and determines the *earliest completion* time for the project. This is obtained by determining the shortest time to arrive at each node, and then summing these. The backward pass is the reverse. It starts with a given completion date, and then determines the last allowable starting time. The computation starts with time required to reach the final node from its predecessor, and then works backward, node by node. A potential use for the forward and backward pass is in determining slack time associated with the non-critical paths. This provides a basis for possible trade-offs among time and effort.

There are some extensions to this basic description of CPM, such as allowing for crash schedules to complete the project by a specified time and separating total slack time into specific sequences of slack time.

PERT differs from CPM primarily in that it permits varying estimates of time required to complete each activity. Thus, for example, if optimistic, pessimistic, and likely activity completion times are given by (t_o, t_p, t_l), we compute the expected time and variance for each activity from the relations $t_e = (t_o + 4t_l + t_p)/6$ and $\sigma^2 = [(t_p - t_o)/6]^2$. The earliest overall expected

---

*In the aerospace and defense industries, network management approaches are usually called by the name PERT. In civil engineering systems areas, such as transportation and construction, the term CPM is often used. Some related names are precedence diagraming method (PDM), and line of balance (LOB). Precedence networking (PN) allows other precedence relationships than "Job B cannot start before job A has been completed," such as: "Job B can only start n time units after job A has started," "job B cannot finish until n time units after job A has finished." Other methods allow for the inclusion of decision points, such as Decision CPM, or network probabilistic elements, such as the Graphic Evaluation and Review Technique (GERT).

time for completion of the project, T_e, is then the sum of all the incremental expected times, t_e, for each activity along the critical path. When several activities lead to the same event, the largest expected sum of activity completion values is used. It is now possible to calculate a forward and backward pass, as with CPM. The slack time is the difference between the longest period permitted for activity completion and the expected time, or $T_S = T_L - T_E$. These values can be computed, as well as the probability of meeting any completion time requirement.

All of these computations are based on some assumptions that are subject to a number of questions. First, it is difficult to provide the three estimates of project completion: optimistic, pessimistic, and likely. There is no truly appropriate way to determine the appropriate statistical distribution for activity duration, even though the classical PERT approach uses (or more appropriately, imposes) a special statistical distribution called a beta distribution. That the longest time path is the most critical one may even be subject to question. What do we do with a longest path of 100 months with a variance of 1 month, as contrasted with a second longest path of 50 months, and a variance of 100 months? Despite these potential concerns, the PERT approach is very useful and often used. As with using any tool, we should be aware of the associated risks and limitations.

The typical products of this network planning and management approach are as follows:

1. Forecasts of project duration.
2. Forecasts of project resource requirements.
3. Identification of activities that are critical to project duration and, as a result, those activities that deserve the greatest systems management focus.
4. An activity schedule that will not violate time, resource, or other constraints.
5. A framework for cost-accounting.
6. A basis for monitoring and control of progress in a large project.
7. Estimates of the minimum additional costs of change in overall project specifications.
8. Increased insight into various critical milestones and identification of possible inconsistencies or conflicts that require additional configuration management efforts.
9. Improved communication among those responsible for the project.
10. Improved documentation of the project.
11. Provision of a basis for configuration management and management control.

Thus, network-based methods are particularly useful for the operational and task level controls that follow from systems management. They can be particularly useful for the configuration management efforts discussed in Chapter 4.

The outputs of this approach include graphic displays and network diagrams that show sequential relations among activities. This graphic display of the critical path normally identifies the earliest and latest possible starting times for each activity. The use of a network for planning and management of a project normally accomplishes the following.

1. *Formulation, Including Identification of Goals and the Substance of the Project to Be Undertaken:* It is essential that a clear definition of the beginning and end of the project and its objectives be stated.
2. *Identification of the Specific Activities to Be Included in the Project:* As one task in configuration management, the project should be disaggregated into packages of tasks that require similar resources, are performed by the same group of people, succeed each other directly, and only on completion of a single precedent package of activities.
3. *Identification of Precedence Relations:* A list is made of the immediate predecessors for each activity. The set of precedence relations is often called the *project logic,* which reflects the fact that precedence requirements are almost always based on logical requirements.
4. *Construction of the Network Diagram:* In activity-on-node networks, each activity is represented by a node and the start of the next activity is denoted by an arrow leading to the successor activity. In activity-on-arrows networks, each activity is represented by an arrow connecting two nodes that represent events. Generally, the events denote the completion of an activity. In activity-on-arrows networks, the event at the tail of the arrow must have occurred before the activity can start, whereas the event at its head can only occur after completion of the activity. Arrows generally point from the left to the right, by convention, thereby indicating one representation of the direction in which time proceeds. Diagrams may be drawn manually, but computer software packages have been developed to aid in handling large projects.* Generally, the software is acceptably easy to use and quickly reveals such inconsistencies in approach as improper precedence relations.
5. *Identification of Time Requirements for Each Activity:* Project personnel or others familiar with the project provide estimates of the time required for completion of each activity. These may include only the most likely, or normal, or the most cost-effective, time estimates. In the case of great uncertainty, it may be desirable to provide 3 estimates: a pessimistic, an optimistic, and a most likely estimate. Often, the estimates are indicated in graphical representations.

---

*Typical software packages include Super Project, ARTEMIS PROJECT, Harvard Total Project Manager, and Microsoft Project. Many integrated operations research software packages, such as STORM, contain a project management module.

6. *Identification of the Critical Path(s):* The critical path is that sequence of activities in the project that takes the longest time to complete. The activities on the critical path are easily determined. We begin this computation at the beginning of the network, where for convenience we set time = 0. An earliest possible start time is computed for each succeeding activity. This leads to determination of an earliest possible completion time for the project, and this time is equal to the length of the critical path. Then, a latest possible start time is assigned to each of the activities, working back from the earliest possible completion time. The latest possible start time is the latest time the activity can be started without delaying the completion of the entire project. The critical activities are those activities for which the earliest and the latest possible start times are equal. The difference between earliest and latest start times for the other, noncritical, activities indicates the amount of float or slack associated with them. This slack represents the time span by which they may be shifted forward or backward without affecting project duration. There may be more than one critical path in a given network. Activities on the critical path deserve most management attention from the perspective of project completion time management. This is the case since delays in progress along the critical path will directly affect project completion time.

7. *Use of the Project Network and Associated Critical Path as a Basis for Planning:* It is possible to make several uses of the just computed critical path and associated project plan.

   7.1. *Estimation of Uncertainty in Forecasted Project Duration:* Uncertainty estimates for each of the critical path activities indicates where statistical methods should be employed to help estimate overall project duration and its standard deviation. When changes in length of the critical path occur, other activities not on the initial critical path may become critical.

   7.2. *Scheduling of Jobs within an Activity:* This can be done directly from the network, given some criterion to schedule jobs with float. The often requested completion time is "as soon as possible" (ASAP). Float occurs whenever jobs on some paths through the network may be delayed without affecting the time to completion on the critical path. The result can be represented graphically as a Gantt or bar chart, as we will discuss in our next subsection

   7.3. *Activity and Job Scheduling under Resource Constraints:* Activities, and the jobs that comprise them, that can be executed at the same time often require similar resources in labor and equipment. The float in activities and jobs can be used to reschedule them to minimize some of the peak demands for these resources. Various scheduling problems can be solved. We can, for example, find the schedule that does not exceed specific peak resource requirements while keeping the total project duration at some minimum value. We can find the schedule that equalizes resource requirements throughout the entire project duration. Mathematical programming algorithms have been developed to solve problems such as these. Most of these problems can and have been formulated and solved using linear, integer, and mixed integer programming approaches.

7.4. *Time–Cost Minimization:* In general the time duration of activities can be shortened, for a price. Obviously, shortening of the task only along the critical path can reduce the overall project duration and potentially justify the increased costs. If the same quality level is maintained, it would be rare that a project completion schedule could be compressed without increasing completion costs. When such critical jobs are compressed, others may become critical and themselves require shortening. The network can be used as a basis to compute the additional costs of reducing total project time. If costs for each job increase linearly with compression, the schedule giving minimum additional costs for a specified overall compression can be found using linear and other mathematical programming methods. The results can be used for planning or bidding purposes, or to help compute cost–benefit trade-offs. Thus, we see that project network management models can be used for a number of the efforts we discussed in Chapter 4, especially for configuration management efforts.

7.5. *Cost Accounting and Budgeting:* The distinction of separate activities provides a basis for cost accounting through such efforts as the cost breakdown structure (CBS) and work breakdown structure (WBS) efforts that we describe in Chapter 8. A network-based management schedule is a good starting point for planning project costs.

8. *Use of the Network as a Basis for Project Management and Control:* Actual and planned project progress, with respect to both schedule and expenditures, can be easily compared. Causes of schedule delays and/or cost overruns can be diagnosed directly. As long as its logic remains unchallenged, an existing project network can also be used to reschedule activities and event completion times to reflect changing circumstances. This enables the project manager to identify changes in criticality of activities and jobs, to resolve emerging resource competitions, and to find out which activities need to be modified to meet impending deadlines.

There are a number of conditions in a project that would normally call for a network approach to management. The most obvious is large size, in which case it would be very helpful to disaggregate the project into a set of predictable, well-defined independent activities with certain precedence relationships. Network management approaches provide a systematic approach that focuses on the overall project configuration effort, while taking relevant and separate project activities into account. This approach provides a unifying framework for planning, scheduling, budgeting, and controlling a project.

Network planning methods can also provide a very useful way to examine feasibility of alternative systems engineering proposals, particularly as regards their costs and schedules.

There are a number of useful by-products that result from use of network planning models. The structure of the project and the interdependencies of activities are generally clarified, as are the critical activities. There are some caveats, however, The validity of time and cost projections depends on the validity of estimates used for the individual tasks. These estimates may be uncertain. Estimates provided by those responsible for execution of tasks may be severely biased. It is difficult to clearly distinguish and assign precedence to all activities before the start of a project. The network planning method is usually not appropriate when iterations or cycles occur in a project. In this special sense, the presence of cycles in a project management network is beneficial. It allows us to detect severe problems that will exist if the project is fielded as configured. It may not always be possible to start or stop activities independently, and activity times may not be independent of each other. Even if they are independent, it may be difficult to identify all activities prior to the start of a project. Finally, network planning methods do not replace human management. As with other systems engineering methods and tools, they can only provide a tool to support human skills.

6.4.2. Bar Charts

A *bar chart*, sometimes called a *Gantt chart** [50] after one of the pioneers in the very early days of "scientific management," is a simple visual chartlike aid for project planning and management. The charts are used to display a schedule of activities as they evolve over time and to compare actual and planned progress in those activities. Traditionally, a time scale is listed along the horizontal axis, and activities are listed along the vertical axis. Horizontal lines or bars show the times during which each activity is planned to be carried out. The length of such a line is a measure of the time required for the activity and can also indicate a quantity of work. Occasionally, milestone dates are illustrated on the vertical axis. These are often indicated by a triangle-like symbol.

Gantt charts are most useful for synthesis and analysis of plans in the project planning and development phases of a systems effort. The Gantt chart is a prime component of the planning for action and implementation step of systems engineering. It can also be helpful for analysis of the feasibility and impacts of alternative plans. Although Gantt charts can be designed and used separately, they are most effective when used in close

---

*It was first developed by Henry Gantt in the beginning of the 20th century. Henry L. Gantt (1861–1919) was a disciple of Frederick Taylor and created a class of charts in which progress is plotted against time. This effort apparently has a major impact on ammunition production and delivery efforts during World War I.

connection with network planning methods. Although Gantt charts empha-size the time required to complete tasks, network models enable us to analyze project logic and precedence requirements. The two may best be combined into an integrated project development and management ap-proach. This is generally the approach taken today.

Inspection of a Gantt chart easily shows which activities are planned to be carried out at each specific point in time. It is particularly useful for displaying the results of a PERT or CPM analysis. Thus, such a chart is helpful in forecasting workload, scheduling the resources required to com-plete a project as a function of time, and other similar uses. Possible shifts in the schedule may also be indicated on the chart and used to reduce peak demands or resolve conflicts over resource allocation issues.

While execution of activities is in progress, the actual portions of work completed at a specific time may also be shown on the chart in some distinctive manner, such as through use of heavy lines. This enables the person viewing the chart to quickly identify items of potential interest, such as those areas where progress differs from that initially planned in the PERT or CPM schedule. Because of their simplicity, use of Gantt charts can considerably enhance communication about, and participation in, planning large projects. The charts have been used for a wide variety of projects over many decades now.

There are actually two closely related but somewhat different types of Gantt charts currently in use. In one type, various departments, crews, pieces of equipment, and so on are listed along the vertical axis. Horizontal lines or bars show the timing and amounts of work to be performed by each working group. In the other type, activities are listed on the vertical axis, and lines or bars show the time during which they are planned to be carried out. Because it is more useful for project planning, we discuss the latter type of chart in some detail here.

The following steps lead to construction of a Gantt chart. Just as with the network management efforts, we describe these efforts as planning, schedul-ing, and control efforts.

1. *Planning:* A decision is made as to which alternative project will be carried out. The plan is worked out in detail and a time scale or completion date specified. The project is disaggregated into a set of independent activities or work packages, perhaps in a cost breakdown structure (CBS) and work breakdown structure (WBS). Precedence relations between the activities are identified, and estimates for time requirements are obtained. Network planning methodology, typically PERT or CPM, is used to compute expected overall project duration, to identify activities critical with respect to project duration, and to derive earliest and latest possible start or completion times for each activity.

2. *Scheduling:* The time scale along the horizontal axis is specified, and the activities are listed along the vertical axis, usually in ascending order to start times from top to bottom. Planned start and completion times for each activity are linked by a horizontal line showing the time during which the particular task is supposed to be *active*. Possible earlier start or later completion times are also indicated, for example by broken extensions of a line to indicate what is planned. The completed chart shows, at a glance, which tasks will be active at any point in time. When estimates of requirements for each job are available, graphs showing total workload, equipment requirements, expenditures, and other items of importance for project scheduling and management may be illustrated as they evolve over time.*

3. *Control:* While execution of the project is under way, the Gantt chart is usually updated regularly in order to show actual accomplishments. Usually, a solid line is drawn parallel to or over the original line to reflect the fraction of the task that has been completed. If, for example, 50 percent of a specific task is completed, a solid line is drawn that is half as long as the line showing total activity duration. This enables comparison of actual and planned progress and revision of plans as necessary. A glance at the updated chart quickly reveals those activities that are on schedule, those ahead of schedule, and those that lag behind the initially planned schedule. Depending on the particular situation, this may lead to a reallocation of priorities, or even to revision of the schedule for those tasks that are not yet completed. Using this information, preparation of a Gantt chart to document the execution of the project becomes possible. If required for documentation, the realized start and completion times of all activities can be shown in a Gantt chart in the same format in which the planned schedule was shown.

We use the same example here that is used in the description of network planning. Inspection of Figures 6.15 and 6.18 easily results in the Gantt chart illustrated in Figure 6.19. It would be easy to show various critical milestones, both original milestones and those rescheduled as the project evolves, on the diagram. Clearly, the notions described here are very simple ones. They are also very useful ones, as are many *simple* notions.

There have been a number of extensions of these early notions. One of these is based on the *Pareto Principle*, which states that a small number of problems are generally responsible for the majority of the ill effects. In terms

---

*Some analytical effort is possible using these charts. For example, peak loads may be reduced by shifting the active periods of tasks over the allowable range. The Gantt chart can be used as a visual aid to directly evaluate the effects of these possible shifts. However, for complicated projects consisting of hundreds of activities, so many combinations of shifts are possible that it becomes infeasible to evaluate them all graphically. Most of the network software described in the previous section have available options that allow display and printing of Gantt charts.

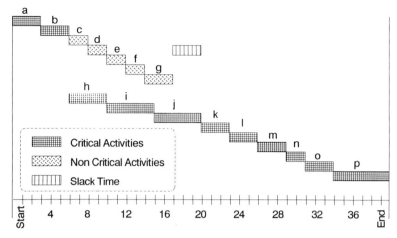

Figure 6.19. Gantt Chart Corresponding to Figures 6.15 and 6.18 Life Cycle

of economic systems analysis, it states that a few of the contributors to costs are responsible for most of the costs. In terms of quality, it states that a few of the contributors to quality loss costs are responsible for most of these costs. Thus, a chart that illustrates the most important areas in terms of costs would have potential value in indicating which problem areas should be addressed first. This is called a Pareto chart [51]; a prototypical Pareto chart is shown in Figure 6.20. It would certainly be possible to plot a Pareto chart showing the number of occurrences of a problem of a given type versus the type. The classical name given to such a chart is a *histogram*. The biggest potential problem with doing this is that the inferred suggestion visually presented by a Pareto chart is that we should work on the most important

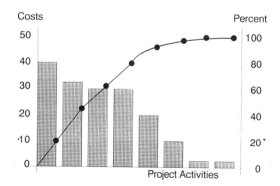

Figure 6.20. Hypothetical Pareto Chart

elements first and that the most important elements are those with the largest number of occurrences. A cost of problem type measure is generally more relevant, or possibly some multiple attribute measure, rather than simple number of occurrences.

The steps involved in constructing a Pareto chart are, conceptually, quite simple. We first identify as many of the relevant influencing elements of interest as possible. This might well be in the form of a matrix of costs. The dimensions of the matrix could be product categories and type of defect, life-cycle phase, and type of error. The entires in the matrix would be the measure of influence, typically cost entires. We would then sum these horizontally or vertically, depending upon the horizontal dimensions selected for the Pareto chart. Then we order these elements according to the selected measure of influence, such as cost. A cumulative distribution of the measure of influence, generally costs, would be illustrated in the form of a line graph and associated with the descending order of the problem elements, such as illustrated in Figure 6.20.

6.4.3. Interaction Matrices

Interaction matrices provide a framework for identification and description of patterns of relations between sets of elements, such as needs, alterables, objectives, and constraints in a problem definition effort. Self-interaction matrices are used to explore and describe relations or interactions between elements of the same set. Cross-interaction matrices provide a framework for study and representation of relations between the elements of two different sets. Basically, a matrix, or table, structure is set up in which each entry represents a possible linkage between two elements. Figures 6.21 through 6.23 illustrate concepts useful in constructing interaction matrices. Each entry is considered, and the existence of a relationship is indicated in the matrix, such as in the adjacency matrices shown in Figure 6.21.

Development of an interaction matrix encourages us to consider every possible linkage, and significantly enhances insight in the connectivity of elements being studied. The resulting matrix may be helpful in identifying clusters of related elements, or elements that appear to be quite isolated from the others. Also, the table may be used to construct a graph or map showing a structure of elements. Interaction matrices are generally applicable in circumstances where the structure of relationships among a set of elements has to be explored.

Formally, there are at least four types of self-interaction matrices. These represent nondirected graphs, directed graphs (or digraphs), signed digraphs, and weighted digraphs. A digraph consists of a set of variables, called nodes, and a set of links between the variables, called edges. The graph is directed, or a digraph, if there is a direction associated with an edge e_{ij} that connects node n_i to node n_j. It is an undirected graph if there is no direction associated with any of the edges in the network. A cognitive map is a signed digraph representation of a structural model.

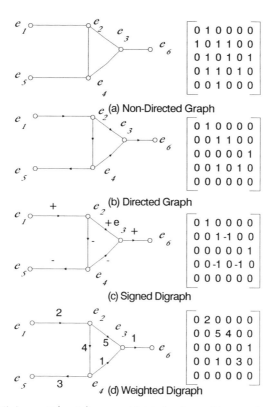

Figure 6.21. Minimum-Edge Adjacency Matrix for Several Interaction Matrix Types

Figure 6.21 illustrates the basic structural relations and graphs associated with these interaction matrices. The theory of digraphs and structural modeling is authoritatively presented by Harary, Norman, and Cartwright [52] and a number of applications to what is called interpretative structural modeling have been described [53–56]. Cognitive map structural models have also been considered [57]. A development of structural modeling concepts based on signed digraphs is discussed by Roberts [58]. Geoffrion has been especially concerned with the development of a structured modeling methodology [59, 60] and environment. He has noted [61] that a modeling environment needs five quality and productivity related properties.

1. A modeling environment should nurture the entire modeling life-cycle, not just a part of it.
2. A modeling environment should be hospitable to decision and policy makers, as well as to modeling professionals.
3. A modeling environment should facilitate the maintenance and on-going evolution of those models and systems that are contained therein.

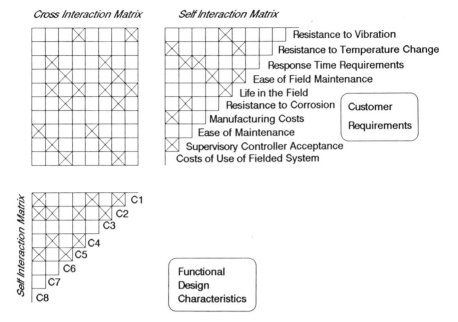

Figure 6.22. Sample Self- and Cross-Interaction Matrices for Quality Function Deployment

4. A modeling environment should encourage and support those who use it to speak the same paradigm neutral language in order to best support the development of modeling applications.

5. A modeling environment should facilitate management of all the resources contained therein.

Structured modeling is set forth as a general conceptual framework for modeling. Associated with it is an executable modeling language that supports the frameworks and a number of software integration approaches [62] to enable it to deal with the wide variety of issues for which modeling is appropriate.

We briefly examine construction of interaction matrices and structural models. It is assumed that one, two, or more sets of elements are given. These may have been obtained by an idea generation method or by using some other approach. Also, a decision has been made as to whether the interactions between the elements of one set (such as needs, alterables, constraints), or the linkages between elements of two or more different sets are to be explored. The following activities are typically followed in setting up an interaction matrix, including a graph theory based structural model.

1. *Determination of the Type of Relationship:* Directed or nondirected graph relationships may be appropriate, depending on the type of

Figure 6.23. Interaction Matrices for Organizational Policy Assessment and Quality Function Deployment

matrix, the type of elements, and the degree of specificity desired. For example, alterables might be related in that they belong to the same subsystem or in that they tend to change in a similar nondirected way. Alternately, it might be that change in one alterable directly affects another alterable. Nondirected relations seem appropriate for cross-interaction efforts. When self-interactions are considered, appropriated directed relations are generally to be preferred. In general, any type of relationship may be used as long as the meaning of the relation considered is clear to all those involved in setting up the interaction matrix.

2. *Setting Up the Interaction Matrix Framework:* The elements are listed and lines are drawn in such a way that the result is a table in which there is a box for each possible combination of elements. Nondirected self-interaction matrices have a triangular form, whereas directed self-interaction matrices are square and cross-interaction matrices are rectangular. We have presented a number of nondirected self- and cross-interaction matrices throughout this text.

3. *Completion of Each Entry in the Interaction Matrix:* The entries in the matrix are completed one-by-one, each time asking the question, "Are these two elements directly linked according to the specified relation, or not?" A positive answer may be indicated by writing a "1" or a "×" in the appropriate box, or by blackening the box, and a negative answer by writing a "0", or by leaving the box blank. Information about the strength of the relation may be included, as noted in Figure 6.21d.

4. *Revision of the List of Elements and Scanning the Displayed Pattern of Relations:* The process of completing the matrix entries may well lead to the identification of important intermediate elements that are missing from the original list. If desired, such an element may be added to the matrix. Also, elements may be redefined, thereby requiring revision of related matrix entries. Certain patterns may appear in the matrix, thereby revealing clusters of interrelated elements, or elements may appear to be quite isolated from others. In an alternatives-objectives cross-interaction matrix, it is important to check whether, for each objective, there is at least one alternative that helps to achieve it.

5. *Translation of the Matrix into an Appropriate Graph:* Since a graph conveys both direct and indirect linkages, it is generally more preferred than a matrix to communicate structure. It may be worthwhile to construct such a graph from a self-interaction matrix provided the matrix is not too large and there are not too many interactions. The graph, containing essentially the same information as the matrix, can be directed or nondirected, depending on the relation used.

6. *Analysis and Formulation of Conclusions:* On the basis of the matrix and possibly the corresponding graph, conclusions may be drawn concerning the major interactions in the problem considered. Graphical analysis methods might be used to show the structure more clearly. The consistency of several interaction matrices of elements relating to the same problem may be checked mathematically.

Interaction matrices are generally useful in the problem or issue formulation step of a systems effort. They provide a simple aid to exploring the structure of all the problem elements, such as needs, stakeholders, alterables, constraints, agencies involved, and so on. The use of both self- and cross-interaction matrices for a comprehensive structural analysis of all elements of problem definition is a central part of very useful approach called Unified Program Planning [63]. Computer assistance may be helpful for setting up and displaying large matrices. Also, computers may be used to perform analysis of (binary) matrices and to help structuring a matrix in such a way that an informative graph may easily be drawn in the form of a structured model. The usual approach is to first represent a mental model of a situation by identifying a number of elements and an appropriate contextual relationship. Questions are posed and answered concerning the binary relations among the elements according to the selected contextual relationships. This enables construction of a reachability matrix or a minimum-edge adjacency matrix according to the approach chosen. Many structural models take the form of a tree, or slightly more complex but hierarchical structure. Also, many structural models take the form of influence diagrams.

Formally, a *minimum-edge adjacency matrix* describes the minimum number of edges needed to describe a directed graph. A reachability matrix describes all possible linkages among digraph elements. It is easily possible to show that the relationship among minimum-edge adjacency matrix $\mathbf{A}$ and the associated reachability matrix $\mathbf{R}$ is $\mathbf{R} = (\mathbf{I} + \mathbf{A})^n$. Here, $\mathbf{I}$ is the identity matrix and n is at least as large as the maximum path length in the digraph network to contain. In all cases, $n \leq N$, where N is the number of elements. Even if $\mathbf{A}$ contains more than the minimum number of edges, the reachability matrix can be obtained in the manner indicated. If transitivity can be assumed, it is possible to uniquely determine a reachability matrix by asking a limited number of contextual relatedness questions about the elements and use the transitivity property to infer many responses. This approach is described in considerable detail in the structural modeling references cited [53–62].

A *quality function deployment matrix* represents an actual or conceptual collection of interaction matrices that provides the means for transitional and functional planning and communication across groups. Basically, it is an approach suggested in the total quality management literature that attempts to encourage an early identification of potential difficulties at an early stage in the life cycle of a system. It encourages those responsible for fielding a large system to focus on customer requirements and to develop a customer orientation and customer-motivated attitude toward everything. Thus, the traditional focus on satisfying technical system specifications is sublimated to, but not replaced by, the notion of total satisfaction of customer requirements.

Quality function deployment (QFD) has been suggested as one successful approach to use in conjunction with implementation of TQM. A quality function deployment interaction matrix is one approach for representation and communication of quality function deployment results. The purpose of QFD is to promote integration of organizational functions to facilitate responsiveness to customer requirements. As described by Clausing [64], QFD is comprised of structured relationships and multifunctional teams. The members of the multifunctional teams attempt to insure that all information regarding customer requirements and how best to satisfy these requirements is identified and used, that there exists a common understanding of decisions, and that there is a consensual commitment to carry out all decisions.

Two applications have been suggested. One is a "house of quality" effort matrix, and the second is a "policy assessment" matrix. Generally, the associated interaction matrices might appear as illustrated conceptually in Figures 6.22, 6.23, and 6.24. Figure 6.22 illustrates some generic interactions among potential elements in the matrices. Figure 6.23 illustrates a more

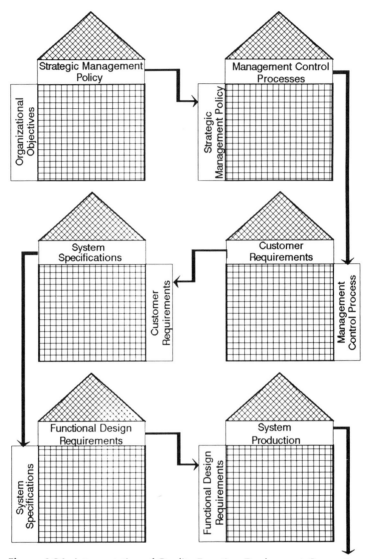

Figure 6.24. Interpretation of Quality Function Deployment Concepts

complete set of QFD matrices. Figure 6.24 presents an illustrative picture of these, as a house of quality and house of policy, that follows illustrative house of quality depiction of these interactions [65, 66]. The generic efforts involved in establishing QFD matrices are as follows.

1. Identify customer requirements and needs. Establish these in the form of weighted requirements expressing their importance.

2. Determine the systems engineering architectural and functional design characteristics that correspond to these requirements.

3. Determine the manner and extent of influence among customer requirements and functional design characteristics and how potential changes in one functional design specification will affect other functional design specifications.

As a related and more strategic matter, it would be necessary to determine the extent to which the specific systems fielding effort under consideration is supportive of the long-term objectives of the organization and the competitive position and advantage to be obtained through undertaking the development. We see that this QFD approach has uses in strategic management to see if the management controls or systems management effort associated with the development of the system under consideration is supportive of the overall development strategies of the organization.

The various interaction matrix entries might well be completed with binary ones and zeros to indicate interaction and no interaction. Alternately, there would be major potential advantages in indicating the strength of interaction. Hypothetically, at least, the QFD matrices provide a link between the semantic prose that often represents customer requirements, and the functional requirements that would be associated with technological system specifications. In a similar manner, an interaction matrix can be used to portray the transitioning from functional requirements to detailed design requirements. Thus, they provide one mechanism for transitioning between these two phases of the systems life cycle. One goal in this would be to maintain independence of the various functional requirements to the extent possible. A measure of this independence can be obtained from the density and location of the interactions among the functional requirements.

6.4.4. Tree Structures and Hierarchical Structures

The term *tree* is used to indicate a special type of graphical representation that consists of elements represented as vertices or nodes and relations between those elements represented as lines or edges. A tree may represent either a single-source or single-sink structure. It starts from a single source or edge, and branches develop along nodes. Each node can have only one in-coming branch, but can have several out-going branches. There can be only one path between every pair of nodes. Trees can be particularly useful for displaying hierarchies of objectives, organizational structures, or various events that can develop from a single starting state. In general, trees are useful aids for representing the structure of elements and for communicating an understanding of such structures. Several types of trees useful in the study of complicated issues include objectives trees, activity trees, event trees, intent trees, decision trees, and worth or attribute trees.

In constructing a tree, it is first necessary to determine the elements that will be structured and the contextual relation to be used in building the structure. In constructing an organizational tree structure, for example, we need to know the functional positions associated with the organizations. An appropriate contextual relation is "reports to." In general, the elements to be structured are available from an issue formulation effort. The type of relation according to which the structuring is done will, in general, determine the feasibility of representing the structure by some type of tree. The various types of trees are so different that comments concerning their construction and use are perhaps best made in terms of separate discussions of the most common tree types.

1. An *objectives tree* represents the hierarchical structure associated with objectives or intents. It is sometimes referred to as an *intent structure* [53, 54].

    1.1. In constructing an objectives tree, it is helpful to formulate the objectives to appear in the objectives tree with an appropriate contextual relation in the form *to (action word) + (object) + (constraint).** It is almost mandatory to use the proper semantic form of contextual relation to aid in developing action-oriented statements. The definition of the contextual relation between the objectives should not change during construction of the tree. Typical contextual relations are *will contribute to* in the case of upward pointing associations and *will be achieved by* for downward pointing associations.

    1.2. Next, elements are arranged in hierarchical order with the more general goals above the more specific goals. If a specific objective seems to fit into several branches, possibly at different levels, such that the structure is not that of a tree, the objective statement should be reworded, or replaced by several different objectives at different hierarchical levels, such that the tree structure is retained.

    1.3. As the tree is constructed, five tests of objective tree logic should be applied. Each test should be made on each statement in the tree as follows:

       a. In going down any branch of the tree, each goal or objective statement must answer how for each immediately superior objective.

       b. In moving up any branch of the tree, each level objective should provide information concerning why the objective below it is needed.

       c. In reading across the objectives at any given level and under any one general goal, it should be possible to see that each of these more specific objectives is needed to accomplish the more general objective.

---

*An example is *to achieve increased happiness through personal health.*

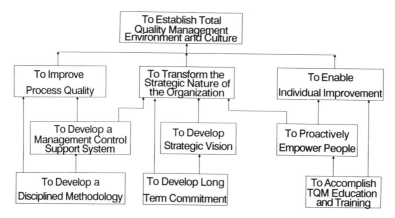

Figure 6.25. Simple Objectives Tree for Total Quality Management

d. In reading across the objectives at a given level and under any one general objective, we should also ask whether there are other specific objectives at this level that are needed to accomplish the more general objective. If there are, they should be included in the tree.

e. Often, it will be desirable to identify the ownership of all or many objectives.

Figure 6.25 indicates a typical objectives tree structure.* This tree represents a set of top-level objectives for total quality management. There is one top-level objective, three second-level objectives, and six objectives in the third level. The contextual relationship is "is desirable or necessary in order," and this connects all the objectives in the manner indicated in Figure 6.25.

2. *Activity trees* can normally be constructed most easily after an objectives tree has been prepared. Frequently, there is a significant correspondence found between activities and objectives. Each vertex entry in an activity tree consists of an activity, and the edge relation is that lower level activities must contribute in some way to the conduct of the higher level activity. See Figure 6.26 for an example. Activity trees are useful in helping to structure a program to determine what is related to what and to examine how responsibilities can be assigned. Applications include developing a responsibility assignment and constructing an activity network. Objectives can be converted into activities by removing the infinitive preposition *to* and the constraint set. It is interesting to note that project

---

*This is often called a hierarchical tree. The structure can easily be redrawn in a classic tree form by defining additional third-level elements, such that a third-level element is directed at one and only one second-level element.

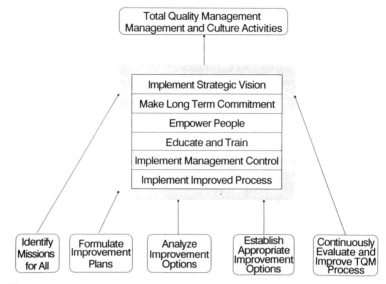

Figure 6.26. Activities Tree for Implementation of Total Quality Management

network modeling methods deal extensively with activities. However, the usual representation framework is not directly that of a tree structure, although the structure of activities can often be recast into a tree structure. Figure 6.26 represents the structure of an activity tree for total quality management. The elements shown represent the activities needed to bring about a specific TQM implementation, and to evaluate and refine the TQM process continually over time. Some elements in the structure are combined in order to simplify the graphical presentation in Figure 6.26. This must often be done when it would be cumbersome to display a vast number of elements in a crowded visual presentation. The elements in Figure 6.26, which are stated in the form of activities, need to be carried out in a planned sequential manner, as established through strategic planning and management controls. Figure 6.27 illustrates a feedback model of these same elements, very close in form to what we call a causal or influence diagram. For most uses, this would be a more appropriate representation than the tree representation. One of the challenges in many systems engineering efforts is choosing an appropriate framework for representation of salient information. This framework has much to do with the manner in which information is thought about and acted upon.

3. *Decision trees* are comprised of decision nodes, event nodes, and event outcome nodes, as illustrated in Figure 6.14. The decision node is generally represented by a square. The lines, or edges, emanating from the

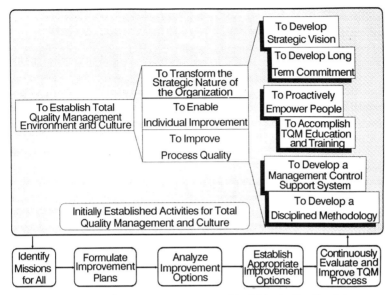

Figure 6.27. Alternate Arrangement of Activities in Feedback Loop

decision node represent the possible alternative courses of action. The event node, or chance node, is generally represented by a circle. Edges emanating from the chance node represent the different possible consequences of choosing the alternative leading to that chance node. Probabilities of occurrence are assigned to the consequences, and costs or profits are usually estimated for all branches of the tree. The probabilities of consequences emanating from a particular chance node must sum to one. Time and activities in a decision tree typically proceed from left to right, but this is not necessary. The final outcome nodes appear at the extreme right-hand end of a decision tree and have incoming edges but no outgoing edges. A decision path represents a possible sequence of decisions and consequences, and leads, in association with uncertainty considerations, to a particular outcome. Chapter 7 develops notions associated with decisions and consequences, in order to make the formal task of decision making more organized and explicit than it otherwise might be.

4. *Attribute trees* are much like objectives trees. One often-found difference, however, is that attributes tend to be more measurable than objectives. Often, it is desirable to cast attributes in the form of objectives measures.

There are many other types of tree structures, including fault trees, event trees, relevance trees, and preference trees. A hierarchical structure is one in which each level is directed upward only to higher levels and downward only to lower levels. Thus, all tree structures are necessarily hierarchical. But, not all hierarchies are trees.

Trees can be used throughout the systems engineering process, but the process of constructing a tree is fundamentally an analysis effort. For example, objectives trees help structure the value elements determined in issue formulation; decision and event trees may be used as aids in the analysis of alternatives; and activity trees may be used as aids to planning for action after a decision has been made.

Various structural modeling approaches are helpful in constructing hierarchies for a large set of elements. An interaction matrix may also be helpful in tree construction. Tree structures may be constructed from the influence diagram approach that we describe next, although it is generally a simple matter, manually, to construct a tree with less than two dozen elements. The more difficult problem is that of identifying the relevant elements for the tree. Additional information concerning hierarchies, tree structures, and their construction may be found in the work of Warfield [53, 67] and others involved in these studies [54–62].

6.4.5. Causal and Influence Diagrams

Causal diagrams or *influence diagrams* [68] are a particular form of structural modeling that represents causal interactions between sets of variables. They are particularly helpful in making explicit one's perception of the causes of change in a system and can serve well as communication aids. In this section, we first discuss causal diagrams and then conclude with a presentation of recent efforts involving influence diagrams.

A *causal diagram* is a graphic representation of those cause–effect mechanisms that are responsible for change, generally change over time although it could be change over other variables such as space. Figures 6.28 and 6.29 are typical causal, or causal loop, diagrams. Lines and arrows between elements show the existence and direction of causal influences. An arrow pointing from element A to element B illustrates that some change in a pertinent aspect of A causes a change in a corresponding pertinent aspect of B. A pertinent change could be a change in value, magnitude, intensity, size, color, beauty, worth, or any of several other possible descriptors. Optionally, we might indicate the nature of the influences by a " + " to indicate that an increase in A causes a larger value of B, or a " − " to indicate that *increases in A cause a smaller value of B to occur*. The relevant sign would be placed at the arrowhead, just as with signed digraphs.

A causal diagram is particularly useful for portraying cyclic interactions, or feedback loops, in which influences act to reinforce or counteract change. It is a powerful communication tool, and aids considerably in structuring discussions. Typically, applications include the identification and portrayal of cycles and feedback, and the construction of a framework for dynamic

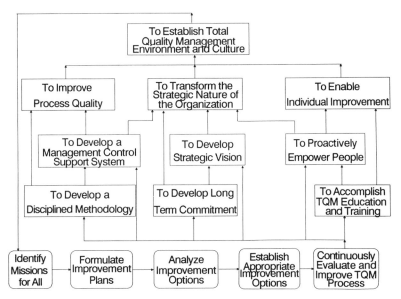

Figure 6.28. Causal Loop Diagram of Activities and Objectives Accomplished in Total Quality Management

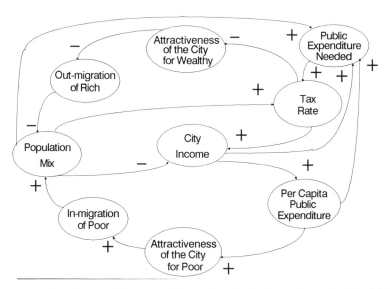

Figure 6.29. Causal Loop Diagram of Some Feedback Loops in a Model of Urban Dynamics

modeling. A causal loop diagram is a special sort of influence diagram or a directed graph or digraph. Causal loop diagrams draw specific attention to interacting loops and to the causal influences these loops portray.

We give a descriptive account of the major activities involved in causal diagram construction and related hypotheses explaining causal mechanisms of change. A causal structure permits tentative conclusions concerning such major influencers of behavior as critical mechanisms, points of leverage, strong cycles, or reasons for resistance of an organization or system to change. These conclusions may indicate why certain policies are, or could be, highly effective or very harmful, and why others are ineffective. Causal diagrams are most appropriate for structuring interactions between relevant issue formulation elements, and to prepare for more thorough analysis efforts. Causal diagrams provide "a bridge" between the formulation and the analysis steps of a systems engineering effort.

We suggest the following steps for construction of a causal, or cause and effect, diagram.

1. Identify the problem or issue for which a causal diagram is needed.
2. Identify the major groupings of causes. Search for more detailed causes, perhaps through brainstorming or use of the nominal group technique. Eliminate inapplicable causes. This needs to be done with some care, of course, What may initially seem frivolous may well turn out to be quite important.
3. Connect causes and effects together by determining predecessor and successor relationships. Refine the causal diagram as appropriate.

Figure 6.28 illustrates a causal loop and an influence diagram representation of the total quality management activities needed for implementation of TQM. There is a single cycle around the overall process to indicate improvement and evolution over time of TQM efforts. The presence of cycles creates difficulties in the use of graphical methods of analysis. Generally these difficulties can be avoided by using a single overall feedback loop, as in Figure 6.28.

Often, causal loop diagrams are developed for situations with multiple feedback loops. As an illustration, consider the erosion of the tax base in a hypothetical city with a fraction of the population characterized as "poor," and the remaining fraction as "wealthy." The tax base erodes through outmigration of the "wealthy," attracted by lower tax rates in the suburbs, and immigration of the "poor" who perceive better public facilities in the city than in the suburbs or rural areas. The city's desired policy is to keep public expenditures at a constant level. With the eroding tax base, due to the outmigration of more wealthy taxpayers and increasing demand for services by immigration of poor people, the city is indeed in trouble. A causal loop diagram that reflects some of the mechanisms causing long-term change is shown in Figure 6.29. This is representative of the sort of dynamic situation described by Forrester in his system dynamics model of urban growth and decay [69], and the causal network representation of Figure 6.29 can be converted to a systems dynamics modeling representation. We note that Figure 6.29 contains a mixture of flows: people, money, and information. It might be desirable to make these distinguishable, and this could easily be done by using branches of a different color or thickness to represent the different flows.

Two major interacting loops can be identified in this figure. The first major loop shows that a continuing outmigration of the "rich" will change the population mix, and reduce city income, all other things being constant. As a result, to keep public expenditures constant, the tax rate will be increased. This causes a decrease in attractiveness of the city to the rich, which leads to an increase in outmigration of the rich, and a further increase in the fraction of the poor in the city. The second loop shows that immigration of "poor" causes similarly deteriorating conditions. A policy that keeps the public welfare expenditures for each poor person at a constant level results in more and more poor people moving into the city, thereby further draining already dwindling revenues through increased social welfare expenditures. A further complication, even if secondary in importance, is the uncontrolled outmigration of the rich. This could potentially be changed by a policy of diverting part of the city income to cultural or other projects that the rich generally find attractive, thereby making it a more attractive place for the rich to live, despite high tax rates. Additional systems dynamics studies and references may be found in [54].

What we have described thus far needs to be put into some sort of methodical framework if it is to be useful. The *influence diagram* concept does just this. It is an approach to structural modeling that was initially developed for decisions analysis and allows representation of probabilistic effects in a relatively direct manner. It was first described by Howard and Matheson [70] and Shachter [71].

An influence diagram is potentially able to represent probabilistic and functional dependencies, and associated information flow patterns, as a directed graph. The basic conventions and notations typically used in influence diagrams are illustrated in Figure 6.30. There are four types of influence diagram nodes: probabilistic or chance nodes, deterministic nodes, decision nodes, and value nodes. Information flows from one node to the other, as indicated in this figure. There are three types of influence: conditioning influence, information influence, and value influence.

A *conditioning influence* indicates the presence, or absence, of probabilistic dependence. Thus conditioning influence and probabilistic influence are equivalent terms. From elementary probability theory, we know that probabilistic independence (sometimes called statistical independence) exists among two random variables, x and y, if and only if

$$p(x, y \mid \&) = p(x \mid \&) p(y \mid \&)$$

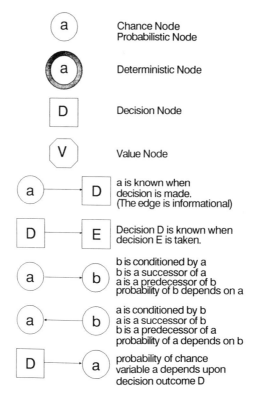

Figure 6.30. Influence Diagram Conventions and Simple Structures

where & represents all other existing prior information or conditioning predecessor that is known to, and available at, the explicit nodes in the network under consideration. A major advantage to probabilistic independence is the considerably reduced effort required in measuring, or eliciting, $p(x, y| \&)$ as two single-dimensional probability density functions rather than one two-dimensional density functions.* Moreover, probabilistic independence does not exist if it can only be stated that

$$p(x, y| \&) = p(x|y, \&)p(y| \&) = p(y|x, \&)p(x| \&).$$

---

*Another major advantage to probabilistic independence is that we can obtain the probability density of a single dependent variable in a much simpler manner. The density of the dependent variable y, which is $p(y| \&) = \int p(x, y| \&) \, dx$, does not depend upon the x variable at all and becomes

$$p(y| \&) = \int p(x, y| \&) \, dx = p(y| \&)\int p(x,| \&) \, dx = p(y| \&).$$

In other words, it must be true that

$$p(x|y, \&) = p(x|\&)$$

$$p(y|x, \&) = p(y|\&).$$

An *information influence*, which is shown by directed relationships or arrows that lead into decision nodes, indicates a causal ordering of elements. Thus any chance outcome leading into the decision node necessarily is known before it is needed to make a decision concerning or associated with that decision node.

A *value influence* variable or node represents outcome influencers or attributes and their interrelations that enable determination of a single value that reflects the underlying interpretations and value system of the decision maker. In the discussion of multiple attribute utility theory (MAUT) and related topics in Chapter 7, value influence is further explored.

The concepts of information or decision influence and value or attribute influence are relatively straightforward. Some further commentary concerning probabilistic, or chance, influence is needed here as there are a number of points that are often not fully appreciated. For the case of three random variables, x, y, and z, it must be true that

$$p(x, y, z|\&) = p(y, z|x, \&)p(x|\&) = p(x|\&)p(y|x, \&)p(z|x, y, \&).$$

But, there are eight possible ways in which $p(x, y, z|\&)$ can be represented as the product of three conditional densities of the form

$$p(x, y, z|\&) = p\left(x|\Gamma_{y, z}, \&\right)p\left(y|\Gamma_{x, z}, \&\right)p\left(z|\Gamma_{x, y}, \&\right).$$

Six of these are shown in Figure 6.31. These are the six possible conditional density representations of $p(x, y, z|\&)$ that are in the form of an acyclic digraph. The other two representations

$$p(x, y, z|\&) = p(x|z, \&)p(y|x, \&)p(z|y, \&)$$

$$p(x, y, z|\&) = p(x|y, \&)p(y|z, \&)p(z|y, \&)$$

involve cyclic digraphs of probabilities and are, therefore, not allowed because of the impossibility of eliciting the requisite information or knowledge needed for the digraph.

Figure 6.32 also represents this situation in the form of alternate influence diagrams. If we wished to know the probability density of the random variable z, conditioned on the common information &, we could obtain this

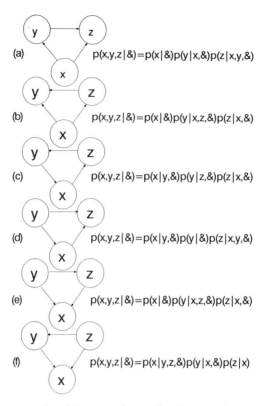

(a)   p(x,y,z|&)=p(x|&)p(y|x,&)p(z|x,y,&)

(b)   p(x,y,z|&)=p(x|&)p(y|x,z,&)p(z|x,&)

(c)   p(x,y,z|&)=p(x|y,&)p(y|z,&)p(z|x,&)

(d)   p(x,y,z|&)=p(x|y,&)p(y|&)p(z|x,y,&)

(e)   p(x,y,z|&)=p(x|&)p(y|x,z,&)p(z|x,&)

(f)   p(x,y,z|&)=p(x|y,z,&)p(y|x,&)p(z|x)

Figure 6.31. Conditioning Relations for Three-Node Diagrams

from an expression such as

$$p(z|\&) = \int\int p(x, y, z|\&) \, dx \, dy$$

$$= \int\int p(x|\&) p(y|x, \&) p(z|x, y, \&) \, dx \, dy.$$

Howard [72] made pertinent suggestions concerning alternate forms of this expression. He considered an illustration for three related variables, x = age, y = education, z = income. Figure 6.31 has illustrated the relevant acyclic knowledge maps or influence diagrams that result from these observations.

Formally, a knowledge map is an influence diagram that has no value nodes and no decision nodes. It is, therefore, equivalent to a probability tree. Some authors call this a *partially formed influence diagram*. The term partial is used because of the lack of an output value node. Inclusion of this would make the influence diagram fully formed. Figures 6.31a and 6.32a indicate

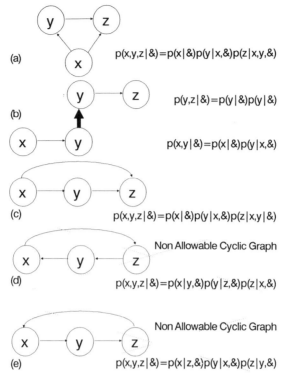

$$p(x,y,z|\&)=p(x|\&)p(y|x,\&)p(z|x,y,\&)$$

$$p(y,z|\&)=p(y|\&)p(y|\&)$$

$$p(x,y|\&)=p(x|\&)p(y|x,\&)$$

$$p(x,y,z|\&)=p(x|\&)p(y|x,\&)p(z|x,y|\&)$$

Non Allowable Cyclic Graph

$$p(x,y,z|\&)=p(x|y,\&)p(y|z,\&)p(z|x,\&)$$

Non Allowable Cyclic Graph

$$p(x,y,z|\&)=p(x|z,\&)p(y|x,\&)p(z|y,\&)$$

Figure 6.32. Additional Conditioning Possibilities for a Three-Chance-Node Set

the map most appropriate for the calculation just indicated. All that is desired is the probability density over z as conditioned on the information $\&$. This can be obtained from

$$p(z|\&) = \int p(y|\&)p(z|y,\&)\,dy,$$

where

$$p(y|\&) = \int p(x|\&)p(y|x,\&)\,dx.$$

To obtain $p(z|\&)$ through use of this two-stage procedure, we may assess a density function over the variable x and another one over the variable y conditioned on the variable x and the common information. We then need to multiply these density functions together and integrate the result to obtain the desired density function $p(y|\&)$. The initial density function estimates suggested by Figure 6.32a involve a one-dimensional assessment, a two-dimensional assessment, and a three-dimensional assessment. Now, we need a one-dimensional assessment and two two-dimensional assessments. Representation by this latter approach requires the sort of disjoint knowledge map represented by Figure 6.32b. This is, indeed, a disjoint representation as probability density of the random variable y is called for twice, with different conditioning variables. We need $p(y|\&)$ and $p(y, x, \&)$, and they are not the same. There might be a temptation to use the influence diagram of Figure 6.32c. This is a well-formed and proper influence diagram, but it is quite the same as that shown in Figure 6.32a.

In a similar manner, we might attempt a representation in the form of Figure 6.32d or Figure 6.28e. However, these are not correct representations of probabilistic information in the sense that they are cyclic. There is simply no way in which the relevant knowledge needed to assess these probability density functions can be assessed because of this cyclic conditioning. Any of the six representations of Figure 6.31 are acceptable, at least in principle.

So, we see that there are a number of perspectives that we need to take concerning knowledge representation by an influence diagram. Doubtlessly, the way knowledge is initially represented and the purposes to which it is put influence this consideration strongly. The various approaches to combining probability distributions [73] are quite relevant to this consideration also, although we will not discuss them here.

It is very important to note that probabilistic, or conditioning, influence is in no way equivalent to the notion of a causal influence. Whereas two variables that are conditionally influenced by a third variable are necessarily correlated, they are not at all necessarily causally influenced.

Shachter [71] has defined an influence diagram to be a single connected network that is comprised of an acyclic directed graph, together with associated node sets, functional dependencies, and information flows. There are three types of nodes: decision nodes, value nodes, and chance nodes. There are two types of arcs: informational arcs and conditioning arcs. An influence diagram is said to be well-formed, fully specified, or fully formed if the following conditions hold.

1. There are no cycles. In other words, an influence diagram is a *directed acyclic graph* or, in other words, a set of nodes or variables and a set of edges or branches that connect the nodes in a directed sense, and where there are no nontrivial paths that begin and end at the same node.

2. There is one, and only one, value node. There is a value function that is defined over the parents of this single value node. A set of nodes P_i are called parents of node n_j if, and only if, there is an edge e_{ji} for each node n_j that is an element of P_i. In a similar manner, a set of nodes C_i are called children of node n_i if, and only if, there is an edge e_{ij} for each node n_j that is an element of P_i. A *barren node* is a node with no

children. A *border node* is a node with no parents. There is much rather specialized graph theory terminology.

3. Each node in the digraph is defined in terms of mutually exclusive and collectively exhaustive states.

4. A joint probability density function is defined over all of the states of the uncertainty nodes.

5. There is at least one path that connects all of the decision nodes both to each other and to the single value node.

6. There are functions over the parents of each deterministic node that are defined over the parents of those nodes.

Even though there exists a unique joint probability density function for a specific well-formed influence diagram, there may be a number of physical influence diagram realizations for any given joint density function.

Three steps enable identification of an appropriate probability distribution from a well-formed influence diagram.

1. *Barren Node Removal:* All decision nodes and chance nodes may be eliminated for the diagram if they do not have successors. If there is nothing that follows a node, that node can have no effect on the outcome value. Such nodes are irrelevant and superfluous and can be eliminated. Formally, this says that a node a, which is barren with respect to nodes b and c can be eliminated from an influence diagram without changing the values these nodes take on, such as $p(b|c, \&)$. Figure 6.33a illustrates this concept. It is important to note that a barren node is not frivolous, but only irrelevant with respect to a particular set of nodes for which it is barren. In other words, $p(x, y, z | \&)$ does not depend at all upon any conditioning W in Figure 6.29a.

2. *Deterministic Node Propagation:* If a well-formed influence diagram contains an arc from a deterministic node a to node b, which may be a chance node or a deterministic node, it is possible to rearrange or transform the influence diagram to one in which there is no edge from node a to b. The new influence diagram will have node b inheriting all conditional predecessors of node a. Furthermore, if node b was deterministic before the transformation, it will remain a deterministic node. Figure 6.33b illustrates this concept.

3. *Arc Reversal:* In a well-specified influence diagram in which there is a single directed path or arc from probabilistic node a to node b, the diagram may be transformed to one in which there is an arc from node b to node a and where the new nodes a and b inherit the conditional predecessors of each node. If node b was initially deterministic, it becomes probabilistic. If it was initially probabilistic, it remains probabilistic. Figure 6.33c illustrates this concept.

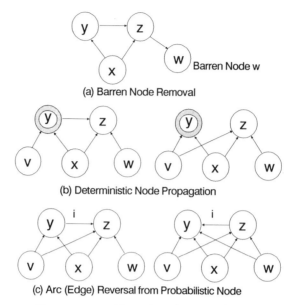

Figure 6.33. Additional Node Manipulations

These simple reductions can be grouped together into a series of transformations that potentially resolve influence problems. This leads to three additional steps.

4. *Deterministic Node, with a Value Node as the Only Possible Successor, Removal:* A given deterministic node may be removed from the network. The given deterministic node is propagated into each of its successors until it has none and is then barren and can be eliminated from the diagram. Figure 6.33b has actually illustrated this. Node y, a deterministic node, is barren after the manipulation leading to Figure 6.33b.

5. *Decision Node, with a Value Node as the Only Possible Successor, Removal:* A given decision node may be removed from the network. When any conditional predecessors of the value node that are not observable at the time of decision, which are typically the successors of the decision node in question, are removed first, and when the decision node is a conditional predecessor of the value node, it may be removed. No new conditional predecessors are inherited by the value node as a result of this operation and the operation ends when all predecessors to value nodes have been removed. Decision nodes are removed through the maximization of expected value, or subjective expected utility. (This concept is discussed further in Chapter 7).

6. *Probabilistic Node, with a Value Node as the Only Possible Successor, Removal:* A probabilistic or chance node that has only a value node as a successor can be removed. In some cases, it is necessary to reverse a conditioning arc between the node and other successors so that the value node inherits conditional predecessors of the node that is removed. Figure 6.33c illustrates this concept. Node y can be removed as it is a barren node after the manipulations leading to Figure 6.33c.

Clearly, these three steps follow from the first three stated. A relatively complete discussion of these steps is contained in Shachter [74, 75] and Call and Miller [76]. Included in these efforts is a discussion of transformations needed to solve inference and decision problems, sufficient information to perform conditional or unconditional inference, and the associated information requirements for decision making, including calculations of the value of (perfect) information.*

One of the questions that naturally arises is whether an influence diagram type of representation or decision tree type of representation is preferable. The answer, of course, is problem and perspective dependent. Figures 6.34 and 6.35 provide some comparisons of alternative representations of decision situations in terms of influence diagrams and decision trees. It is relatively easy to construct a situation for which one representational framework is the best. In any case, both decision trees and influence diagrams are equivalent to spreadsheet-like matrix representations.

One particularly impressive demonstration of potential superiority of the influence diagram representation is in situations where probabilistic independence exists. Figure 6.36 illustrates this sort of situation in decision tree format and influence diagram format. What is displayed here is a case where $p(c_2|c_1, \&) = p(c_2|\&)$. This independence is clearly illustrated in the influence diagram, but it is not at all evident in the decision tree structure unless we actually examine the probabilities shown in the tree. This effective representation of probabilistic independence also makes it easier to enforce the distinction between probabilistic influences and values influences.

A relatively good illustration of this is provided by the influence diagram of Figure 6.37. There are three decision nodes in this figure: detection (DE), diagnosis (DI), and correction (CO). The figure shows directly unobservable components which are influenced by some chance mechanism, C, which is also a border node. The detection decision outcome is influenced by the chance mechanism and some additional random mechanism CDE. Depending on the detection result, we enter a diagnostic decision phase, DI. Again,

---

*The Call and Miller paper also contains a very useful discussion of available influence diagram software. INDIA, from Decision Focus, and Decision Programming Language (DPL), are two commercially available MS-DOS products for implementation of influence diagram type structuring approaches. Although INDIA is able to deal only with influence diagrams, DPL provides a fourth-generation command language that is able to result in either influence diagrams or decision tree structures.

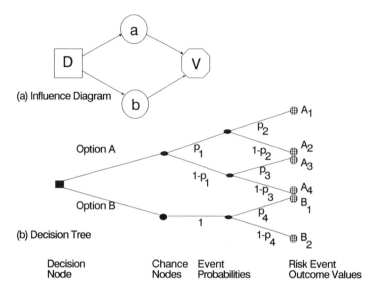

Figure 6.34. Simple Influence Diagram and Associated Decision Tree

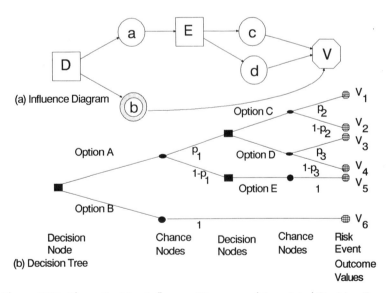

Figure 6.35. Three-Decision Influence Diagram and Associated Decision Tree

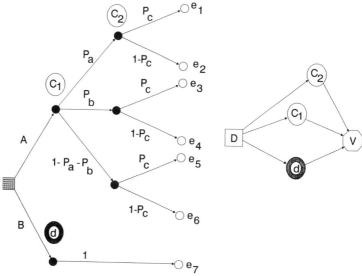

Figure 6.36. Decision Tree and Equivalent Influence Diagram Illustrating Conditional Independence of Two Chance Nodes

there are chance mechanisms involved that influence the actual diagnostic outcome. Finally, the correction decision phase (CO) produces an outcome that depends on the failure state of the system and the chance result of the corrective effort. This influence diagram is a relatively illuminating and straightforward representation of the decision situation. The decision tree is not comparably illuminating and straightforward. This difficulty usually increases for more complex decisions and provides some potential and real advantages to the influence diagram approach.

Recent efforts [77] have introduced the concept of super value nodes that enable the representation of separable value functions in influence diagrams. Other recent extensions to influence diagram efforts [78] remove restrictions

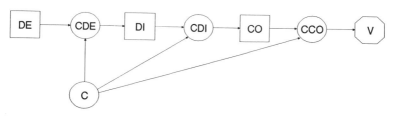

Figure 6.37. Influence Diagram for Fault Detection (DE), Diagnosis (DI), and Correction (CO)

2, 3, 4, and 6 on pages 299–301 from the definition of a well-formed influence diagram. This enables the expansion of the influence diagram concept to decision processes in which value aspects are critical. In particular, it enables the determination of value driven clusters and decision driven clusters of elements so that a relatively complex influence diagram may be viewed in an abstracted form, which simplifies its representation and presentation.

6.5. SUMMARY

In this chapter, we have discussed a number of relevant issues. One of the major concerns in earlier chapters has been the identification and representation of requirements. We specifically address this concern in the first part of this chapter. Clearly, activities that are undertaken to produce an intended result will not always produce precisely that result, and only that result. Inevitably, risks are associated with intentional efforts. We also examine a variety of risk management issues here. Finally, there are a number of methods and tools that can be used to both represent and communicate requirements and risks. We conclude the chapter with a presentation of some of these important methods and tools for systems engineering.

PROBLEMS

6.1. Determine a cross-interaction matrix between the five risk management facets that we identified and Deutsch's project success factors. Also prepare a discussion of the relationships between requirements identification and these two sets of factors.

6.2. Define "requirements engineering." Describe four approaches to identifying client requirements for a system. Contrast and compare these in terms of anticipated costs and benefits associated with each approach.

6.3. In systems engineering, one often speaks of process and product. Describe the requirements engineering process. Describe the requirements engineering product. How are these related? What is an appropriate requirements engineering process? How do you identify an appropriate requirements engineering process?

6.4. It is possible, and generally desirable, to speak of structured requirements, functional requirements, and purposeful requirements. How are these related, and how are they incorporated in a (typical and appropriate) requirements engineering process and product.

6.5. Please discuss the role of rapid (rabid) prototyping in requirements engineering.

6.6. Evaluation, and related efforts in audits, reviews, verification, validation, etc., of the results of the requirements engineering process might generally be expected to lead to better requirements engineering. How?

6.7. How does one deal with potentially imperfect information in the requirements engineering process? Give some specific examples of imperfect information and illustrate how this might arise.

6.8. How is requirements engineering in a systems integration house similar to and different from requirements engineering in system hardware development house? Please discuss the impact of the requirements engineering process on the likely cost effectiveness of the resulting process and product for each house.

6.9. What are possible roles for requirements in the spiral life cycle process?

6.10. How is requirements engineering a part of the concurrent engineering process?

6.11. What role does requirements engineering play in configuration management? Please discuss, in particular, reviews and audits relative to requirements.

6.12. John Rockart [15] of MIT has coined the term *critical success factors* (CSF) to denote those few critical areas where things must go right for an organization to flourish and for goals to be attained and objectives achieved. We would consider the hierarchical structure of an existing corporation. This might consist of a President or CEO, VPs for research, finance, manufacturing, marketing, etc. We could attempt to identify the CSFs of each of these and thereby obtain a hierarchy of CSF for the organization. Suppose that the intent of a systemhouse is to discern requirements for an information system. How does the process of identifying CSFs differ from the process of requirements engineering? How is the product of the two similar? Different?

6.13. Suppose that the precedence and duration relations for a project are as follows.

Activity	Duration	Predecessor	Successor
a	2	Start	b
b	3	a	c, g, e
c	4	b	d
d	6	c	e
e	4	d	f
f	4	e, i, k	g
g	5	f	1
h	3	g	i
i	2	h	f
j	6	e	k
k	7	j	f

What is the critical path for the network, and the minimum time to project completion? What is the slack time(s) for project completion?

6.14. Please construct a project management network that has an error in one network loop, a cycle in which activity c depends upon activity b which depends upon activity c. How might this be detected?

6.15. What are the implications of the sort of feedback iterations that might occur around phases of the life cycle on PERT- and CPM-based scheduling algorithms?

6.16. Is it absolutely necessary to have the successor relations, or is knowledge of the predecessor relations sufficient?

6.17. Please draw a PERT diagram and perform an analysis for a project with the following precedence and duration relations.

Activity	t_o	t_a	t_p	Pre.	Suc.
a	1	3	4	a	b, c
b	1	3	5	a	d, e
c	2	3	4	b	f, i
d	6	8	9	b	g
e	4	7	9	e	h
f	3	6	7	d	h
g	4	7	8	e	j
h	2	5	8	e	j
i	4	4	5	g, h, i	j
j	4	5	6		End

Construct a chart showing the various statistical relations obtained from this project activities.

6.18. Identify the elements for and construct a tree structure of intents or objectives for your university or company. What is the self-interaction matrix that corresponds to these intents. Postulate a process where, for a very complicated and large set of elements, you might elicit responses that enable construction of a tree (or other) structure from elicitation of a reachability matrix that indicates which elements are influenced by, or influence, other elements according to some contextual relationship.

6.19. Take the objectives tree identified in Problem 6.18 or some other identified tree of objectives. Develop an activities tree that corresponds to the objectives tree. Is this a simple effort to accomplish? Why or why not?

6.20. Take the objectives tree identified in Problem 6.18 or some other identified tree of objectives. Develop an attributes tree that corresponds to the objectives tree. Is this a simple effort to accomplish? Why?

6.21. What efforts would be needed to construct a PERT or CPM activities flow from the activity tree obtained in Problem 6.19?

6.22. What efforts would be needed to construct a Gantt chart and a Pareto chart from the activity tree obtained in Problem 6.19?

6.23. Please identify a problem area of interest to you. Construct a hierarchy of objectives and intents for resolution of the problem. Identify an activity tree of elements that might aid in achieving these intents. Construct a PERT chart and a Gantt chart for these activities. Identify appropriate milestones on the Gantt chart.

6.24. Please develop the spreadsheet-like decision situation model, of options and potential outcomes for the influence diagrams and decision trees of Figures 6.34 and 6.35.

6.25. Are there more chance nodes in the decision tree representation of Figure 6.34 than the influence diagram representation? What has been accomplished to allow this?

6.26. One of the cycles described in the TQM literature is the Shewart cycle. It involves (1) planning a change; (b) doing or implementing the change, generally on a small scale initially; (c) checking and observing the effects brought about by the change; and (d) acting on the basis of what was learned, by either implementing the change on a broad basis or iterating through the cycle with an improved set of plans. Discuss this Shewart cycle in terms of the methods discussed in Section 6.5.

6.27. Please write a brief discussion on how you might go about knowledge elicitation to identify the probabilities needed for the cyclic representation of Figure 6.32d.

6.28. Discuss cross-impact analysis and its relation to influence diagrams. Would an influence diagram be of use for cross-impact analysis?

6.29. Discuss the equivalence, or lack thereof, between event trees and influence diagrams.

6.30. Please draw the decision tree equivalent to Figure 6.37.

6.31. The TQM literature often discusses a number of TQM tools. Included among these are cause-and-effect diagrams, checklists, control charts, decision matrix, Pareto charts, flowcharts, histograms, and scatter diagrams. Provide a brief description of each of these and illustrate how they may be obtained from the approaches discussed in this chapter.

6.32. Provide definitions of a good decision, bad decision, good outcome, and bad outcome. Provide an illustration of each of these.

6.33. Discuss trustworthiness and TQM impacts on the factors on page 258.

REFERENCES

[1] Volkema, R. J., "Problem Formulation," in Sage, A. P. [Ed.], *Concise Encyclopedia of Information Processing in Systems and Organizations*, Pergamon Press, Oxford, 1990, pp. 377–382.

[2] Smith, G. F., "Defining Managerial Problems: A Framework for Prescriptive Theorizing," *Management Science*, Vol. 35, No. 8, 1989, pp. 963–981.

[3] Thayer, R. H., and Dorfman, M. (Eds.), *System and Software Requirements Engineering*, IEEE Computer Society Press, Los Alamitos CA, 1990.

[4] Dorfman, M., and Thayer, R. H., (Eds.), *Standards, Guidelines, and Examples on System and Software Requirements Engineering*, IEEE Computer Society Press, Los Alamitos CA, 1990.

[5] Simon, H., "Applying Information Technology to Organization Design," *Public Administrative Review*, May/June 1973, pp. 268–278.

[6] Shannon, C. E., and Weaver, W., *The Mathematical Theory of Communication*, University of Illinois Press, Urbana IL, 1949.

[7] Taggart, W. M., and Tharp, M. O., A Survey of Information Requirements Analysis Techniques," *Computing Surveys*, Vol. 9, No. 4, 1977, pp. 273–290.

[8] Tiechroew, D., "A Survey of Languages for Stating Requirements for Computer Based Information Systems," *Proceedings AFIPS National Computer Conference*, Vol. 41, 1972, pp. 1203–1224.

[9] Tiechroew, D., and Hershey, E. A., "PSL/PSA: A Computer Aided Technique for Structured Documentation and Analysis of Computer Based Information Systems," *IEEE Transactions on Software Engineering*, Vol. 3, No. 1, 1977, pp. 41–48.

[10] Yadav, B. B., "Determining an Organization's Information Requirements: A State of the Art Survey," *Data Base*, Spring 1983, pp. 3–20.

[11] Davis, G. B., "Strategies for Information Requirements Determination," *IBM Systems Journal*, Vol. 21, No. 1, 1982, pp. 4–30.

[12] Naumann, J. D., Davis, G. B., and McKeen, J. D., "Determining Information Requirements: A Contingency Method for Selection of a Requirements Assurance Strategy," *The Journal of Systems and Software*, Vol. 1, 1980, pp. 273–281.

[13] Davis, G. B., and Olson, M., *Management Information Systems*, McGraw Hill, New York, 1985.

[14] Sage, A. P., and Palmer, J. D., *Software Systems Engineering*, John Wiley, New York, 1990.

[15] Rockart, J. F., "Chief Executives Define Their Own Data Needs: Critical Success Factors," *Harvard Business Review*, Vol. 57, No. 2, 1979, pp. 81–93.

[16] Rockart, J. F., and Bullen, C. V. (Eds.), *The Rise of Managerial Computing*, Dow Jones Irwin, Homewood IL, 1986.

[17] Pinto, J. K., and Prescott, J. E., "Variations in Critical Success Factors Over the Stages in the Project Life Cycle," *Journal of Management*, Vol. 14, No. 1, 1988, pp. 5–18.

[18] Selvin, D. P., and Pinto, J. K., "The Project Implementation Profile: New Tool for Project Managers," *Project Management Journal*, Vol. 18, 1986, pp. 57–70.

[19] Schultz, R. L., Slevin, D. P., and Pinto, J. K., "Strategy and Tactics in a Process Model of Project Implementation," *Interfaces*, Vol. 17, No. 3, 1987, pp. 34–46.

[20] Heninger, K. L., "Specifying Software Requirements for Complex Systems: New Techniques and Their Applications," *IEEE Transactions on Software Engineering*, Vol. 6, No. 1, 1980, pp. 2–13.

[21] Sommerville, I., *Software Engineering*, Addison-Wesley Publishing Co., Reading MA, 1982.

[22] Davis, A. M., *Software Requirements: Analysis and Specification*, Prentice Hall, Englewood Cliffs NJ, 1990.

[23] Ross, B. H., Ryan, W. J., and Tenpenny, P. L., "The Analysis of Relevant Information for Solving Problems," *Memory and Cognition*, Vol. 17, No. 5, 1989, pp. 639–651.

[24] Yadav, S. B., and Chand, D. R., "An Expert Modeling Support System for Modeling an Object System to Specify Its Information Requirements," *Decision Support Systems*, Vol. 5, 1989, pp. 29–45.

[25] Jain, H. K., Tanninu, M. R., and Faziollahi, B., "MCDM Approaches for Generating and Evaluating Alternatives in Requirement Analysis," *Information Systems Research*, Vol. 2, No. 3, 1991, pp. 223–229.

[26] Saunders, C., and Jones, J. W., "Temporal Sequences in Information Acquisition for Decision Making: A Focus on Source and Medium," *Academy of Management Review*, Vol. 15, No. 1, 1990, pp. 29–46.

[27] March, J. G., "Ambiguity and Accounting: The Elusive Link Between Information and Decision Making," *Accounting: Organizations and Society*, Vol. 12, No. 2, 1987, pp. 153–168.

[28] Daft, R. L., and Lengel, R. H., "Organizational Information Requirements, Media Richness and Structural Design," *Management Science*, Vol. 32, No. 5, 1986, pp. 554–571.

[29] Slovic, P., and Lichtenstein, S., "Relative Importance of Probabilities and Payoffs in Risk Taking," *Journal of Experimental Psychology*, Vol. 78, Part 2, 1968, pp. 1–18.

[30] Libby, R., and Fishburn, P. C., "Behavioral Models of Risk Taking in Business Decisions: A Survey and Evaluation," *Journal of Accounting Research*, Vol. 14, 1977, pp. 272–292.

[31] Lowrence, W. W., *Of Acceptable Risk*. Kaufman, Los Altos CA, 1976.

[32] Camerer, C. F., and Kunreuther, H., "Decision Processes for Low Probability Events: Policy Implications," *Journal of Policy Analysis and Management*, Vol. 8, No. 4, 1989, pp. 565–592.

[33] Deutsch, M. S., "An Explanatory Analysis Relating the Software Project Management Process of Project Success," *IEEE Transactions on Engineering Management*, Vol. 38, No. 4, 1991, pp. 365–375.

[34] Singh, M. G. (Ed.)., *Systems and Control Encyclopedia*, Pergamon Press, Oxford, 1987.

[35] Covello, V. T., Menkes, J., and Mumpower, J. (Eds.), *Risk Evaluation and Management*, Plenum, New York, 1986.

[36] Merkhoffer, M. W., *Decision Science and Social Risk Management*, Reidel, Dordrecht, 1987.

[37] Porter, A., Rossini, F. A., Carpenter, S. R., and Roper, A. T., *A Guidebook for Technology Assessment and Impact Analysis*, North Holland, New York, 1980.

[38] Porter, A. L., Roper, A. T., Mason, T. W., Rossini, F. A., and Banks, J., *Forecasting and Management of Technology*, John Wiley, New York, 1991.

[39] Fischoff, B., "Informed Consent in Societal Risk-Benefit Decisions," *Technological Forecasting and Social Change*, Vol. 15, 1979, pp. 347–357.

[40] Sage, A. P., and White, E. W., "Methodologies for Risk and Hazard Assessment: A Survey and Status Report," *IEEE Transactions on Systems, Man, and Cybernetics*, Vol. 10, No. 8, 1980, pp. 425–446.

[41] Fischhoff, B., Lichtenstein, S., Slovic, P., Derby, S. L., and Keeney, R. L., *Acceptable Risk*, Cambridge University Press, New York, 1981.

[42] Covello, V. T., von Winterfeldt, D., and Slovic, P., "Risk Communication: A Review of the Literature," *Risk Abstracts*, Vol. 3, No. 4, 1986, pp. 171–182.

[43] Wiest, J. D., and Levy, F. K., *Management Guide to PERT/CPM*, Prentice Hall, Englewood Cliffs NJ, 1977.

[44] Archibald, R., and Villoria, R., *Network-Based Management Systems*, (*PERT/CPM*), John Wiley, New York, 1967.

[45] Woodgate, H. S., *Planning by Network*, Brandon/Systems, New York, 1964.

[46] Badiru, A. B., *Project Management in Manufacturing and High Technology Operations*, John Wiley, New York, 1988.

[47] Blanchard, B. S., and Fabrycky, W. J., *Systems Engineering and Analysis*, Prentice Hall, Englewood Cliffs NJ, 1990.

[48] Kezsbom, D. S., Schilling, D. L., and Edward, K. S., *Dynamic Project Management*, John Wiley, New York, 1989.

[49] Michaels, J. V., and Wood, W. P., *Design to Cost*, John Wiley, New York, 1989.

[50] Clark, W., *The Gantt Chart*, 3rd ed., Pitman, London, 1957.

[51] Grant, E. L., and Leavenworth, R. S., *Statistical Quality Control*, 6th ed., McGraw Hill, New York, 1988.

[52] Harary, F., Norman, R. Z., and Cartwright, D., *Structural Models: An Introduction to the Theory of Directed Graphs*, John Wiley, New York, 1965.

[53] Warfield, J. N., *Societal Systems: Planning, Policy, and Complexity*, John Wiley, New York, 1976.

[54] Sage, A. P., *Methodology for Large Scale Systems*, McGraw Hill, New York, 1977.

[55] Steward, D. V., *Systems Analysis and Management: Structure, Strategy, and Design*, Petrocelli Books, New York, 1981.

[56] Lendaris, G. G., "Structural Modeling: A Tutorial Guide," *IEEE Transactions on Systems, Man, and Cybernetics*, Vol. SMC 10, No. 12, 1980, pp. 807–840.

[57] Eden, C., Jones, S., and Sims, D., *Messing About in Problems*, Pergamon Press, Oxford, 1983.

[58] Roberts, F. M., *Discrete Mathematical Models*, Prentice Hall, Englewood Cliffs NJ, 1976.

[59] Geoffrion, A. M., "An Introduction to Structured Modelling," *Management Science*, Vol. 33, No. 5, 1987, pp. 547–588.

[60] Geoffrion, A. M., "The Formal Aspects of Structured Modeling," *Operations Research*, Vol. 37, No. 1, 1989, pp. 30–51.

[61] Geoffrion, A. M., "Computer Based Modeling Environments," *European Journal of Operations Research*, Vol. 41, No. 1, 1989, pp. 33–43.

[62] Geoffrion, A. M., "FW/SM: A Prototype Structured Modeling Environment," *Management Science*, Vol. 37, No. 12, 1991, pp. 1513–1538.

[63] Hill, J. D., and Warfield, J. N., "Unified Program Planning," *IEEE Transactions on Systems, Man, and Cybernetics*, Vol. SMC-2, 1972, pp. 610–622.

[64] Clausing, D., "Quality Function Deployment: Applied Systems Engineering," *Proceedings 1989 Quality and Productivity Research Conference*, University of Waterloo, June 1989.

[65] Hauser, J. R., and Clausing, D., "The House of Quality," *Harvard Business Review*, Vol. 66, No. 3, 1988, pp. 63–73.

[66] Clausing, D., and Pugh, S., "Enhanced Quality Function Deployment," *Proceedings of the Design and Productivity International Conference*, Honolulu, February 1991.

[67] Warfield, J. N., "On Arranging Elements of a Hierarchy in Graphic Form," *IEEE Transactions on Systems, Man, and Cybernetics*, Vol. SMC-3, No. 2, 1973, pp. 121–132.

[68] Howard, R., and Matheson, J. E., "Influence Diagrams," in Howard, R., and Matheson, J. E. (Eds.), *The Principles and Applications of Decision Analysis*, Stanford University Press, Stanford, CA, 1984.

[69] Forrester, J. W., *Urban Dynamics*, MIT Press, Cambridge MA, 1969.

[70] Howard, R. A., and Matheson, J. E., "Influence Diagrams," *Readings on the Principles and Applications of Decision Analysis, Vol. II*, Strategic Decisions Group, Palo Alto CA, 1984, pp. 719–762.

[71] Shachter, R. D., "Evaluating Influence Diagrams," *Operations Research*, Vol. 34, No. 6, 1986, pp. 871–882.

[72] Howard, R. A., "Knowledge Maps," *Management Science*, Vol. 35, No. 8, 1989, pp. 903–922.

[73] Genest, C., and Zidek, J. V., "Combining Probability Distributions: A Critique and an Annotated Bibliography," *Statistical Science*, Vol. 1, No. 1, 1986, pp. 114–148.

[74] Shachter, R. D., "Probabilistic Influence and Influence Diagrams," *Operations Research*, Vol. 36, No. 4, 1988, pp. 589–604.

[75] Shachter, R. D., "An Ordered Examination of Influence Diagrams," *Networks*, Vol. 20, 1990, pp. 535–563.

[76] Call, H. J., and Miller, W. A., "A Comparison of Approaches and Implementations for Automating Decision Analysis," *Reliability Engineering and System Safety*, Vol. 30, 1990, pp. 115–162.

[77] Tatman, J. A., and Shachter, R. D., "Dynamic Programming and Influence Diagrams," *IEEE Transactions Systems, Man, and Cybernetics*, Vol. 20, No. 2, 1990, pp. 365–379.

[78] Buede, D. M., and Ferrell, D. O., "Convergence in Problem Solving: A Prelude to Quantitative Analysis," *IEEE Transactions on Systems, Man, and Cybernetics*, Vol. 22, No. 6, December 1992.

Chapter 7

Decision Assessment

The systems engineering life cycle is intended to enable evolution of high-quality, trustworthy systems that have appropriate structure and function support for identified purposeful objectives. Many major efforts called for throughout all phases of a systems engineering life cycle involve the making of decisions. This chapter continues our discussions of methodological issues in systems engineering. The focus, however, shifts to normative, descriptive, and prescriptive aspects of decision making. Much has been said on this subject.* We attempt to present the salient features of some of the work in this area that is most relevant to systems engineering. As our coverage is necessarily broad, it helps to first provide an overview and perspective. Then we examine normative decision making, descriptive decisionmaking, and prescriptive decisionmaking.

7.1. INTRODUCTION: TYPES OF DECISIONS

Decisions range from very simple to very complex and from very narrow to very broad. We make personal and organizational or business decisions every day. We make very simple decisions without much explicit consideration of the factors affecting the decision or those individuals who may be affected by the decision. At least we often think that this is what we do. In reality, a decision often seems *simple* because it concerns an issue, environment, and relevant considerations with which we are familiar. Intuition or other skill-based forms of reasoning may not only be perfectly acceptable in these

---

* "Much" is a very large understatement. This is, by far, the largest chapter in this book. We have not even begun to scratch the surface of this important subject.

situations, but perhaps actually preferred to a very formal and lengthy process. In the case of more complex decisions, we give much more thought and consider more of the factors involved, especially when we do not have a wealth of initial experience with the decision situation. The term *decision situation* is intended to be a relatively broad term that includes the objectives to be achieved, the needs to be fulfilled, constraints and alterables associated with the decision, those affected by the decision, the decision options themselves, the environment in which each of these are imbedded, and the experiential familiarity of the decision maker with all of these. Thus, the decision situation is dependent upon many contingency variables.

Depending on the complexity and scope involved, the thought first given to an impending decision may include a brief mental comparison of experientially familiar alternatives, or it may be a thorough analysis appropriate to a complex, less-understood situation in which there are significant differences in the impacts of various alternative courses of action.

The decision assessment process may be described as follows. The decision maker is presented with a problem that requires a decision. Certain objectives may be provided by those to whom the decision maker is responsible. Also available, or identified as part of the effort, are certain alternative courses of action, each of which satisfies the objectives in some way and to some extent. The problem is to choose the alternative course of action that best meets or satisfies the objectives and is fully responsive to any constraints. If there is more than one factor contributing to the satisfaction of the objectives, the decision maker should generally find some way to combine the effects of several factors in some best or most appropriate way.

Descriptive, or behavioral, decision assessment is concerned with the way in which human beings in real situations actually make decisions. The term *way* is intended to include both the process used to make a decision and the actual decision that is made.

Normative decision analysis is often called rational decision analysis or axiomatic decision analysis. In the more formal approaches, a set of *axioms* that a *rational* person would surely agree with is postulated. From this results the *normative*, or most desirable, or optimum, behavior which the rational decisionmaker should seek to follow in order to conform to the accepted axioms. *Prescriptive* decision assessment involves suggestions of appropriate decision behavior that tempers the formal normative approaches to insure that the process reflects the needs of real people in real decision situations. For the most part, our concern here is with prescriptive approaches. Since the prescriptive includes some blend of the normative and the descriptive, as indicated in Figure 7.1, we must really be aware of all three approaches. Experimental and other evidence indicate that the normative theory is often unacceptable as a descriptive theory of decision making. The normative theory is intended as a standard guide to what people would have to do to accommodate very desirable axioms for judgment. Since unmodified normative theory does not describe unaided descriptive decision

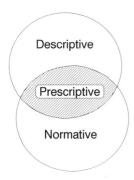

Figure 7.1. Prescriptive Decision Support as the Confluence of the Normative and the Descriptive

making, considerable caution in the use of the theory for aided prescriptive behavior should be exercised. Otherwise, very serious cognitive stress may result from potential acceptance of a normative theory that may be at variance with unaided descriptive behavior. The fact that unaided decision behavior (what people do) differs from the normative (what people ideally could do) and the prescriptive (what real people should do) appears not a criticism of the normative theory but, rather, a strong indication of need for the theory.

In providing prescriptive support for decision making, great care must be taken to understand the decision situation extant. Unless there is a good assessment of the decision situation, efforts at decision selection may not produce the intended results. The primary objectives of a prescriptive decision assessment are as follows:

1. To formulate the issue in terms of objectives to be obtained, needs to be satisfied, and the identification of potential alternative courses of action;
2. To analyze the impacts of the alternatives upon the needs and objectives of the appropriate group of stakeholders; and
3. To interpret these impacts in terms of the objectives so as to enable selection of a most appropriate* alternative course of action, given all of the realities of the decision situation.

These are just the formal steps of the systems engineering process, as described earlier, and there appears to be little difference between decision

---

*Often, this is called the *best*, or *optimum* decision. Although we have no problem with this terminology, it has sometimes led to the *paralysis through analysis* effect noted in Chapter 5. It is quite possible to have models that are too complex for the realities of the decision situation and that are otherwise inappropriate, and to then sublimate the real task of aiding the decision maker in selecting the best decision with solution of the artificial optimization problem that has been posed.

assessment and a general systems engineering approach to problem resolution. Often the term *decision analysis* is used for the complete effort. In the way that we describe it, analysis is one of the three fundamental steps of an assessment effort.

Many would argue that these objectives are precisely the objectives of normative, or formal-rational, decision analysis. Then, the identification of a prescriptive approach that is distinct from the normative approach would be unneeded. Another way of classifying decision making is identifying whether it is technique oriented, problem oriented, or user oriented. Related commentaries may be found in Brown [1], Howard [2], and Watson [3].

We may subdivide decision assessment efforts into five types:

1. *Decision under Certainty:* Those issues in which each alternative action results in one and only one outcome and where that outcome is sure to occur. The decision situation structural model is established and is correct. There are no parametric uncertainties.

2. *Decision under Probabilistic Uncertainty:* Those issues in which one of several outcomes can result from a given action depending on the state of nature; these states occur with known probabilities. The decision situation structural model is established and is correct. There are outcome uncertainties, and the probabilities associated with these are known precisely. The utility of the decision maker for the various event outcomes can be quite precisely established.

3. *Decision under Probabilistic Imprecision:* Those issues in which one of several outcomes can result from a given action depending on the state of nature, and these states occur with unknown or imprecisely specified probabilities. The decision situation structural model is established and is correct. There are outcome uncertainties, and the probabilities associated with the uncertainty parameters are not all known precisely. The utility of the decision maker for the events outcomes can be quite precisely established.

4. *Decision under Information Imperfection:* Those issues in which one of several outcomes can result from a given action depending on the state of nature, and these states occur with imperfectly specified possibilities.* The decision situation model is established but may not be fully specified. There are outcome uncertainties, and the possibilities associated with these are not all known precisely. Imperfections in knowledge of the utility of the decision maker for the various event outcomes may exist as well.

5. *Decision under Conflict:* Those issues in which there is more than a single decision maker, and where the objectives and activities of one

---

*Here, we use the term *possibility* to refer to a chance occurrence, and do not wish to necessarily state that the conventional Bayesian probabilistic rules are to be used.

decision maker are not necessarily known to all decision makers. Decision under conflict issues may also be viewed as those in which *nature* is augmented or replaced, at least in part, by a not necessarily hostile challenger. We could further subdivide these issues according to information imperfection concerns. We shall not do this here, as we will not consider decision under conflict issues here.

Problems in any of these groupings may be approached from a normative, descriptive, or prescriptive perspective. Problems in category 1 are those for which deterministic principles may be applied. This condition of known states of nature ignores the overwhelming majority of issues faced in typical private or public sector decision making, including such factors as how will my customer base or constituency be affected, what will happen to the cost or quality of my product or service, how will the morale and organization of my workforce be affected, how does my ever-changing institutional and societal environment bear on this decision, what are the implications of this decision on other decisions I must make elsewhere in my organization, and so on.

Problems in category 5 are game theoretic based problems and will not be considered in this text. The majority of decision assessment efforts have been applied to issues in grouping or category 2, although current approaches to decisions under information imperfections allow solutions to some problems in categories 3 and 4 as well. We initially concentrate on a description of problems in category 2, and will then provide an overview of some solution methods for problems in categories 3 and 4.

Figure 7.2 illustrates a simple decision tree structure that can be used to illustrate the differences among these five categories. For category 1, a decision under certainty problem, all of the probabilities are equal to either 1.0 or 0.0, such that there is a single known outcome from each alternative. In general, Figure 7.2 is most suitable to a category 2 problem. It is also valid for a category 3 problem, except that some of the probabilities themselves, are, unknown. In a category 4 problem, there may be event outcomes that have not been identified. In addition, probabilities of the outcomes may not necessarily be specified. In a category 5 problem, the probabilities take on the nature of variables that are to some extent alterable by a competitor.

Decision assessment provides a framework for describing how people do choose, or ideally could best choose, or should choose among alternative courses of action when the outcomes resulting from these alternatives are clouded by uncertainties and information imperfections. These are the descriptive, normative, and prescriptive approaches to decision assessment noted earlier.

Many foundations for decision assessment are available. Behavioral psychology provides us with information concerning both the process and the product of judgment and choice efforts. In principle, systems modeling provides us with methods that can be used to represent decision situations. Probability theory allows the decision maker to make maximum use of the

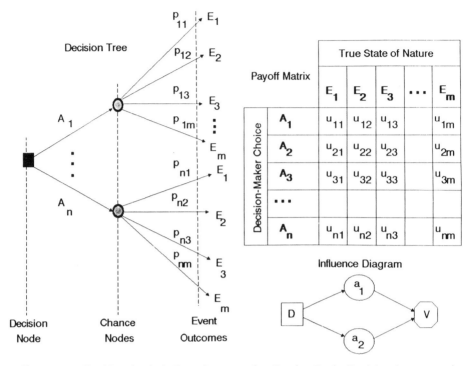

Figure 7.2. Decision Analysis Tree Structure for Simple, Single Decision Lottery and Associated Payoff Matrix

uncertain information that is available. Utility theory guarantees that the choice will reflect the decision maker's preferences if all aspects of the decision situation structural model have been modeled correctly.

In order to develop a descriptive theory that agrees with how decisions could be made in a normative sense, we need to assume the following.

1. Past preferences are valid indicators of present and future preferences.
2. People correctly perceive the values of the uncertainties that are associated with the outcomes of decision alternatives.
3. People are able to assess decision situations correctly, and the resulting decision situation structural model is well formed and complete.
4. People make decisions that accurately reflect their true preferences over alternatives that have uncertain outcomes.
5. People are able to process decision information correctly.
6. Real decision situations automatically provide people with decision alternatives that allow them to express their true preferences.
7. People accept the axioms that are assumed to develop the various normative theories.

8. People make decisions without being so overwhelmed by the complexity of actual decision situations that they must necessarily use suboptimal decision strategies.

Given these necessary assumptions for a valid descriptive and a valid normative theory, it is not surprising that departures exist between normative and descriptive theories of decision making.* A principal task and responsibility of those systems engineering professionals who seek to aid others in decision assessment and decision aiding is to retain those features from the descriptive approach that enable an acceptable transition from normative approaches to prescriptive approaches. The prescriptive features should eliminate potentially undesirable features of descriptive approaches, such as flawed judgment heuristics and information processing biases, while retaining acceptable features of the normative approaches.

There are a variety of approaches to normative, prescriptive, and descriptive decision assessment. Most of these rely on what is called utility theory. Fishburn [4, 5] has identified 25 different variations, or generalizations, of what we call expected utility theory. Most of the variations are minor and most lead to relatively similar normative recommendations. We discuss some of the major generic approaches here.

7.2. FORMAL DECISIONS

In this section, we examine approaches intended to aid in the process of making decisions, or, more precisely, making effective decisions.

7.2.1. Prescriptive and Normative Decision Assessments

A prescriptive decision assessment provides a systematic framework of search, deliberation, evaluation, and selection that facilitates selection of a most preferred course of action for a decision maker in a complex decision situation. In a formal sense, this may be accomplished by a seven-step process of formulation, analysis, and interpretation.

I. *Formulation:* This results in knowledge of the decision issue to be resolved.
   1. Assess the decision situation. This includes the identification of objectives, needs, constraints, alterables, and potential alternative courses

---

*It turns out that assumption 7 is not a real difficulty in that most of the axiomatic basis for the normative theory is usually above reproach. There have been, to the author's knowledge, no really successful attacks on the axiomatic foundations of the theory, although there have been a great many observations that the axioms are not necessarily implementable by real people in realistic settings.

of action. These are the ingredients of the decision situation structural model.

II. *Analysis:* This results in a complete decision tree model of the situation.

2. Structure the relationship between the decision alternatives and outcomes, typically in the form of a decision tree or some other appropriate decision situation structural model, such as an influence diagram.

3. Encode the information known by the decision maker, or by others, on the decision situation structural model in order to describe the probability of occurrence of the outcomes of uncertain situations. With steps two and three completed, we have an impact analysis model of the decision situation that enables us to predict the likelihood of various outcomes resulting from alternative courses of action.

4. Determine the impacts of the identified alternative courses of action.

III. *Interpretation:* This results in identification of the *best* decision alternative.

5. Assess the decision maker's preferences for the various attributes that characterize the possible outcomes and then formally include the decision maker's attitudes toward the risk associated with uncertain event outcomes. This generally results in a number of different utility functions.

6. Use the utility function of the decision outcomes and the probabilistic information to evaluate an overall expected utility score for the set of event outcomes that are associated with each alternative.

7. Rank the alternatives in terms of the expected utility of each alternative course of action. Discuss the implication of these results with the decision maker.

8. Use sensitivity analysis and other approaches to refine the decision assessment process and product. Ideally, this is done so that the decision maker more fully appreciates, accepts, and is supportive of the final results obtained. Communicate these as appropriate.

Figure 7.3 indicates this sequence of steps in a formal prescriptive decision assessment* process. Ideally, the normative theory is prescribed. Our description here is actually a typical normative decision analysis process, which evolved in the 1960s from the efforts of a number of investigators [6–9]. A very interesting and readable and recent overview of the historical foundations of decision analysis is provided by Fishburn [10].

---

*The term *decision analysis* is more common than *decision assessment*. One of the steps involves analysis of the impacts of alternative decisions. For this reason, we prefer the slightly more generic term, decision assessment. Clearly, this is a very small point. We often use the terms interchangeably.

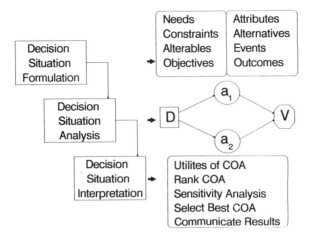

Figure 7.3. Fundamental Steps in Prescriptive Decision Assessment

The fact that the process just described is suggested for implementation qualifies it as a prescriptive process. It also happens to be the product of a normative theory. Thus, this is a prescriptive–normative process. Even if a formal normative approach is optimal under hypothetical circumstances, modifications may be needed to make it more appropriate for a specific decision situation. For example, the decision tree may be extraordinarily complex, and it may be better to treat subsequent acts as if they were events [1], thereby restricting the tendency of decision trees to become extraordinarily complex. This would replace uncertain act–event combinations by uncertain events only. In a very complex situation, obtaining solutions might be more realistic of accomplishment than if the approximation were not made. An approximate and useful answer is generally much better than a potentially more correct answer that is not obtained until the time for decision enactment is past.

The typical final products of formal normative decision assessment generally involve the following:

1. A description of the decision situation structural model.
2. A decision tree that represents the real decision situation and indicates probabilities assigned to outcomes.
3. A utility function that describes the decision makers' preferences for relevant attributes. This function might be useful not only for the present decision being taken but also to guide future decisions of a similar nature.
4. An evaluation and ranking of alternative courses of action according to their utility to the decision maker.

5. A sensitivity interpretation of how the results might change with changes in such decision parameters as subjective probabilities, personal preferences, and attribute measures.

The specific steps of the decision assessment effort, including the decision situation structural model, are configured to enable successful completion of the decision assessment effort and achievement of these five results. It is worth noting that these elements of a decision situation do not correspond to popular perceptions, although there have been efforts to bring the formal approaches into more widespread usage [11–16]. One of the tasks of those doing decision assessments, therefore, is to be sure that the customer or client is aware that these formal descriptions of decisions are in fact quite similar to the more common descriptors.

It is also important to note some of the important differences. One of these is that the formal decision assessment effort generally produces no final usable product until the entire process has been concluded. Sometimes this may create difficulties, especially for an impatient decision maker who wishes to see some intermediate product of immediately recognizable value.

Quite a bit of information is needed to accomplish the decision assessment process just outlined. This includes information about

1. The decision situation, including an identification of needs, constraints, alterables, objectives and measures by which outcomes are evaluated.
2. Alternative courses of action that are available to the decision maker.
3. Possible outcomes that may be realized from implementation of each alternative course of action.
4. The probability of occurrence of each outcome.
5. The attributes or objectives measures of all possible outcomes.
6. The costs of purchasing additional information about the decision situation.

Quite clearly, this information will not always be completely known with precision. If we have all of this information, we can incorporate it into a structural model of a decision situation, use this model to analyze the impact of the alternative decisions that have been identified, and interpret these according to the value system of the decision maker to enable formal rational choice.

There are a number of questions that should be answered to appraise the worth of a projected decision assessment effort, including the following:

Was the decision situation identified in a satisfactory manner?

Did the decision situation structural model satisfactorily capture the reality of the decision situation to give the decision maker confidence in the indicated results?

Were a sufficiently robust set of alternative courses of action identified?

Were the possible outcomes of the alternatives made explicit, and was knowledge about the probability of occurrence of each sufficient for the intended purpose?

Were the attributes of the outcomes identified and explained?

Did the decision assessment effort produce decisions of sufficient quality to justify its cost?

Did the decision maker have confidence in the decision situation structural model and the validity of the utility function?

Since good decisions do not necessarily lead to good outcomes,* and bad decisions may well result in good outcomes, there is no totally objective method of directly assessing the benefits of a decision assessment effort by observing the outcomes alone. Thus an indirect evaluation approach, such as that suggested in most studies of decision support systems [17, 18], will have to be used. In any case, we are interested in valuing the decision support process itself and not just the product of the decision support process, important as that may be.

7.2.2. A Formal Normative Model for Decision Assessment

Figure 7.2 illustrates a way to represent alternatives, event outcomes, and utilities. This forms the basis for the model that we use for our study of normative decision analysis, and it leads to the decision analysis process represented in Figure 7.3.

A decision analysis model is generally comprised of six elements:

1. The set of n alternative actions $A = \{a(1), a(2), \ldots, a(n)\}$.
2. The set of m states of nature $E = \{e(1), e(2), \ldots, e(m)\}$.
3. The set of nm outcomes $Q = \{q(11), \ldots, q(ij), q(nm)\}$, where $q(ij)$ corresponds to the pair of action–event possibilities $\{a(i)e(j)\}$.
4. A set of probabilities of these outcomes, which is actually a set of conditional probabilities $P[e(j)|a(i)]$ or, in shortened form, $P(e_j|a_i)$.
5. A utility function $U = \{u(11), u(ij), \ldots, u(nm)\}$, where $u(ij) = u_{ij} = U[q(ij)]$, that expresses the decision maker's utility at having selected alternative i and receiving state of nature j as a result.
6. The objective function, representing what is desired to be accomplished through selection of an alternative.

---

*The decision to pay $1 for a lottery ticket in which we might win $100 with probability 0.99 and receive nothing with probability 0.01 is, almost assuredly, a good decision. We are able to purchase something with an expected return of $99 for only $1. But, even though the decision may be a good one, we may lose our $1 in that we obtain an unlucky outcome. Good decision! Bad outcome!

The term *decision analysis*, or *decision assessment*, has come to be a generic term that refers to a wide variety of problems in such decision-related areas as systems engineering. The one common bond among all these problems is that they involve some kind of optimization, in terms of selecting a best alternative. Often "best" will mean "maximum." But the resulting u_{ij} will also depend upon the particular value of the random variable e_j, so the best the decision maker can do here is to maximize some function such as the expected value of the random variable u_{ij}. This is given by the maximum expected utility expression

$$\text{MEU} = \max_i \text{EU}\{a_i\} \sum_{j=1}^{\infty} u_{ij} P(e_j | a_i)$$

where $P(e_j | a_i)$ is the probability mass function that the state of nature that results or follows from selection of alternative a_i is e_j. More precisely, this is the probability of state e_j occurring, given that alternative a_i is selected. We use notation $\text{EU}\{a_i\}$ to mean the expected utility of taking the alternative course of action a_i.

7.2.3. Decision Assessment with No Prior Information on Uncertainties

To solve decision problems where there is information concerning the probability of various outcomes e_j that may result from selection of an alternative a_i, we need to know the associated conditional probability mass function $P(e_j | a_i)$.

Generally, we deal with finite event situations in decision analysis. So, the probabilities are probabilities of events, or probability distributions, or probability mass functions. It is not uncommon to find the terms probability density function used as well. Let us suppose for the moment that nothing is known about this probability density function. Although we may argue that there are few if any practical problems where we know absolutely nothing about the probabilities of the states of nature, there is nothing to prevent us from hypothesizing this total lack of knowledge. We now turn our attention to several approaches, described in many early operations research books [19], which may be used in such a situation.

1. *The Laplace Criterion:* Assuming all $P(e_j | a_i)$ are equal, that is, where $P(e_j) = P(e_j | a_i) = 1/m$, we obtain $\text{EU}\{a_i\}\sum_{j=1}^{m} = u_{ij} P(e_j | a_i) = \sum_{j=1}^{m} u_{ij}/m$. Since the common factor $1/m$ is invariant over the summation, it is not significant in influencing the optimization result and can be omitted. The optimal decision rule is to pick the a_i that maximizes $\sum_{j=1}^{m} u_{ij}$. We should note that the Laplace criterion is not a very good decision rule in general, and unless there is some reason for believing that all $P(e_j | a_i)$ are equal, or nearly so, the application of this rule may potentially yield very undesirable results.

2. *The Max–Min, or Pessimist, Criterion:* This is an extremely pessimistic rule which holds that regardless of which action a_i is chosen, nature will somehow malevolently cause the state e_j, which yields the very smallest utility possible for the action taken. We partially compensate for this pessimism by choosing the alternative course of action a_i which has the largest possible value of the smallest such utility. So, we have a criterion that allows us to minimize the maximum damage that can be done.*

3. *The Max–Max, or Optimist, Criterion:* Here we assume that nature takes on a benevolent character such that regardless of which a_i is chosen, nature will act kindly to produce the state e_j that yields the largest utility, or u_i^+. We then choose the action which results in the largest such u_i^+, where $u_i^+ = \max_j u_{ij}$. An alternative explanation of this criterion is that the decision maker chooses that action whose row in the payoff matrix contains the largest element and hopes for the best.†

4. *The Optimum Index α:* In this approach we combine the max–min criterion with the max–max criterion. We let $u_i^- = \min_j u_{ij}$ and $u_i^+ = \max_j u_{ij}$. Next we define a function of the optimism index $u_i(\alpha) = u_i^+ + (1 - \alpha)u_i^-$ for $(0 \leq \alpha \leq 1)$. This describes a linear function of α, which can be plotted for each value of i. The maximum over i of this function defines a convex function of α, which gives the optimal utility according to this rule. For the two extreme cases of $\alpha = 0$ and $\alpha = 1$, this rule reduces to the max–min rule and the max–max rule, respectively.

5. *The Min–Max Regret Criterion:* From the payoff matrix we generate a regret matrix R according to the rule $R_{ij} = u^+_j - u_{ij}$, where $u^+_j = \max_i u_{ij}$. Regret is the sorrow we experience about not getting a better outcome. For each alternative a_i we find the largest element in each row of the R matrix. This is $R_i^+ = \max_j R_{ij}$. Finally, we choose the alternative course of action a_i that yields the smallest such regret, which is $R_i^+ = \min_i R_i^+ = \min_i \max_j R_{ij}$.

---

*This criterion will allow us to refuse to pay $1 for the lottery noted earlier in which one receives a $100 outcome with probability 0.99 and a $0 outcome with probability 0.01. One might well extend the decision situation structural model such that refusal to pay is the desirable alternative. We are stuck on an island that is scheduled to be destroyed tomorrow. All we have on the island is our clothing and $1. Someone offers to take us off of the island for $1. Clearly, we would keep the $1, reject the lottery, and buy the escape ticket. Obviously, however, we have altered the initially stated decision situation.

†Such an approach would result in paying $1 for a lottery in which the payoff is $100 with probability 0.01 and $0 with probability 0.99. This bad decision might result in a good outcome but I would not bet on it. Clearly there is a potential problem if advice leads to paying for the lottery, we win, and think that we received good judgmental advice. Again, we could consider being stuck on an island that is going to be destroyed tomorrow. It costs $100 to buy a ticket on the last boat off of the island. Now, the lottery becomes quite desirable. But, we have altered the initially stated decision situation structural model through this additional information!

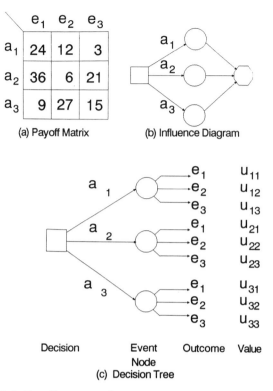

	e_1	e_2	e_3
a_1	24	12	3
a_2	36	6	21
a_3	9	27	15

(a) Payoff Matrix

(b) Influence Diagram

Decision   Event   Outcome   Value
Node
(c) Decision Tree

Figure 7.4. Payoff Matrix and Decision Situation Structural Models

Example. To observe how these methods may be applied in a practical decision problem, consider this simple example. The payoff matrix, decision tree, and influence diagram for this simple decision problem are shown in Figure 7.4.

First, we apply the Laplace criterion. Course of action a_1 results in an expected utility of 13. Course of action a_2 results in an expected utility of 21. For course of action a_3, we have an expected utility of 17. All of these expected utilities are based on the assumption that the probability of each outcome state being realized, $P(e_j|a_i) = 1/m$, is 0.333. Since $21 > 17 > 13$, we see that course of action a_2 is the best course of action from a Laplace criterion perspective. The expected utility of this course of action is 21 *if* all probabilities are equal.

Next we apply the max–min criterion and obtain $u_1^- = 3$, $u_2^- = 6$, and $u_3^- = 9$ as the minimum utility value associated with each course of action.

Since $u_3^- > u_2^- > u_1^-$, a_3 is the best course of action, because it yields the maximum value of all of the minimum return values.

Applying the max–max criterion results in $u_1^+ = 24$, $u_2^+ = 36$, and $u_3^+ = 27$, since we assume that nature gives us the maximum utility for each action. From this we see that the best course of action is a_2, because this yields the maximum utility.

Combining the max–min and max–max techniques using the optimism index α, we obtain $\hat{u}_1(\alpha) = 24\alpha + 3(1 - \alpha) = 3 + 21\alpha$, $\hat{u}_2(\alpha) = 36\alpha + 6(1 - \alpha) = 6 + 30\alpha$, and $\hat{u}_1(\alpha) = 27\alpha + 9(1 - \alpha) = 9 + 18\alpha$. It would be easy to sketch the utility versus α curves. Such an illustration would show that for $\alpha < \frac{1}{4}$, course of action a_3 is the best course of action. For $\alpha > \frac{1}{4}$, the optimal course of action alternative, or decision, is a_2.

Finally, applying the min–max regret criterion yields $u^+_1 = 36$, $u_2^+ = 27$, $u_3^+ = 21$, and thus we have for the regret matrix

$$R = \begin{bmatrix} 12 & 15 & 18 \\ 0 & 21 & 0 \\ 27 & 0 & 6 \end{bmatrix}$$

We now compute the maximum regret for each alternative. These are obtained as $R_1^+ = 18$, $R_2^+ = 21$, $R_3^+ = 27$. Thus the optimal decision is the course of action a_1, since we obtain the minimum of the maximum regret by this selection.

In the absence of any prior knowledge of the event probability function for the various e_j, we find that the optimal decision may, and generally will, depend upon the particular criterion selected. For the example considered, we can by judicious choice of the decision criterion cause any action to be optimal. There are other approaches that might also be used. For example, we might observe that for two of the three outcomes, alternative a_1 is better than alternative a_2. Also, we observe that for two of the three outcomes, alternative a_2 is better than alternative a_3. If we stop here, we might be tempted to conclude that since $a_1 \rightarrow a_2$ and since $a_2 \rightarrow a_3$, with each preference on two out of three outcomes, then we should have $a_1 \rightarrow a_3$. Sadly, however, it turns out that a_3 is better than alternative a_1 on two out of three alternatives. It turns out that we often are in deep trouble in trying to make pairwise comparisons of alternatives, as we are here. Intransitive preferences may well result, as they do here, if we are not very careful in formulating the comparisons.

Actually, this example is relatively academic in the sense that it would be rare that we had no knowledge whatever about the probabilities of outcome occurrence. Usually, we do have some information. Nevertheless, this somewhat academic example does provide some insight concerning how we might obtain solution to problems using much less than complete information. In real situations, information imperfections generally exist.

7.2.4. Decision Assessment Under Conditions of Event Outcome Uncertainty

We now turn to the case where the probability density function of the states of nature is known. For many decades, basic notions of probability have intuitively come to be associated with events that could be statistically replicated many times, such as flipping a coin or rolling a die, thereby giving the events a relative frequency interpretation. However, some events may occur only once or not at all. In these situations, a relative frequency interpretation may not be fully meaningful. Subjective factors need to be included in such estimates. Often, different people associate different personal probability assessments associated with the same action. The early works of von Neuman and Morgenstern [20] and Savage [21] were concerned with notions of subjective probability and subjective utility.

In a standard decision situation, once the appropriate probability density functions and outcome values have been determined such that the decision tree is well structured, all the information necessary for determining the expected utility resulting from any decision action is then potentially available. This is the decision under probabilistic uncertainty situation described earlier. Most efforts in normative decision analysis have been concerned with this problem. In such a situation, a "rational" decision maker attempts to maximize subjective expected utility.* How this is done can perhaps best be shown by an example. Numerous pedagogical examples of this type of decision problem exist in the literature. The "anniversary" problem described by North [22] in 1968, and the "wildcatter" problem outlined by Raiffa [9] are especially good. A number of very useful examples are provided by Holloway [23]. To keep the discussion relatively context-free, our approach in this section roughly parallels the "ball-urn" approaches that were initially described by Raiffa [9].

We assume a collection of opaque urns, each identical in external appearance and each containing a number of red balls and black balls. The decision maker knows in advance that there are n_1 urns, called type Θ_1, containing r_1 red balls and b_1 black balls. There are n_2 urns that are called type Θ_2. They contain r_2 red balls and b_2 black balls. The person conducting the experiment chooses an urn at random and gives it to the decision maker, who must, on the basis of the information given, decide whether the urn is of type Θ_1 or of type Θ_2. The reward or return to the decision maker depends upon the type of urn actually obtained by the decision maker (called the *true state of nature*) and the choice of decision alternative a_1 or a_2. The payoff matrix is given in Figure 7.5, where we use W and L as mnemonics for win and lose, respectively. Any of the four numbers W_1, W_2, L_1, and L_2 could be positive

---

*Actually, this is rational behavior, but only one form of rational behavior. Generally, it is called something like *techno-economic rationality* to indicate the rationality context.

or negative, but it is assumed that $W_1 > L_1$ and $W_2 > L_2$. Also it is assumed that $W_1 > L_2$ and $W_2 > L_1$; otherwise one decision would always dominate the other, and the decision choice would be obvious—select the nondominated decision alternative.

For the moment let us assume that our decision maker's utility curve for money is a linear function of the amount of money involved. Then the utilities W_1, W_2, L_1, and L_2 may be represented numerically by the equivalent amount of money. Such a decision maker is called an expected monetary value player, or EMVer. More is said about the nonexpected monetary value player. or non-EMVer, in Sections 7.2.5 and 7.2.6, which specifically discuss utility theory.

The decision maker has two choices. Either decision a_1 or a_2 may be chosen. If the decision maker chooses a_1 the expected reward is given by $E\{a_1\} = W_1 P(e_1) + L_1 P(e_2)$. On the other hand, if the decision maker chooses a_2, the expected reward is given by $E\{a_2\} = L_2 P(e_1) + W_2 P(e_2)$, where $P(e_i)$ is the probability of being given an urn of type Θ_i. These probabilities are easily determined, using the relative frequency concept, as $P(e_1) = n_1/(n_1 + n_2)$ and $P(\Theta_2) = n_2/(n_1 + n_2)$. These probabilities are called a priori probabilities since they represent the prior knowledge or the knowledge before sampling that exists concerning the relative proportions of type e_1 and type e_2 urns.

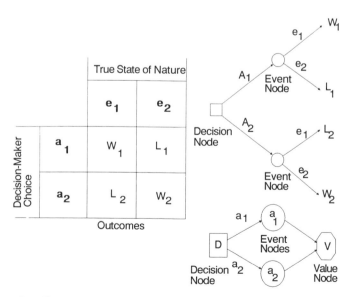

Figure 7.5. Payoff Matrix, Decision Tree, and Influence Diagram for Simple Decision Options and Outcomes

The appropriate decision rule is simply stated:

If $EU\{a_1\} > EU\{a_2\}$, select a_1.
If $EU\{a_2\} > EU\{a_1\}$, select a_2.
If $EU\{a_1\} = ER\{a_2\}$, either a_1 or a_2 may be selected.

There are two basic approaches that can be taken to actually determine the best decision, given a well-structured decision tree. These are described as the *extensive form* and the *normal form* approaches. Decision assessment in the extensive form has just been described. Solution of problems in extensive form is comprised of six steps, which are illustrated in the next example.

1. Structure the decision tree diagram.
2. Model the deterministic outcome payoffs.
3. Model the probabilistic outcomes by identifying the probabilities that exist at all chance forks.
4. Determine the utilities of all possible outcomes.
5. Analyze the problem, generally through averaging out and folding or rolling back.
6. Determine the optimum decision and communicate the results.

The extensive form is often called the *roll-back form*. Figure 7.6 illustrates these steps and the associated decision analysis process. The extensive or

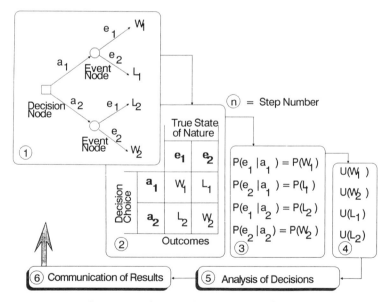

Figure 7.6. The Decision Assessment Process

roll-back form is generally used when all probabilities are known and the problem involves decision making under risk.

Example. Suppose a decision maker is presented with the following decision situation model. An experimenter owns 1000 urns, 750 of which are called type 1 and 250 of which are type 2. Each type 1 urn contains seven red and three black rolls. Each type 2 contains two red and eight black balls. The experimenter will select an urn at random, and the decision maker must guess whether the urn is type 1 or type 2. If the DM guesses type 1 and the urn is actually a type 1 urn, a \$500 return is received. But if the urn is actually a type 2 urn, the DM must pay \$100. If the DM says it is a type 2 urn and the urn is actually a type 2 urn, an \$800 return is received. But if the urn is actually a type 1 urn, the DM must pay \$150. Thus, the DM has three alternative courses of action, or decisions.

$$a_0\text{: Refuse to play.}$$
$$a_1\text{: Guess urn type 1.}$$
$$a_2\text{: Guess urn type 2.}$$

If the DM refuses to play, nothing is gained and nothing is lost; the payoff is zero. If the DM plays this little game, the payoff or consequence of play is depicted by the payoff matrix shown in Figure 7.7, which also illustrates the complete decision tree and influence diagram that represents the evolving decision situation structural model.

We can easily compute the expected return from each decision. We obtain $E\{a_0\} = 0$, $E\{a_1\} = 500(0.75) + (-100)(0.25) = \$350$, and $E\{a_3\} = -150(0.75) + 800(0.25) = \$87.50$. In general, we should obtain the expected utility of the various decisions. In this case, the utility of a given amount of money is just that amount of money. So, $EU(a_i) = E(a_i)$. Since the expected return is the greatest if decision a_1 is chosen, the DM should play the game and should choose a_1 for an expected return of \$350. For this simple example, there is no real opportunity to average out and fold back in the process of making the optimum decision. In more complicated problems, which we will now consider, the operations of averaging out and folding back are quite useful.

Now suppose that before the decision maker decides to select option a_0, a_1, or a_2, it is possible, for a price S, to draw one ball out of one of the urns, to be selected at random, and observe its color. The *rational* decision maker still seeks to maximize the expected utility resulting from the decision. Also, the decision maker should desire to use the information gained from sampling one ball from an urn to update the assessment of $P(e_1)$ and $P(e_2)$. Thus, the DM needs to determine $P(e_1|R)$ and $P(e_2|R)$ if a red ball is drawn and $P(e_1|B)$ and $P(e_2|B)$ if a black ball is drawn. These are calculated from

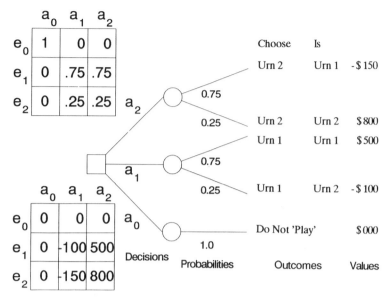

Figure 7.7. Payoff Matrix, Probability Matrix, and Decision Tree for Simple Urn Model Example

Bayes' rule as

$$P(e_i|R) = P(R|e_i)P(e_i)/P(R)$$
$$P(e_i|B) = P(B|e_i)P(e_i)/P(B)$$

where

$$P(R) = P(R|e_1)P(e_1) + P(R|e_2)P(e_2)$$
$$P(B) = P(B|e_1)P(e_1) + P(B|e_2)P(e_2)$$
$$P(R|e_i) = \frac{r_i}{r_i + b_i}$$
$$P(B|e_i) = \frac{b_i}{r_i + b_i}$$
$$P(e_i) = \frac{n_i}{n_1 + n_2}$$

The probabilities $P(e_i|R)$ and $P(e_i|B)$ are called a posteriori probabilities, because they represent the probabilities of the outcome states of nature, e_i, that result after the sampling that occurs by picking a ball from an urn. We assume that there are r_i red balls and b_i black balls in the type i urn.

Using $E\{a_i|R\}$ to mean the expected reward from choosing option a_i, given that a red ball was the ball was drawn from the urn, the appropriate expected values are

$$E\{a_1|R\} = (W_1 - S)P(e_1|R) + (L_1 - S)P(e_2|R)$$
$$E\{a_2|R\} = (L_2 - S)P(e_1|B) + (L_1 - S)P(e_2|B)$$
$$E\{a_1|B\} = (W_1 - S)P(e_1|B) + (W_2 - S)P(e_2|B)$$
$$E\{a_2|B\} = (L_2 - S)P(e_1|B) + (W_2 - S)P(e_2|B)$$

It is important that we note that the cost of sampling, S, has been subtracted from each reward. Thus we see that after sampling, the optimal decision rule is

Red ball drawn:
   If $E\{a_1|R\} > E\{a_2|R\}$, select alternative a_1.
   If $E\{a_2|R\} > E\{a_1|R\}$, select alternative a_2.
Black ball drawn:
   If $E\{a_1|B\} > E\{a_2|B\}$, select alternative a_1.
   If $E\{a_2|B\} > E\{a_1|B\}$, select alternative a_2.

These expected values provide us with the expected reward after sampling. The expected reward before we sample, and given that we do sample, is given by

$$E_s\{a_i\} = \left[\max_i E\{a_i|R\}\right] P(R) + \left[\max_i E\{a_i|B\}\right] P(B)$$

Now we need to determine whether it is worthwhile to sample. To determine this we compare the expected reward without sampling to the expected reward with (but before sampling). The optimal decision rule here is simply

If $\max_i E\{a_i\} > E_s\{a_i\}$,   do not sample.
If $E_s\{a_i\} > \max_i E\{a_i\}$,   sample.

It is important to note that the decision makers (DM) must now first decide whether to sample or not to sample. If the DM decides to sample, then based upon the result of the sampling outcome as a red ball or a black ball, the DM must make another decision. This concerns whether to choose a_1 or a_2. So, we see that two decisions must be made, with an intervening chance event. This type of decision problem is generally called a *sequential decision problem*. Many realistic decision problems are sequential in nature. The decision situation structure for the problem with sampling as described here is represented by the decision tree as shown in Figure 7.8.

If we use the extensive or fold-back approach, the optimal policy is developed by starting at the terminal nodes on the right-hand side of the tree and working back through the tree by assigning to each decision node the maximum of all the expected utilities emanating from that node. This approach is called by Raiffa [8] *averaging out and folding back*. It is often a

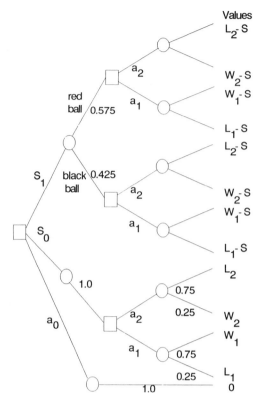

Values

Figure 7.8. Decision Tree with Sampling for Simple Urn Model Example

tedious approach to track, because we must have all the information regarding probabilities, sampling costs, and rewards before we can begin working our way through the tree. This makes appropriate software for decision analysis very desirable.

Example. Let us consider an extension to our previous example. Suppose that we are permitted to sample the selected urn by drawing a ball and observing its color before making our decision. We are granted that privilege in return for a sampling fee of $60. Of course we may elect not to sample and then would pay no sampling fee. The question is: "Should the DM choose to sample or not to sample?" And then, "Should the DM select urn 1 or urn 2?" Decision analysis or decision assessment provides the answer. We again assume that the decision maker is an expected monetary value player. That is to say, the DM will select the course of action that yields the highest expected value monetary payoff.

The first step in the decision assessment, as we see from Figure 7.3 or Figure 7.6, is to structure the decision flow diagram. The result of doing this is illustrated in Fig. 7.8. This figure indicates all the possible courses of action the decision maker may take. The square nodes are those points at which the decision maker selects the branch to follow, whereas the round nodes are those points at which the branch to follow is selected by the chance outcome of an experiment that occurs as the decision situation evolves in real time.

Our next step is to assign payoffs at the tips of the branches. These payoffs are those indicated in the payoff matrix of the previous example. The decision flow diagram with payoffs is shown in Figure 7.9. The $60 fee for sampling is indicated in this figure. We note again that the payoff for alternative a_0, the option of refusing to play, is $0.00.

The ensuing step calls for the assignment of probabilities at the chance forks. These follow each uncertainty node. In the branch that follows decision a_0, there are no chance events. Hence, there are no probabilities to be assigned as we have encountered a deterministic node. The branches for decisions a_1 and a_2 involve only the chance node outcome of whether the urn is a type 1 urn or a type 2 urn. The only information available to the DM

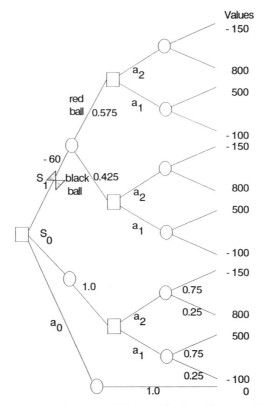

Figure 7.9. Decision Tree Outcome Values with Sampling for Simple Urn Model Example

is that there are 750 type 1 urns and 250 type 2 urns. Since the random selection of an urn implies that individual type 1 or type 2 urn is equally likely to be selected, we assess probabilities at the chance nodes in the appropriate branches as

$$P(R) = P(R|\text{urn } 1)P(\text{urn } 1) + P(R|\text{urn } 2)P(\text{urn } 2)$$
$$= (0.7)(0.75) + (0.2)(0.25) = 0.575$$
$$P(B) = P(B|\text{urn } 1)P(\text{urn } 1) + P(B|\text{urn } 2)P(\text{urn } 2)$$
$$= (0.3)(0.75) + (0.8)(0.25) = 0.425$$

Figure 7.10 illustrates the completed decision tree showing these probabilities. We have deducted the $60 sampling cost from the payoff in each case in which we sample, as also indicated in Figure 7.10. The information from the sample can now be used to calculate the probability of each type of urn. We use Bayes' rule to find the probability that the urn is type 1, given that a red

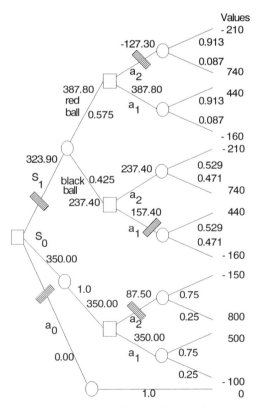

Figure 7.10. Decision Tree Outcome Values with Sampling for Simple Urn Model Example

ball was drawn. This is given by the expression

$$P(\text{urn } 1|R) = P(R|\text{urn } 1)P(\text{urn } 1)/P(R) = (0.7)(0.75)/0.575 = 0.913.$$

We determine the other probabilities in the same manner and have indicated them in Figure 7.10.

The next step involves the averaging out and folding back effort. In it, the DM evaluates the expected value of the payoff at each chance node and selects the alternative that leads to the largest expected value. This averaging process begins at the points closest to the end of the tree and works back to the start to indicate the alternative with the highest expected value. This is doubtlessly why an appropriate name for this portion of the effort is averaging out and folding back.

The expected value of the events at a chance node is the sum of the payoffs at that chance node multiplied by their respective probabilities. If we elect not to sample, the expected value of decision a_1 is, as was obtained previously, $E(a_1) = (0.75)(500) + (0.25)(-100) = \$350$. Similarly, $E(a_2) = \$87.50$. The same kind of computation must be made at the tips of nodes that spread out from the branch associated with the sample decision. For the decision node following the situation where the DM obtains a red ball sample, we have $E(a_1) = (0.913)(440) + (0.087)(-160) = \$387.80$ and $E(a_2) = -\$127.30$. For the decision branch that follows a black ball sample, we obtain $E(a_1) = \$157.40$ and $E(a_2) = \$237.40$.

Each of these expected values are shown in Figure 7.10. If we examine the red ball sample obtained branch, we see that decision a_1 has a higher expected monetary value outcome than does decision a_2. The implication of this is that a DM who has draw a red ball should decide to accept option a_1. Since decision option a_2 is less desirable in this situation, a double slash is drawn through the alternative option a_2 to indicate its removal from any further consideration. The decision node on the red sample obtained branch then has an expected value of $\$387.80$ provided the decision maker chooses a_1 after drawing a red ball. Our efforts in this example have just led us to a determination of a part of the optimal decision.

In a similar manner, the DM observes that a_2 has the greatest expected value on the black ball sample obtained branch. Thus, the DM should remove a_1 from any further consideration there. The DM assigns an expected value to the decision node on the black ball sample obtained outcome node of $\$237.40$. The DM finally folds back one more time and is then able to find the expected value of the sample chance node e_1 as $E(e_1) = (0.575)(387.80) + (0.425)(237.40) = \$323.88$. Now each of the alternatives has an expected monetary value assigned to it, and the decision maker can select the one with the largest expected monetary value. For this problem, the alternative with the greatest expected value is the no sample alternative a_1. That is to say, we should simply select urn type 1 without a sample. If the sampling cost were lower, say $\$30$ rather than $\$60$, the sample branch would have the higher expected value, and a different decision would then be indicated. We wouldthen sample and choose a_1 if we draw a red ball and a_2 if we drew a black ball. This concludes this relatively simple example. While nothing at all is difficult, either mathematically or conceptually, the example is rather tedious to go through. And we have not accomplished any sort of sensitivity analysis. Clearly, software support is desirable.

In this simple example, we assumed that the DM made judgments on the basis of the expected value of money. This expected monetary value approach is open to some criticism. Not all decision makers would be either content or comfortable with an expected monetary value approach. Concepts of utility theory are introduced in Section 7.2.5 to permit risk attitudes to be considered, such that the DM may then be other than expected monetary value-oriented.

The expected value of perfect information and the expected value of sampling information are two important concepts that arise in examples of the sort we have just considered. To continue in the context of the ball–urn problem, suppose that the decision maker knows in advance, perhaps by consulting an oracle of some sort, which type of urn will be given for purposes of taking the sample. For each state of nature, e_1 or e_2, the DM should choose that decision alternative, a_1 or a_2, that maximizes the resulting reward. Each such reward is multiplied by the corresponding a priori probability of obtaining e_1 or e_2. These weighted rewards are then summed over all the states, and from this sum is subtracted the optimal expected outcome without sampling. So for the problem posed here, we obtain the expected value of perfect information, $\text{EVPI} = W_1 P(e_1) + W_2 P(e_2) - \max E\{e_i\}$. A related term, the expected value of sampling information, is simply the difference between the optimal expected reward with sampling and the optimal expected reward without sampling. We easily obtain this value, $\text{EVSI} = E_s(a_i) - \max E(a_i)$.

The important point here is that perfect information represents an upper limit to the amount of knowledge that can be gained by sampling. Consequently we should never pay more than EVPI, the expected value of perfect information, for sampling information. The expected value of sampling information must be no greater than EVPI, and the amount by which it is less than this is indicative of greater value inherent in the sampled information.

Example. We consider now the determination of the expected value of perfect information and the expected value of actual sampled information for our previous example. Application of the relevant equations easily yields, with the probabilities and returns shown in Figure 7.10, $\text{EVPI} = 500(0.75) + 800(0.25) - 350 = \$225.00$. Also, we obtain $\text{EVSI} = 323.88 - 350.00 = -\$26.12$. Thus we see that perfect information would be worth quite a bit here. The expected return without sampling is $\$350.00$. We could increase this return by $\$225.00$ to $\$575.00$, for a 64 percent increase in the anticipated return, if we had perfect information. Actual sampled information would increase our expected return by only $33.88, even when this information is free. It costs the DM $60 to sample, however, and it is not worth it for the DM to pay $60 for information worth $33.88, as there is a loss of $26.12 by doing this.

This sort of episode can go on and on! Suppose that, instead of drawing the ball from the unknown urn and observing its color, the DM decides to hire an assistant to perform this sampling task and report back the color of the ball to the DM. Unfortunately, many assistants are not completely reliable and occasionally report results inaccurately (see Chapter 9 for a discussion of human errors). The conditional probabilities describing such a situation would then be given by an appropriate matrix of probabilities. For example we would have the conditional probability that the assistant reports *red* when the true color is *black*. Ideally the elements along the diagonal should be much larger than the off-diagonal elements. In the case of such "noisy" measurements, the a posteriori probabilities of interest are of the form $P(e_i|r)$, that is, the probability that the urn is of type e_i given that the assistant reported that a red ball was drawn from it. Also, the DM will need the probabilities $P(r)$ and $P(b)$, the probabilities of what the assistant reports concerning the color of the balls, rather than the previously needed $P(R)$ and $P(B)$. These needed expressions may be obtained from

$$P(r) = P(r|R)P(R) + P(r|B)P(B)$$

$$P(e_1|r) = P(r|e_1)P(e_1)/P(r)$$

$$P(r|e_1) = P(r|R)P(R|e_1) + P(r|B)P(B|e_1)$$

Similar expressions result for the other needed probabilities. We could now specify all of these probabilities for and continue the example of this section. We will leave this as an example for the interested reader (see Problem 7.4).

What has been accomplished in this section may appear, at least at first glance, not especially related to systems engineering problems, and decision assessment issues. It may appear more attuned to gambling strategy. However, the urn model example used here may easily be used as a model for a large number of important decision related issues. It is especially useful as a basic probability model in choice under uncertainty situations.

7.2.5. Utility Theory

When choosing among alternatives, we must be able to indicate preferences among decisions that may result in a diversity of outcomes. With the monetary rewards in the just-considered ball-in-the-urn examples, this was relatively straightforward. We assumed an expected value approach and preferred a larger amount of money to a smaller amount.

Utility theory is needed for at least two reasons. The decision maker may have an aversion toward risk. On the other hand, a DM may be risk-prone and make decisions that are associated with high impact unfavorable consequences simply for the "fun" of it, or because of very large rewards if the chance outcome does turn out favorable. Not all consequences are easily quantified in terms of monetary value, especially when there are multiple and potentially incommensurate objectives to be achieved. A scaler utility function makes it, in principle, possible to attempt to resolve questions that involve uncertain outcomes with multiple attributes such as these. Before discussing utility theory, however, we need to define and discuss a lottery, as this provides a basis for the utility concept.

It is desirable to reflect briefly on what we mean by the concept of utility in decision analysis. Formally, a *utility function* is a mathematical transformation that maps the set of outcomes of a decision problem into some interval of the real line. In other words, *a utility function assigns a numerical value to each outcome of a decision problem*. The basic assumptions, or axioms, needed to establish utility theory may be described as follows:

1. Any two outcomes resulting from a decision may be compared and placed in a preference order. If A and B are two such outcomes, then one must either prefer A to B ($A \to B$), prefer B to A ($B \to A$), or be indifferent between A and B ($A \leftrightarrow B$). An extension of this assumption leads to the concept of transitivity. This requires that if a decision maker prefers A to B and B to C, then this decision maker must also prefer A to C. If a decision maker is not transitive, all sorts of maladies can result, as can easily be demonstrated.* In the simple way posed here, every individual should normatively seek to be transitive relative to preferential expressions. Sadly, what applies to individuals does not apply to groups, as group preferences may well be intransitive, even though all individuals in the group have transitive preferences. Many decision analysis texts demonstrate this, a generalization of which won a Nobel prize for Kenneth Arrow.

2. Utilities may be assigned to lotteries involving outcomes as well as to the outcomes themselves. The term *lottery* is defined as a chance mechanism that yields outcomes $e_1, e_2, \ldots e_n$ with probabilities $P_1, P_2, \ldots, P_n$, respectively, and where each $P_i \geq 0$ and the sum over all P_i must be equal to 1. (This insures that we have a valid probability

---

*For example, one can become a money pump. If for example, I have the preference structure $A \to B$, then there should be some amount of money, say \$1, such that $A \leftrightarrow B + \$1$. Now suppose that $B \to C$ and $B \leftrightarrow C + \$1$. If I have alternative C to start with, then I should be willing to trade in that alternative and \$1 and obtain alternative B. If I have alternative B, then I should be willing to trade it and get \$1 for alternative A. This new relation, $A \leftrightarrow C + \$2$ is quite consistent with a transitive preference structure $A \to C$, but very much a problem if I am intransitive and have $C \to A$. If this were my preference structure, then I should be willing to pay someone to take away A and give me back C. Then the cycle can start all over again and continue *ad infinitum*!

function, but not necessarily that the probabilities we use are appropriate for the problem at hand.) A lottery, as described here, is denoted by $L = (e_1, P_1;\ e_2, P_2; \ldots;\ e_n, P_n)$ and is easily modeled by a treelike diagram, such as in Figure 7.2. Obviously, a lottery is just a particular outcome alternative pair. It describes the alternatives that may be selected and the possible outcomes. From this it follows that if one prefers outcome A to outcome B, then one should also prefer A to the lottery $(A, P;\ B, 1 - P)$, and one should prefer this lottery to outcome B. Furthermore, if we prefer A to B, then we should prefer the lottery $L = (A, P;\ B, 1 - P)$ to the lottery $L' = (A, P';\ B, 1 - P')$ if and only if $P > P'$.

3. If A is preferred to C and C is preferred to B, then for sufficiently large P, the lottery $(A, P;\ B, 1 - P)$ is preferred to C. Similarly for sufficiently small P, C is preferred to the lottery $(A, P;\ B, 1 - P)$. Thus there exists some P $(0 < P < 1)$, such that one is indifferent between receiving risk free outcome C and the lottery $(A, P;\ B, 1 - P)$. This is an especially important condition in that it establishes the certain money equivalent (CME). This is the lottery which is believed to be quite exactly equivalent to a specific sum of money. It is obtained by letting $B = 0$.

4. There is no intrinsic reward in the lotteries themselves, or as it is more commonly expressed, there is "no fun in gambling."* The DM is ambivalent about deciding between lotteries of comparable value, and complex lotteries that may be, appropriately, simplified.

Condition 1 is often called the *orderability* condition or axiom. It insures that the preferences of the DM impose a complete ordering and a transitive preference ordering across outcomes. Condition 2 is a *monotonicity* condition. It insures that the preference function is smooth and that increasing preferences are associated with increasing rewards. Condition 3 is a *continuity* condition that insures comparability across outcomes with equivalent preferences. Condition 4 insures decomposability of lotteries and substitutability of equivalent lotteries.

Condition 3 contains the key to resolving a problem where the decision yields uncertain outcomes by using decision analysis techniques. In actual practice, a situation like the one described in condition 3 is used to measure the utility that a decision maker has for various outcomes. These four assumptions form a basic set for the establishment of a utility theory.

Although this theory is not complicated, putting the theory into practice successfully is very complicated. A number of important theoretical issues are

---

*In other words the decision maker's preferences should not be affected by the particular means chosen to resolve the uncertainty. If there is indeed "fun" in gambling, this should be considered as one of the attributes of the alternative outcomes and multiattribute evaluation methods used. This is easily accomplished and the result is a situation in which there is no reward associated with the lotteries themselves.

easily overlooked. It is important, for example, to keep in mind that the concept of preference precedes the concept of utility. In other words, A may be preferred to B $(A \rightarrow B)$, but not because the utility of A is greater than that of B, or $U(A) > U(B)$; on the contrary, we have the relation $U(A) > U(B)$ simply and directly because of the fact that A is preferred to B. Also, it is important to note that the numerical values assigned by the utility function are not unique. Given a utility function $U_1(x)$, it is possible to generate a new utility function $U_2(x) = aU_1(x) + b$, where a is a positive constant, and b is any constant, such that the preferences are not altered. Thus in a sense, a utility function may be viewed as a kind of preference thermometer. The utility function must also be monotonically nondecreasing in the sense that the highest utility is assigned to the most preferred outcome and the lowest utility is assigned to the least preferred outcome. Intermediate utility values and preferences are assigned in an analogous ordering.

There are many potential behavioral pitfalls as well. Most of them are more insidious than the theoretical and quantitative pitfalls. Some of these are discussed later in this chapter and in Chapter 9.

According to expected utility theory, the decision maker should seek to choose the alternative $a(i)$ that makes the resulting expected utility the largest possible. The utility $u(ij)$ of choosing decision a_i and obtaining outcome event e_j also depends upon the particular value of the probabilistically uncertain random variable $e(j)$, as conditioned on the decision path that is selected. So, the best the decision maker can and should do here is to maximize some function, such as the expected value of utility

$$\max_i EU\{a_i\} = \max_i \sum_{j=1}^{N} u_{ij} P(e_j|a_i)$$

where the maximization is carried out over all alternative decisions, and $P[e(j)|a_i]$ is the probability that the state of nature is $e(j)$. Often, we use the abbreviated notation $EU\{a(i)\}$ to mean the expected utility of taking action $a(i)$. Generally, this is also called the *subjective expected utility* (SEU), which denotes the fact that the probabilities and utilities may be personal or subjective, that is, belief probabilities and personal utilities for consequences.

A lottery is a chance mechanism that results in an outcome with a prize or consequence $e(1)$ with probability $P(1)$, an outcome with prize or consequence $e(2)$ with probability $P(2), \ldots$, and an outcome with prize or consequence $e(r)$ with probability $P(r)$. The probabilities $P(i)$ must be such that they sum to one,

$$\sum_{j=1}^{M} P(j) = P(1) + P(2) + \cdots + P(M) = 1$$

where the lottery is denoted, as in Axiom 2 on page 341, by

$$L = [e_1, P_1; e_2, P_2; \ldots; e_N, P_N]$$

We can now formally define a utility function as a transformation that maps the set of consequences into an interval on the real line. We denote the utility of a consequence $a(i)$ as $U[a(i)]$, even though this will usually require that we calculate expected utility or subjective expected utility and should therefore be written as $EU[a(i)]$ or $SEU[a(i)]$. We now utilize the consequence of the four utility relevant assumptions just made. The number of axiomatic assumptions varies from four to six, depending on which approach is taken. The basic approach taken by Luce and Raiffa [24], for example, in a classic and seminal work, utilized six assumptions. We have just described four assumptions, conditions, or axioms, which enable establishment of subjective expected utility as the criterion of choice for rational decision making. The result of these assumptions is that the utility function $U(ij)$ satisfies several properties. These utilities turn out to be indicators of preference and not absolute measurements. They are unique only up to a general linear transformation or, more properly stated, affine transformation. A considerable part of this undertaking is related to a need for concepts of risk aversion and relative risk aversion [25]. Let us examine some of these concepts here.

A utility curve for money can be generated in the following manner. We arrange all the outcomes of a decision in order of preference. We denote the most preferred outcome W and the least preferred outcome L. We arbitrarily assign utility $U(W) = 1$ and $U(L) = 0$. Next, the decision maker is encouraged to answer the question: Suppose I owned the rights to a lottery which pays W with probability $1/2$ and L with probability $1/2$. For what amount, say $X(0.5)$, would I be willing to sell the rights to this lottery? Since the decision maker is indifferent between definitely receiving $X(0.5)$ and participating in the lottery $[W, 0.5; L, 0.5]$, then we must have $U[X(0.5)] =$

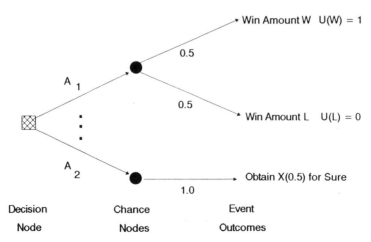

Figure 7.11. Decision Analysis Tree Structure to Determine Decision-Maker Utility for $X(0.5)$

$0.5U(W) + 0.5U(L) = 0.5$. Figure 7.11 illustrates this lottery in decision tree format.

We could have picked any other lottery with outcomes W and L, such as, $[W, 0.01; L, 0.99]$. There are several analytical reasons, however, for choosing a lottery which gives W or L with equal probability, or $[W, 0.5; L, 0.5]$. This may, however, be unrealistic. For example, W might mean some enormous increase in profit to the firm and L might mean *go bankrupt*. It might well be that a business person would so abhor a situation in which there is a 50% chance of going bankrupt that it becomes impossible to realistically think of utilities in these terms. Nonetheless, it may be conceptually quite simple for the decision maker to imagine a lottery that gives W or L with equal probability. In this case, the decision maker need only be concerned with the relative preferences of W and L. Also, the value of the lottery $[W, 0.5; L, 0.5]$ occurs midway on the U axis between W and L and therefore facilitates plotting the utility curve.

Next the decision maker is encouraged by a human analyst or a decision support system under the guidance of a facilitator to answer the question: Suppose I owned the rights to a lottery that paid W with probability $1/2$ and $X_{0.5}$, which has a utility of 0.5, with probability $1/2$. For what amount, call it $X_{0.75}$, would I just be willing to sell the rights to this lottery? Here, the decision maker is indifferent between receiving $X_{0.75}$ for certain and participating in the lottery $[W, 0.5; X_{0.5}, 0.5]$. Therefore, $U(X_{0.75}) = 0.5U(X_{0.5}) + 0.5U(L) = 0.75$. Next, we consider the lottery $[X_{0.5}, 0.5; L, 0.5]$, and the amount to be determined is $X_{0.25}$. In this case, we obtain $U(X_{0.25}) = 0.5U(X_{0.5}) + 0.5U(L) = 0.25$. This process could be continued indefinitely. However, the three points generated here and the two end points are often sufficient to give an idea of the shape of the utility curve, which will usually

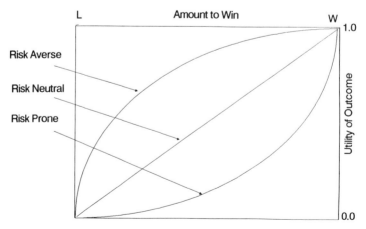

Figure 7.12. Hypothetical Utility Curve

take on the general shapes shown in Figure 7.12. It is rare that a decision maker is risk averse. Thus, the typical utility curve is concave downward. There are some notable exceptions to this risk averse nature of the DM, especially when the utility function is not required to be symmetric about losses and gains (see our later discussions on regret and prospect theory for illustrations of this). The utilities we have obtained are indicators of preference, and are not absolute measurements. As a result, they are unique only up to a linear transformation of the form $U'(L) = \alpha U(L) + \beta$, where $\alpha > 0$.

The first step in a utility assessment process results in a definition of the scope of the assessment procedure. Elements of the value system have been identified and often structured in the issue formulation step. Attributes, according to which alternative output values are to be assessed, are identified and structured. Much of this is accomplished in the formulation and analysis steps, and the results should be organized to match the requirements of the utility assessment portion of the interpretation step. The final part of the utility assessment process concerns either elicitation of and aggregation of the information to facilitate selecting the best alternative, or alternatively, use of a number of holistic judgments and regression analysis to identify these utilities.

A number of techniques exist for eliciting utility functions in terms of attribute weights and alternative scores [26–28]. There are four very common and useful approaches that are of value both for the single attribute case assumed here and the multiple attribute situation of Section 7.2.6.

1. *Direct Elicitation:* The decision maker is asked to assign a relative, or ordinal, utility score for outcomes. Generally, these value scores would be based on experiential familiarity with previous situations. The utility might be assigned into categories such as fair, good, or excellent, or might be based on numerical values, perhaps anchored on a scale. The range of the scale might typically be 0–100, or it could be left indeterminate until after the elicitation is completed. Direct methods can provide precise numerical scores and therefore are very attractive for their speed in application. However, providing precise, consistent, and meaningful numerical values is a very difficult task. It is also quite difficult to consider risk aversion in an approach of this sort.

2. *Ranking Methods:* The decision maker orders the levels of a specific set of attributes from most preferred to least preferred. The ordering may consist of levels of a single attribute, or combinations of the levels of two or more attributes, as in the case of investigating value trade-offs, or assessing the utility of interrelated attributes. This type of utility assessment measures the relative strength of one preference relation with respect to another preference relation. Outcomes are assumed to occur with certainty for purposes of outcome preference ranking.

3. *Indifference Methods:* The decision maker identifies indifference points in a decision space. Indifference methods may involve the assessment

of several possible combinations of attribute levels and determining indifference among these combinations. Judgments related to trade-offs between the utilities of various attribute level combinations must be made. When uncertainty is involved, indifference methods rely on the decision maker's ability to choose among uncertain outcomes with known probabilities. In this case, we obtain the certainty equivalent of a lottery, that is, the level of attributes for which the decision maker is indifferent to that lottery. Generally, indifference methods are preferred by most practicing decision analysts, especially those deeply committed to axiomatic approaches. A number of heuristic and approximate approaches to utility assessment exist using indifference methods. In principle, indifference judgments involve lotteries. Thus, they provide a combined assessment of strength of preference and risk attitude. It is important to note that indifference methods require prior knowledge of the decision situation model, including the outcome attribute tree. In standard multiattribute utility decision assessment, which we soon discuss, ranking methods are often used prior to the use of indifference methods. This provides a rough, imprecise, assessment of the possible values of the parameters of the model. It facilitates coordination of the subsequent precise assessments by means of indifference methods.

4. *Assessment through Decision Observation:* This approach consists in observing decision behavior in real-world or simulated real-world situations and inferring from these observations the parameters of a prespecified model. The work of Hammond and his colleagues on social judgment theory, discussed in Section 7.3.4, makes use of this concept. Techniques based on this approach are sometimes called bootstrapping techniques. Use of approaches based on regression-analysis to estimate parameters in the assumed structural model is central to the decision observation approach.

There are three generally used utility scale types in common use. The *ordinal scale* ranks items in preference order. The statement $A_1 \rightarrow A_2 \rightarrow A_3 \rightarrow A_4 \rightarrow A_5 \rightarrow A_6$ is a statement of ordinal preferences. The *interval scale* enters preferences according to their distance from some arbitrary end points. We might say for example that an outstanding result receives a score of 100 and a very poor result receives a score of 0. The statement $A_1 = 90$ and $A_2 = 30$ indicates that the first alternative is quite good and the second one rather poor. But it would not be meaningful to say that the first alternative is three times as good as the second one. A classic example of an interval scale would represent a Faranheit temperature scale, where $40°$ is warmer than $20°$, but not twice as warm. A *ratio scale* is a scale in which alternatives are associated with a cardinal ranking according to their relative distance from some calibrated and nonarbitrary end points. The indifference methods just discussed lead to ratio scales. The assessment-through-direct-observation approach may do this. The other approaches generally lead to ordinal or interval scale measurements.

Utility functions are only fully useful when they are intelligently obtained and thoughtfully applied. Structuring of a decision assessment issue is a crucial part of an overall decision assessment process. The final result obtained, in practice, depends upon the veridicality of both the decision situation structural model and the parameters within this structure.

The process of eliciting probabilities and associated utilities such that they truthfully represent individual judgments is one of the major tasks in decision analysis. The seminal text by von Winterfelt and Edwards [29] describes many of the approaches that have been taken to elicit these values. The efforts of Merkhofer [30], Wallsten and Budescu [31], Borcherding et al. [32], and Keeney et al. [33] are also pertinent, as is a set of papers edited by Lehner and Adelman [34]. Harvey has documented a number of useful findings in this area [35, 36]. Risk assessment and the identification of rare-event probabilities is considered by Sampson and Smith [37]. A definitive three-volume set provides a compendium of a great variety of approaches to judgment and decision related measurement [38–40], including measurements of probabilities and utilities for decision assessment.

Example. Let us give a simple example of utility determination. To give a specific numerical example, let $W = \$100$ and $L = -\$50$. Then we might find $X_{0.5} = \$10$, $X_{0.75} = \$47$, and $X_{0.25} = -\$25$. In other words, the decision maker owns the rights to the lottery $[\$100, 0.5; -\$50, 0.5]$ and would just be willing to sell the rights to this lottery for $10. An EMVer, of course, would

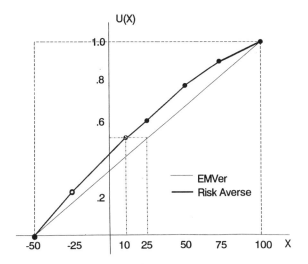

Figure 7.13. Utility For Money Curves

be willing to sell the rights to the lottery for $0.5(\$100) + 0.5(-\$50) = \$25$. This utility curve may be expressed graphically as shown in Figure 7.13. The units on the abscissa $X_0, \ldots, X_1$ represent increments from the decision maker's present assets.

The concave shape of the utility curve in Figure 7.13 results from the risk-averse nature of the decision maker. An EMVer would have the utility curve shown by the straight line in Figure 7.13. Because of the concavity of the curve, the decision maker places a value on a lottery somewhat lower than its expected monetary value. For example, suppose our decision maker owned the rights to a lottery that pays $75 with probability 0.7 and $30 with probability 0.3. The EMV of such a lottery is $0.7(\$75) + 0.3(\$30) = \$61.50$. However, the expected utility is given by $0.7U(\$75) + 0.3U(\$30) = 0.7(0.9) + 0.3(0.65) = 0.825$. This corresponds to a CME (certain monetary equivalent) of $60. The fact that the CME is less than the EMV is indicative of the conservative or risk-averse nature of our decision maker.

Once the decision maker's utility curve has been defined, the standard tenets of axiomatic decision analysis suggest that it may be employed in any decision problem as long as all the outcomes fall between L and W in desirability. In principle, the decision maker can now delegate authority to an agent, who can employ this utility curve to arrive at the same decisions that the decision maker would make. The utility curve conveys all the information necessary for making decisions that reflect the decision maker's preferences for the problem under consideration. Clearly, all of this is very idealistic. Although what is said may not be untrue, there are a lot of contingencies that need to be associated with the statements.

These are highly provocative statements. Substantively, there are several major problems. First, and probably most obvious, is that delegating authority has all sorts of ego, status, and authority problems that have nothing to do with utility curves and the substance of the decision in question. Second, delegation of authority would presume that the subordinate could be trusted fully to understand and adhere to the utility curves of someone else.

At a more theoretical level, we have assumed that, once determined, a utility function is immutable. Neither of these assumptions is valid. A person's utility curves, regarding any number of aspects of life, continually change, as the person learns new things, has new experiences, or develops new perspectives. Similarly, a utility curve that applies to one kind of problem, say, preferences for apples over oranges, may offer very little insight into another problem, say preferences for apples over pears. Thus, it may simply make no sense to suggest a subordinate (or anyone else) can learn someone's utility curves, and then operate as if these can be rigidly applied in changing circumstances.

Example. Thus far, we have assumed that our decision maker owned the rights to the lottery $[W, P; L, 1 - P]$ and was trying to determine a selling price, or a CME, for the lottery. Now suppose the decision maker does not own the rights to such a lottery, but is instead seeking to buy these rights. What should be the buying price? For an EMVer the buying price and the selling price would be the same, but for a non-EMVer they would not necessarily be the same, since we are, in reality, considering two different lotteries. This is true because in the case of buying a lottery, the buying price must be subtracted from the outcomes of the lottery. Thus if b is the buying price of a lottery, we are trying to find b such that $U(0) = PU(W - b) + (1 - P)U(L - b)$. This is also illustrated by the decision tree of Figure 7.14a. Similarly, the selling price s can be determined either from the relevant equation or from the decision tree of Figure 7.14b. Here, we assume that $W = \$100$, $L = 0$, and that the probability of winning is 0.25. Thus an EMVer should be willing to either buy or sell this lottery for \$25. With the risk-averse behavior we assumed in Figure 7.13, the buying price is such that $0.44 = 0.25U(\$100 - b) + 0.75U(-b)$. Thus, the buying price for the lottery is approximately \$19. The selling price for a risk-averse owner is determined from $U(s) = 0.25(1) + 0.75(0.44) = 0.58$ and is, from the utility function of Figure 7.13, approximately \$22.00.

Suppose we are faced with an equal chance of winning or losing \$10,000. Rather than risk losing \$10,000, we might be willing to pay, say, \$1,000 to avoid having to take this gamble, especially if we did not have \$10,000 to lose! This is, of course, the basis for the insurance industry. An example of this is described in Fishburn [41]. The shape of an individual's utility curve clearly depends upon that individual's present assets. Since an individual with a high asset position would be less averse to taking risks that might result in a loss, we might expect that the utility curve would approach that of an EMVer (i.e., a straight line) as the present asset position increased. Such a decision

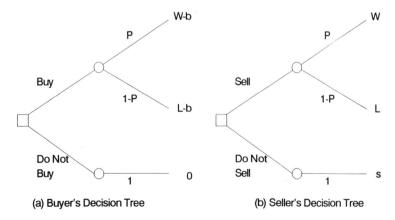

Figure 7.14. Decision Trees for Buyer and Seller

maker exhibits what is called decreasing risk aversion. In other words, when faced with a risky gamble, say an even chance of winning or losing $100,000, the amount which a risk-averse decision maker would pay to avoid this gamble, called the *insurance premium*, is a function of current assets, and should therefore decrease to zero as assets increase indefinitely.

Example. Let us examine briefly the interesting result that life insurance policies may be good for both the buyer and the seller in that they increase the utility of both the seller and the buyer of the insurance. We assume that we are considering a $100,000 term life insurance policy which costs $1000 per year. We assume that the individual is risk-averse. We assume that the probability of death in a given year is $P_{Death} = 0.009$, the present assets of the buyer are $10,000, and the value of future income discounted to the present is $100,000. Figure 7.15 details the payoff matrix representations of this life insurance problem for both the buyer and the seller. The expected value of a buy decision for the buyer is $109,000, which is less than the expected value of the do not buy decision, which is $109,100. If this were not the case the insurance company average profit would be negative. Even if the utility curves for both buyer and seller are risk-averse and precisely the same, the regions of operation are so different that the buyer is much more risk-averse than the seller.

Consider for example a utility for money of the form $U = $ in (assets $\times$ 10^{-3}) that is valid for both buyer and seller. The buyer is operating over the region $10,000 to $110,000 for determination of this utility curve, whereas the

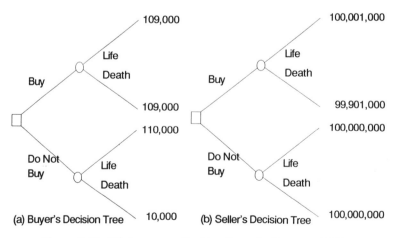

Figure 7.15. Life Insurance Decision Trees for Buyer and Seller

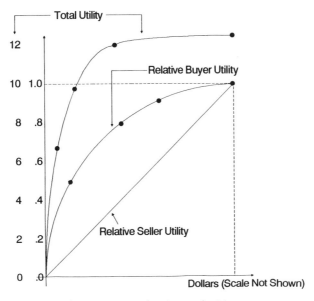

Figure 7.16. Utility Curves for Money

insurance company is operating over the region $99,901,000 to 100,001,000. We may use the relation $U' = aU + b$ to scale the utility curves for the buyer and seller over the range 0 to 1. The original and scaled utility curves are illustrated in Figure 7.16.

The total utility curve is concave, since both buyer and seller are risk-averse, but the buyer is relatively more risk-averse, since the seller can afford to "play the averages" but the individual buyer cannot. The buyer is willing to reduce total assets to less than the EMV, $109,000 versus $109,100, of the lottery to prevent a financial hardship for the surviving members of the family in the event of death. The insurance company is willing to assume this risk for the buyer, since the EMV is now higher, $100,001,000 as contrasted with $100,000,000, than its assets before selling the policy. In the payoff matrices for this example, or the equivalent decision tree of Figure 7.15, none of the outcomes for the seller differs significantly from the initial assets. This is not at all the case for the buyer. So, it is not too surprising that the insurance company is more of an EMVer than the individual buyer.

It is quite interesting to note that these results were obtained assuming that the utility curves for both the buyer and the seller were the same. However, the two parties operate over such different portions of the total utility curve that the buyer is quite risk-averse, whereas the seller is effectively an expected monetary value player.

7.2.6. Multiple Attribute Utility Theory

In this section, we describe an extension of the basic decision analysis paradigm to include situations where the outcomes have multiple attributes, we assume that a set of feasible alternatives $A = (a, b, \ldots)$ and a set $(X_1, \ldots, X_n)$ of attributes or evaluators of the alternatives can be identified. Initially, we assume that a single known outcome follows with certainty, from selection of an alternative. We comment on the decision under outcome uncertainty case later. Associated with each alternative course of action a in A, there is a corresponding consequence $X_1(a), X_2(a), \ldots, X_n(a)$ in the n-dimensional attribute (consequence) space $X = X_1 X_2 \ldots X_n$.

The problem faced by the decision maker is to choose an alternative a in A so that the maximum pleasure with the payoff or consequence $[X_1(a), \ldots, X_n(a)]$ results. It is always possible to compare the values of each $X_i(a)$ for different alternatives, but, in most situations, the values $X_i(a)$ and $X_j(a)$ for $i \neq j$ cannot be easily compared, because they may be measured in totally different units. It would be very convenient if we had a scaler utility value, just as in our earlier efforts in this section. Thus, a scaler-valued function defined on the attributes $(X_1, \ldots, X_n)$ is sought that will allow comparison of the alternatives across the attributes. The interested reader is referred to extended discussions by Keeney and Raiffa [42]. Only the highlights of this extensive theory are presented here.

A primary interest in multiattribute utility theory (MAUT) is to structure and assess a utility function of the general form

$$U[X_1(a), \ldots, X_n(a)] = f\{U_1[X_1(a)], \ldots, U_n[X_n(a)]\},$$

where U_i is a utility function over the single attribute X_i and f aggregates the values of the single attribute utility functions to enable one to compute the scaler utility of the alternatives. We assume that the utility functions U and U_i are continuous, monotonic, and bounded. Usually, they are scaled by $U(x^+) = 1$, $U(x^-) = 0$, $U_i(x_i^+) = 1$, and $U_i(x_i^-) = 0$ for all i. Here $x^+ = (x_1^+, x_2^+, \ldots, x_n^+)$ designates the most desirable consequence and the expression $x^- = (x_1^-, x_2^-, \ldots, x_n^-)$ denotes the least desirable consequence. The symbols x_i^+ and x_i^- refer to the best and worst consequence, respectively, for each attribute X_i. Thus, we have $x_i^+ = X_i(a^+)$ where a^+ is the best alternative for attribute i, and $x_i^- = X_i(a^-)$ where a^- is the worst alternative for attribute i. In the simplest situations, additive independence of attributes [42] exists such that the MAUT function may be written as

$$U(a_i) = w_1 U_1(a_i) + w_2 U_2(a_i) + \cdots + w_n U_n(a_i) = \sum_{j=1}^{n} w_j U_j(a_i)$$

Here the w_j are the weights of the various attributes of the decision alternative a_i and the U_j are the attribute scores for that alternative.

Mutual preference independence is required for this utility function to be valid in the case of certain outcomes. There are many cases where this independence relationship holds. It simply requires that preferences over specific outcomes on attribute $X_i(a)$ do not depend upon the outcomes on attribute $X_j(a)$ for all i, j. Somewhat stronger than the notion of preferential independence is that of utility independence. The utility independence condition is basically the same as that of preference independence, except that the preferences for uncertain choices involving different realizations of $x_i(a)$ do not depend on a fixed value that might be set for $X_j(a)$ for all i, j.

It turns out that precisely this same linear form of utility expression is valid in the case where decision outcome uncertainties are involved. In cases where there is probabilistic event outcome uncertainty, we determine the expected utility of the multiattributed outcomes as suggested by Figure 7.17.

We have just described very briefly the case of certainty in a multiattribute decision-making framework. Associated with each alternative there is a known consequence that follows with certainty from implementation of the alternative. These results also carry over to the case of decisions under uncertainty of event outcomes. The foundations for decision making under risk were discussed in an earlier portion of this section. The implications of this work are that probabilities and utilities can be used to calculate the expected utility of each alternative and that alternatives with higher expected utilities should be preferred over alternatives with lower ones. In the multiple

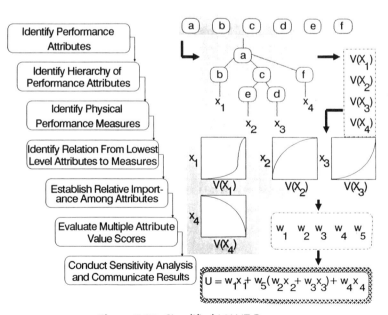

Figure 7.17. Simplified MAUT Process

attribute case, we simply calculate the scaler utility function of each multiat-tribute outcome and use this scaler utility function to calculate the subjective expected utility function as we have done previously.

Multiattribute utility theory is an approach for evaluating, prioritizing, and ranking the outcomes associated with different action alternatives in complex decision situations. The outcomes of alternative policies may occur with either certainty or uncertainty, and have multiple attributes that indicate the degree to which the outcomes meet prespecified objectives.

Even though the theory of multiattribute decision analysis is conceptually straightforward, especially when the linear additive independence of at-tributes condition holds, there are circumstances that make its implementa-tion very complex. Putting the methodology into practice is much more involved than we might initially believe. Each of the foregoing decision analysis steps requires substantial interaction between the analyst and the decision maker if useful results are to be obtained. There are a number of subtleties associated with scaling and, of course, a number of simplified approaches that can be utilized [43].

Application of MAUT presumes that the decision situation has been structured, for example, in a decision tree, that a hierarchy of objectives or attributes has been developed in the form of an attribute tree, and that the lowest level attributes are capable of being observed, as indicated in Figure 7.17. In the case of uncertainty, probabilities should be assigned to each of the possible outcomes of a decision alternative. Then, a utility or value function is derived, based on preference information provided by the decision maker. Using the multiple attribute utility function, the probability estimates, and utility values for the lowest level attributes associated with each out-come, an overall utility expected is computed for each decision alternative. These utility values reflect the decision maker's preferences. Therefore, they may be used for prioritization of alternative decisions.

The use of MAUT should be preceded by an effort in which objectives are defined, alternative decisions or policies are identified, and their impacts investigated. Following this decision issue formulation effort, the decision situation is structured, usually in the form of a decision tree or influence diagram that illustrates the possible outcomes of each action alternative and its associated probability. The attribute tree should always be structured in such a way that the lowest level attributes of the decision outcome events are measurable. Also, attributes should be defined in such a way that it is easily possible to elicit the decision maker's preferences over different possible outcomes. This requires that the attributes be defined in such a way that the linear independence conditions hold. Then, assessment of the decision maker's utility function results in a single number that represents the preference of the decision maker over the expected outcome of each decision alternative. The lowest level attribute measures for each outcome and the probability estimates for the outcomes are used as input data to this calcula-tion. The resulting cardinal multiple attribute utility function represents the decision maker's value system, assuming that we have been faithful to the tenets of MAUT and that the theory is applicable in the decision situation at hand.

It will, most often, be highly desirable to obtain a sensitivity assessment of the effort to investigate the effects of changes in probability estimates, elicited multiple attribute utility parametric coefficients, and other imprecise information that may have been used in the analysis. The results of a sensitivity assessment should give a good idea of the robustness of the alternative preference rankings that have been obtained and indicate the critical sensitivity factors and areas of the decision assessment that are in need of further attention.

The following simplified approach, an adaptation of a procedure called worth assessment [44–46], is suggested as a heuristic for construction of a multiple attribute utility function. It is illustrated in Figure 7.17.

1. *Identify the Overall Performance Attributes:* The list of identified attributes should be restricted to those performance attributes with the highest degree of importance and should include all relevant attributes. Ideally, the attributes in the list should be mutually exclusive and independent so that no one attribute encompasses any other attribute. The attributes should also be linearly independent in the sense that the decision maker is willing to trade partial satisfaction on one attribute for reduced satisfaction on another attribute.
2. *Construct a Hierarchy of Performance Attributes:* Once the overall attributes have been established, they are subdivided until the decision maker feels there is enough detail in each attribute that its value can be measured. This dividing and subdividing process results in a tree-type hierarchy.
3. *Select Appropriate Physical Performance Measures:* In constructing a hierarchy, the decision maker defines a set of lowest level attributes that are combined in some fashion to define the overall performance attributes. Some characteristic of performance must be assigned to each lowest level attribute to measure the degree of criterion satisfaction. It is important to distinguish between an attribute measure and a performance objective. A performance objective reflects what a decision maker desires from a set of decision consequences, whereas attribute measures reflect what a decision consequence can actually deliver across the identified attributes. In considering a possible standard of measure, the following question can be asked: Would changes in the value of this measure bring about significant increases or decreases in the extent to which each lowest level attribute is satisfied? If the answer is yes, the measure is appropriate.
4. *Define the Relationship between Lowest Level Attributes and Performance Measures and Deal with the Scoring Problem:* Selecting the performance measures establishes the connection between the measure and the worth indicated by that measure, but it does not completely specify the connec-

tion. The connection is established by a scoring function, which assigns an attribute score to all possible values of a given performance measure. The domain of the scoring function is the set of all possible values of the performance measure, and its range is the closed interval [0, 1]. The scoring function may be stated explicitly or graphically.

5. *Establish Relative Importance within Each Attribute Subcriteria Set:* In constructing the attribute hierarchy in step 2, each attribute is successively subdivided. At each point of division, an attribute criterion is defined by its subcriteria. Some of those subcriteria may be more important than others, and a weighting function is needed to indicate the relative importance of attributes. When this step is completed, the hierarchy can be used to create an overall attribute score for each alternative being considered.

Attribute weights are assigned to each attribute set such that the sum of the attribute weights is unity. The first step in this process is to rank the subcriteria by relative importance of attribute satisfaction. Next, the most important criterion is assigned a temporary value of 1.0. Then the decision maker must be more precise. How much less important is the second criterion than the first? If the answer is "half as important," the criterion is assigned a temporary value of 0.5. Next, the second- and third-most important are compared. If the third is one-fifth as important as the second, we assign it a temporary value of $1/5 \times 1/2 = 1/10$. Note that the third is also one-tenth as important as the first.

This process is continued until all attributes have been assigned temporary weights. Then those temporary weights are scaled so that their sum is unity. The final weight for the ith subcriterion is $w_i = \beta_i/(\beta_1 + \beta_2 + \cdots + \beta_p)$. It is important to note that the relative importance of any two attributes is reflected in the ratio of their assigned weights. Thus, we have obtained a ratio scale.

The issues of attribute independence and subdivision of attributes are important. Clearly, if the overall attribute performance objectives are independent of each other, the elements of an attribute set must also be independent, and they must be included within the meaning of the attribute they subdivide. Theoretically, it makes little difference whether we consider a single level attribute tree or a multiple level attribute tree. In practice, the effort involved and the results obtained may differ quite a bit across various approaches taken. There are really two problems here. The first is how to define the attribute sets, and the second is how to determine attribute independence. Let us examine these separately.

In comparing an attribute with its related higher level attributes, the following two statements should be examined to ascertain which better describes the relationship. A given attribute may be included within the meaning of, or be an integral part of, the higher level criterion. Alternately, an attribute may represent one of several alternative means of satisfying the higher level criterion and may be important only insofar as it contributes to it. If the first statement is the more accurate description, the candidate subcriterion may belong in the hierarchy under the higher level criterion. If the second is more accurate, worth independence does not exist, and so the attribute does not belong in the hierarchy.

Attribute independence can be defined by three interrelated statements:

5.1 The relative importance of satisfying separate performance criteria does not depend on the degree to which each criterion has in fact been satisfied.

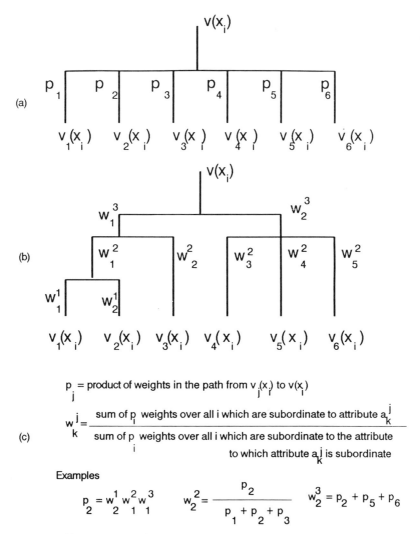

(a)

(b)

(c)

p_j = product of weights in the path from $v_j(x_i)$ to $v(x_i)$

$$w_k^j = \frac{\text{sum of } p_i \text{ weights over all } i \text{ which are subordinate to attribute } a_k^j}{\text{sum of } p_i \text{ weights over all } i \text{ which are subordinate to the attribute}}$$
to which attribute a_k^j is subordinate

Examples

$$p_2 = w_2^1 w_1^2 w_1^3 \qquad w_2^2 = \frac{p_2}{p_1 + p_2 + p_3} \qquad w_2^3 = p_2 + p_5 + p_6$$

Figure 7.18. (a) Single and (b) Multiple Level Attribute Structure and (c) Weight Relations

5.2 The rate at which increased satisfaction of any given criterion contributes to the overall worth is independent of the levels of satisfaction
already achieved on that and other criteria.

5.3 The rate at which decision makers would be willing to trade off
decreased satisfaction on one criterion for increased satisfaction on
another is independent of the levels of satisfaction already achieved by
the criteria.

6. *Calculate the Multiple Attribute Utility:* Using the attribute weights and
performance scores on lowest level attributes, the performance of each
alternative being evaluated is calculated.

7. *Conduct Sensitivity Studies as Needed and Communicate the Results of the
MAUT Evaluation.*

The activities associated with this effort are illustrated in Figure 7.17. The
approach is sufficiently general that it can accommodate either single-level
tree structures or multiple-level structures, as illustrated in Figure 7.18.

Example. We present below some sample criteria for database management
system (DBMS) software selection. These provide us with an appropriate
summary of the critical and important factors in conducting a MAUT-based
study. First, it is important to present the sequence of steps with which we
approach DBMS software evaluation:

1. Identify user requirements for DBMS software.
2. Define critical attributes for the DBMS software. Identify the range of
performance expected for each attribute from best performance to
worst performance. Also identify the importance of this performance
range across each attribute.
3. Obtain literature describing potential software packages and arrange
for an evaluation/demonstration of the software. If needed, iterate
back to step two and refine attributes and attribute scoring measures.
4. Evaluate each software package according to its score on the various
performance attributes.
5. Conduct a sensitivity analysis of the results of the analysis, if needed.
6. Select the highest scoring software package.

In the first step, we identify critical user needs to be fulfilled by the DBMS
software. When we speak of user requirements, we refer to the purposes and
associated functions that the user wishes or needs to perform. These user
requirements can be converted into *requirements specifications*, which are the
more technical and structural characteristics of the DBMS. Two closely

related, but different approaches, are possible. One results in a ratio scale, and the other in an interval scale. The first approach, called *relative weight* or *swing weight*, is anchored on the best performing and worst performing attribute scores for the actual alternatives and yields a ratio scale. The second approach, called *absolute weight*, is based on definition of a hypothetical best and worst attribute score. In the way accomplished here, it results in an interval scale.

1. Relative weight lists desired attributes in order of importance; evaluates the degree to which each candidate software package has each such attribute; weights these evaluations according to relative importance of each attribute; and sums to come up with a final evaluation of each candidate alternative.

2. Absolute weight lists attributes in order of importance; establishes maximum desired and minimum expected performance level for each attribute; evaluates each candidate alternate by the degree to which it satisfies this maximum value for each attribute; weights these evaluations according to relative importance of each attribute; and sums to come up with a final evaluation of each candidate alternative.

We now describe these two multiattribute-based approaches to performance evaluation. The first of these is based on relative weights and relative performance scores across attributes of performance and performance alternatives.

In the swing weight based approach, we identify the maximum performance expected on each of the identified performance attributes and assign these performance levels a score of 1.0. The actual score used to indicate the maximum is arbitrary and the overall evaluation does not depend upon this value. We could just as well use 100 as the maximum score possible. We identify the minimum performance expected on each attribute and assign these performance levels a score of zero.

Next we assign a level of importance to each attribute. Clearly, some attributes are very important and some relatively unimportant. A possible way to identify weights for the attributes is to identify the most important performance attribute in terms of the worst to best performance scores expected. It is necessary to identify this importance in terms of the performance ranges expected. If all software packages have virtually the same performance on some attribute, then the *relative* importance of this attribute is zero. Recall that there will be a performance score of 1.0 for the best performing software package and a performance score of 0.0 for the worst performing package. This attribute is assigned a value of one. The next most important performance attribute, in terms of the range in performance expected, is identified. If the most important attribute has a weight of 1.0, then we should associate a lower weight, which is called a swing weight for this second most important attribute. This is continued until we have identified swing weights for all attributes of importance. These swing weights are then normalized such that they sum to one. The normalized swing weight, w_j, becomes $w_j = W_j/(\sum_{j=1}^{N} W_j)$, where N is the number of attributes considered and W_j is the (unnormalized) swing weight of the jth attribute.

Following this, we identify performance scores for each alternative on each attribute. We have already identified the 0.00 and 1.00 performance scores and their meaning and so the performance evaluation task is just that of placing the performance of each DBMS alternative within this range for each of the performance attributes. The final performance score is then determined from the weighted sum $\mathcal{S}_i = \sum_{j=1}^{N} (w_j S_{ji})$, where w_j is the swing weight associated with the jth attribute, S_{ji} is the performance score of the ith software package on the jth attribute of performance.

The absolute weight approach is based on identifying an ideal performance alternative, one whose performance is ideal across all performance attributes and then associating a performance score of 1.0 with this for all attributes. The following associations of numerical scores with performance characteristics are somewhat mercurial, but may be of value in identifying an ideally performing DBMS package and in setting the standards for judgment of actual DBMS packages.

1.00 Package completely meets performance requirements on this attribute
0.75 Package is very acceptable on this performance attribute, in all but very minor ways
0.50 Package generally satisfies maximum performance requirements on this attribute, but is deficient in some important aspects
0.25 Package fails to meet performance specifications associated with this attribute
0.00 Package is almost totally deficient in performance on this performance attribute

Now that we have defined the performance of an ideal standard DBMS package, and established performance scores across each attribute for less than ideal performance, we need to identify the weight to assign with each attribute. Clearly, some attributes are important and some unimportant. These importance weights can be expected to vary, perhaps considerably, across specific software evaluations. A possible way to identify the (absolute) attribute weights is to identify the most important performance attribute. This attribute is assigned a value of one. The next most important performance attribute, in terms of the range in performance expected, is identified. We might associate weights in accordance with the following associations:

1.00 The feature described by the attribute in question is mandatory and very important

0.75   The feature described by the attribute in question is quite important
0.50   The attribute in question is moderately important as is performance at the ideal level specified by a 1.0 performance score
0.25   Performance at the level specified by the ideal standard for this attribute is desirable, but not at all needed
0.00   Performance of this attribute is unimportant and this attribute can be disregarded in terms of performance evaluation

The process of listing the attributes in terms of importance continues. After they are identified, the absolute weight values are identified. These must be veridical to the interval scale listing. A danger with this particular approach is identifying too many performance attributes and assigning most or all of them too high a worth. A worth of 0.35 many not seem very high. However, if there are 10 attributes with this weight, then it is possible to obtain a performance score of as much as 3.5 on these attributes alone. This is 3.5 times the maximum score of 1.0 that can be obtained due to performance on the most important attribute. For this reason, primarily, this author prefers the relative, or swing weight, approach.

These absolute weights are then normalized such that they sum to one. The normalized absolute weight w_j becomes $w_j = W_j/(\sum_{j=1}^{N}W_j)$ where N is the number of attributes considered and W_j is the unnormalized weight of the jth attribute. The final performance score is then determined from the weighted sum $\mathscr{S}_i = \sum_{j=1}^{N}(w_j S_{ji})$ where w_j is the absolute weight associated with the jth attribute, S_{ji} is the performance score of the ith software package on the jth attribute of performance.

Even though the formula used to compute the final software package worth appears to be the same for the two approaches, the interpretation placed on the attribute weights and alternate scores are quite different. Nevertheless, each method is subjective and appropriate. The swing weight approach has the greatest theoretical basis for use. It would be appropriate to evaluate the performance of candidate software packages using each approach and, if there is a discrepancy in the final result, then explore the causes for this, perhaps by means of a sensitivity analysis and more refined definition of terms. The critical difference between the two methods is whether one wants to think critically about trade-offs among attributes, implying the value of the swing weight method on relative attributes, or whether one wants to focus on the overall performance of the system, implying the value of the absolute method on some absolute standard.

As a specific numerical example of DBMS software package evaluation, let us consider that four attributes are initially thought to be of importance:

1. Software company trustworthiness (W_1).
2. Database security and integrity (W_2).
3. Operating effectiveness (W_3).
4. Data dictionary (W_4).

Assume that three software packages are being reviewed for purchase. To use the swing weight approach, first determine the best and worst performing software packages across each alternative. This might result in the following partially filled in performance scoring matrix:

$$\text{Software Package}$$

$$
\begin{array}{ccc}
1 & 2 & 3
\end{array}
$$

$$
S_{ji} = \begin{bmatrix}
0.00 & 1.00 & \\
0.00 & 1.00 & \\
1.00 & & 0.00 \\
& 1.00 & 0.00
\end{bmatrix}
$$

This says that package 1, represented by the matrix column S_{j1} is the worst performer on the first two attributes in the evaluation and the best performer on the third attribute. On the other hand, package 2 is the best performer on attributes 1, 2, and 4. It is not a worst performer on anything. Package 4 is neither a best nor a worst performing package on any attribute. It is important to note that there is a single 0.00 and a single 1.00 in each row.

Based on these assessments, evaluate the performance of the other alternative attribute pairs. Suppose that we obtain the following elicited results.

$$
S_{ji} = \begin{bmatrix}
0.00 & 1.00 & 0.60 \\
0.00 & 1.00 & 0.90 \\
1.00 & 0.00 & 0.00 \\
0.40 & 1.00 & 0.00
\end{bmatrix}
$$

Based on this definition of best and worst, we now need to determine the attribute swing weights. Suppose that the difference between worst and best performance is such that attribute 4 is the most important. Then assign it an unnormalized weight of 1.00. Given this, we might say that attribute 3 is only 0.80 as important as attribute 1, that attribute 1 is 0.40 and attribute 2 is 0.30 as important as attribute 4. We then calculate the normalized swing weights and obtain

$$W_1 = 0.40 \qquad W_2 = 0.30 \qquad W_3 = 0.80 \qquad W_4 = 1.00$$

$$w_1 = 0.16 \qquad w_2 = 0.12 \qquad w_3 = 0.32 \qquad w_4 = 0.40$$

The final evaluation scores are obtained using the equation $\mathscr{S}_i = \sum_{j=1}^{N}(w_j S_{ji})$ as $\mathscr{S}_1 = 0.46$, $\mathscr{S}_2 = 0.62$, and $\mathscr{S}_3 = 0.20$. Thus, we see that alternative two is the best alternative, actually by quite a fair amount.

To do this same evaluation using the absolute weight approach, we might proceed as follows. We might obtain as weights, for example,

$$W_1 = 0.60 \qquad W_2 = 0.50 \qquad W_3 = 1.00 \qquad W_4 = 1.00$$

and then calculate relative weights as

$$w_1 = 0.19 \qquad w_2 = 0.16 \qquad w_3 = 0.32 \qquad w_4 = 0.32$$

Now we assign performance scores.* These might be

$$S_{ji} = \begin{bmatrix} 0.50 & 1.00 & 0.70 \\ 0.50 & 1.00 & 0.90 \\ 1.00 & 0.60 & 0.50 \\ 0.60 & 1.00 & 0.30 \end{bmatrix}$$

such that the evaluation scores are $\mathscr{S}_1 = 0.56$, $\mathscr{S}_2 = 0.86$, and $\mathscr{S}_3 = 0.48$. Thus, again alternative two is the best alternative according to the criteria and approach used. The others appear to be close competitors. In part this is caused by the fact that we are working over a smaller range of performance scores in that no score is now less than 0.30 and many are close to 1.00.

7.3. DESCRIPTIVE DECISION MODELS

In this section, we examine a number of models that lie in the nexus of the normative and the descriptive. First, we examine some decision framing situations and show how these may lead to a number of anomalies. In the course of our journey, we discuss regret models of judgment and choice, cognitive heuristics and biases, and prospect theory. We also describe a number of descriptive judgment and choice strategies, many of which are flawed. Our presentation concludes with a discussion of social judgment theory (SJT) which, as we indicated earlier, may serve as an appropriate way to identify utility functions for use in prescriptive and/or descriptive approaches.

7.3.1. The Framing of Decision Situations to Include Regret

Violations of consistency and coherence in choice may often be traced to cognitive limitations, which govern the perceptions of decision situations, the processing of information, and the evaluation of opinions. A strong determinant of the frame or structure adopted for a given decision is the decision maker's familiarity with the issue under consideration. In this section, we examine some of the ways in which our perspective strongly influences how we assess a decision situation. In large part, our efforts are based on previous work by Sage and White [47, 48].

---

*Ideally, it would be best to do this across all alternatives at a single attribute, and then proceed to a new attribute. To make this assignment across attributes for a single alternative is generally more cognitively demanding and invites cognitive bias.

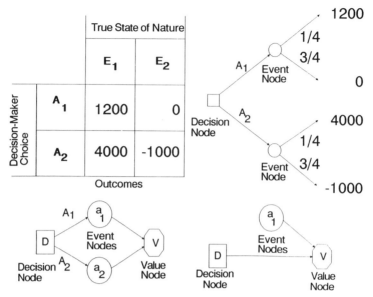

Figure 7.19. Payoff Matrix, Decision Tree, and Influence Diagrams for Simple Decision Options and Outcomes

Because of imperfections in human cognition, it turns out that alternative framing representations, and the associated differences in perspective from which situations are assessed, may often act to cause a reversal in the relative desirability of achieving various objectives. This, in turn, may alter the relative desirability of the decision options. Sometimes the order in which we assess information often unduly influences the initial estimate (or anchor or base rate) used to assess a situation. It is essential that issues be framed and situations assessed in a very careful fashion, such that either affective mental models or any analytical models that the decision maker may choose for the decision situations are truly representative of the essential features of the situation at hand. This is needed to avoid or at least minimize possibilities for cognitive bias and the use of improper heuristics.

Example. Consider the representative decision situation illustrated in Figure 7.19. This is precisely the same generic decision model considered earlier as Figure 7.5. Now, specific probabilities and outcomes are associated with the two decision options. Also shown in this figure is a decision tree for the issue considered and two influence diagrams. Both are correct, but the one on the right is preferable, because it explicitly recognizes that the event outcome probabilities are independent of the decision option selected.

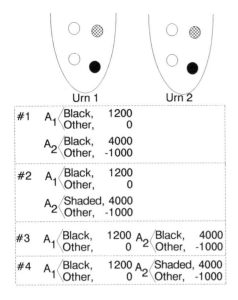

Urn 1   Urn 2

#1 $A_1\begin{cases}\text{Black,} & 1200 \\ \text{Other,} & 0\end{cases}$

$A_2\begin{cases}\text{Black,} & 4000 \\ \text{Other,} & -1000\end{cases}$

#2 $A_1\begin{cases}\text{Black,} & 1200 \\ \text{Other,} & 0\end{cases}$

$A_2\begin{cases}\text{Shaded,} & 4000 \\ \text{Other,} & -1000\end{cases}$

#3 $A_1\begin{cases}\text{Black,} & 1200 \\ \text{Other,} & 0\end{cases}$ $A_2\begin{cases}\text{Black,} & 4000 \\ \text{Other,} & -1000\end{cases}$

#4 $A_1\begin{cases}\text{Black,} & 1200 \\ \text{Other,} & 0\end{cases}$ $A_2\begin{cases}\text{Shaded,} & 4000 \\ \text{Other,} & -1000\end{cases}$

Figure 7.20. Urn Models for Decision Shown in Figure 7.19

From the information presented in Figure 7.19, it is reasonably clear that alternative A_1 should be chosen over alternative A_2, assuming that our decision rule is to select the alternative with the greatest expected return. The expected return from A_1 is $300, whereas the expected return from option A_2 is $250. Generally, the value of money is not a linear function of the amount of money, however, due both to the satiation effect of money on value and to an attitude toward risk that further shapes that value function into a utility function. If the decision maker is risk averse, the strength of preference for option A_1 is further increased. These separate and distinct issues of preferences and risk aversion are discussed in our commentary on prospect theory in Section 7.3.4.*

The models presented in Figure 7.19 are not necessarily fully representative of the decision situation, even though they are correct models. Let us examine the various urn representations shown in Figure 7.20 from which the decision situation graphically illustrated in Figure 7.19 could have been obtained. Let us also speculate on how the decision maker might react in each case. There are four choice situations illustrated. In each case, the decision maker is able to view the ball drawn from the urn that is used for the four decision situations. In some cases, additional information is available and then should be used.

---

*It turns out that negative outcomes are often valued in a different fashion than positive outcomes of the same amount and that value functions are often convex for gains and concave for losses. These are not especially important for our present discussions.

Choice Situation 1. Suppose that we must choose between the options described in choice situation 1. If we choose option A_1 and a white or shaded ball is drawn, then we obtain nothing. However, it does not appear that we would feel badly about not having chosen option A_2, because, in this latter case, we would surely have lost a nice sum of money ($1,000). So, we might actually feel quite good at having chosen option A_1 and not A_2. Thus, we see that the value felt from a decision outcome may be a combination of value for what we did obtain as well as regret or joy for what we could have obtained had we chosen the other option.

If after having selected decision option A_1 a black ball is drawn, we win $1,200. While this is a rather desirable outcome, we might or might not have regret associated with our not having selected option A_2, where we would have won $4,000. It is possible that we would just feel good about having played it safe and won the $1,200. Alternatively, there could be regret at not having chosen option A_2 as we did suffer a loss of $2,800 by selecting A_1. It appears that many people would express a postdecision regret at selecting option A_1 and having obtained a black ball. Therefore, the potential need to include regret as an attribute of the choice situation becomes apparent. We examine some of the implications of this regret here.

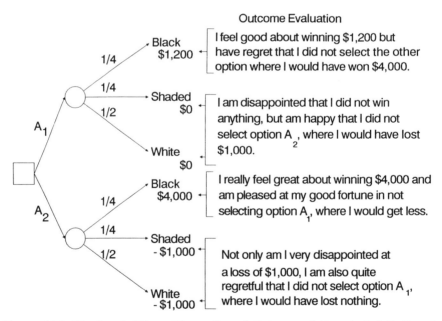

Figure 7.21. Situation 1 Where Observation of Outcome of Nonselected Options is Permitted

Suppose that we choose option A_2. If a white or shaded ball is drawn, it is very possible that we would feel quite badly about having been greedy in choosing option A_2 over the less risk laden option of choosing A_1. Thus, there exists regret associated with this choice. Here, again, we experience a postdecision regret over what we might have obtained. If a black ball is drawn, however, we would be quite happy with having selected option A_2. Again, we see a postdecision effect, bliss or sadness, that is perhaps modified from the predecision situation. Figure 7.21 describes the expanded sense of value that we feel from the possible realized outcomes associated with this decision situation.

Choice Situation 2. Suppose that we must choose between the options described in choice situation 2 of Figure 7.20. We note that the same decision tree as in problem 1 may be used. But, the decision tree of Figure 7.19 is not fully indicative of the decision situation. The present decision situation differs from that of problem 1 in that the "win" state for option A_2 is associated with a black ball in choice situation 1. Here, it is associated with a shaded ball.

If a black ball is drawn, we should really feel quite good about having selected option A_1 instead of option A_2. Had option A_2 been selected $1,000 would have been lost. If a shaded ball is drawn, no money is lost under decision option A_1 but $4,000 would have been won had option A_2 been selected. Here, we may feel quite badly about not having selected decision A_2 in the first place as we know, postdecision, that we would have won if we had but selected option A_2. There should be no predecision regret, but we might well be quite sorry about our choice after we see the outcome here.

Suppose that we choose decision option A_2. If a white ball is drawn, we would most likely feel quite badly having been greedy in selecting this option instead of option A_1. We loose $1,000 now, whereas nothing would be lost or gained had we chosen the other option. Finally, if a shaded ball is drawn, then we would probably feel extraordinarily happy about having both drawn the shaded ball and selected option A_2. We win $4,000 with the decision selected and the outcome obtained. We would obtain nothing under the other option. Figure 7.22 illustrates some of the possible decision-outcome values associated with this problem.

Choice Situation 3. We assume that we are only informed that we have won or lost according to the color of the ball drawn. This may be represented by a two-urn model for choice situation number 3 or choice situation 4, as indicated in Figure 7.20. The major difference between this problem and the other two is that we now have no idea what would have been obtained as the outcome event under the option not selected.

Suppose that we choose option A_1. If a white ball or a shaded ball is drawn, we are simply informed of the loss. Since there is no way of determining what would have been the outcome had the other option been selected, we have no basis for regretting not having chosen the other option.

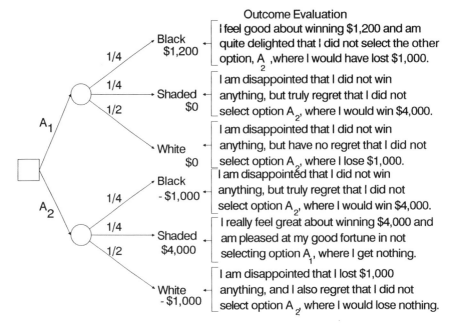

Outcome Evaluation

I feel good about winning $1,200 and am quite delighted that I did not select the other option, A_2, where I would have lost $1,000.

I am disappointed that I did not win anything, but truly regret that I did not select option A_2, where I would win $4,000.

I am disappointed that I did not win anything, but have no regret that I did not select option A_2, where I lose $1,000.

I am disappointed that I did not win anything, but truly regret that I did not select option A_2, where I would win $4,000.

I really feel great about winning $4,000 and am pleased at my good fortune in not selecting option A_1, where I get nothing.

I am disappointed that I lost $1,000 anything, and I also regret that I did not select option A_2 where I would lose nothing.

Figure 7.22. Situation 2 Where Observation of Outcome of Nonselected Options Is Permitted

If a black ball is drawn, we are informed that we have won. Since there is no way of determining what would have occurred had the other option been selected, we can only be happy about having won. We surely should not assume that we would have obtained a black ball if we selected option A_2 in choice situation 3 where we would be even happier.

Suppose that we do choose option A_2. If a white or shaded ball is drawn, we surely feel badly about having lost $1,000. We should have no regret, however, at not choosing the other option. If a black ball is drawn, we should be quite satisfied with having gone for the "big win" of $4,000 and won this amount. Figure 7.23 illustrates the decision-outcome values for this problem 3 situation model. Choice situation 4 is so very similar to this one that we immediately see the reasonableness of the decision-outcome values illustrated in Figure 7.23.

In our discussion, we assumed that the decision maker had explicit information available concerning the outcomes from the option not selected in choice situations 1 and 2. These are the one-urn situation model problems. We assumed that no information was available concerning the outcomes from the option not selected in choice situations 3 and 4, the two-urn problems. Even though we may perceive subtle differences between the choice situations, they are sometimes modeled in the same way in the decision tree

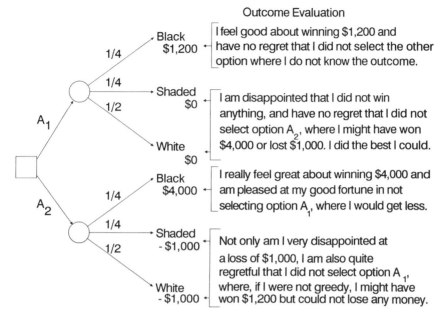

Figure 7.23. Situation 3(a) Where Observation of Outcome of Nonselected Options Is Not Permitted

representation shown in Figure 7.19. This is surely a good representation of the decision tree portion of the decision situation. However it is not a good representation of the attributes of the *decision outcome combinations*. For this, we might choose to use an attribute tree, or a more complete labeling of the outcomes in terms of their attributes, as shown by the decision trees of Figure 7.21 through 7.23.

Even in the very simple situations, we see that we have identified two attributes associated with the outcomes. These are the amount of money, won or lost, and the regret or rejoicing associated with what could have been obtained from the decision not selected. Figure 7.24 illustrates this attribute tree for the three choice situations considered. It includes both the notions of

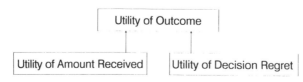

Figure 7.24. Outcome Utility Comprised of Utility for Outcome and Regret

utility for the outcome that we do receive as well as *regret* over what we would have obtained had we chosen the other option.

In the single-urn model, we may generally have postdecision regret or postdecision happiness associated with outcomes. In the two-urn model, we do not generally have postdecision regret since we do not know the outcome for the option not selected. The central issue in this is that the decision situation model should not be the same for all of these choice situations. The outcomes and the associated values are not truly the same, and our feelings toward them are not the same since the available information is different in each case. In other words, Figure 7.19 is a valid decision tree, decision matrix, and influence diagram; however, the decision tree is incomplete.

Let us examine this further. Postoutcome information about outcomes from decisions not selected is both context dependent and different from predecision information. Thus, we need to examine models that allow inclusion of the available predecision and postdecision information on choice situations. In particular, we wish to resolve such questions as the following.

1. Is there a difference between choice situations 1, 2, 3, and 4? Should there be a difference with respect to judgment concerning the most preferred alternatives in each situation?
2. What constructs are available to explain, in descriptive, prescriptive, or normative senses, any differences in these?
3. Are there possibilities for nontransitivities or other anomalous results from using these constructs?

We will now provide some answers to these questions.

7.3.2. Regret

We have just introduced some notions of postdecision regret. Bell and his colleagues [49–53] were perhaps the first to discuss the inclusion of regret as an additional attribute sometimes needed to more fully capture the decision maker's values. It has often been suggested that regret be used as a surrogate for value and that an option alternative be selected that minimizes regret.

The famous Allais Paradox, perhaps first brought to popular attention in the work of Raiffa [8], is an especially good example of this. In gamble 1, the DM has a choice of option A_1 which provides \$1,000,000 for sure, and option A_2, which is a lottery described by [\$5,000,000, 0.10; \$1,000,000, 0.89; \$0, 0.01]. Most people will reason that a million dollars is quite a bit of money and it is well worth taking it, as contrasted with a second option where there is a 0.10 chance of getting five times this much, but a very small chance of getting nothing. After all, how would you feel if you selected option A_2 and just happened to get nothing. In the second gamble, option B_2 results in a lottery

[$5,000,000, 0.10; $0, 0.90] and alternative B_2 results in the lottery [$1,000,000, 0.11; 0, 0.89]. It turns out that most people will now select the first alternative. This is not unreasonable since it has a much larger expected value and the probabilities of winning are about the same in each case. It turns out that this behavior is quite inconsistent with the precepts of subjective utility theory. Raiffa's explanation for this is quite interesting. He reframes the decision situation structural model in such a way that the postdecision regret can not be sensed. We will address some of the fundamental ingredients associated with this phenomenon in the remainder of this section.

Often, regret is measured against some ideal best and worst outcomes, associated with the specific choice situation at hand, and these same anchors are used to measure regret for all outcomes. Bell's effort, and related results due to Loomes and Sugden [54, 55], consider value and regret as simultaneously present attributes of decision outcomes. Two-option situations are considered, and regret is a differential concept that is measured within the outcome states across the options that could have led to these states, as in Bell's work. Loomis presents experimental evidence to test the prediction that certain alterations in the pairwise comparison of alternatives will affect pairwise comparisons [56]. An effort by Sugden [57] relates these efforts to alternative approaches.

These developments appear more applicable in those situations where the selection of an option that leads to a winner or loser would result in a certain postdecision error-free identification of a win or loss situation than had the other option been selected. There are, of course, many examples of decision situations where one may obtain, postdecision, full knowledge about what would have happened under the option(s) not selected. There are, however, other decision situations where complete postdecision knowledge about what would have happened under the option not selected is not available. Also there are many decision situations that involve more than the two alternative courses of action options. Our effort here [48, 49] extends regret theory to incorporate these considerations. Again, an example is a very suitable method of illustrating more general results.

Example. Consider the illustrative decision situation represented by the decision tree of Figure 7.25. Here, selecting option A_1 yields $100,000 with probability 0.9 and $0 otherwise. Selecting option A_2 will result in obtaining $80,000 for sure. Most people, very likely including you and me, will select options A_2. The choice of this option, which has a certain outcome, over the option with the uncertain but potentially larger outcome, may be attributed to this certainty effect. We prefer the certain $80,000 to a 0.9 chance of winning $100,000 and a 0.1 chance of winning $0. In other words, we are quite willing to pay an effective premium of $10,000 in expected value loss, which represents foregone winnings of 12.5% of the certainty winnings, to

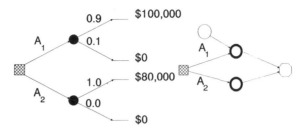

Figure 7.25. Illustrative Situation Depicting Regret Due to Certainty Effect

avoid the possibility of this postdecision regret of getting nothing when we could have received $80,000.

One full information framing situation, from which the decision maker might view this choice situation, is shown in Figure 7.26. This is a somewhat more general situation than depicted in Figure 7.25 and includes it as a special case. We consider that there are two essential attributes here. The first attribute $v_i(A)$ represents the utility value for the ith outcome for what we actually gain or lose. The second attribute $f_i(A)$ represents the utility value for the ith outcome for the regret that we have over assets foregone by not having selected the other alternative. We assume a linear multiattribute utility function form such that we have $EU(A_i) = \sum_j p(e_j|A_i)[v_j(A_i) + r_j(A_i)]$. Here $p(e_j|A_i)$ is the probability of obtaining outcome e_j given that option A_i is selected. $v_j(A_i)$ is the value associated with obtaining outcome e_j as a result of selection of option A_i. $r_j(A_i)$ is the regret associated with obtaining outcome e_j from option A_i and not obtaining some other outcome that would have been known to occur had another option been selected.

Clearly, we need to be careful in anchoring this other outcome. We assume that M is the largest possible outcome, O is the smallest, and that the certainty outcome C is intermediate between these two. In order to have

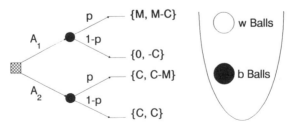

Figure 7.26. Illustrative Situation Depicting Regret Due to Certainty Effect for Given Urn Situation Model

option A_1 preferred to option A_2, $A_1 \to A_2$, we must have

$$p[1 + r(M - C)] + (1 - p)r(-C)$$
$$< v(C) + pr(C - M) + (1 - p)r(C) \qquad (7.1)$$

where we assume, for convenience and without any loss of generality, that $v(M) = 1$, $v(0) = 0$.

The regret attribute is just the regret of the foregone assets, because it is just the difference between the outcome actually realized and the outcome that knowingly would have been obtained in the other lottery. It is very important to note that value here is a cardinal value function measured with some presumed anchor whereas regret is a differential cardinal measure anchored on another outcome that *"could have been."* This approach, assuming that regret and utility are additive, is correct for the special case where the problem being evaluated may be represented by the one-urn, two-outcome full information model shown in Figure 7.26. In this case, if one wins or loses after having selected one option, then it can be determined whether one would have won or lost had the other option been selected.

However, let us suppose that Figure 7.25 was intended to be representative of the two-urn situation model shown in Figure 7.27. There are only two outcomes associated with the decision situation model of Figure 7.27, a situation model in which regret enters the decision situation as a second attribute. These involve option A_3, the uncertain outcome option. If we select this option, then we know for sure what we could have obtained under the uncertain option. If we select decision option A_4 and the certain return, there is no way of knowing whether we would have won or lost had the other urn been selected. Thus, there can be no regret associated with this option.

The significant question here concerns whether or not postdecision knowledge of the outcomes increases or decreases the overall utility of the certain

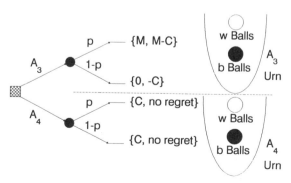

Figure 7.27. Illustrative Situation Depicting Regret Due to Certainty Effect for Given Urn Situation Models

outcome alternative A_4. The expected utility of A_4, with no postdecision information about the outcomes obtained from alternative A_3 available, is just $v(C)$.

However, in the contrasting situation with postdecision regret information available in the model of Figure 7.26, we have $EU(A_2) = v(C) + pr(C - M) + (1 - p)r(C)$. We can obtain the answer to our question through investigation of this equation. The answer to this question appears to be associated with the nature of the regret function. We know that the utility of alternatives A_1 and A_3 are the same since they have precisely the same outcome values and regrets, as indicated in Figures 7.26 and 7.27. This is not the case for the expected utilities of the other two decisions. Postdecision outcome regret information will increase the utility alternative of A_4, relative to that of alternative A_2, such that $A_4 \rightarrow A_2$ if

$$pr(C - M) + (1 - p)r(C) < 0 \qquad (7.2)$$

This same postdecision regret information will decrease the expected utility of alternative A_4 relative to that of alternative A_2, such that $A_2 \rightarrow A_4$ if

$$pr(C - M) + (1 - p)r(C) > 0 \qquad (7.3)$$

The alternatives A_1 and A_3 should have the same expected utility here for these pairwise preference comparison examples, because we know in each case what outcome occurs from the option not selected. For the second gamble considered and illustrated in Figure 7.27, decision alternative A_3 is preferred to decision alternative A_4, or $A_3 \rightarrow A_4$, if we have

$$p[1 + r(M - C)] + (1 - p)r(-C) > v(C) \qquad (7.4)$$

where we assume that the regret associated with zero money is zero, $r(0) = 0$ for convenience and without loss of generality.*

There may exist preference reversals, in the sense that $A_2 \rightarrow A_1 \leftrightarrow A_3 \rightarrow A_4$ if the inequalities of Eqs. (7.1) and (7.4) are satisfied. If this occurs, then the inequality of Eq. (7.3) holds. In a similar way, we will obtain the preference reversals indicating that $A_4 \rightarrow A_3 \leftrightarrow A_1 \rightarrow A_2$ if neither Eq. (7.1) nor Eq. (7.4) is satisfied. If this occurs, Eq. (7.2) is satisfied. Thus, we see that, under appropriate circumstances, preference reversals may be made to occur by changing the information set that is made available to the decision maker. Changing the probabilities of obtaining the outcomes may also result in preference reversals. In most cases we expect that greater postdecision outcome information will increase the utility of the certain outcome alternative, assuming it is favorable, relative to the uncertain outcome alternative.

---

*This will necessitate having negative regrets. These amount to not a regret in a negative sense but a regret in a rejoicing sense over how badly we would have done if we chose the other alternative.

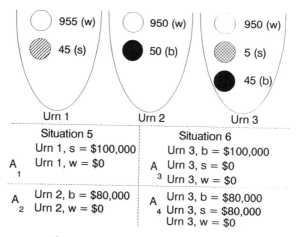

	Situation 5	Situation 6
A_1	Urn 1, s = $100,000 Urn 1, w = $0	Urn 3, b = $100,000 Urn 3, s = $0 Urn 3, w = $0
A_2	Urn 2, b = $80,000 Urn 2, w = $0	
A_3		
A_4		Urn 3, b = $80,000 Urn 3, s = $80,000 Urn 3, w = $0

Figure 7.28. Decision Situation Model

An interesting modification to the urn-model decision situations can be made by inserting s no-win shaded balls in each urn for the decision situations initially depicted by Figures 7.26 and 7.27. The resulting situations are just those depicted in Figure 7.28. In making this change, a little reflection convinces us that we should do it to preserve the general regret features initially associated with the initial choice situations. The certainty effect initially associated with choice situation six now vanishes, however. This is so because there is now no certain way of knowing what would have occurred under outcome A_4 if we choose decision option A_3.

Of course, if the number of shaded balls, s, is small we have a relatively good idea of what would have occurred, but there is no way in which we can know for sure. Suppose, for example, that we let M = $100,000, C = $80,000, b = 45, w = 5. It seems not at all unreasonable that we prefer choice A_3 to choice A_4 in choice situation 5 and choice A_4 to A_3 in choice situation 6.

Let us now suppose that s = 950 shaded balls are added to each urn for these two decision situations. We then obtain the decision situation structural models of Figure 7.29. Our preferences should remain the same regardless of s, the number of shaded balls, if we assume that the conventional utility theory that we have described earlier is fully applicable. The sure thing principle of Savage and the strong independence axiom of Samuelson each require this. These early seminal results in decision analysis are based on the assumption that predecision and postdecision regret information is the same. Raiffa [8] provides some elaboration on this point, as do a number of other texts. It is very likely, however, that we now prefer the more risky options A_1 and A_3 in decision situations 5 and 6 since there is potentially a 25 percent greater monetary return with only slightly greater probability of not receiving

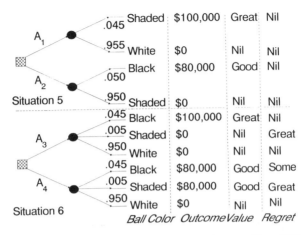

Figure 7.29. Illustrative Situation Depicting Regret Due to Two Different Situation Models

the money. By adding the shaded balls to the urns, we have changed the decision situation considerably. The significance of this observation is that a knowledge of the information patterns of choice and outcomes in the decision situation model is essential. Further, the notion of regret is not at all independent of uncertainty levels, especially when there are postdecision information uncertainties. There have been a number of attempts to illustrate the nonrationality of choices that violate one or more of the classic axioms of decision theory. The prospect theory discussions to follow provide additional illustrations. Our purpose in doing this is to demonstrate the need to carefully construct decision situation structural models, especially with respect to information flow patterns, including the nature of any regret that is associated with the decision situation.

It is well known that sets of pairwise preference comparisons are often unintentionally nontransitive. This may occur using the regret concepts presented here, because the regret associated with selecting an alternative must necessarily be associated with the alternative not selected. Thus it is not fully meaningful to speak of the expected utility of decision alternative A_1 when regret associated with not selecting decision alternative A_2 is involved. We should instead use

$$\mathrm{EU}(A_1, \neg A_2) = \sum_{i=1}^{N} p(e_i|A_1)[v(e_i, A_1) + r(e_i, A_i, \neg A_2)]$$

$$\mathrm{EU}(A_2, \neg A_1) = \sum_{i=1}^{N} p(e_i|A_2)[v(e_i, A_2) + r(e_i, A_2, \neg A_1)]$$

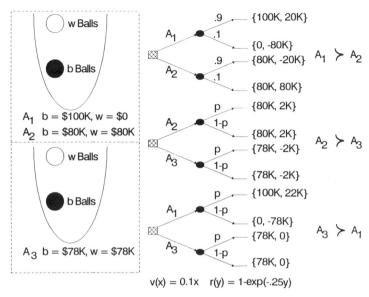

Figure 7.30. Illustrative Situation Showing Preference Intransitivity Associated with Regret

where the $v(e_i, A_1)$ are the, perhaps multiattributed, components of the utility of the ith outcome of option A_1; the $r(e_i, A_1, \neg A_2)$ are the regrets associated with the ith outcome that results from selecting option A_1 and not selecting A_2. In cases such as these, the regrets may be negative for true regret and positive for those cases where we are more pleased with the actual choice that is received than we are with the choice that is foregone.

We will say $A_1 \rightarrow A_2$ if $EU(A_1, \neg A_2) > EU(A_2, \neg A_1)$. From this, we easily see that there is no reason to infer that if $A_1 \rightarrow A_2$ and $A_2 \rightarrow A_3$, we must necessarily have $A_1 \rightarrow A_3$. Figure 7.30 illustrates a three-choice situation with preference intransitivities that occur because of the different information sets available in the three pairwise preference comparisons. This would suggest much caution in the use of any prioritization approach that is based on pairwise preference comparisons and assumed, but nonverified, transitivity among preference relations. It becomes extraordinarily easy to produce agenda-dependent results in these situations.

In this subsection, we have examined the concept of pairwise comparison regret. The regret concept does not necessarily lead to transitive preference comparisons. We have demonstrated the strong need to incorporate decision process descriptions in framing the regret situations. Several illustrative examples indicate that the framing of decision output states and the information available concerning the outputs, resulting from decision options not selected, greatly influences the regret calculations. The central conclusion from this is that careful decision situation structuring must be associated with the regret concept if the results obtained are to have prescriptive value for decision making.

In a descriptive sense, many people obviously experience notions of regret and hesitate rejecting alternatives because of some desirable properties of the rejected alternatives that will then be foregone. We must be careful, however, to note that the preference relation implied through use of regret by $A_1 \to A_2$ is not just that A_1 is preferred to A_2 but that selecting A_1 and rejecting A_2 is preferable to selecting A_2 and rejecting A_1. Thus the two preference relations $A_1 \to A_2$ and $A_2 \to A_3$ just simply do not provide sufficient information to infer or conclude that $A_1 \to A_3$, when regret is included in the contextual relation prefer. The contextual preference relation is not uniquely defined here. It should really be written to infer the true contextuality of the situation at hand. What we have really stated here is that $A_1 \to_{12} A_2$ and $A_2 \to_{23} A_3$ do not necessarily allow us, when comparing A_1 and A_3, to say anything about the comparison, because this will require knowledge of what really is the contextual preference relation $\to_{13}$, and not just a preference relation $\to$.

Kahneman and Tversky have commented that departures from subjective expected utility theory, as can be described by their prospect theory notions, and regret theory as well, "must lead to normatively unacceptable consequences" [58]. A well-conceived decision assessment process should assist the decision maker in understanding the potential decision process errors involved in the use of flawed judgment heuristics. For these reasons, among others, we strongly encourage decision situation assessment [59], structuring, and elicitation processes that encourage comparison within attributes and across outcome states for alternatives. It would seem that pairwise preference comparison of alternatives is fraught with difficulties and unless great care is exercised can lead to potentially misleading results. At the very least, we must be fully aware of the potential context and associated framing-dependent definition of the pairwise contextual relation, $\to$, and that this relation is, strictly speaking, incomparable in the expressions $A_1 \to_{12} A_2$ and $A_2 \to_{23} A_3$ when these expressions imply pairwise comparison with no ideal alternatives to serve as a reference point, anchor, or basis for the comparison. In other words, we must be careful to insure that contextual relation used for comparison purposes are defined in such a fashion that it is context independent across compared alternatives and outcomes. If this is accomplished, it becomes possible to make meaningful multiple pairwise comparisons across alternatives.

This cannot, however, be used as an invocation against using the regret concept or the more general prospect theory of Section 2.3.4 to describe behavior. For descriptive purposes these approaches have much to recommend them. Understanding of descriptive reality is a necessary first step toward a meaningful prescriptive process. In this sense, among others, these approaches have considerable value. Of course, it will ideally turn out that regret and prospect theory approaches, with suitable information framing, do not lead to preference nontransitivities.

7.3.3. Heuristics and Biases in Descriptive Decision Making

Many recent studies indicate that people are not perfectly consistent in their choices and often unknowingly violate even the preference transitivity axiom, which states that if $x \rightarrow y$, or that x is preferred to y, and $y \rightarrow z$, then $x \rightarrow z$. We have seen this difficulty in our previous subsection. Problems of this sort may be due to use of any number of poor heuristics, such as choice models based on elimination by aspects* [60] and lexicographic ordering. It has also been shown that people, even including "experts," express a very strong belief in a "law of small numbers [61]." They sometimes seriously underestimate the error and unreliability inherent in a very small sample of data. This is accomplished by placing unreasonably high expectations concerning repeatability of results from a very small sample. Further, undue confidence is often placed in these small sample size estimates, and hypotheses are often identified as correct on the basis of small samples, without appreciating that small samples can mask numerous factors in a situation. Often people do not understand that the variance of a sample decreases as the sample size becomes larger and, in its place, use the "law of small numbers" in which they rarely attribute unexpected results to sampling variability. Consequently, many fail to regress predictions toward a correct central value, which is generally the prior mean, as information validity decreases or as information redundancy decreases.

A large number of contemporary studies have indicated that the attempts of people, including experts, to apply various intuitive strategies to acquire and analyze information for purposes such as prediction, forecasting, and planning, are often flawed. Many studies have been conducted to describe and explain the way information is acquired and analyzed and the results of faulty acquisition and analysis. Generally the descriptive behavior of subjects in tasks involving information acquisition and analysis is compared to the normative results that would prevail if people followed an "optimal" procedure. Most of these information processing biases involve the processing of statistical information and a Bayesian framework is usually assumed in order to obtain normative results. There have been a number of recent discussions,

---

*Elimination by aspects is a simple selection aid in which those alternatives not fulfilling certain minimum requirements on every aspect or attribute are eliminated from further consideration. Alternatively, only alternatives that exceed a minimum aspiration level on each attribute may be retained. This is an extensively used heuristic in practice. It is used in many areas as a screening method to select only those options for further consideration that meet a number of minimum requirements. It may be very appropriate for this purpose. When used, as it often is in practice, to select a single "best" alternative, it can be very flawed.

from several perspectives, of information processing heuristics and cognitive biases.* The text by Hogarth [62] concerning strategies and biases associated with judgment and choice is especially noteworthy as are the collection of papers edited by Kahneman, Slovic, and Tversky [63], Hogarth and Reder [64], and Caverni et al. [65]. A text by Bazerman [66] provides a very readable overview of this area as does the book by Yates [67]

Example. We now examine some ways in which humans actually process information that is acquired through observation. After this, we provide some guidelines that potentially enable better information processing. We describe only a single bias in any detail here. It is known as the *base rate bias* or *base rate fallacy*. Suppose that it is known that 7.69 percent of the cards in a standard 52-card deck are aces. A single card is drawn, face down, and we are asked the question: What is the probability that the card drawn is not an ace? It turns out that almost everyone will say 0.9231 or 92.31 percent. In almost all cases like this, the base rate information is used as representative information, and a particular sample is assumed to have the same statistical properties as the entire set of items. This is, of course, quite reasonable.

Now suppose that the subject is allowed to ask the opinion of a spy. The spy says that the actual card drawn was an ace. But, spies have been known to be inaccurate. So, a vision test is given the spy. The vision test is precise and it determines that the spy can correctly identify the face value of a card 90% of the time. This means that the probability that the card will be identified as an ace given that it is an ace is 0.90, that is, $P(SA|A) = 0.90$. The visual ability of the spy also means that the probability that the spy will say that it is not an ace, given that it is not an ace is also 0.90, that is, $P(SNA|NA) = 0.90$. The other two conditional probabilities are, of course, $P(SNA|A) = 0.10$ and $P(SA|NA) = 0.10$. Often the last two sentences are not initially provided the subject; in many cases they will be, and they would generally be provided if requested by the subject.

The question now posed to the subject is: Given the uncertainty associated with the response of the spy, what is the probability that the card is an ace, given that the spy says that it is an ace? Most people will reason that: Since the spy says it is an ace and since the spy is accurate 90 percent of the time, the probability that the card drawn from the deck is actually an ace is 0.90!

---

*Among the many flawed heuristics and information processing biases that have been identified are anchoring and adjustment, availability, base rate, confirmation, conservatism, data presentation context, data saturation, desire for self fulfilling prophecies, ease of recall, expectations, fact-value confusion, fundamental attribution error (success/failure error), gamblers fallacy, habit, hindsight, illusion of control, illusion of correlation, law of small numbers, order effects, outcome irrelevant learning system, overconfidence, redundancy, reference effect, regression effects, representativeness, selective perceptions, spurious cues, substitution of correlation for causation, and wishful thinking.

A significant problem is associated with this answer as the subject has ignored base rates. What we really wish to know is the probability of the event—card is an ace, given that the spy says that it is an ace. Generally, this sort of neglect of important information, base rates in this case, leads to errors.

It is quite possible that subjects misinterpret the conditional probabilities that are given and perceive these as probabilities of the actual number of the card given the claimed number seen by the spy. An experimenter, in investigating these responses, is generally very explicit in indicating to the subjects that this is not the case and that the spy is really telling the number that the spy actually believes was observed.

The actual probability of the card being an ace, conditioned upon the observation by the spy that it was an ace, may be obtained from Bayes rule as $P(A|SA) = P(SA|A)P(A)/P(SA)$. We have from the joint-conditional probability law, $P(SA) = P(SA|A)P(A) + P(SA|NA)P(NA)$, the result $P(SA) = 0.90(0.0769) + 0.10(0.9231) = 0.1640$. So, we see that $P(A|SA) = 0.90(0.0769)/(0.1640) = 0.42$.

Rather than the spy being correct, in that the card is an ace, with probably 0.90, the spy is only correct with probability 0.42. The difficulty here, generally unrecognized by those who do not do the calculation, is that the spy has a lot of *false alarms*, that is to say misdiagnoses a large percentage of nonaces and calls them aces. However, it appears that not only do subjects do calculations like this, they neglect base rates completely when they are given individuating information.

This is a prototypical illustration of the sort of information processing bias that is observed, generally in experimentally controlled laboratory settings. The many studies that have been made suggest that the errors that occur are systemic errors and not just random errors that might be due to factors such as guessing. Several reasons have been cited that justify study of these errors and inferential biases:

1. They reveal the psychological processes that govern human inference and judgment.
2. They indicate those portions of statistical theory that are not intuitive or counterintuitive.
3. This identification of human intellectual limitations may suggests ways to improve the quality of human reasoning and information processing through adoption of appropriate corrective strategies.

Thus a study of flawed heuristics and information processing biases may lead to approaches that compliment the standard Bayesian approach and are potentially more appropriate for use in specific situations. It may provide guidelines for the design of decision support systems [17] and other aids that support minimization of human cognitive errors.

A central result of the plethora of efforts in this area is the conclusion that biases are systematic and very strongly prevalent in a great many cognitive activities. A fundamental hypothesis of much of these works is that decisions and judgments are influenced by differential weights of information, and that these differential weights are often improperly selected. There are a number of information processing biases that have been identified. Several of them will now be discussed.

Base rates are, in principle, objectively defined in the statistics literature as prior statistics. Failure to properly consider relevant background information, in the form of prior statistics, is called the base rate fallacy. Generally the base rate fallacy results, not from the order in which information is presented or responses elicited nor from a simple misinterpretation of a problem. Rather, it results from the apparent fact that people will often simply consider base rates irrelevant and not causally related to events and judgments, especially after they have been presented with individuating information. These are the points emphasized in our foregoing discussion and example.

Representativeness refers to the diagnostic attributes of an object. Often, these are defined subjectively. Those unduly influenced by representativeness are insensitive to base rates. They estimate event likelihoods on the basis of the degree to which events are similar or representative of the primary features of the parent population which generated the event. Use of stereotypes, failure to search for potentially disconfirming evidence, and confusion of such probabilities are $P(A|B)$ and $P(B|A)$ are characteristics of representativeness.

Often, people assess the probability of an event given some evidence as the probability that makes the evidence most likely. As a result, predictions are insensitive to reliability of evidence and expected prediction accuracy. The confidence placed in these predictions depends primarily on the degree of representativeness. There is little or no regard for the many factors that limit predictive accuracy, an effect which is called the *illusion of validity*. Generally people underestimate the amount they learn from information and have strong beliefs that they knew all along what would happen. This conservatism causes premature cessation of information acquisition and processing and premature onset of choice making. As a result of representativeness, hindsight judgments are invariably "better" than foresight judgments. The distinction between a good decision and a good outcome is very important but is often overlooked or confused when we practice the representativeness bias and believe, in effect, that "we knew it all along."

Availability, *adjustment*, and *anchoring*, represent the types of cognitive processes or mental constructs used to describe, compare, forecast, and explain events. In many situations, people make estimates by starting from an initial value that is adjusted to yield a final answer. The initial value or starting point may be suggested by the formulation of the problem or may be the result of a partial computation. In either case, the adjustments that are made on the basis of individuating information are typically inappropriate.

This phenomenon is called anchoring. Studies of judgments of probability indicate that people tend to overestimate the probability of conjunctive events and underestimate the probability of disjunctive events. As a consequence of anchoring, the overall probability is overestimated in conjunctive problems and underestimated in disjunctive problems.

The successful completion of an undertaking, such as the development of a new product, typically has a conjunctive character. For the undertaking to succeed, each of a series of events must occur. Even when each of these events is very likely, the overall probability of success can be quite low if the number of events is large. Overestimation of the probability of the conjunctive events leads to unwarranted optimism in evaluation of success probabilities. Disjunctive events are typically encountered in the evaluation of risks. A complex system, such as a nuclear reactor or a human body, malfunctions if any of a large number of components fail. Even when the likelihood of failure in each component is slight, probabilities of overall failure can be high if many components are involved. Because of the anchoring heuristic, people tend to underestimate the probabilities of failure in complex systems. Directions of anchoring biases can sometimes be inferred from a knowledge of the event structures.

Typically, people assess the frequency of a class or the probability of an event by the ease with which the past instances can be brought to mind. Availability is a useful clue for assessing relative frequency or probability to be used for adjustment and anchoring. However, availability is affected by factors other than frequency and probability. When the size of a class is judged by availability, a class whose instances are easily retrieved will appear more numerous than a class of equal frequency whose instances are less retrievable. In addition to familiarity, there are other factors, such as imaginability or salience, which affect the retrievability of instances. For example, the impact of seeing a horrible automobile accident or the subjective probability of such accidents occurring in the future is probability greater than the impact of reading about an automobile accident in the local paper. Furthermore, recent occurrences are likely to be relatively more available or influential than earlier occurrences.

Most of the flawed heuristics and information processing biases that have been identified are not independent of other biases. One of the most interesting biases, for example, is the *conjunction fallacy*. In this, people consistently make judgments that require $P(A, B) > P(A)P(B)$ or perhaps even $P(A, B) > P(A)$ or some other incorrect variant of the joint probability rule. A relatively standard version of the conjunction fallacy results from an initial stereotypical statement S of some person or situation. This is followed by some statement of an event A that is inconsistent with the stereotype and another event B that is compatible with it. Generally, when shown the statements, people conclude that $P(A, B|S) > P(A|S)$.

Our earlier example is associated with the base rate and/or representativeness information processing bias. It is of interest also to comment briefly on work on adjustment and anchoring. A judgment and choice model based on belief updating (or an anchoring adjustment model) has been developed by Einhorn and Hogarth [68–70]. It predicts that there will be an information order bias under some conditions, and not in others. This model extrapolates the descriptive observation that improper anchoring and adjustment may occur by providing a prediction of situations when various heuristics will be in use. The model predicts this on the basis of six characteristics of the contingency task structure situation model:

1. Whether the task is simple or complex.
2. Whether the task involves evaluation or estimation.
3. The amount of situational information that is presented.
4. Whether the information is consistent or inconsistent (contradictory).
5. The order of presentation of the information.
6. Whether the response mode adopted in the decision situation requires that belief updating occur after each information item is presented, called *step by step* updating, or whether updating occurs after all available information has been presented, which is called *end of sequence* updating.

When information is presented in the step by step (SbS) mode, and a new probability estimate obtained, people anchor on the current information and adjust on the new information in such a way as to associate recent information with more weight than it deserves relative to prior information. This results in the *recency effect*: as the anchor becomes larger, a given piece of negative or inconsistent information will have greater impact than if the anchor were smaller. In a similar way, as the anchor becomes smaller, a given item of consistent information will have a greater impact than if the anchor were larger.

The step by step model for anchoring and adjustment is

$$S_k = S_{k-1} + w_k\left[s(x_k) - R\right]$$

$$w_k = \alpha S_{k-1} \qquad\qquad \text{for } s(x_k) \leq R$$

$$w_k = \beta(1 - S_{k-1}) \qquad\qquad \text{for } s(x_k) > R$$

In this equation, S_k is the degree of belief in some hypothesis, impression, or attitude formed after evaluating k pieces of information. S_{k-1} is the anchor of prior belief, and the initial strength of belief is denoted by some S_0. The decision maker's subjective evaluation of the kth item of information is denoted by $s(x_k)$. R represents the reference point or background against which the impact of the kth item of information is evaluated. Finally, the weight w_k represents the weight for the kth item of information. The different expressions for w_k insure that disconfirming information, where

$s(x_k) < R$, has a larger impact for a larger anchor than for a smaller anchor. Confirming information, where $s(x_k) > R$, will have a larger impact for a small anchor than for a large anchor.

When people evaluate information at the end of sequence mode, and probability updating occurs at that time, people aggregate all information in such a way that the order of confirming and disconfirming evidence is unimportant. The end of sequence model for information updating of probability is

$$S_k = S_0 + w_k \big[s(x_1, x_2, \ldots, x_k) - R \big]$$

in which $s(x_1, x_2, \ldots, x_k)$ is some appropriate function, such as a weighted average, of the evaluation of the items of information that follow after the initial anchor has been obtained.

Recent experimental studies by Adelman et al. [71, 72] generally confirm the applicability of this model as a valid descriptive model. They do indicate, however, that disconfirming information may not have as much of an impact as might be predicted from the model. This is, of course, another of the cognitive information processing biases: the neglect of potentially confirming information.

A central goal of much of this research in psychological decision theory is not only that of describing flawed information processing but also prescriptive debiasing efforts. Of particular interest are circumstances under which these biases occur; their effects on activities such as decision making and corrective efforts that might result in debiasing or amelioration of the effects of cognitive information processing bias. Many of the cognitive biases that have been found to exist have been found in the unfamiliar surroundings of the experimental laboratory, and generalization of this work to real-world situations is a contemporary research area of much interest. However, most of the laboratory experiments have concerned very simple, if unfamiliar, tasks. Kahneman and Tversky [73] offer the following additional observations concerning the influential factors affecting human cognitive information processing biases and flawed heuristics.

1. There is often a highly consistent bias in identifying confidence intervals and probability distributions. Often, people are overconfident. They estimate a personal probability of an estimate being correct as greater than is actually the case.
2. A surprise occurs if an error in a personal estimate of some event falls outside the confidence interval.
3. If confidence does indeed reflect knowledge, then the true value should fall outside the k fractional confidence interval on an approximate $(1 - k)$ fraction of the estimates made.
4. People may be overconfident or underconfident, depending on how the percentage of surprises compares with that anticipated.

5. The degree of overconfidence generally increases with ignorance.
6. Overconfidence does not generally occur when a person has considerable knowledge of conditional outcome distributions, probably due to the repetitive situations and outcome feedback associated with this knowledge.
7. Sample size and information reliability do not significantly influence human confidence in judgments.
8. Insensitivity to evidence quality may explain overconfidence effects.
9. Oversensitivity to the consistency of available data is a second cause of overconfidence.
10. In a search for coherence, people often
   10.1. See patterns where none exist.
   10.2. Reinterpret data to increase consistency.
   10.3. Ignore evidence that does not fit their view.
   10.4. Overestimate the consistency of available information.
   10.5. Derive too much confidence from information that is available.

Ford et al. [74] have conducted a recent survey and assessment of decision strategies, conducted by means of process tracing studies. They focused principally on the use of protocol assessment and compensatory models versus noncompensatory models for making decisions. The reason for this selection was that compensatory models (such as those based on multiple attribute utility theory) can compensate for poor scoring or performance of an alternative on one attribute or dimension by good performance on another. Noncompensatory models (such as lexicographic ordering models or conjunctive models*) cannot accomplish this trade-off. Among the findings were the following.

1. Increasing task complexity increases the likelihood that the decision maker will use simplifying noncompensatory strategies to make a decision more manageable.
2. Increasing task complexity leads to a decrease in the range and quantity of information searched.
3. The display format affects the amount of information searched, the temporal pattern in which information is acquired and assessed, and the

---

*A lexicographic order is the ordering used in compiling a dictionary. In a lexicographic evaluation, the most important attribute would be chosen. The alternatives would be ranked according to their scores on this attribute and the alternative with the largest score on the most important attribute is selected as the best. A conjunctive model is one in which alternatives are retained for further selection if performance on each attribute of importance is above some threshold. Combinations of these models are possible. We could, for example, scan a large set of alternatives and eliminate those on a first pass that fall below acceptable performance on some threshold. Then we could select the single alternative that has the best performance on the most important attribute. These models may be quite acceptable, but may perform quite poorly as compared to compensatory models.

amount of time spent examining a particular piece of information or to make a final decision.

4. Labeling resulted in less information being searched.
5. Time constraints affect the depth and the focus of search.
6. High performers tend to examine the same amount of information across trials, whereas low performers looked at less information over trials.
7. Empirical evidence indicates the extensive use of noncompensatory strategies and that task complexity is strongly related to strategy use.
8. Preconceived notions about one or more alternatives affects information search.
9. Decision makers pay more attention to cues with extreme values.
10. The type of decision affects the search process.
11. Selection of a decision strategy is seen as a complex function of a compromise between a desire to make a correct decision and a desire to minimize cognitive effort. This conclusion assumes that the more formalized or analytic the decision strategy, the higher the probability that a "correct" solution can be selected.
12. Compensatory strategies cost more to use in terms of cognitive demands than noncompensatory strategies.
13. Variables that have the greatest effect on decision behavior are
    13.1. Number of alternative courses of action.
    13.2. Ambiguity or clarity of goals.
    13.3. Familiarity with the decision task.
14. Decision makers tend to use simplifying noncompensatory strategies when there are few incentives, such as low task significance, no accountability, no monetary rewards, and no risks for optimal decision making, in the decision environment.
15. Experienced, knowledgeable, and highly-motivated decision makers are more likely to select a complex, analytical, compensatory strategy than someone on the other end of the continuum.
16. When a decision taken is necessarily irreversible, decision makers use more analytic techniques, perceived more task pressure, and rated the problem as more important than when the decision is reversible.

Ford and his colleagues found that most of the research using process tracing approaches has focused almost exclusively on the effects of task, rather than on the combined effects of the decision situation, the environment into which it was imbedded, and the experiential familiarity of the decision maker with these. They observe that a theory of problem solving cannot predict behavior unless it encompasses both an assessment of the task and the limits of rational adaptation to task requirements. This observation relates strongly to requirements for prescriptive decision making.

A particularly cogent summary of those principles that encourage use of proper information processing, particular in the statistical settings in which many of the information processing biases have been developed is contained in Nisbett, Krantz, Jepson, and Kanda [75]. Three task variables are identified as being particularly important in influencing appropriate statistical reasoning:

1. The degree to which randomness in data sensing devices is evident.
2. Experiential familiarity with analogous situations.
3. Cultural disposition for statistical reasoning in the particular task being considered.

The results of much of this work, although there is some controversy, show that simple quantitative models perform better in human judgment and decision-making tasks, including information processing, than wholistic expert performance in similar tasks. Einhorn and Hogarth [76] provide guidelines that should be useful in assisting people to process statistical information better. Klayman and Ha [77, 78] are particularly concerned with the provision of appropriate feedback, such that humans can process information more correctly in stochastic environments.

There are a number of prescriptions that might be given to encourage avoidance of possible cognitive biases and to debias those that do occur. Some suggestions to avoid cognitive bias, gleaned from a number of studies, include the following:

1. Sample information from a broad data base and be especially careful to include data bases that might contain disconfirming information.
2. Include sample size, confidence intervals, and other measures of information validity in addition to mean values.
3. Encourage use of models and quantitative aids to improve upon information analysis through proper aggregation of acquired information.
4. Avoid the hindsight bias by providing access to information at critical past times.
5. Encourage decision makers to distinguish good and bad decisions from good and bad outcomes in order to avoid various forms of selective perception such as, for example, the illusion of control.
6. Encourage effective learning from experience. Encourage understanding of the decision situation and methods and rules used in practice to process information and make decisions, so as to avoid outcome irrelevant learning systems.
7. Use structured frameworks based on logical reasoning in order to avoid confusing facts and values, and wishful thinking; and to assist in processing information updates.
8. Collect both qualitative and quantitative data and be sure that all data is regarded with "appropriate" emphasis. None of the data should be

overweighted or underweighted in accordance with personal views, beliefs, or values only.

9. People should be reminded, from time to time, concerning what type or size of sample from which data are being gathered, so as to avoid the representativeness bias.

10. Information should be presented in several orderings so as to avoid recency and primacy order effects and the data presentation context and data saturation biases.

Kahneman and Tversky [79] discuss procedures to enhance debiasing of information processing activities. A five-step procedure designed to produce properly regressive procedures by experts who are familiar with the subject area of the investigation is proposed.

1. Select a proper reference class from among the many that are potentially available.

2. Make a statistical estimate to obtain the probability distribution of the reference class.

3. Make an intuitive, generally nonregressive, estimate.

4. Assess probabilities and the degree of predictability that is possible.

5. Correct the intuitive estimate, generally by regressing toward the reference class or reducing the class average by a factor to represent the correlation coefficient.

Of course, not everyone agrees with the conclusions just reached about cognitive human information processing and inferential behavior. Several arguments have been advanced for a decidedly less pessimistic view of human inference and decision. In one of these, Jonathan Cohen [80, 81] argues that all of this research is based upon a conventional model for probabilistic reasoning, which Cohen calls the *Pascalian* probability calculus. He expresses the view that human behavior does not appear biased at all when it is viewed in terms of other equally appropriate schemes for probabilistic reasoning, such as his own inductive probability system. Cohen states that human irrationality can never be demonstrated in laboratory experiments, especially experiments based upon the use of what he calls "probabilistic conundrums."

There are a number of other contrasting viewpoints, as well. There is a major body of literature that deals with alternatives to the Bayesian perspective on uncertain information processing. For an overview of some of these approaches, in terms of alternative representational systems, the reader may refer to [82–85] and the references contained therein. An interesting and appropriate area for current effort relates to the notion that information processing biases may be more due to the attempt to force fit a Bayesian perspective on reasoning, than due to fundamental human error.

In their definitive study of behavioral and normative decision analysis, von Winterfelt and Edwards [86] refer to the information processing biases we identified earlier as cognitive illusions. They indicate that there are four fundamental elements to every cognitive illusion:

1. A formal operational rule that determines the correct solution to an intellectual question.
2. An intellectual question that almost invariably includes all of the information required to obtain the correct answer through use of the formal rule.
3. A human judgment, generally made without the use of these analytical tools, that is intended to answer the posed question.
4. A systematic and generally large and unforgivable discrepancy between the correct answer and the human judgment.

Many of the concerns relative to the heuristics and biases studied are based upon what are believed to be inadequacies of the studies upon which the conclusions are based. Von Winterfeldt and Edwards [86] and Phillips [87] describe some of the ways in which subjects might have been put at a disadvantage in this research on cognitive heuristics and information processing biases. Much of this centers around the fact that the subjects have little experiential familiarity with the tasks that they are asked to perform. Other suggestions intended to make cognitive illusions vanish [88] have been made.

7.3.4. Prospect Theory: A Descriptive Model of Human Judgment and Choice

Even when the outcome event, or state, probabilities are known objectively, it has been noted that unaided descriptive decision behavior may not be in accord with the aggregation rule of expected utility theory. Kahneman and Tversky have described *prospect theory* as a descriptive model of human judgment and choice [89, 90]. We summarize prospect theory here, based in part on [47].

A *prospect* is defined as a complete set of event-outcomes associated with some choice. Figure 7.31 illustrates a simple *prospect*, Γ. The expected utility of prospect Γ is defined in the usual way, according to the tenets of subjective utility theory as

$$\text{EU}(\Gamma) = P_1 U(e_1) + P_2 U(e_2) + \ldots + P_N U(E_n)$$

where P_i is the objective probability of event outcome e_i and $U(e_i)$ is the utility of this outcome. Prospect Γ is acceptable to the decision maker with present asset position W if, and only if, $\text{EU}(\Gamma + W) > U(W)$. Furthermore, it is assumed that the decision maker is risk adverse, such that the utility

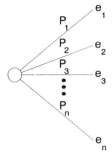

Figure 7.31. Model of Event with Stanford Prospect Form

function is concave. In other words, $d^2U(x)/dx^2 < 0$. There are several additional restrictions and assumptions.

1. Money is essentially the only outcome of value.
2. There is essentially a single attribute—money.
3. Probabilities are objective.
4. Prospect scenarios are fully descriptive.

Kahneman and Tversky describe three effects that lead unaided experimental subjects to violate the Von Neumann, or Savage, axioms of subjective expected utility theory. These three effects are doubtlessly based on their many studies of flawed human information processing heuristics and biases.

1. *The certainty effect,* in which people often overweigh the value of outcomes that are considered certain relative to those that are merely probable.
2. *The reflection effect*, in which people reverse their preference order compared to that predicted from the certainty effect when the monetary amounts of a prospect is reflected about zero. Thus, the certainty effect increases the repugnance of losses as well. This is incompatible with the notion, often expressed concerning risks in business situations, that certainty is generally desirable. It appears desirable only when the outcomes are beneficial.
3. *The isolation effect*, in which people often disregard components common to all alternative outcomes or prospects, and instead focus only on incremental components that distinguish the outcomes from each other. Thus, value or utility is determined by changes of wealth rather than final asset position including current wealth. The isolation effect may produce inconsistent preferences since it is possible to decompose prospects in several ways.

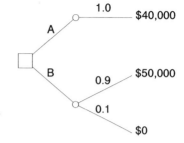

Figure 7.32. Decision Tree Associated with Choice Under Uncertainty Situation 1

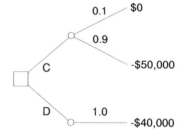

Figure 7.33. Decision Tree Associated with Choice Under Uncertainty Situation 2

Some simple examples are instructive in illustrating these effects. Figures 7.32 through 7.38 illustrate decision trees for eight simple decision situations that involve gambles. The probabilities of various monetary outcomes are known and shown in the figures. Stated simply, we are asking: Which gambles, or prospects, do we prefer? It would be unreasonable to expect that everyone would have precisely the same preferences across the eight illustrated gambles. However, *most* people, including those who obey the descriptive prospect theory, will state that $A \to B$, $C \to D$, $E \to F$, $G \to H$, $I \to J$, $K \to L$, $M \to N$, $Q \to R$. By stating several pairs of these preferences,

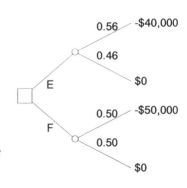

Figure 7.34. Decision Tree Associated with Choice Under Uncertainty Situation 3

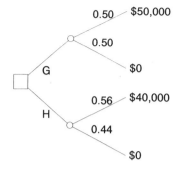

Figure 7.35. Decision Tree Associated with Choice Under Uncertainty Situation 4

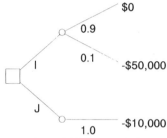

You have accepted this lottery. I will give you $50,000 in addition to anything you win through selection of an alternative.

Figure 7.36. Decision Tree Associated with Choice Under Uncertainty Situation 5

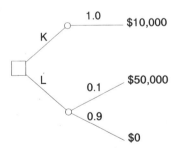

You have paid $50,000 and I will give you anything you win through selection of an alternative.

Figure 7.37. Decision Tree Associated with Choice Under Uncertainty Situation 6

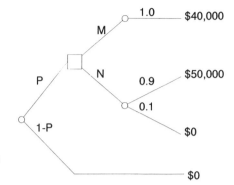

Figure 7.38. Decision Tree Associated with Choice Under Uncertainty Situation 7

aud assuming that our utility is solely a function of the money received, a number of violations of the axioms of expected utility theory occur. Among these are the following.

1. $A \rightarrow B$ and $G \rightarrow H$. The certainty effect is present. We have violated Savage's substitution axiom, which states, in effect, that if we prefer A to B, or $A \rightarrow B$, then we must also have the same preference in a lottery involving these decisions, or $(A, P) \rightarrow (B, P)$. Here, if we let $P = 0.56$, then we have $(A, P) = (\$40,000, 0.56) = H$ and $(B, P) = (\$50,000, 0.5) = G$, and thus we do indeed violate the substitution axiom with these choices.

2. $C \rightarrow D$ and $E \rightarrow F$. The certainty effect is again present. Compared with the immediately foregoing situation, we see that the reflection effect has reversed all preferences from those where the outcomes were nonnegative. The reflection effect alone does not necessarily involve any violation of expected utility theory. The combination of the reflection effect and the certainty effect does often produce a violation.

3. $A \rightarrow B$ and $I \rightarrow J$, or $C \rightarrow D$ and $K \rightarrow L$. The isolation effect is present. The decision situation of Figure 7.32 is precisely the same as that of Figure 7.36, and the decision situation of Figure 7.33 is precisely the same as that of Figure 7.37. Only the method used for display of the outcomes that result from the possible choices is different.

4. $G \rightarrow H$ and $M \rightarrow N$, or $E \rightarrow F$ and $Q \rightarrow R$. The decision situations of figures 7.35 and 7.38, and the decision situations of Figure 7.34 and 7.39, are the same if we have $P = 0.56$ and if we do not consider the time sequences implied by rearranging the decision trees. This is one of the explanations used by Raiffa [8] in his resolution of the famous Allais paradox (see page 370) which was an early demonstration of this certainty effect. Actually, the decision situation of Figure 7.38 is

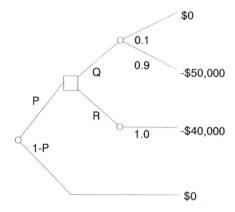

Figure 7.39. Decision Tree Associated with Choice Under Uncertainty Situation 8

equivalent to that illustrated in Figure 7.32 for $P = 1$ and to that decision situation illustrated in Figure 7.35 for $P = 0.56$. The fact that we may be tempted to state $M \rightarrow N$ for the decision situation in Figure 7.38, regardless of the value of P, indicates the potential to ignore the first stage of this gamble, probably because one has no control over its outcome. There may, in reality, never be an opportunity to state $M \rightarrow N$. However, if the opportunity does occur, it will generally be stated as a preference.

You are encouraged to carefully examine each of these decision situations and see if you do not agree with these preferences. In almost all cases, you will either agree with the results as stated or see how slight changes in either the probabilities or outcome monies would lead you to these same conclusions. One informal explanation is that we are risk averse for gains and risk averse for losses. Let us look at an explanation for this behavior using the formal prospect theory.

Prospect theory consists of two phases: editing and evaluation. Editing, as described by Kahneman and Tversky, consists of six tasks.

1. *Coding,* in which gains and losses are defined relative to a reference point.
2. *Combination,* in which probabilities associated with identical outcomes are aggregated.
3. *Segregation,* in which risky components are separated from riskless components.
4. *Cancellation,* in which components shared by all offered prospects are discarded.
5. *Simplification,* in which probabilities and outcomes are simplified in appropriate fashions, such as rounding.

6. *Elimination by domination,* in which dominated prospect alternatives are eliminated.

Since the sequence of editing operations will vary with the structure and content of the decision situation and since the results obtained depend on this, there will often not be a unique result obtained from use of prospect theory. The editing phase of prospect theory results in the presentation of a modified prospect which is then evaluated by a set of formal aggregation rules much like those of expected utility theory.

In the evaluation phase of prospect theory, the utility of an edited prospect is expressed in terms of two components, π and v. The decision weight of $\pi(P)$ reflects the impact of the objective probability P on the overall value of a prospect, $V(x, P, y, Q)$. It is important to note that $\pi(P)$ is not a probability measure and generally $\pi(P) + \pi(1 - P)$ is less than one. This indicates what is called a *subcertainty effect.* The value function v assigns to each outcome x a number called $v(x)$, which reflects the value of that outcome. Outcomes x are defined relative to some reference point. This is the zero point on the value scale. Thus, a value function like $v(x)$ measures gains and losses relative to this reference point.

In prospect theory, the value function $v(x)$ has three interesting and important properties. It measures the deviation of values from some reference point. The value function is concave for gains and is steeper for losses than for gains.* The probabilitylike function π, which measures the subjective impact of objective probabilities upon choices has a number of interesting properties also.

1. It is a monotone increasing function of P and is anchored at $\pi(0) = 0$, $\pi(1) = 1$.
2. It has the subcertainty property that $\pi(P) + \pi(1 - P) < 1$.
3. It has a subproportionality property, $\pi(PQ)/\pi(P) \leq R\pi(PQ)/\pi(PR)$, that requires that decision weights are closer to one when probabilities are small than when probabilities are large.
4. It has an overweighting property, $\pi(P) > P$, for small probabilities.
5. It has a subadditivity property, $\pi(PQ) > Q\pi(P)$, for small enough probabilities.
6. It has an underweighting property, $\pi(P) < P$, for sufficiently large probabilities.

---

*An interesting consequence of this is that a prospect theory value function may not be concave, or *s*-shaped, with respect to a moved reference point. If W is the initial asset position and $W + x_2$ is the reference point, then $v(W + x)$ is concave for $x_1 \leq x \leq x_2$. If the reference point is shifted to $W + x_1$, the value function $v(W + x)$ is not necessarily concave about the reference point x_1. This indicates a fundamental change between the notions of a value function in prospect theory and the corresponding utility function in the normative SEU theory.

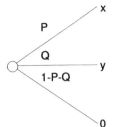

Figure 7.40. The Definition of Prospect $(x, P; y, Q)$

7. The probability weighting function $\pi(P)$ is not necessarily well behaved at $P = 0$ and $P = 1$, as people often have great difficulty in dealing acceptably with very small probability events, especially when they are associated with significant outcomes, and certainty effects in situations that also involve uncertainties.

A three-outcome prospect $(x, P; y, Q)$ is defined as we illustrate in Figure 7.40. No more than two nonzero value outcomes can be considered in the present prospect theory. Extension to more than two nonzero outcomes is conceptually easy, however. A prospect is said to be strictly positive if $x > 0$, $y > 0$ and if $P + Q = 1$. A prospect is said to be strictly negative if $x < 0$, $y < 0$ and $P + Q = 1$. A prospect is said to be regular if it is neither strictly positive nor strictly negative. It is interesting and potentially important that probabilities do not necessarily sum to one in prospect theory. The sum of all probabilities can be less than one or greater than one. Obviously, for mutually exclusive and collectively exhaustive uncertain events, the sum of the marginal probability mass functions must total one in any normative, or realistic prescriptive, theory. Prospect theory is a descriptive model of judgment and choice, however. Probabilities of event outcomes, as identified by real people in complex situations, do not necessarily sum to one!

The value of prospect $(x, P; y, Q)$ is given by one of two possible expressions. For regular prospects, we have

$$V(x, P; y, Q) = \pi(P)v(x) + \pi(Q)v(y)$$

For strictly positive $(x > y > 0)$ or strictly negative $(x < y < 0)$ prospects the value of a prospect is given by

$$V(x, P; y, Q) = v(y) + \pi(P)[v(x) - v(y)]$$

Figure 7.41 illustrates a typical probability decision weight and outcome value function. Parameters appropriate for the example choice prospects described

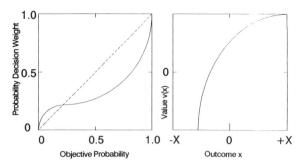

Figure 7.41. Typical Prospect Theory Decision Weight and Value for Outcome

in Figures 7.32 through 7.39 are $\pi(0.1) = 0.1$, $\pi(0.5) = 0.4$, $\pi(0.56) = 0.46$, $\pi(0.9) = 0.8$, $V(50000) = 1.0$, $V(40000) = 0.85$, $V(10000) = 0.30$, $V(-10000) = -0.60$, $V(-40000) = -1.7$, $V(-50000) = -2.0$, and these values do result in the prospect theory associated preferences we have just presented.

Prospect theory is not proposed as a normative theory or as a prescriptive theory. Departures from the expected utility theory may lead to normatively unacceptable consequences, such as inconsistency or intransitivity. If observed, by the decision maker, these will generally be corrected. However, if the decision maker does not have the opportunity to discover and correct these violations, then anomalous behavior such as that implied by the descriptive prospect theory may be expected to occur.

A major question of interest, however, is whether or not it is possible to reach an accommodation of prospect theory and subjective utility theory such that a prospect-theory-like set of decision situation models can be constructed and such that these models are prescriptively reasonable. There are many questions in need of resolution to bring this to full fruition, however. What are efficient ways to elicit weights and values such as those in Figure 7.41? Do people interpret objective probabilities in subjective ways such as that implied by the decision weight curve of Figure 7.41? Can effects such as the certainty effect also be explained by introduction of multiattribute utility functions, and is this a reasonable compliment to prospect theory? Is it reasonable, for instance, to assign the same value or utility to the equal monetary outcomes of Figures 7.32 and 7.36 when the action alternatives leading to these outcomes represent quite different gambling probabilities?

A related question concerns whether probability should be considered as one of the attributes of an outcome? We can easily posit multiattribute value elicitations that appear quite reasonable and resolve problems with many, seeming, inconsistencies explainable by the certainty effect without recourse to prospect theory. Thus $B \rightarrow A$ and $G \rightarrow H$ can be accommodated within the normative theory at the expense of a considerably more complex utility function. It is known that values are often highly labile and that people are sometimes nearly inchoate [91]. Does this account for the isolation effect of prospect theory and are there effective approaches to elicitation which will avoid these? Or is the effect more than an anomaly? Should not the time occurrence of events implied by Figure 7.38 suggest a fundamentally different situation model than that of Figure 7.35, even when $P = 0.56$? Do people discount costs, benefits, and perhaps even probabilities over time in fundamentally different ways, and what are the possible implications of this for descriptive, prescriptive, and normative decision assessment? How do and how should people choose when to stop receiving information and when to act; what could a prescriptive version of prospect theory say about this important concern?

There have been a considerable number of related studies. One of these extends prospect theory to include *expression theory* as an approach to explicitly separate the notions of judgment and choice [92] and the potential differences in preference reversal tendencies across judgment and choice activities. Another extension involves a modeling concept for probabilities as decision weights that is called *venture theory* [93]. In this effort, gains are distinguished from losses, as in prospect theory, in terms of an asymmetric influence on utility and value functions. Attitudes toward probabilistic uncertainty (called risk in Ref. 92) and attitudes towards ambiguity of imprecision in knowledge of probabilities or utilities are each considered. Machina [94, 95] has developed a model that more formally attempts to relax the independence of the irrelevant alternatives axiom, which is the central axiom precluding the Allais Paradox and preference reversals, from the normative theory such that it might be used as a descriptive theory. This leads to a utility model of descriptive preference, not expected utility base, of the generic form

$$\text{NEU}[A] = \sum_i v(x_i)P(x_i) = \sum_i [\tau(x_i)P(x_i)]^2$$

The research of Harvey on structured prescriptive models of risk attitudes [35, 96] and that of Schoemaker on the nature of risk attitudes across different payoff domains [97] are also of considerable value to prospect theory and related, concepts. Peterson and Lawson [98] investigate the incorporation of political perspectives in prospect theory models. Starmer and Sugden [99] test and evaluate prospect theory, especially with respect to the independence axiom, and compare it to alternate approaches.

A number of prospect-theory-related issues have been addressed by Currim and Sarin [100]. One major objective in this interesting research is the calibration of a prospect theory model for an individual. A second major objective is a contrast and comparison of the resulting model with a normative SEU model in terms of verification of the postulates of each model and predictive accuracy. Two types of experiments were conducted. One of these is a nonparadoxical situation in which both a prospect model and a SEU model should lead to the same correct result. This type of experiment considered the classical SEU utility situation, as well as one in which the utility for gains and losses could be different. The second experiment type involved paradoxical choice situations in which the prospect-theory-based model would generally be preferred as a descriptive model. The conclusions are that both the prospect model and the SEU model performed well in terms of predicting actual behavior with the prospect model performing slightly better in the paradoxical choice situations.

7.3.5. Social Judgment Theory

Social judgment theory (SJT), called policy capture in the early literature, is an approach to obtaining a descriptive multiattribute model of judgment and choice on the basis of actual judgment. It is based on the assumption that the best way to obtain an accurate description of the parameters of a preference structure in terms of the relative importance given to various attributes or objectives is through empirical analysis of actual judgments. Initially, lists of alternatives, objectives of the individual or group, and relevant attributes of the alternatives are determined. Then the individual or group is asked to specify an overall preference ranking for each of the alternatives. Regression analysis is used to derive a value or scoring function that would lead to similar rankings for all alternatives. The coefficients in the function represent the weights implicitly given by the individual or group to the various attributes taken into account.

The basis for social judgment theory is the philosophy of *probabilistic functionalism* as initially set forth by Brunswick [101] and later extended by Hammond [102, 103]. In this theory, uncertainty was regarded as a characteristic of both the environment and the organism. This results in the lens model that is illustrated in Figure 7.42. This lens model assumes that people are guided by rational programs in their attempt to adapt to the environment. The two fundamental systems of interest in this model are the environmental system and the subject's response system.

The lens model itself is represented by an equation of the form

$$r_a = GR_eR_s + C\left[\left(1 - R_e^2\right)\left(1 - R_s^2\right)\right]^{0.5}$$

In this equation, r_a is the correlation between the judgment of a subject and the criterion of accuracy for the judgment. Thus, r_a represents achievement. G is a term used to represent the achievement realized when the assumed linearly predictable uncertainty in the environmental system and the response system of the subject is removed. The term R_e is a term that is equal to the linearly predictable variance in the environment. R_s represents the linearly predictable variance in the subject, and C represents the correlation between the nonlinear components of each system. Thus, C is the correlation

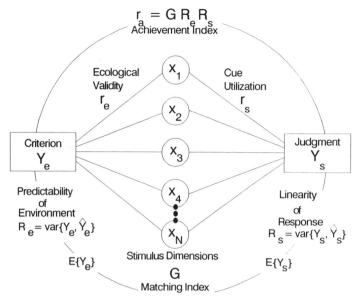

Figure 7.42. Lens Model

between the residual variance in the environmental system and the prediction system of the subject.

The left portion of the lens model of Figure 7.42 represents *ecological validity*. These are the correlations, or achievement r_e, between the cues x_i and the criterion value Y_e. The subject is represented on the right, or organismic, side of the figure. A subject will base a response, or cue judgment or cue utilization r_s, on the correlation between the cue values and judgments. The achievement r_a is then the correlation between the judgments and the criterion value.

A particularly interesting and useful form of this lens model equation (LME) results when $C = 0$, such that there is no correlation between the residual terms. The nonlinear contribution to the lens equation represents the extent to which a person uses nonlinear situational and personal cue-related aspects of the evaluation effort. This lack of correlation results in a linear achievement or lack of achievement of the form

$$r_a = GR_eR_s$$

The task predictability term R_e represents multiple correlations among the criterion and the cues. It provides an upper limit to the achievement that is possible. R_s represents the consistency of a person's judgment across evaluations. Finally, the term G represents the relative weights subjectively provided by an individual to match the optimum weights as determined through a regression analysis.

We assume a criterion value Y_e for the environmental variables and a subject response of judgment as to this value Y_s. The relationship between these are directly comparable if a linear combination of the N cues, or attributes, is assumed. We then have

$$Y_e = \sum_{i=1}^{N} h_{ei} x_i + v_e$$

$$Y_s = \sum_{i=1}^{N} h_{si} x_i + v_s$$

where the cue values x_i may be nonlinear functions of the actual cues

$$x_i = v(X_i)$$

so that we have the expected values and variances

$$E\{Y_e\} = \sum_{i=1}^{N} h_{ei} x_i = \sum_{i=1}^{N} h_{ei} V(X_i)$$

$$E\{Y_s\} = \sum_{i=1}^{N} h_{si} x_i = \sum_{i=1}^{N} h_{si} V(X_i)$$

$$r_{ei} = \text{var}\{Y_e, x_i\}$$

$$r_{si} = \text{var}\{Y_s, x_i\}$$

Here, the h_{ei} and h_{si} are the optimum regression weights for the independent cues or x_i. The terms v_e and v_s represent error terms due to inadequacy of the linear model. The expressions Y_e and Y_s represent the true criterion value and subject response or estimate. The expected value of these terms represents the estimates of the criterion value and subject response.

This explicit value structure can be useful in a variety of ways: it helps understanding past and present judgments, increases the individual's or group's understanding of its own judgment policy, leads sometimes to revision in rankings or decisions, provides a means for guiding future judgments to replicate or be consistent with past judgments, clarifies different value structures that cause different judgments, and promotes conditions favorable to conflict reduction and management, and associated efforts at conflict resolution.

The technique has been used for elicitation of value of structures of stakeholder groups in a variety of areas such as public transportation, labor management, and land/water resource allocation. Having this explicit description of individual or group preferences ideally leads to better communication between individuals or groups through improved understanding of their respective value structures. Much additional detail concerning SJT, and other decision-making approaches, may be found in the book by Hammond, McClelland, Mumpower [104].

The following steps lead to determination of a social judgment theory model of decision making.

1. *Formulate and Structure the Decision Situation:* This activity includes the issue formulation step of a systems engineering effort. It results in defining the judgment of interest and the situational context for judgment, identifying the cues or attributes of importance, and identifying the range of attribute score values to be considered. The following tasks are accomplished.

   1.1. The objectives for the SJT endeavor are specified.
   1.2. A list of feasible alternative choices is identified.
   1.3. The major attributes of the outcomes resulting from the alternative choices are identified.
   1.4. The attribute tree is structured.
   1.5. Attribute measures are specified.

   The number of different attribute or objectives measures should be minimized in order to keep the model as simple as possible. Generally, the attributes must be linearly independent of one another. If they are not, anomalous results will generally occur. One very good illustration of this is cited in the SJT literature. It would be possible to have people indicate preference for family composition by regarding the two attributes of the decision situation as the number of boys and the number of girls. Alternately, it would be possible to regard the two attributes as the total number of children, and the difference between the number of girls and the number of boys. It turns out that the latter structure is much more appropriate and leads to far more trustworthy results. It should be remembered that a structural model is proposed and we are only identifying the best parameters, or weights, for that structure.

2. *Estimate the Parameters of the Preference Structure Model:* A number of subactivities comprise this step.

   2.1. Numerical measures are determined of the degree to which each of the alternatives meets each of the objectives or performs with respect to each of the attributes.
   2.2. The group or individual whose value structure is to be modeled is asked to indicate preference for each of the individual attribute measures over an assumed relevant range. This leads to single-attribute utility functions that may have different shapes.
   2.3. The group or individual whose value structure is to be modeled is asked to rank order the alternatives with respect to their overall desirability. In the case of multiple rankings by different individuals,

the rankings could be weighted and combined into an overall preference ordering for estimation of the group's preference structure.

2.4. Estimation, or capture, of the attribute weights for the individual or
group is accomplished through use of regression analysis. Use of
regression analysis results in identification of those weights that most
appropriately explain the preference rating of the alternatives as
specified by the individual or group, given the attribute measures for
each alternative and the single-attribute utility functions. Essentially,
weighting coefficient values are determined such that the combined,
weighted attribute utilities lead to the same overall ranking as specified. More specifically, the weight coefficient w_i in the equation
$Y = w_1U_1 + w_2U_2 + w_3U_3 + \cdots + w_nU_n$ are estimated. Y is a cardinal number that represents the overall rating of an alternative.
$U_1, U_2, \ldots, U_n$ represent the contribution of each attribute measure to
the overall rating of an alternative. Each alternative represents one
data point in the regression.

3. *Interpretation of SJT Results:* The individual or group is presented the
derived preference weights preferably in a clear, graphical way. Feedback
is solicited. The decision maker may wish to modify the initial ranking of
alternatives when it is felt that the attribute weights derived by regression
are not consistent with real preferences. This may be done by changing
attribute weights and reevaluating alternatives, as in multiple attribute
utility assessment. The iteration may continue until the decision maker
agrees that both the weights that have been obtained and the corresponding overall ranking of the alternatives are appropriate for the task at hand.

Figure 7.43 illustrates one possible implementation of these steps and generic
results that might be obtained at each step.

Social judgment theory can be used whenever there is a need to determine
the relative importance that an individual or group adhere to different
attributes of a decision situation. It can be particularly appropriate as an aid
for evaluation and comparison of alternatives for decision making and also
for value system design in issue formulation. This approach has been used in
a rather large number of applications [105–109]. A recent single volume that
provides a definite overview of this approach, as well as some discussion of
other approaches, is also available [110].

It is very useful to have as many realistic alternatives as possible included
in the SJT analysis. Although the failure to include a number of alternative
choices will not doom the approach, it will make the weight parameters that
are estimated through its use potentially much less accurate than they might
be. A more important concern is inclusion of relevant attributes and identification of independent attributes.

The SJT technique results in the identification of the best weights that
most closely model actual value judgments as reflected by alternative rank-

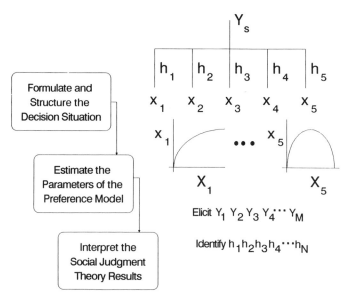

Figure 7.43. Steps in Social Judgment Theory

ings provided by the users of the approach. It cannot predict future judgments or behavior except as replicas of present behavior. When the decision maker is very inconsistent the weights may not mean much. This is generally indicated by the identified regression error variance.

7.4. SUMMARY

In this lengthy chapter, we have examined a large number, selected from an even larger number, of approaches to decision assessment. We began our efforts with a discussion of normative approaches. Then, we examined some descriptive approaches to the ways in which real humans in real situations evaluate alternatives and make judgments and decisions. A goal in this was to provide an appropriate basis for the selection of appropriate prescriptive approaches to aid people in evaluating alternatives and making decisions.

While it may seem at first glance that the normative theory is entirely reasonable and plausible, it suffers a number of disadvantages in imperfect information situations and in situations where time stress (see Chapter 9 for a special interpretation of time stress) is important. There are some realistic behavioral considerations that appear to warrant a prescriptive approach that incorporates features that do not immediately follow as a consequence of the normative theory. Some behavioral characteristics of the decision maker that strongly influence aiding requirements and considerations are as follows:

1. DMs are often impatient with time-consuming and stressful assessment procedures that seem unrelated to the decision task at hand.
2. DMs want to see some preliminary results promptly when these are needed quickly, although most decision analytic approaches produce no output until completion of the entire analysis.
3. DMs may lack interest in interacting directly with complex quantitative procedures for decision aiding that do not seem tailored to the specific contingency task structure of the issue at hand.
4. As a consequence, DMs may require a decision-aiding approach that adapts to the decision-making style appropriate for the DM in the given contingency task structure.

For each of these reasons, use of an interactive approach that allows imprecise and incomplete information is desirable. These sorts of concerns have led to a number of suggested implementations. A number of these are discussed in an excellent reprint volume [111].

Figure 7.44 presents some salient features of a dominance-process-model-based approach for search, discovery, judgment, and choice that attempts an integration of several approaches. The support system design paradigm is based upon a process model of decision making in which a person perceives an issue that may require a change in the existing course of action. On the basis of a framing of the decision situation, one or more alternative courses of action, in addition to the present option which may be continued, are identified. A preliminary screening of the alternatives, using conjunctive and disjunctive scanning, may eliminate all but one alternative course of action. Unconflicted adherence to the present course of action or unconflicted change to a new option may well be the metastrategy for judgment and choice that is adopted if the decision maker perceives that the decision situation is a familiar one and that the stakes are not so high that a more thorough search and deliberation is needed.

This decision evaluation model [112] was based upon the hypothesis that people are able to evaluate alternative plans and decisions efficiently and effectively and with low stress when a clear dominance pattern exists among alternatives that allow the establishment of a sufficiently discriminatory priority structure. It is called *A*lternative *R*anking *I*nteractive *A*id based on *D*omi*n*ance Structural Information *E*licitation or [ARIADNE].* This simplest decision-under-certainty version of ARIADNE was discussed by White and Sage [113]. Extensions to the stochastic case are presented by White and colleagues [114, 115]. An extension to incorporate rule-based knowledge is

*Ariadne, a character in Greek mythology, was the daughter of Minos. She gave Theseus a thread, thereby helping him escape from the labyrinth.

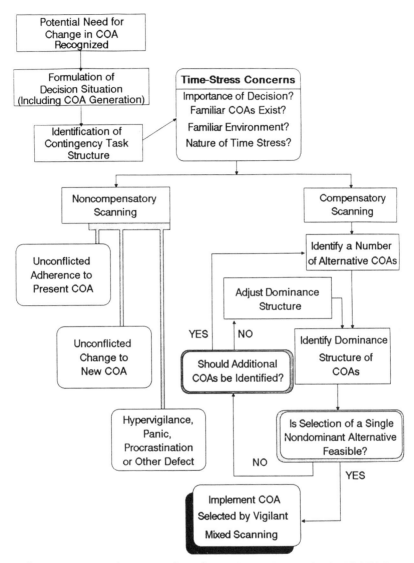

Figure 7.44. Mixed Scanning through Dominance Structuring in ARIADNE

presented by White and Sykes [116]. This effort addresses the notion of a model as a set of rule-based statements about the world.

Successful efforts to evolve high-quality and trustworthy support for evaluation, judgment, and choice doubtlessly need to blend the descriptive and the normative. Klein, Orasanu, and Calderwood [117] have published a volume in which eight characteristics of what is called naturalistic decision making are identified, each of which may be at least partially ignored in abstract axiomatic approaches that lead to normative theories. One of the initiatives of this research is augmentation of the descriptive and normative perspectives with a naturalistic perspective that is potentially able to prescribe appropriate behavior in high-velocity environments with complex decision situations. These settings are characterized by

1. Ill-structured problems.
2. Uncertain and dynamic environments.
3. Ill-defined, competing, and inconsistent objectives.
4. Causal action feedback loops.
5. Time stress.
6. High-risk, high-stake outcomes.
7. Multiple decision makers.
8. Poorly understood and rapidly changing organizational goals and norms.

Each leads to the need for a prescriptive theory. In this work, a naturalistic decision-making approach is suggested. In part, this approach calls for a rapprochement of the normative theory of decision making with meta-level cognitive realities. We examine some of these in Chapter 9.

Caution must, however, be exercised relative to the notion that the normative theory will, or should, be dispensed with. In an insightful paper, Howard [118] illustrates a decision situation much like that shown by two representational formats in Figure 7.45. This is just a particular implementation of the decision situation model illustrated in Figure 7.2. A single die is thrown and we may select options a_1 or a_2. The probabilities of each of the possible outcomes are the same. On five of the six outcomes, which occur with equal ($P = 1/6$) probability, alternative a_2 is better than alternative a_1. Yet, the expected return from selecting alternative a_1 is easily shown to be higher than that from alternative a_2. (In the specific example chosen by Howard, the expected values are the same.) Given this, we might ask whether there is any difference between the decision situation shown in Figure 7.45

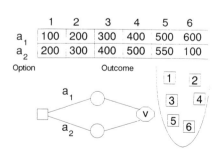

	1	2	3	4	5	6
a_1	100	200	300	400	500	600
a_2	200	300	400	500	550	100
Option			Outcome			

Figure 7.45. Decision Situation in which Regret May Cause Regret

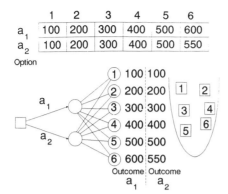

	1	2	3	4	5	6
a_1	100	200	300	400	500	600
a_2	100	200	300	400	500	550

Option

Figure 7.46. Decision Situation in which Regret Does Not Arise

and that shown in Figure 7.46, which easily suggests selection of alternative a_1 on an absolute basis and with no calculation whatever!

The only possible way in which decision alternative a_2 could be preferred to a_1 is because of some postoutcome regret. The extent to which this regret is visualized depends upon the specific way in which the outcome lotteries are posed. There are questions concerning the extent to which decision regret is a reasonable prescriptive notion. These have been addressed earlier, and we will not comment further on them here, except to note that we may well regret having regret, and we may well regret not paying sufficient attention to the normative axioms!

There have been a number of variants of the approaches that we have discussed here. A MAUT-like approach has, for example, been devised by Saaty [119, 120], called the Analytical Hierarchy Process (AHP). A substantial number of applications have been considered, and the process has been studied intently by a number of researchers [121–123].

Thus, we have indeed come upon a most interesting subject that has major theoretical and pragmatic implications. These vary from some excellent current pedagogical works on normative decision analysis [124] to design procedured for decision support systems [18], the evaluation of decision analytic software features [125], a host of realistic applications [126], decision making in groups and organizations [127], and associated equity considerations [128]. Clearly, the next 100 years in this important arena will be very interesting [129]. But, the last 30 years have been of major interest and importance as well!

PROBLEMS

7.1. Suppose that a city planner must decide whether to do nothing A_1 or to build in a given uninhabited area a single-lane highway A_2, a

dual-lane highway A_3, or a giant expressway A_4. The land-use characteristic of the area in the future may be: uninhabited as it is now E_1,
sparsely populated E_2, residential usage E_3, or heavily industrialized
E_4. The planner will receive blame or praise for foresightedness in the
decision made. The utilities of the outcomes are given by the payoff
matrix

	E_1	E_2	E_3	E_4
A_1	4	4	3	3
A_2	3	5	4	3
A_3	2	3	5	4
A_4	1	2	4	6

If the city planner uses (a) the max-max criterion, (b) the max-min
criterion, (c) the minimum regret criterion, or (d) the Laplace criterion,
what is the optimum strategy? Maximum payoff from the decision taken
is desirable.

7.2. Suppose that the probabilities of the various states of nature in Problem 7.1 are $P(E_1) = 0.1$, $P(E_2) = 0.2$, $P(E_3) = 0.5$, $P(E_4) = 0.2$.
Which alternative yields maximum expected utility?

7.3. Two medical doctors must decide whether to treat, T, a patient for a
particular diagnosed disease or wait, W. The utilities for the outcomes,
cure, paralysis, or death may be assumed independent of the corrective
strategy or decision adopted. They are

	$U(C)$	$U(P)$	$U(D)$
$MD1$	1.0	0.4	0
$MD2$	1.0	0.7	0

The conditional probabilities of the outcomes are given by $P(C|T) =$
0.6, $P(P|T) = 0.1$, $P(D|T) = 0.3$, $P(C|W) = 0.2$, $P(P|W) = 0.6$,
$P(D|W) = 0.2$. Each doctor desires to maximize the utility of their
decision strategy for the patient. What will be each doctor's best
strategy? Is it reasonable for the utilities of the outcomes to be
independent of the alternatives, so that these utilities are really utilities
of the state of nature? Draw several models of the decision situation
and analyze each. How could the doctors reach agreement if they
attempted to arrive at a single utility curve for the two of them?

7.4. Please reconsider the example discussed in the text on page 338.
Suppose that the conditional probabilities associated with the judgment
of the assistant are $P(r|R) = 0.7$, $P(r|B) = 0.3$, $P(b|R) = 0.2$, $P(b|B)$
$= 0.8$. What are the best decisions for this example?

7.5. Show that the buyer should indeed buy insurance in the example on page 330.

7.6. Reconsider the insurance example on page 330, where the initial assets of the buyer are $500,000. What is the smallest value of the buyer's assets such that he should decide not to buy the insurance? Does the value of the buyer's assets make any difference to the seller?

7.7. A manufacturer desires to ship goods worth $100,000 by truck. There is a probability P that the shipment may be lost, stolen, or otherwise destroyed in transit. It costs $5,000 to insure the shipment. What must the probability P be so that the business executive should buy the insurance. Consider the cases where the manufacturer is an expected value operator, risk-averse, and a risk-prone gamble. Also consider several cases for total assets of the manufacturer, say $100,000, $1,000,000 and $100,000,000.

7.8. A city planner must decide whether to recommend formulated plans A_1 or A_2 to the mayor, who in turn may submit it to the city council for approval. One plan must be recommended and both cannot be recommended. In order for the plan to become policy, both mayor and city council must prove it. The city planner believes A_1 to be the best plan but believes that there is a 20 percent chance the mayor will not approve it and pass it on to the city council. If the mayor approves it, there is a 50 percent possibility that the city council will not approve it. The city council will, in the opinion of the planner, surely approve plan A_2, but the probability of the mayor passing it on to them is only 0.5. The planner would prefer that rejection, if it occurs, occur at the hands of the city council. The planner's utilities are

Alternative	Mayor Outcome	City Council Outcome	Utility
A_1	Accepts	Accepts	1.00
A_1	Accepts	Rejects	0.50
A_1	Rejects		0.30
A_2	Accepts	Accepts	0.80
A_2	Rejects		0.00

What should the planner recommend? Please illustrate your solution according to the tenets of SEU theory.

7.9. An investor who is wealthy enough to be an expected value operator may buy shares of a stock for $25,000. The net return r_i on the investment, the selling price minus the cost price of $25,000, will be one of three values $E_1 = \$40,000$, $E_2 = \$10,000$, or $E_3 = \$25,000$ and the investment returns will be influenced by the state of the economy according to e_1 = bull market, e_2 = fair market, e_3 = depression. The joint probabilities of the return and the state of the economy $P(E_i|e_j)$ are given by

		State of Economy		
		e_1	e_2	e_3
Return	r_1	0.12	0.06	0.02
	r_2	0.10	0.30	0.10
	r_3	0.04	0.12	0.14

For a price of $1,000 an economic consultant may be employed to predict the state of the economy. The conditional probabilities of the economist's estimate of the state of the economy given the true state, $P(\hat{e}_i|e_j)$ are

		True Economic State		
		e_1	e_2	e_3
Estimated	$\hat{e}_1$	0.7	0.1	0.1
Economic	$\hat{e}_2$	0.2	0.8	0.2
State	$\hat{e}_3$	0.1	0.1	0.7

Find the optimal decision regarding hiring the economist and buying the stock. Also find the expected value of perfect information and the expected value of the economist's information.

7.10. A quantity called the risk aversion function is defined as the second derivative of the utility function, or $r(e) = U''(e)$ and indicates how large a risk premium a decision maker will pay to eliminate uncertainty in a given situation. For cases in which the risk aversion function $r(e)$ is a constant R, we may find the utility function as

$$U(e) = \alpha - (\text{sgn } R)\beta \exp(-Re), \qquad R \neq 0$$
$$U(e) = \alpha + \beta e, \qquad\qquad\qquad\quad R = 0$$

Discuss possible uses for a risk aversion factor such as this. How could you conduct a lottery to assess U?

7.11. Howard has suggested use of the exponential as an adequate approximation to many utility function. Figure 7.47 illustrates the suggested exponential utility function. How could you assess R for this utility function? How would you use this for multiattribute utility determination?

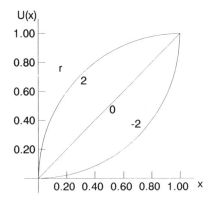

Figure 7.47. Illustration for $U(x) = [1 - \exp(-rx)] / [1 - \exp(-r)]$

7.12. Suppose you have your choice of two lotteries. The first lottery costs $1.00 to play, the probability of winning is 0.5, and the payoff if you win is $2.10. There is no payoff if you lose. You may play this lottery 10^8 times. The second lottery may be played only once. It costs 10^8 to play this lottery. The probability of winning is 0.5, and the payoff is 2.1×10^8 if you win, and nothing if you lose. Use the theory of normative decision analysis to thoroughly explore the varied aspects of these two lotteries. Which would you choose to play?

7.13. Suppose that Problem 7.12 is changed such that the probability of loosing the second lottery is 10^{-6}. The payoff for winning the second lottery is given precisely by $[5(10^{12} - 10^8)]/(10^6 - 1)$. How are the results of the previous problem changed? Are there any difficulties in assessment of utilities in this problem or the previous one that are not present in many decision analysis problems? Both this problem and the previous one have interesting relations with Problem 7.14.

7.14. A simplistic view of the coal versus nuclear fuel for electric power plant controversy views the probability of environmental damage from coal as much higher than that from nuclear energy. However, the likely destruction from a nuclear accident causing environmental damage would be far greater than that from coal. How can decision analysis be used to help resolve this controversy? Please relate your discussion to your readings concerning risk and hazard and Problems 7.12 and 7.13.

7.15. There are firms in the oil exploration industry that drill for oil with little scientific knowledge concerning the seismic characteristics of a given area. Instead, these firms, known as wildcatters, base their drilling decisions on what other firms have discovered while drilling in similar areas. We consider a simplified version of the wildcatter prob-

lem and add complexity and a degree of practicality as we analyze the
problem.

 a. Initially suppose that a wildcatter is considering the decision to drill
in a section where oil has been found on 20 percent of the past
efforts of other firms. The cost of drilling initially is assumed to be
$100,000. If oil is discovered, the gross return over the life of the
producing well is $400,000. Assume that the wildcatter is an ex-
pected monetary value player. Should the wildcatter drill? What is
the expected net return from drilling?

 b. Suppose that our wildcatter has a perfect seismic device that will
predict perfectly the existence or nonexistence of oil beneath the
surface of the earth. What is the expected return using the perfect
seismic predictor? What is the expected value of this perfect information?

 c. We consider a more realistic case in which we have an imperfect
seismic predictor. We assume that performance records of this
device have been kept. The output of the imperfect seismic detector
is a set of two lights. A green light is turned on when the seismic
indicator predicts that oil is present and a red light is turned on
when it predicts that no oil is present. A summary of the performance record of the seismic device is given by the following conditional probabilities

if oil	if no oil	then
0.75	0.40	green light on
0.25	0.60	red light is on

 What is the expected net return using this seismic device? What is
the expected value of this seismic device?

 d. If the imperfect seismic predictor costs a certain rental fee to use,
what is the maximum fee that the wildcatter would be willing to pay
in order to use the device?

7.16. Suppose that the price of crude oil increases such that the gross return
over the life of the well is increased from $400,000 to $600,000. Repeat
the calculations in Problem 7.15 and contrast the result obtained here
with those in the previous problem.

7.17. Reexamine Problem 7.15 for the case where the prior probability of
locating oil (without tests) is set equal to P. How sensitive are the
results of Problem 7.15 to the value of P?

7.18. Suppose that the wildcatter of Problem 7.15 is risk averse. How will this
affect the answers in Problem 7.15? What will be the wildcatter's

decisions when the gross return from a successful well is changed as in Problem 7.16?

7.19. What is your certainty equivalent for the following lotteries:

$$L_1 = (\$1{,}000, 0.50; \$0, 0.50)$$

$$L_2 = (\$2{,}000, 0.25; \$0, 0.75)$$

$$L_3 = (\$5{,}000, 0.10; \$0, 0.90)$$

$$L_4 = (\$667, 0.75; \$0, 0.25)$$

$$L_5 = (\$556, 0.90; \$0, 0.10)$$

On the basis of this, please plot your utility function for money.

7.20. How would Problem 7.19 be modified if you consider the various urn model representations shown in Figure 7.20?

7.21. You are in charge of ordering shoes for resale in the store that you manage. You can order these in unit quantities of 50 boxes of shoe pairs. If you order 50 boxes, the cost is $10. If you order 100 boxes, the cost is $9 per box. If you order 150 boxes, the costs are $8 per box. If you order 200 or more boxes, the cost is $7 per box. The selling price per box is $14. Any unsold shoes at the end of the period in question will have to be sold at your discount outlet at a price of $6 per box. Suppose that the probability for various demands for shoes is $P(0) = 0$, $P(50) = 0.1$, $P(100) = 0.3$, $P(150) = 0.2$, $P(200) = 0.2$, $P(250) = 0.1$, $P(300) = 0.1$. What is the order quantity that will maximize the expected profit (neglecting any merchandising costs)? What is the value of perfect information for this problem?

7.22. Investigate the possible appropriateness of the following heuristic for choosing among brands of beer. Only consider from among brands that are fundamentally acceptable in taste. Make pairwise comparisons of beers. If the price difference among two competing brands is less than $0.20, select the better beer. If the price difference among two competing brands is $0.20, or greater, select the cheaper beer.

7.23. You are a geologist. You estimate that there is a probability 0.3 that the area in which you are presently drilling contains a significant quantity of oil. You examine a seismograph and get a reading S. From past experience, or through consulting an expert, you find that the probability of obtaining this reading given that there is oil is rather high, say $P(S|Oil) = 0.85$. Should your belief in the probability of oil be changed, and by how much, from your initial estimate of $P(Oil) = 0.50$? Please justify your response.

7.24. Prepare a multiple attribute assessment procedure that will enable the evaluation of which of a number of homes to purchase.

7.25. A not uncommon situation for a civil defense emergency preparedness effort might read as follows. A civil defense committee in a large metropolitan area met recently to discuss emergency plans in the event of various crisis situations. One emergency under discussion was the following: A train carrying a very toxic chemical substance derails and the storage tanks begin to leak. The threat of explosion and lethal discharge of poisonous chemicals is imminent. Two possible actions were considered by the committee. Indicate your opinion about the relative merits of each. Option A carries with it a 0.5 probability of containing the threat without any loss of life and a 0.5 probability of losing 100 lives. Option B would produce a certain loss of 50 lives.

Which option is preferred? How would this issue be resolved using the various decision models and rationality perspectives discussed here? Please provide a brief discussion in your commentary concerning how an interface design might support or might have difficulty in supporting each of these.

7.26. Identify one human information processing bias that appears to be of interest and write a case study scenario of how a human might correctly or incorrectly process information using it.

7.27. Write a case study report on a choice effort you have faced in which regret over outcomes selected appeared to be present. Please conduct a decision assessment of this issue. In particular, please discuss the reasonableness of regret in the situation you describe.

7.28. How would we use social judgment theory to determine the preference of users for particular features of a DBMS software package? Of what use would the results obtained be in determining requirements for a software package?

7.29. Prepare a discussion from consulting furnished references of the steps involved in using the analytical hierarchy process. Please illustrate how this approach could be used to resolve the DBMS evaluation issue discussed in the text. Please contrast and compare this approach versus direct use of MAUT.

7.30. Prepare a prospect-theory-based explanation of the regret issues illustrated in Figure 7.29.

7.31. There exists an urn with a mixture of 1,000 balls with color red, blue, and white. 400 of the balls are red and the other 600 are some unknown mixture of blue and white balls. A single ball is taken from the urn. You are first offered a choice between option A_1, in which you win $10,000 if a red ball is chosen and A_1, in which case you win $10,000 if a blue ball is chosen. Which option would you prefer? Why? Please explain your choice on the basis of theories of decision assessment developed here. You are next offered a choice between option A_3, in which you win $10,000 if a red ball or a white ball is chosen and A_4, in which case you win $10,000 if a blue ball or a white ball is chosen. Which option would you prefer? Why? Please explain your choice on the basis of theories of decision assessment developed here. Now please contrast and compare your results for these two choice situations using SEU theory, prospect theory, and regret theory.

7.32. Many people believe that a favorable outcome becomes quite sure to occur after a long sequence of failures. This is known as the *gambler's fallacy*. Please describe it, and illustrate it with a few simple examples.

7.33. Prepare a case study report on the use of MAUT to aid a person in selection of a job.

7.34. Prepare a discussion of group decision analysis, that is to say decisions that impact a group of people. Discuss the thought that group decision analysis is just a special case of MAUT in which each individual, or special interest group, is represented by one attribute.

REFERENCES

[1] Brown, R. V. "The State of the Art of Decision Analysis: A Personal Perspective," *Interfaces*, in press.

[2] Howard, R. A., "Heathens, Heretics, and Cults," *Interfaces*, in press.

[3] Watson, S. R., "The Presumptions of Prescription," *Acta Psychologica*, in press.

[4] Fishburn, P. C., *Nonlinear Preference and Utility Theory*, Johns Hopkins University Press, Baltimore, 1988.

[5] Fishburn, P. C., "Generalization of Expected Utility Theories: A Survey of Recent Proposals," *Annals of Operations Research*, Vol. 19, No. 1, 1989, pp. 3–28.

[6] Brown, R., Kahr, A., and Peterson, C., *Decision Analysis for the Manager*, Holt, Rinehard, and Winston, New York, 1974.

[7] Howard, R. A., "Foundations of Decision Analysis," *IEEE Transactions on Systems Science and Cybernetics*, Vol. SSC-4, No. 3, 1968, pp. 211–219.

[8] Raiffa, H., *Decision Analysis*, Addison-Wesley, Reading MA, 1968

[9] Schlaifer, R., *Analysis of Decision Under Uncertainty*, McGraw-Hill, New York, 1969.

[10] Fishburn, P. C., "Foundations of Decision Analysis: Along the Way," *Management Science*, Vol. 15, No. 4, 1989, pp. 387–405.

[11] Wheeler, D. D., and Janis, I. L., *A Practical Guide for Making Decisions*, Free Press, New York, 1980.

[12] Beyth-Marom, R., and Dekel, S., *An Elementary Approach to Thinking Under Uncertainty*, Lawrence Erlbaum Associates, Hillsdale NJ, 1985.

[13] Norman, D. A., *The Psychology of Everyday Things*, Basic Books, New York, 1988.

[14] Behn, R. D., and Vaupel, J. W., *Quick Analysis for Busy Decision Makers*, Basic Books, New York, 1982.

[15] Russo, J. E., and Schoemaker, P. H., *Decision Traps: The Ten Barriers to Brilliant Decision Making and How to Overcome Them*, Simon and Schuster, New York, 1989.

[16] Gilovich, T., *How We Know What Isn't So: The Falibility of Human Reason in Everyday Life*, Free Press, New York, 1991.

[17] Sage, A. P., *Decision Support Systems Engineering*, John Wiley, New York, 1991.

[18] Adelman, L., *Evaluating Decision Support and Expert Systems*, John Wiley, New York, 1992.

[19] Hadley, G., *Introduction to Probability and Statistical Decision Theory*, Holden-Day, San Francisco, 1967.

[20] von Neumann, J., and Morgenstern, O., *Theory of Games and Economic Behavior*, Princeton University Press, Princeton NJ, 1944.

[21] Savage, L. J., *The Foundations of Statistics*, John Wiley, New York, 1954.

[22] North, D. W., "A Tutorial Introduction to Decision Theory," *IEEE Transactions on Systems Science and Cybernetics*, Vol. 4, No. 3, 1968, pp. 200–210.

[23] Holloway, C. A., *Decision Making Under Uncertainty: Models and Choices*, Prentice Hall, Englewood Cliffs NJ, 1979.

[24] Luce, R. D., and Raiffa, H., *Games and Decisions: Introduction and Critical Survey*, John Wiley, New York, 1957.

[25] Dyer, J. S., and Sarin, R. K., "Relative Risk Aversion," *Management Science*, Vol. 28, 1982, pp. 875–886.

[26] Huber, G. P., "Multi-Attribute Utility Models: A Review of Field and Field-like Studies," *Management Science*, Vol. 20, 1974, pp. 1393–1402.

[27] Johnson, E. M., and Huber, G. P., "The Technology of Utility Assessment," *IEEE Transactions on Systems, Man, and Cybernetics*, Vol. SMC-7, No. 5, 1977, pp. 311–325.

[28] Farquhar, P. H., "Utility Assessment Methods," *Management Science*, Vol. 30, 1984, pp. 1283–1300.

[29] von Winterfeldt, D., and Edwards, W., *Decision Analysis and Behavioral Research*, Cambridge University Press, Cambridge MA, 1986.

[30] Merkhofer, M. W., "Quantifying Judgmental Uncertainty: Methodological, Experiences and Insights," *IEEE Transactions on Systems, Man, and Cybernetics*, Vol. SMC 17, No. 5, 1987, pp. 741–752.

[31] Wallsten, T. S., and Budescu, D. V., "Encoding Subjective Probabilities: A Psychological and Psychometric Review," *Management Science*, Vol. 29, No. 2, 1983, pp. 151–173.

[32] Borcherding, K., Eppel, T., and von Winterfeldt, D., "Comparison of Weighting Judgments in Multiattribute Utility Measurements," *Management Science*, Vol. 37, No. 2, 1991, pp. 1603–1619.

[33] Keeney, R. L., von Winterfeldt, D., and Eppel, T., "Eliciting Public Values for Complex Policy Decisions," *Management Science*, Vol. 36, No. 9, 1990, pp. 1011–1030.

[34] Lehner, P. E., and Adelman, L. (Eds.), "Special Issue on Perspectives in Knowledge Engineering," *IEEE Transactions on Systems, Man, and Cybernetics*, Vol. 19, No. 3, 1989, pp. 443–662.

[35] Harvey, C. M., "Structured Prescriptive Models of Risk Attitudes," *Management Science*, Vol. 36, No. 12, 1990, pp. 1479–1501.

[36] Harvey, C. M., "Model of Tradeoffs in a Hierarchical Structure of Objectives," *Management Science*, Vol. 37, No. 8, 1991, pp. 1030–1042.

[37] Sampson, A. R., and Smith, R. L., "Assessing Risks through the Determination of Rare Event Probabilities," *Operations Research*, Vol. 30, No. 5, 1982, pp. 839–866.

[38] Krantz, D. H., Luce, R. D., Suppes, P., and Tversky, A., *Foundations of Measurement, Volume I: Representational Theory of Measurement*, Academic Press, Orlando FL, 1971.

[39] Suppes, P., Krantz, D. H., Luce, R. D., and Tversky, A., *Foundations of Measurement, Volume II: Geometric, Threshold, and Probabilistic Representations*, Academic Press, Orlando FL, 1989.

[40] Luce, R. D., Krantz, D. H., Suppes, P., and Tversky, A., *Foundations of Measurement, Volume III: Representation, Axiomatization, and Invariance*, Academic Press, Orlando FL, 1990.

[41] Fishburn, P. C., "Utility Theory," *Management Science*, Vol. 14, No. 5, 1968, pp. 335–378.

[42] Keeney, R., and Raiffa, H., *Decisions with Multiple Objectives*, John Wiley, New York, 1976.

[43] Edwards, W., "How to Use Multiattribute Utility Measurement for Social Decisionmaking," *IEEE Transactions on Systems, Man, and Cybernetics*, Vol. SMC-7, 1977, pp. 326–340.

[44] Miller, J. R., "A Systematic Procedure for Assessing the Worth of Complex Alternatives," MITRE Corporation Report AD 662001, New Bedford MA, 1967.

[45] Farris, D. R., and Sage, A. P., "On Decision Making and Worth Assessment," *International Journal of System Sciences*, Vol. 6, No. 12, 1975, pp. 1135–1178.

[46] Sage, A. P., *Methodology for Large Scale Systems*, McGraw Hill, New York, 1977.

[47] Sage, A. P., and White, E. B., "Methodologies for Risk and Hazard Assessment: A Survey and Status Report," *IEEE Transactions on Systems, Man, and Cybernetics*, Vol. 10, No. 8, 1980, pp. 425–446.

[48] Sage, A. P., and White, E. B., "Decision and Information Structures in Regret Models of Judgment and Choice," *IEEE Transactions on Systems, Man, and Cybernetics*, Vol. 13, No. 3, 1983, pp. 136–145.

[49] Bell, D. E., "Components of Risk Aversion," *Operational Research '81*, J. P. Brans, (Ed.), North-Holland, New York, 1981, pp. 371–378.

[50] Bell, D. E., Dyer, J. S., and Sarin, R. K., "Relative Risk Aversion," *Management Science*, Vol. 28, No. 8, 1982, pp. 875–886.

[51] Bell, D. E., "Regret in Decision Making Under Uncertainty," *Operations Research*, Vol. 30, 1982, pp. 961–981.

[52] Bell, D. E., "Risk Premiums for Decision Regret," *Management Science*, Vol. 29, 1983, pp. 1156–1166.

[53] Bell, D. E., "Disappointment in Decision Making Under Uncertainty," *Operations Research*, Vol. 33, No. 1, 1985, pp. 1–27.

[54] Loomes, G., and Sugden, R., "Regret Theory: An Alternative Theory of Rational Choice Under Uncertainty," *Economic Journal*, Vol. 92, 1982, pp. 805–824.

[55] Loomes, G., and Sugden, R., "Some Implications of a More General Form of Regret Theory," *Journal of Economic Theory*, Vol. 41, 1987, pp. 270–278.

[56] Loomes, G., "Predicted Violations of the Invariance Principle," *Annals of Operations Research*, Vol. 19, 1989, No. 1–4, 1989, pp. 103–113.

[57] Sugden, R., "New Developments in the Theory of Choice Under Uncertainty," *Bulletin of Economic Research*, Vol. 38, No. 1, 1986, pp. 1–24.

[58] Kahneman, D., and Tversky, A., "Prospect Theory: An Analysis of Decision Under Risk," *Econometrica*, Vol. 47, 1979, pp. 263–291.

[59] Smith, C. L., and Sage, A. P., "A Theory of Situation Assessment for Decision Support," *Information and Decision Technologies*, Vol. 17, No. 1, 1991, pp. 91–124.

[60] Tversky, A., "Elimination by Aspects: A Theory of Choice," *Psychological Review*, Vol. 79, 1972, pp. 281–299.

[61] Tversky, A., and Kahneman, D., "Belief in the Law of Small Numbers," *Psychological Bulletin*, Vol. 76, No. 2, 1971, pp. 105–110.

[62] Hogarth, R. M., *Judgment and Choice*, John Wiley, New York, 1987.

[63] Kahneman, D., Slovic, P., and Tversky, A. (Eds.), *Judgment Under Uncertainty: Heuristics and Biases*, Cambridge University Press, New York, 1981.

[64] Hogarth, R. M., and Reder, M. W. (Eds.), *Rational Choice: The Contrast Between Economics and Psychology*, University of Chicago Press, Chicago, 1987.

[65] Caverni, J. P., Fabre, J. M., and Gonzalez, M. (Eds.), *Cognitive Biases*, North Holland Elsevier, Amsterdam, 1990.

[66] Bazerman, M. H., *Judgment in Managerial Decision Making*, John Wiley, New York, 1986.

[67] Yates, J. F., *Judgment and Decision Making*, Prentice Hall, Englewood Cliffs NJ, 1990.

[68] Einhorn, H. J., and Hogarth, R. M., "Ambiguity and Uncertainty in Probabilistic Inference," *Psychological Review*, Vol. 92, No. 4, 1985, pp. 433–461.

[69] Einhorn, H. J., and Hogarth, R. M., "Adaptation and Inertia in Belief Updating: The Contrast Inertia Model," Technical Report, University of Chicago, 1987.

[70] Hogarth, R. M., and Einhorn, H. J., "Order Effects in Belief Updating: The Belief Adjustment Model," University of Chicago Report, 1989.

[71] Adelman, L., Tolcott, M. A., and Bresnik, T. A., "Examining the Effects of Information Order on Expert Judgment," *Organizational Behavior and Human Decision Processes*, Vol. 46, 1992.

[72] Adelman, L., and Bresnik, T. A., "Examining the Effect of Information Sequence on Patriot Air Defense Officers' Judgment," *Organizational Behavior and Human Decision Processes*, Vol. 46, 1992.

[73] Kahneman, D., and Tversky, A., "Intuitive Prediction: Biases and Corrective Procedures," in *TIMS Studies in Management Science*, Vol. 12, 1979, pp. 313–327.

[74] Ford, J., Schmitt, N., Schectman, S., Hults B., and Doherty, M., "Process Tracing Methods: Contributions, Problems, and Neglected Research Questions," *Organizational Behavior and Human Decision Processes*, Vol. 43, 1989, pp. 75–117.

[75] Nisbett, R. E., Krantz, D. H., Jepson, C., and Kunda, Z., "The Use of Statistical Heuristics in Everyday Reasoning," *Psychological Review*, Vol. 90, No. 4, 1983, pp. 339–363.

[76] Einhorn, H. J., and Hogarth, R. M., "Decision Making Under Ambiguity," in Hogarth, R. M. and Reder, M. W. (Eds.), *Rational Choice*, University of Chicago Press, Chicago, 1987, pp. 41–66.

[77] Klayman, J., "Learning from Feedback in Probabilistic Environments," *Acta Psychologica*, Vol. 56, 1984, pp. 81–92.

[78] Klayman, J., and Ha, Y. W., "Confirmation, Disconfirmation, and Information in Hypothesis-Testing," *Psychological Review*, Vol. 94, No. 2, 1987, pp. 211–228.

[79] Kahneman, D., and Tversky, A., "Intuitive Prediction, Biases and Corrective Procedures," in Makridakis, S., and Wheelwright, S. C. (Eds.), *Forecasting*, North Holland, Amsterdam 1979, pp. 313–327.

[80] Cohen, L. J., "On the Psychology of Prediction: Whose is the Fallacy," *Cognition*, Vol. 7, No. 4, 1979, pp. 385–407.

[81] Cohen, L. J., "Can Human Irrationality Be Experimentally Demonstrated?," *The Behavioral and Brain Sciences*, Vol. 4, 1981, pp. 317–370.

[82] Stephanou, H. E., and Sage, A. P., "Perspectives on Imperfect Information Processing," *IEEE Transactions on Systems, Man, and Cybernetics*, Vol. 17, No. 5, 1987, pp. 780–798. (Also in *Knowledge Based Systems: Fundamentals and Tools*, Garcia, O. N., and Chien, Y. T. (Eds.), IEEE Computer Society Press, Los Alamitos CA, 1992, pp. 294–311.)

[83] Post, S., and Sage, A. P., "An Overview of Automated Reasoning," *IEEE Transactions on Systems, Man, and Cybernetics*, Vol. 20, No. 1, 1990, pp. 202–224.

[84] Neapolitan, R. E., *Probabilistic Reasoning in Expert Systems*, John Wiley, New York, 1990.

[85] Smith, G. F., Benson, P. G., and Curley, S. P., "Belief, Knowledge and Uncertainty: A Cognitive Perspective on Subjective Probability," *Organizational Behavior and Human Decision Processes*, in press.

[86] von Winterfeldt, D., and Edwards, W., *Decision Analysis and Behavioral Research*, Cambridge University Press, New York, 1986.

[87] Phillips, L., "Theoretical Perspectives On Heuristics And Biases In Probabilistic Thinking," in Humphries, P. C., Svenson, O., and Vari, O. (Eds.), *Analyzing And Aiding Decision Problems*, North Holland, Amsterdam, 1984.

[88] Gigerenzer, G., "How to Make Cognitive Illusions Disappear: Beyond 'Heuristics and Biases'," *European Review of Social Psychology*, Vol. 2, 1991, pp. 83–115.

[89] Kahneman, D., and Tversky, A., "Prospect Theory: An Analysis of Decision under Risk," *Econometrica*, Vol. 47, 1979, pp. 263–291.

[90] Tversky, A., and D. Kahneman, "The Framing of Decisions and the Psychology of Choice," *Science*, Vol. 211, 1981, pp. 453–458.

[91] Fischhoff, B., Slovic, P., and Lichtenstein, S., "Knowing What You Want: Measuring Labile Values," in Wallsten, T. S. (Ed.), *Cognitive Processes in Choice and Decision Behavior*, Lawrence Erlbaum Assoc., Hillsdale NJ, 1980, pp. 117–141.

[92] Goldstein, W. M., and Einhorn, H. J., "Expression Theory and the Preference Reversal Phenomena," *Psychological Review*, Vol. 94, No. 2, 1987, pp. 236–254.

[93] Hogarth, R. M., and Einhorn, H. J., "Venture Theory: A Model of Decision Weights," *Management Science*, Vol. 36, No. 7, 1990, pp. 780–803.

[94] Machina, M. J., "'Expected Utility' Analysis Without the Independence Axiom," *Econometrica*, Vol. 50, No. 2, 1982, pp. 277–323.

[95] Machina, M. J., "Choice Under Uncertainty: Problems Solved and Unsolved," *Economic Perspectives*, Vol. 1, No. 1, 1987, pp. 121–154.

[96] Harvey, C. M., "Prescriptive Models of Psychological Effects on Risk Attitudes," *Annals of Operations Research*, Vol. 19, No. 1–4, 1989, pp. 143–170.

[97] Schoemaker, P. J. H., "Are Risk-Attitudes Related Across Domains and Response Modes," *Management Science*, Vol. 36, No. 12, 1990, pp. 1451–1463.

[98] Peterson, S. A., and Lawson, R., "Risky Business: Prospect Theory and Politics," *Political Psychology*, Vol. 10, No. 2, 1989, pp. 325–339.

[99] Starmer, C., and Sugden, R., "Violations of the Independence Axiom in Common Ratio Problems: An Experimental Test of Some Competing Hypotheses," *Annals of Operations Research*, Vol. 19, No. 1–4, 1989, pp. 79–102.

[100] Currim, I. S., and Sarin, R. K., "Prospect Versus Utility," *Management Science*, Vol. 33, No. 1, 1989, p. 2241.

[101] Brunswick, E., *The Conceptual Framework of Psychology*, University of Chicago Press, Chicago, 1952.

[102] Hammond, K. R., "Probabilistic Functioning and the Clinical Method," *Psychological Review*, Vol. 62, 1955, pp. 255–262.

[103] Hammond, K. R., *The Psychology of Egon Brunswick*, Holt, Rinehart, and Winston, New York, 1966.

[104] Hammond, K. R., McClelland, G. H. and Mumpower, J., *Human Judgment and Decision Making: Theories, Methods, and Procedures*, Praeger Publishers, New York, 1980.

[105] Hammond, K. R., Stewart, T. R., Brehmer, B., and D. O. Steinmann: "Social Judgment Theory", in *Human Judgment and Decision Processes*, M. F. Kaplan, G. Schwartz (Eds.), Academic Press, New York, 1975, pp. 271–312.

[106] Hammond, K. R., Rohrbaugh, J., Mumpower J., and Adelman, L., "Social Judgment Theory: Applications in Policy Formation", in Kaplan, J. M. F., and Schwartz, S. (Eds.), *Human Judgment and Decision Processes*: *Applications in Problems Setting*, Academic Press, New York, 1977.

[107] Hammond, K. R., Mumpower, J. L., and Cook, R. L., "Linking Environmental Models with Models of Human Judgment: A Symmetrical Decision Aid," *IEEE Transactions on Systems, Man, and Cybernetics*, Vol. 7, 1977, pp. 358–367.

[108] Mumpower, J. L., Veirs, V., and Hammond, K. R., "Scientific Information, Social Values, and Policy Formation: The Application of Simulation Models and Judgment Analysis to the Denver Regional Air Pollution Problem," *IEEE Transactions on Systems, Man, and Cybernetics*, Vol. 9, 1979, pp. 464–476.

[109] Anderson, D. F. and Rohrbaugh, J., "Some Conceptual and Technical Problems in Integrating Models of Judgment with Simulation Models," *IEEE Transactions on Systems, Man, and Cybernetics*, Vol. 22, No. 1, 1992.

[110] Brehmer, B., and Joyce, C. R. B. (Eds.), *Human Judgment*: *The STJ View*, Elsevier, Amsterdam, 1988.

[111] Arkes, H. R., and Hammond, K. R., *Judgment and Decision Making*: *An Interdisciplinary Reader*, Cambridge University Press, New York, 1986.

[112] Sage, A. P., and White, C. C., "ARIADNE: A Knowledge Based Interactive System for Planning and Decision Support," *IEEE Transactions on Systems, Man, and Cybernetics*, Vol. SMC-14, No. 1, 1984, pp. 35–47.

[113] White, C. C., and Sage, A. P., "A Multiple Objective Optimization Based Approach to Choicemaking," *IEEE Transactions on Systems, Man, and Cybernetics*, Vol. SMC-10, No. 4, 1980, pp. 315–326.

[114] White, C. C., and Sage, A. P., "Multiple Objective Evaluation and Choicemaking under Risk with Partial Preference Information," *International Journal of Systems Science*, Vol. 14, 1983, pp. 467–485.

[115] White, C. C., Sage, A. P., and Scherer, W. T., "Decision Support with Partially Identified Parameters," *Large Scale Systems*, Vol. 3, 1982, pp. 177–190.

[116] White, C. C., and Sykes, E. A., "A User Preference Guided Approach to Conflict Resolution in Rule Based Expert Systems," *IEEE Transactions on Systems, Man, and Cybernetics*, Vol. 16, 1986, pp. 276–278.

[117] Klein, G., Orasanu, J., and Calderwood, R. (Eds.), *Decision Making in Action*: *Models and Methods*, Ablex Press, Norwood NJ, 1992.

[118] Howard, R. A., "Decision Analysis: Practice and Promise," *Management Science*, Vol. 34, No. 6, 1988, pp. 679–695.

[119] Saaty, T. L., *The Analytical Hierarchy Process*: *Planning, Priority Setting, Resource Allocation*, McGraw Hill, New York, 1980.

[120] Saaty, T. L., and Kearns, K. P., *Analytical Planning*: *The Organization of Systems*, Pergamon Press, Oxford, 1985.

[121] Dyer, J. S., "Remarks on the Analytical Hierarchy Process," *Management Science*, Vol. 30, 1990, pp. 259–268. (Also see Dyer, J. S., "A Clarification of

Remarks on the Analytical Hierarchy Process," *Management Science*, Vol. 30, 1990, pp. 274–275.

[122] Harker, P. T., and Vargas, L. G., "Reply to Remarks on the Analytical Hierarchy Process," *Management Science*, Vol. 30, 1990, pp. 269–273.

[123] Winkler, R. L., "Decision Modeling and Rational Choice: AHP and Utility Theory," *Management Science*, Vol. 30, 1990, pp. 247–248.

[124] Clemen, R. T., *Making Hard Decisions*: *An Introduction to Decision Analysis*, PWS-Kent Publishing, Boston MA, 1991.

[125] Buede, D. M., "Superior Design Features of Decision Analytic Software," *Computers and Operations Research*, in press.

[126] Corner, J. L., and Kirkwood, C. W., "Decision Analysis Applications in the Operations Research Literature, 1970–1989," *Operations Research*, Vol. 39, No. 2, 1991, pp. 206–219.

[127] Fishburn, P. C., "Multiperson Decision Making: A Selective Review," in Kacprzyk, J., and Fedrizzi, M. (Eds.), *Multiperson Decision Making Using Fuzzy Sets and Possibility Theory*, Kluwer Academic Publishers, Amsterdam, 1990, pp. 3–27.

[128] Fishburn, P. C., and Sarin, R. K., "Dispersive Equity and Social Risk," *Management Science*, Vol. 37, No. 7, 1991, pp. 751–769.

[129] Fishburn, P. C., "Decision Theory: The Next Hundred Years," *The Economic Journal*, Vol. 101, 1991, pp. 27–32.

Chapter **8**

Microeconomic Systems and Cost and Operational Effectiveness Analysis

In the preceding chapter, we provided a discussion of relatively common systems engineering approaches to evaluation, judgment and choice, and decision making. Although we noted a variety of approaches, the primary approaches developed were decision analysis methods and the regression-based methods of policy capture or social judgment theory. In this chapter, we examine cost-benefit analysis (CBA) and cost-effectiveness analysis (CEA) as alternative and complimentary approaches to achieving many of these same objectives. The use of CBA and CEA concepts requires an understanding of microeconomics, so we provide a very brief overview of this discipline. We also discuss some concepts of economic discounting, or the time value of money, in order that we be able to meaningfully compare costs and benefits at one time with those at another.

In Chapter 7, we considered uncertainty effects associated with decision analysis issues, but this chapter, for the sake of simplicity, will generally not introduce such effects. For the most part, these may be dealt with as discussed in Chapter 7. In this chapter, we cover a relatively broad range of topics, dealt with in greater depth in Ref. 1, the references therein, and the references cited in this chapter. Only in the latter portions of the chapter do we comment on approaches to cost-effectiveness evaluation that include uncertainty and risk management considerations.

8.1. MICROECONOMIC SYSTEMS ANALYSIS

Markets produce a diverse array of economic decision-making activities. Various decision models have been developed that describe the behavior of such participants in the marketplace as firms, consumers, and resource owners. These models deal primarily with markets in equilibrium. Most are designed to yield maximum economic satisfaction for one or another category of participants. We present some essential features of such models. There are a rather large number of references to which the interested reader can turn [2–5] for additional details concerning this most interesting and important subject.

An economic system is said to be in equilibrium when total supply equals total demand. A situation with no market imperfections is typically assumed in most elementary considerations of economic systems. This means that there are no monopolies, externalities, or other peculiarities. A microeconomic model is a set of relations describing consumer sector desires, production sector desires, and the general conditions that result in maximum satisfaction for all consumers and firms in the system. Consumers desire to maximize happiness by purchasing a mix of products and services. Firms desire to maximize profit. Specific application areas for microeconomic models in the real world include design of production facilities, marketing, price regulation, taxation policies, and a variety of resource management considerations. Extensions of microeconomic theory result in cost-benefit and cost-effectiveness models that are very useful for systems management decisions. We examine many of these uses throughout this chapter.

8.1.1. Theories of the Firm and the Consumer

In an economic system, the role of the household consists of selling such resources as labor to firms and purchasing products and services with the wages that are received for this labor. The labor input to the firms from the households is called a *factor input* in the economics literature, and household purchases are an element of what is known as *final demand* (the other elements of final demand are government purchases and exports). It is assumed that each household attempts to maximize their satisfaction or bliss through their purchasing decisions. It is also assumed that a utility function, which mathematically describes the amount of satisfaction that a household derives from possession of a given set of produced goods and factor inputs, can be developed. Each household has a maximum budget or spending restraint. This is the total amount of income a household has to spend on final goods.* The relationship that describes this limitation is known as the

---

*Sometimes the term commodity is used to describe the products and services purchased by consumers. But, commodities also describe the raw materials that generally are inputs to the production process. We will use the term *goods* as a simple term for products and services.

budget constraint. The income of a household is derived from the selling of factor inputs for wages, and in some cases from the ownership of firms. The goal of each household is assumed to be the maximization of its utility function subject to the budget constraint. The solution to this optimization problem yields the optimum quantities of factor inputs that the household will provide and the goods that will be purchased. The prices, wages, and costs associated with factor inputs and goods are assumed fixed in the classic theory of the consumer. The optimum factor input and goods bundle will be different for each household.

The role of a firm in an economic system generally consists in buying factor inputs in the form of land, capital, and labor. It produces goods and services from these resources, and then sells these goods to households for consumption purposes. It will be necessary for some firms to use goods and services produced by other firms as factor inputs to their own production. For example, a firm that manufactures television sets may purchase components from a firm that manufactures electronic parts. Thus, firms sell their products both to other firms and to consumers. It is assumed that associated with each firm is a *production function*, which describes the maximum amount of final goods that a firm can produce for a given quantity of factor inputs. The form of this production function will depend upon the firm's technology and productive capacity.

The *profit* of a particular firm is the total value of products and services sold, or revenues, minus the cost of producing those goods. It is assumed that the basic goal of the firm is to maximize profits. This maximization is necessarily subject to the constraints on the technology and production capacity of a firm, as reflected in its production function. The strategy that maximizes the profit of a firm results in determination of the quantity of goods that should be produced at various levels of prices. This is the supply curve for the firm in question. Solution of this profit maximization equation also results in a factor demand equation. We may obtain a relation that gives the demand for factor inputs in terms of their prices. We develop these relations more fully here.

The fundamental model that comprises the classical *theory of the consumer* is easily stated. We assume that the consumer has a utility function that expresses the bliss, or utility, that is received from consumption of a bundle of goods and services. The consumer is assumed to have a utility function given by

$$U = U(x_1, x_2, \ldots, x_N) = U(\mathbf{x})$$

where $\mathbf{x} = [x_1, x_2, \ldots, x_N]^T$ is an N-dimensional vector that denotes bundle of goods and services (or goods bundle). We assume also that there is a set of prices represented by the vector $\mathbf{p} = [p_1, p_2, \ldots, p_N]^T$. This price vector is assumed to represent the price that has to be paid for a unit of each of the N goods and services.

Consumers are assumed to believe that more of any given good or service is always better than less. Of course, the consumer has limited resources. As a consequence of this, the consumer cannot pay more than some fixed income, here denoted by the symbol *I*, for these. The fundamental problem of the consumer is to maximize the utility function subject to the budget constraint on income. This budget expenditure constraint is, in equation form,

$$I \geq \sum_{i=1}^{N} p_i x_i = \mathbf{p}^T \mathbf{x}$$

We can explore various aspects of consumer behavior that maximize bliss using consumer utility as the objective function and subjecting the maximization to the constraint on consumer income and expenses. This maximization of utility, combined with the constraint on disposable income, produces the demand curve for a consumer. The form of this utility function will depend upon the tastes and preferences of each consumer for the variety of products available in the marketplace.* In effect, this creates an optimization problem —identifying an optimum quantity of commodities demanded for consumption and the optimum input of income for this commodity. The income of a consumer may be produced through labor, through interest earned on savings (capital), or through savings.

Within the overall bundle of goods, the consumer can substitute one good for another. The comparative economic value of such goods can be shown by a utility function. If such a utility function is, for example, given by $U = x_1^{0.5} x_2^2$, then the utility of bundle $\mathbf{x}^T = [1 \ 2]$ is the same as that of the bundle $\mathbf{x}^T = [16 \ 1]$. This does not say that 15 units of x_1 are worth 1 unit of x_2. Incontrovertibly, all that we can conclude is that the utility of this consumer is the same for each bundle. All that has been said is that at two units of x_2, and one unit of x_1, a decrease of one unit of x_2 would be balanced out by a gain of 15 units of x_1. It would seem that only in the additive utility function, such as $U(\mathbf{x}) = a_1 x_1 + a_2 x_2$, could we talk about so many units of x_1 being worth one unit of x_2 independently of the level of the goods. Often, this subtle distinction in the meaning of substitution effects is forgotten.

The notion of economic utility is not unlike utility concepts that we have already examined in our previous chapter. There is some satisfaction or utility from unsold factor inputs. Most often, this means labor hours available but not spent working are used for leisure time. Generally, the utility associated with increasing the total bundle of goods tends to decline with each incremental purchase, since the labor required for such increased purchases takes up more and more leisure time. The optimum mix of

---

*A number of interesting and relevant questions exist. Some consumers might, for example, have developed a considerable taste for some special product if they knew of its existence. There are many questions like this that are not a formal part of basic microeconomic theory but follow as extensions of it.

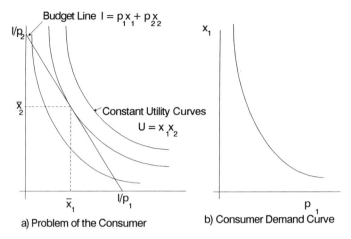

Figure 8.1. Simple Illustration of the Problem of the Consumer

consumption and leisure for a particular consumer will depend upon the consumer's resources and preferences. Under the theory of *revealed prefer-ences*, identification of consumer utility functions can be made from actual behavior.

The notion of consumer utility is, however, somewhat different from the notion of subjective expected utility examined in Chapter 7. There are no notions of risk aversion in consumer utility, for example. It is more like the notion of multiple attribute value.

Figure 8.1 illustrates a typical consumer demand function. This consumer demand function results from the utility function $U = x_1 x_2$ and the income equation $I = p_1 x_1 + p_2 x_2$. This yields the demand curve $x_1 = I/2p_1$ and $x_2 = I/2p_2$. The major feature here is the monotonically decreasing demand with price. There are very few goods, called Giffin goods, where demand for these goods increases with price increases (Lambergini automobiles might be in this category).

The primary economic activities of firms, or productive units, in an economic system consists of

1. Buying factor inputs from households.
2. Producing products.
3. Selling products or goods to households or to other firms.

The profit function of a particular firm relates the total income received for products sold, or revenues, and the cost of producing the product in terms of the wages paid, the price of raw materials, and the price of capital. In these terms, we see that profit is just revenue minus production cost. This is an elementary statement of the microeconomic theory of the firm.

This microeconomic *theory of the firm* is, conceptually, very simple. We adopt, as a fundamental hypothesis of this theory, that the goal of a firm is to maximize profit. To do this, the firm will need to know its costs of production. These costs will depend upon the market for the three fundamental types of economic resources: land, capital, and labor. If we use T, K, and L to represent land (or perhaps more appropriately natural resources), capital, and labor; and if the known costs of these factor inputs are c_T, c_K, and c_L, then the costs of production are given by the expression

$$C(T, K, L) = c_T T + c_K K + c_L L + F \qquad (8.1)$$

where F denotes fixed costs of production. The revenue to the firm for selling a production quantity q at fixed price p is assumed to be given by

$$R = pq \qquad (8.2)$$

The quantity of goods produced by the firm is related to the input factors of production. Traditionally, the input factors of production are assumed to be land, labor, and capital. We might well argue that this classic relation needs to be augmented such that it contains an information component. On the other hand, we can incorporate information into a model such as this by assuming that it is one of the goods that can be purchased by capital. The expression which gives the production of the firm is given by

$$q = f(T, K, L) \qquad (8.3)$$

There are many possible production functions in economic systems analysis. Two of the more common ones are the linear production function

$$q = f(x) = b^T x = \sum_{i=1}^{n} b_i x_i$$

and the Cobb–Douglas (multiplicative) production function

$$q = f(x) = b_0 \prod_{i=1}^{n} x_i^{b_i}$$

which has a log-linear property

$$\log q = \log b_0 + \sum_{i=1}^{n} b_i \log x_i$$

that is quite useful. For example, use of regression analysis to estimate the parameters in the Cobb–Douglas production function is relatively straightforward.

The profit of the firm is the difference between revenue and production costs, or

$$\pi = R - C \qquad (8.4)$$

Obviously, this is a very simple static model; it neglects taxes and shareholder profit considerations among other elements. There are a large number of other accommodations that would have to be made to make this a valid model of organizational behavior. This is not our intent here.

Three key questions are relevant to the economic productivity of the firm.

1. How can we maximize the profit of the firm?
2. How can we minimize production costs for a given quantity of any product?
3. Are there circumstances under which we will not produce a product at all?

Every well-managed firm seeks an answer to these questions. The answers to the first two questions are related. To obtain maximum profit, we maximize the profit π, as given by Eq. (8.4), subject to the equality constraints Eqs. (8.1) through (8.3). The result of doing this is that we obtain production, or supply, curves that give the quantity of goods that will be produced as a function of their prices. To minimize the production costs, we minimize C of Eq. (8.1), subject to the equality constraints of Eqs. (8.2) and (8.3). The result of doing this is a relation for the minimum production cost, $C(q)$, for producing a quantity of goods, q. The answer to question 3 is that we should produce as long as we can obtain a nonnegative profit.

The production function of a firm describes the maximum amount of final goods that the firm can produce given a quantity of factor inputs. The form of this expression will depend on the firm's technology and productive capacity. It is assumed that the firm makes use of the production inputs as efficiently as possible for a given production technology. It is also assumed that the basic goal of a firm is to maximize profit, subject to constraints or limitations on the technology and production capacity of the firm. The solution to the optimization problem for economic efficiency, in the sense of maximum profit, is obtained in terms of the quantity of goods produced and sold for a given set of input prices. As in the theory of the consumer, factor input prices (wages) and product selling prices are assumed to be fixed. Almost all of these restrictions can be removed, and are in more advanced treatments of this subject.

As a simple example of optimization of the firm, we consider a simple firm with production function

$$q = L^a K^{1-a}$$

The profit relation is then given by Eq. (8.4) as

$$\pi = pL^a K^{1-a} - c_L L - c_K K$$

and the necessary conditions for profit maximization are that the partial derivatives of profit with respect to L and K be equal to zero. These requirements give rise to the expressions

$$O = apL^{a-1}K^{1-a} - c_L$$
$$O = (1-a)pL^a K^{-a} - c_K$$

We obtain the price relation through solution of these two simultaneous equations. It is

$$p = (c_L/a)^a [c_K/(1-a)]^{1-a}$$

For the particular case where $a = 0.5$, we have

$$p = 2(c_L c_K)^{0.5}$$
$$Kc_K = c_L L$$
$$q = (LK)^{0.5} = K(c_K/c_L)^{0.5} = p/(2c_L)$$

As we might expect, the quantity of goods produced will increase with the price that can be obtained for them. This is the usual shape that one would

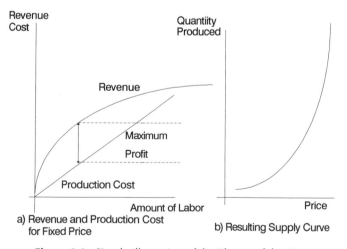

Figure 8.2. Simple Illustration of the Theory of the Firm

expect to obtain for a production curve. Figure 8.2 illustrates a typical supply curve for a firm. This illustration is obtained for a production function $q = 10L - L^2$ and costs, or wages, of 10.

8.1.2. Microeconomic Supply–Demand Models

We now extend these concepts to simple microeconomic models that describe the behavior of economic agents such as firms, consumers, and resource owners in a market economy. We are primarily concerned with conditions that prevail in a market system that is in equilibrium, and in which no market imperfections are present. Market imperfections occur because of monopolies, cartels, and other externalities that our introductory discussion here cannot discuss in any detail. Microeconomic models such as these serve primarily as guides to the behavior that results in maximum satisfaction for each economic agent. They also provide a basis for incorporation of more advanced features.

The interaction of firms and consumers leads to a microeconomic model that describes

1. The price and quantity of products and services desired by a consumer who is maximizing utility.
2. The price and quantity of products or services offered by a firm that is maximizing profits.
3. The general conditions characterizing the markets in which firms and consumers interact.

These relations may be combined to determine equilibrium market conditions resulting in maximum satisfaction for firms and consumers. This equilibrium is found at the intersection of the supply and demand curves for products, and the intersection of the supply and demand curves for the factor inputs to production.

Microeconomic systems analysis models provide insight into the workings and effects of ideal market systems and can be used to evaluate policies designed to regulate economic behavior or alter economic conditions. They can also be used to investigate the effects of changes in such elements as consumer preferences, firms technology, and the availability and costs of the various factor resource inputs to production.

There are two foundational supports for the formulation of market equilibrium equations. First, the total quantity of goods supplied by the firms in the economy is assumed to be equal to the total quantity that is demanded by all consumers and all firms. This neglects exports. These, and such other factors as taxes, can be included and are included in more advanced treatments. The products remaining after purchases by the firms and consumers would equal zero. Second, fluctuating prices and wages are assumed, with the equilibrium value of these being determined by freely competitive market conditions.

The *factor clearing equations* specify that the total demand for factor inputs exhibited by all firms equals the amount supplied by all individuals and all firms. These equations, at least in their most basic formulation, do not consider such market imperfections as monopolies and cartels, as we have noted earlier.

It is important to emphasize again that microeconomic models, just as with any model, may produce results that are mathematically correct, but the correctness is only guaranteed within the contexts of the assumed model. They cannot, of course, answer questions about the equity or fairness of a particular proposed distribution of resources, market structure, or other such economic configurations that were not a part of the model requirements used to formulate the model. Normative, or welfare, economic systems analysis models attempt to provide answers to some of these questions. We examine these in a later subsection of this chapter.

The first step in building a microeconomic model of the supply demand relations describing economic activity is to identify the basic constituents of the economic system under consideration. These constituents generally include the following.

1. A consumption sector, generally represented by a set of consumers or households.
2. A production sector, generally represented by a set of firms.
3. A set of final goods, commodities, or services.
4. A set of economic resources that are the factor inputs to production and that generally consist of capital K, land T, and labor L.

These constituents follow from a set of fundamental and foundational relations for a microeconomic model. They result directly from theoretical considerations of the economic behavior of households, the economic behavior of firms, and the resulting equilibrium conditions that prevail in freely competitive markets where households and firms exchange resources and commodities.

An economic market is in equilibrium when the quantity of all goods and services demanded is equal to the quantity of that good or service that is supplied. Equilibrium conditions require that two conditions are fulfilled. The first is that for each final good, commodity, or product that is produced, the total produced quantity supplied by all firms is equal to the total quantity demanded by all households. This gives rise to a set of product market-clearing equations. The second requirement is that, for each economic factor input to production, the total quantity supplied by all households is equal to the total quantity demanded by all firms. This gives rise to a set of factor

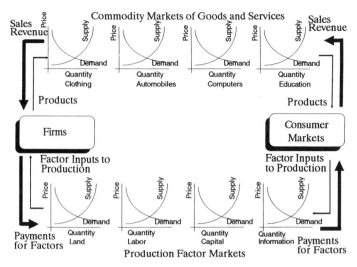

Figure 8.3. Market Equilibrium

market clearing equations. These statements describe the basic semantic structure of the model.

After the basic structure of a specific model has been identified, the precise forms of the production functions for firms and the utility functions of consumers need to be determined. Utility functions might be reconstructed from observed past behavior or might be elicited directly. Such statistical techniques as regression analysis and estimation theory are also useful in constructing models of production functions for firms and utility curves for consumers.

Figure 8.3 illustrates, conceptually, the results that we might obtain from construction of a simple microeconomic model. Shown in this figure are two fundamental feedback loops—one involving the flow of products and the second involving the flow of capital. It is important to note the flows into and out of the firms and consumer sectors. The input to the firm is the income due to sales and the factors for production. The outputs from the firm are the products of goods and services, and payments for the factors are inputs to production.

A simple example will provide some illustration of the concepts involved in constructing a microeconomic model. For this example, we construct a model based on a very simple economy, consisting of two firms and two consumers. In this small market system, firm 1 buys labor from the consumers to make computers, and firm 2 buys labor from the consumers to make automobiles. The generic question we wish to answer is: how many automobiles and how many computers will each consumer purchase, and how much labor will each firm use? The question naturally arises as to the time interval during which these prices, quantities, and profits are valid. The answer is that such an interval must be identified by the people building the model. Actually, the introduction of dynamic evolution over time into microeconomic models is not quite as simple a task as might initially be suspected, but we do not explore this interesting topic here.

To construct our model, we first identify the relevant structural variables. We let q_1 equal the number of computers produced by firm 1, q_2 equal the number of automobiles produced by firm 2, L_1 equal the labor bought by firm 1, L_2 equal the amount of labor bought by firm 2, q_{11} equal the number of computers bought by consumer 1, q_{12} equal the number of automobiles bought by consumer 1, q_{21} equal the number of computers bought by consumer 2, q_{22} equal the number of automobiles bought by consumer 2, p_1 equal the price of one computer, p_2 equal the price of one automobile, c equal the wage for each unit of labor, π_1 equal the profit of firm 1, π_2 equal the profit of firm 2, U_1 equal the utility of consumer 1, and U_2 equal the utility of consumer 2.

Now, we need a brief description of the production, consumption, and market mechanisms. For production, we assume that firm 1 can make units of computers for units of labor purchased in such a way that its production function is $q_1 = (320L_1)^{0.5}$.

We assume a similar production function for automobiles. It is $q_2 = (640L_2)^{0.5}$. Furthermore, we assume that each consumer initially owns 40 units of labor, all of which is sold to the firms. The pay rate, c, is the same for each firm, so that the income of each consumer is given by $I = 40c$. We assume that consumer 1 is partial to computers. Consumer 1 derives satisfaction from consumption of computers and automobiles according to the utility function $U_1 = (q_{11})^{0.75}(q_{12})^{0.25}$. Also, we assume that consumer 1 spends all of the income earned. Therefore, the purchase of computers and automobiles will be subject to the budget constraint for consumer 1. This is $I = 40c = p_1 q_{11} + p_2 q_{12}$. Consumer 2 prefers automobiles and has utility and budget equations that are given by $U_2 = (q_{21})^{0.25}(q_{22})^{0.75}$ and $I = 40c = p_1 q_{21} + p_2 q_{22}$.

In equilibrium, the quantities of labor, computers, and automobiles supplied must equal the respective quantities that are demanded. This is a simple statement of the definition of market, or supply–demand, equilibrium. In the market for labor, the total quantity purchased by the two firms must be equal to the quantity provided by the consumers. This is represented by the factor market clearing equation, which is given by $L_1 + L_2 = 40 + 40 = 80$.

In a similar manner, the quantities of computers and automobiles purchased by the consumers must equal the quantities produced by firms 1 and 2. Thus the commodity market clearing equations are $q_1 = q_{11} + q_{21}$ and $q_2 = q_{12} + q_{22}$.

Now, we can easily set up the optimization problems that must be solved to determine the quantities of goods and resources that will be exchanged when each firm and consumer is maximizing its own profit and utility. Each firm will maximize profit subject to the relevant production constraint for that firm. Each consumer will maximize his own utility, subject to the relevant budget constraint. We have four optimization problems.

The problem of firm 1 is to produce computers in such a quantity as to

$$\begin{array}{ll} \text{Maximize} & \pi_1 = p_1 q_1 - cL_1 \\ \text{Subject to} & q_1 = (320 L_1)^{0.5} \end{array}$$

The profits of firm 1 are the value of the computers sold minus the cost of the labor required to produce them. In a similar manner, firm 2 can make units of automobiles from each unit of labor in such a way that its objective is to

$$\begin{array}{ll} \text{Maximize} & \pi_2 = p_2 q_2 - cL_2 \\ \text{Subject to} & q_2 = (640 L_2)^{0.5} \end{array}$$

The problem of consumer 1 is to

$$\begin{array}{ll} \text{Maximize} & U_1 = (q_{11})^{0.75}(q_{12})^{0.25} \\ \text{Subject to} & p_1 q_{11} + p_2 q_{12} = 40c \end{array}$$

and the problem of consumer 2 is, similarly,

$$\begin{array}{ll} \text{Maximize} & U_2 = (q_{21})^{0.25}(q_{22})^{0.75} \\ \text{Subject to} & p_1 q_{21} + p_2 q_{22} = 40c \end{array}$$

Also we have two sets of clearing equations. These are the commodity market clearing equations $q_1 = q_{11} + q_{21}$ and $q_2 = q_{12} + q_{22}$. The factor market clearing equation is given by $L_1 + L_2 = 80$. Taken together, the last four maximization relations and the associated seven equality constraints comprise the optimization problem. Solution of this optimization problem yields the microeconomic equilibrium results for this simple economy in terms of the model used for the economy.

Solution of this optimization problem is relatively straightforward. We first find the goods bundles that maximize the utility of consumers 1 and 2. We obtain, where capital letters are used for optimum values, $Q_{11} = 30c/P_1$, $Q_{12} = 10c/P_1$, $Q_{21} = 10c/P_2$, and $Q_{22} = 30c/P_2$. We use the commodity market clearing equations to obtain, from the foregoing equations for the total amounts of computers and automobiles that are produced, the relations $Q_1 = Q_{11} + Q_{21} = 40c/P_1$ and $Q_2 = Q_{21} + Q_{22} = 40c/P_2$.

As we should intuitively expect, these consumer demand relations show that the quantity of each product that is demanded by the consumers will decrease with increasing price of the product. To obtain maximum profit for each firm, we find the quantity produced such that the partial derivatives of the profit of each firm with respect to the price for the product of that firm is zero. We obtain, in a very straightforward manner, the optimum relations for the total amount of computers and automobiles that are produced. This is $Q_1 = 160 P_1/c$ and $Q_2 = 640 P_2/c$. As we would likely have intuited, the quantity of each product that is produced increases with the price that the firm can get for it.

We note that the first two relations that we obtain for Q_1 and Q_2 are the demand relations, whereas the latter two relations are supply relations. Economic equilibrium requires that the two equations for each quantity produced lead to the same quantities of the two individual goods. We obtain, then, as conditions for economic equilibrium $c = 2p_1 = 4p_2$. This simply tells us that either one of the prices or the labor wage can serve as a numeraire or normalization term. We should not expect otherwise, as a little thought will show. We may as well let the wages for labor, c, equal 50,000, such that we obtain $P_1 = 25{,}000$ and $P_2 = 12{,}500$.

We have already incorporated the factor market clearing equation into the equilibrium relations. This occurred when we wrote the budget constraint for each consumer. All of the necessary relations in the following list have now been obtained.

1. The equations for optimality of the firm.
2. The equations for optimality of the consumers.
3. The commodity market clearing equations.
4. The factor market clearing equations.

The final result is that we obtain the equilibrium relations $Q_{11} = 60$, $Q_{21} = 20$, $Q_{12} = 40$, $Q_{22} = 120$. So, it turns out in this rather simple and contrived example that we produce a total of 80 computers and 160 automobiles. 60 computers are purchased by consumer 1, and 20 by consumer 2. On the other hand, consumer 1 purchases 40 automobiles whereas consumer 2 purchases 120 of them. In this particular example, we might assume that each consumer is, in reality, a consumer group of some 2,000 people each with an income of $25 per hour. This would suggest a weekly sales volume for computers and automobiles and consumption rates that are realistic.

It is interesting to compute the profit of each firm. It turns out that the profit is zero for each firm. This may, at first, be a surprising result. In reality, it is not surprising since in a more realistic example, we would have included such important ingredients as the cost of capital and the related effect on profit. This is really just another way of saying that we need to consider the cost of equity capital. Also, we would need to consider taxes, exports, and a host of other factors. We could, for example, use a minimum finite profit as an operating condition. Nevertheless, this simple model and the results obtained from it are quite useful in providing some insight into economic behavior.

8.1.3. Perfectly Competitive Economic Conditions

Many models and results in economic systems analysis assume what are called perfectly competitive economic conditions. There are many microeconomic books that deal with this topic, including [2–6]. Six critical assumptions establish the requirements for a perfectly competitive economy.

1. There are a large number of relatively small producers and consumers. The assumption that producers are small is needed to insure that no single firm is big enough to establish control of a market and establish a monopoly.

2. All firms in the same industry produce similar goods of comparable quality. No consumer has any reason to prefer the output of one firm over that of another. Products and services are completely standardized among firms making the product or service. There is no brand loyalty among consumers.

3. Resources are completely mobile. The owners of the productive resources of land, capital, labor, or other inputs to production, are free to put them to whatever use they believe will yield maximum return. People can work in, or sell their resources to, any industry they wish. No barriers exist to prevent any firm from leaving an industry, or becoming a part of another one.

4. Each economic agent has perfect information and knowledge. All firms and all consumers know with certainty all present and future prices. Each knows its own characteristic production functions and utilities. There are no uncertainties.

5. Each economic agent, regardless of whether the agent is a producer or a consumer, is an optimizer. Each agent acts to maximize their own satisfaction. Thus all consumers act to maximize their utility, and all firms act to maximize their profits.

6. There are no price controls. Prices may move up or down freely, subject only to the market pressures of a perfectly competitive economy.

If the above six conditions hold, it is easy to show that prices and the quantity of goods produced and consumed are determined by the market supply-demand equilibrium and that all goods are produced and sold at the lowest possible price.

In practice, not all of these six conditions will always exist. When condition 1 does not exist, a monopoly or monopsony will result. Firms will have knowledge of how prices will vary with demand and they can control the entire quantity produced (monopoly), or suppliers of labor can exert a similar control (monopsony). When condition 1 does not hold, we need to consider similar products as if they are separate, and develop different supply and demand relationships for the different brands of a product. The nonexistence of condition 2 gives rise to the profession of marketing.

Taken together, conditions 1 and 2 paint a picture of a large number of producers, such as family farms, producing and selling goods to a vast market. This is an ideal situation for the flourishing of perfectly free competition, but it is not essential to competitive marketplaces. For example, huge agribusinesses like Cargill and ConAgra currently operate in intensely competitive world grain markets. And highly advanced product differentiation, such as occurs in numerous consumer products markets, is perfectly compatible with free markets. As long as the consumer sees competitively priced products (automobiles, shoes, cosmetics, etc.) in the particular consumer category of interest, the consumer is operating in a competitive market, even if there are only two producers of such shoes.

Monopoly situations can frustrate economic efficiency in several ways. First, competitive pressures to hold down costs or desire to improve product or service quality are often lightened or eliminated, thereby driving up prices. (Many would suggest that the U.S. Post Office is one example of such a monopoly.) Second, a monopoly or oligopoly may consciously choose to cut production below the level that would maximize its profits for any number of reasons. For example, a labor union may go on strike to enhance its bargaining power during contract negotiations. An oil cartel or oligopoly may choose to limit total oil production in an effort to stabilize prices or to shift advantage away from cartel members with low-cost production (e.g., Saudi Arabia) toward cartel members with high-cost production (e.g., Egypt).

As an alternative to limiting output, monopolies and oligopolies may fix prices at levels higher levels than would be implied by the intersection of supply and demand curves. Although such tactics can often achieve short-run surplus profit, in the long run the consumers will redirect their attention to more competitively priced products.

Eight additional assumptions characterize the abstract concepts of a perfectly competitive marketplace.

7. Individual consumers are totally self-interested, and seek to maximize their bliss or utility. Self-interest is a more subtle, far-reaching motivation than is often appreciated, however. For example, buying a new car or adding a new rear porch to keep up with the Jones' is based on what some people might call the irrational motive of envy. Yet it certainly is also an expression of economic behavior that is driven by self-interest. To give another example, charitable giving may be thought of as a form of self-interest because it affords a feeling of self-satisfaction and may be perceived to contribute to one's physical or economic security, as in the case of giving aid to unemployed workers in an effort to discourage rioting. This is a multiattributed issue.

8. Individuals are greedy. More is always better or, at least, never worse. A person never reaches the satisfaction point and always feels better

off by consuming more. We have already pointed out that people may well substitute leisure time for consumption of goods. And, of course, common observation shows that as people's incomes rise, they tend to save more and may contribute more to worthy causes, instead of continuing to increase their consumption from choices among a re-stricted set of goods.

9. Preferences are such that the rate at which additional amounts of product *x* may be substituted for another product *y* to retain the same level of utility diminishes with increasing amounts of the product *x*. That is to say, diminishing marginal rates of substitution between goods or products exist. Indifference curves of constant utility are convex to the origin.

10. There are no externalities, such as in cases where the increased production of one firm results in decreased production ability for another firm.

11. There are no production processes that exhibit increasing returns to scale in the sense that increased production quantity will always result in increased profits. In other words, production of a given product is increased to the point where the marginal return is zero. Thus, the demand curve is just as important as the supply curve. A firm continues to produce a good until the marginal revenue equals the marginal production cost. Marginal revenue tends to decline with each additional unit, almost by the definition of a demand curve. Marginal production costs are usually more difficult to identify, however. Up to a point, economies of scale tend to drive down the marginal cost of a product. Beyond that point, diseconomies of scale may set in, and these will drive up the marginal cost of each additional unit.

12. All goods and services are exchanged in markets, and all markets are in steady-state equilibrium.

13. There are no public goods. There are only private goods that are consumed by a single individual or household.

14. There is neither government taxation nor government subsidization of the production or consumption of any goods or services.

If each of these 14 conditions are satisfied, we may establish the very important result that all goods and services have market prices and that the market prices are exactly equal to the corresponding shadow prices, such that the shadow price and the market price have true social values. This is a very important result of a perfectly competitive economy. *Shadow price* is a relatively important concept in theoretical microeconomics. It is somewhat difficult to provide a simple definition of a shadow price. They are the differential amounts by which consumer utility changes with respect to increases in consumption qualities. Mathematically, they are Lagrange multipliers associated with the consumer income constraints in the optimization of consumer utility in a welfare economic optimization.

A major assumption of microeconomic theory is that buyers and sellers have behavior that tends toward an equilibrium. For example, if the demand for a good is strong, sellers tend to raise its price, and, if demand is weak, sellers tend to reduce the price. This mechanism of groping for stable prices is known as *tatonnement*. A fundamental issue is whether such an equilibrium can be stable and will it be a unique equilibrium, unique in that there will only be one equilibrium point. Even if the equilibrium is stable and unique, will it be reached in any finite time? Dynamic economic models are needed to answer these questions. The generally accepted answer is that any realistic economic model will exhibit a stable equilibrium.

Another issue raised is whether the equilibrium of an economy evolves over time, or tends to jump from one stable equilibrium point to another. For example, did the 1970s oil scares jerk our economy from one equilibrium to another, or did they simply represent faster than normal change in a single stable equilibrium point over time? Either view is possible, although the latter seems more in keeping with most economic thought.

8.1.4. Welfare Economics

Welfare economics is that branch of microeconomics that addresses resource allocation, commodity distribution, and economic organization with an aim to maximize economic welfare for the economic system as a whole [6–8]. As such, welfare economics might also be called *normative* economics. In deciding what distribution of resources among the members of a society is most desirable or equitable, value judgments must be made. Microeconomics is primarily concerned with economic efficiency and economic effectiveness. Normative or welfare economics is concerned also with equity and value judgments that, presumably, reflect equity considerations in economic systems.

In its purest, or perhaps impurest, form, welfare economics is an effort to quantify and compare a diverse set of preferences reflecting the different perspectives of diverse segments of the population, particularly as regards public sector expenditures, regulations, and taxes. Whereas normative, or equity, considerations are often involved, they do not have to be the defining element of welfare economics. For example, a cynical politician could use a welfare economics model to calculate how to maximize the political values of a particular spending program, without having the remotest interest in equity or fairness to anyone. Presumably, however, this is not the intent of normative economics.

The economic systems analysis basis for welfare economics consists of sets of mathematical relations that describe the following.

1. The satisfaction derived by each consumer from possession or consumption of given quantities of goods and resources.

2. The productive capacity of the economic system for transforming economic resources into consumable goods.
3. The level of general economic welfare associated with each distribution of goods and resources among the individual participants in the economic system.

These relations can be imbedded into an optimization problem. The solution of this problem will determine a resource allocation and commodity distribution pattern that yields maximum economic efficiency, maximum economic effectiveness, and maximum equity according to the specified welfare function. It turns out that it is not possible to correctly optimize equity independently of efficiency and effectiveness concerns. This is important and, sadly, not always acknowledged in practice by welfare policies.

The intent of welfare economics is to determine policies that will enhance or increase the general economic welfare. Examples of application areas include policies pertaining to income distribution among citizens, allocation of raw materials among alternative productive uses, sustainable development, and government funding of public projects. A welfare, or normative, economics study includes these typical final results.

1. A quantitative model describing the allocation, utilization, and consumption of economic resources.
2. A mathematical relation that assigns an index of social desirability or general economic welfare to each resource allocation and commodity distribution pattern in the economic system.
3. A determination of the economic configuration for which the social welfare function will attain a maximum value.
4. A determination of economic decisions that will result in maximum satisfaction for individual consumers.
5. Determination of how each consumer's satisfaction contributes to overall social welfare.
6. Increased understanding of the workings of the economic system under consideration, and increased social welfare.

All of this seems very desirable. However, there are major requirements for large-scale welfare economic models that are sometimes quite difficult to accomplish in practice.

1. A set of utility functions that represent the tastes and preferences of each consumer with respect to possession or consumption of given quantities of services and products.
2. An aggregate production function that represents the total productive capacity of the economic system.

3. A social welfare function that places the various levels of utility attained by individual consumers into an overall measure of social, or normative, economic welfare.

4. Specification of information sources on which the welfare economic model is based.

Unfortunately, these needs combine to make it very difficult to formulate or solve realistic normative economics problems through optimization-based approaches. The fact that the values of individuals in society are very different, and are often labile, adds to these difficulties. Nevertheless, the concept is an appealing one.

There are several steps that might be undertaken to construct a normative or welfare economics model. The first step is to identify the components of the economic system under consideration. These components will generally include the following.

1. A set of commodities or final consumable goods.

2. A set of economic resources used as factor inputs to the production of final goods.

3. A set of consumers who sell resources or labor to firms and who buy final goods for consumption.

4. A set of firms who produce final goods using factor inputs.

The construction of a normative economics model will also require determination of consumer preferences, production functions of the firms in the economy, and an overall economic welfare function.

In such a model, each firm will produce final goods from factor inputs, subject to constraints on the firm's technology and productive capacity. The production functions of all the firms in the economic system can, in principle, be combined to form an aggregate production function that describes the total productive capacity of the system. This aggregate production function gives the maximum quantity of consumable goods that the system can produce for given amounts of available resources. The utility functions of all consumers in the economic system are used to determine a *social welfare function*, which incorporates the diverse values and preferences of all consumers, and thus depends upon very subjective value judgments. The premise is that social welfare is maximized and the distribution of resources and goods among consumers is the most *equitable*. These augmentations to the model of the firm and the consumer are needed to develop the welfare economics model.

The next step is to use this welfare economics model and solve the resulting mathematical optimization problem that has been formulated. We desire to maximize the social welfare function, subject to the productive capacity constraint represented in the aggregate production function. The result is often called the *welfare optimum*.

8.2. ECONOMIC DISCOUNTING FOR SYSTEMS ENGINEERING ECONOMICS ANALYSIS

In the first section of this chapter, we gave a brief overview of microeconomic theory, including the theory of how firms seek to maximize profits. Profit maximization involves numerous facets of enterprise management, including such matters as product development, sales, marketing, personnel, and operations. In previous chapters, we have addressed some of these issues, such as maintaining quality in the production process and promoting commitment to excellence among employees. Here we look at some issues necessary for making investment decisions. Section 8.2 discusses techniques for present-value analysis of investment options. In Section 8.3, we look at the related and broader issue of cost–benefit and cost–effectiveness assessment.

Present-value analysis is a technique for evaluating the merit of a potential investment. For example, suppose the purchase by a firm of a $60,000 machine would generate annual returns of $24,000 for 4 years. The firm estimates that an annual percentage return of 18% is required to justify the investment, given its risks. In present-value analysis, the firm calculates what amount would need to be invested at 18% to yield the four-year income flow of $24,000. A fairly simple calculation indicates that the figure is $64,560. This is the amount that is at the present time, able to generate the income stream produced from the machine. This amount is more than the $60,000 cost of the machine. Such an investment is therefore worthwhile from the perspective of this simple analysis.

The remainder of this section is devoted to an algebraic exposition of present value analysis, and related engineering economic systems subjects.

8.2.1. Present and Future Worth

When we deposit an amount P_0 into a savings account at year 0 and the annual interest rate is i, then we earn iP_0 in interest in 1 year, and have available at the end of the first year the principal plus interest, or

$$F_1 = P_0 + iP_0 = P_0(1 + i),$$

where F_1 is the future value at year 1. If this amount is invested for the next year, we have

$$F_2 = P_1(1 + i) = F_1(1 + i) = P_0(1 + i)^2$$

at the end of the second year. At the end of N years we have accumulated a grand total amount of

$$F_N = P_0(1 + i)^N$$

Stated in present value terms, denoted by $P_{0,n}$, this may be written as

$$P_{0,n} = A_n(1 + i)^{-n}$$

where A_n is the amount of money invested at different points in time and the constant interest rate over the time interval is i. We simply add these amounts for all n under consideration and then obtain for the present value of a series of investments

$$P_0 = \sum_{n=0}^{N} P_{0,n} = \sum_{n=0}^{N} A_n(1 + i)^{-n}$$

Of course there is no need for the interest to remain constant from year to year, just as there is no need for the amount invested to remain constant. Should the interest rate from year k to year $k + 1$ be i_k, we then have

$$P_{0,n} = A_n \prod_{k=0}^{n-1} (1 + i_k)^{-1}$$

$$P_0 = \sum_{n=0}^{N} P_{0,n}$$

This expression represents the present worth of the several amounts A_n invested for n years, where the interest rate varies from year to year. Figure 8.4 illustrates three typical cash flow situations.

In some cases the annual amounts of a benefit or a cost are constant over time. Then, if the discount rate is constant over the period under consideration, relatively simple closed-form expressions for present worth result. Many ways can be used to obtain these. One simple way is to note that the present worth, at time 0 of an amount 1 that is initially invested at annual interest rate i beginning at year 0 can be obtained from the future worth equations

$$(1 + i)^{-1} + (1 + i)^{-2} + \cdots + (1 + i)^{-n} + \cdots = i^{-1}$$

This follows in a very simple manner from the identity

$$a/(1 - a) = a + a^2 + a^3 + \cdots$$

which will converge for $a < 1$. Here, we let $a = (1 + i)^{-1}$. The present value at time N of an amount 1 invested annually from time $N + 1$ is also i^{-1} and the present value at time 0 of this amount is $i^{-1}(1 + i)^{-N}$. We subtract this from the expression i^{-1} and obtain the expression for the present worth, at time 0, of an amount 1 invested annually for N years.

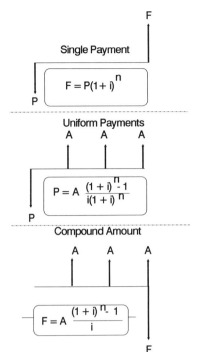

Figure 8.4. Illustrations of Simple Economic Discounting

We find that the present discounted worth P_0 of a constant amount A invested at each of N periods from $n = 1$ to N is given by

$$P = A\left[(1 + i)^N - 1\right]i^{-1}(1 + i)^{-N}$$

We may easily calculate the future worth, at the Nth period, of annual payments from this relation as the expression

$$F = P(1 + i)^N = A\left[(1 + i)^N - 1\right]i^{-1}$$

where we now drop the subscripts on P and F for convenience.

Relations such as these are especially useful for calculating the amount of annuity payments that an initial principal investment will purchase or for calculating constant-amount mortgage payments. We must, however, remember that other factors such as inflation and depreciation, or appreciation, need to be considered when evaluating the present or future worth of a project. For the most part these can be considered just as if they were interest. For example, the worth at year $n + 1$ of an investment at year n of P_n that is subject to interest i, inflation r, and depreciation d is given by the expression

$$P_{n+1,n} = P_{n,n}(1 + i_n)(1 - r_n)(1 - d_n)$$

where i_n is the interest in year n, r_n is the inflation in year n, and d_n is the depreciation in year n. If we assume that this is augmented by an amount A_{n+1} in year $n + 1$, then the worth of the investment at the start of the next period is

$$P_{n+1,n+1} = P_{n+1,n} + A_{n+1}$$

or

$$P_{n+1,n+1} = P_{n,n}(1 + i_n)(1 - r_n)(1 - d_n) + A_{n+1}$$

We can solve this difference equation with arbitrary i_n, r_n, d_n, and A_{n+1} to yield the future worth of any given investment path and annual investment over time.

In the simplest case, the initial investment $P_{n,n}$ is zero and the annual investment and rates are constant. Then it is a simple matter to show that the result of making an investment of amount A over N years is given by the future monetary amount at year N

$$F = P_{N,N} = \left[(1 + I)^N - 1\right] A I^{-1}$$

where I is the effective interest rate which is given by

$$I = i - r - d + rd - ir - id - ird$$

If the percentage rates are all quite small, little error results from using the approximation obtained by dropping the products of small terms. We then have for the foregoing effective interest rate

$$I \approx i - r - d$$

such that we simply use an effective interest rate that is the true interest rate less inflation and depreciation.

8.2.2. Economic Appraisal Methods for Investments Over Time

Several methods can be used to yield economic values of a project. One is present-value analysis, as described above. Another method, payback period calculation, does not consider the time value of money at all. It simply determines the time required from the start of a project before total revenues flowing from the project start to exceed the total cost of the project. As can be easily shown, the payback period is a rather naive criterion to use in evaluating projects.

Another common method of project evaluation is called the *internal rate of return* (IRR). This approach determines the interest rate implied by anticipated returns on an investment, and then compares that rate with some predetermined standard of measure. It is that interest rate that will result in economic benefits from the project being equal to economic costs, assuming that all cash flows can be invested at the internal rate of return. The assumption that it is possible to invest cash flows from the project at the internal rate of return will often be impossible to fulfill. Thus the number that results from the IRR calculation may give an unfortunate impression of the actual return on investment. It tends to favor short-term projects that yield benefits quickly, as contrasted with projects that yield long-term benefits only, because of this reinvestment at the constant internal rate of return assumption that is inherent in this equation.

In some cases, IRR is simply the inverse relation of *net-present-value* (NPV) analysis. NPV starts with an assumed interest rate, solves for the present value of the income stream, and compares this figure with up-front cost. IRR starts with an estimated income flow, solves for the interest rate this flow implies, and compares this interest rate with a predetermined standard of comparison. Assuming the forecasts are done honestly, the calculations are done accurately, and that monetary returns received during the period that the investment is active can be reinvested at the IRR, which is to be calculated over the remaining life of the project, the two approaches represent equally legitimate tools of analysis. The choice between them should depend on the peculiar circumstances of the firm and the investment it is considering and whether this reinvestment at the IRR assumption is acceptable.

The IRR may be calculated fairly easily. If we assume that a project, of duration N years, has benefits B_n in year n and costs C_n, then the present value of the benefits and costs are, assuming that the interest rate is constant,

$$PB_0 = \sum_{n=0}^{N} B_n(1 + i)^{-n}$$

and

$$PC_0 = \sum_{n=0}^{N} C_n(1 + i)^{-n}$$

The net present value of the project is given by

$$\mathrm{NPV} = PB_0 - PC_0 = \sum_{n=0}^{N} (B_n - C_n)(1 + i)^{-n}$$

To obtain the internal rate of return, we set the net present worth equal to zero and solve the resulting Nth-order algebraic equation for the IRR.

These equations lead naturally to a consideration of the *benefit–cost ratio* which is the ratio of benefits to cost and is given by

$$\text{BCR} = \frac{PB_0}{PC_0} = \frac{\sum_{n=0}^{N} B_n (1+i)^{-n}}{\sum_{n=0}^{N} C_n (1+i)^{-n}}$$

The *return on investment* (ROI) is the ratio of the net present value to the net present costs and is therefore given by

$$\text{ROI} = (PB_0 - PC_0)/PC_0 = \text{BCR} - 1$$

Sometimes the ROI criterion is called the *net benefit to cost ratio* (NBCR) or the net present value, or net present worth to cost ratio.

It is of considerable interest to contrast and compare these techniques to determine those most appropriate in particular circumstances. We first consider the selection of a single project from several. An immediate problem that arises is that NPV and IRR may lead to conflicting results. The BCR criterion may result in a different ranking also. Unless costs are constrained, there is no real reason to use a BCR criterion, however.

Two projects, *A* and *B*, may have the NPV-versus-interest rate curves shown in Figure 8.5. Here we see that if the actual interest rate *i* is less than *I*, the rate at which the NPV of the two investments are the same, we prefer investment *A* over investment *B* if we use the NPW criterion. If the interest rate *i* is greater than *I*, we prefer investment *B*. However, if the actual interest rate is greater than the IRR for each investment, we would prefer not to invest at all, if this is possible, since the NPW of each investment is negative. The IRR of project *A* is IRR_A and that of project *B* is IRR_B. Since IRR_B is greater than IRR_A, we would prefer project *B* to project *A* by

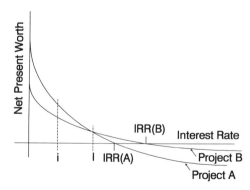

Figure 8.5. Illustration of Potential Difficulties with IRR as a Criterion

the simplest IRR criterion. This simple example illustrates some of the potential difficulties with careless use of IRR as an investment criterion.

There are several essential properties or assumptions that any rational investment rule or criterion should satisfy. All cash flows should be considered. The cash flows should be discounted at the opportunity cost of capital, that is, the interest rate that can be obtained on an investment or some specified discount rate. The rule should select, from a set of mutually exclusive projects, the one that maximizes benefit or value as appropriately defined. Also, we should be able to valuate one project independently of all others. This requires that the investment criterion be complete. By complete, we require that we can always say that **x** is not preferred to **y*** or **y** is not preferred to **x** on the basis of any two sequences of cash flows **x** and **y**. We also require that the criterion be transitive. With three cash flow vectors **x**, **y**, and **z**, if **x** → **y** and if **y** → **z**, then transitivity requires that **x** → **z**.

There are five desirable properties that should be possessed by any reasonable preference ordering of cash flow vectors. The first of these is the property of *continuity*; that is if **x** is preferred to **y**, which we write as **x** → **y**, then for sufficiently small $\epsilon > 0$, cash flow vector **x** − ϵ is also preferred to **y**, or **x** − ϵ → **y**. This continuity property is very reasonable in that it assures us that the preference criterion and associated preferences are not schizophrenic for small arbitrary changes in the return of the investments. Also, we need a property associated with *dominance*. If cash flow **x** is at least as large as **y** at every period in the overall investment interval $(0, N)$ and is strictly greater in at least one period, then investment **x** is preferred to **y**, or **x** → **y**. This dominance property is equivalent to greed, since more is always preferred to less, or consumer satiation.

Preferences should also be based on the *time value of money*. This suggests that if two investments **x** and **y** are identical except that an incremental cash flow obtained by investment **x** is at some period n does not result until period $n + 1$ for investment **y**, then investment **x** is preferred to investment **y**. The time value of money investment property is equivalent to impatience. We would prefer to have an incremental amount of money now than to have precisely this same amount at some future time.

Also, we prefer cash flow **x** to **y** if and only if the differential cash flow **x** − **y** = $[x_0 - y_0, x_1 - y_{12}, \ldots, x_N - Y_N]^T$ is preferred to a cash flow of **0** = $[0, 0, \ldots, 0]^T$. This property is called *consistency at the margin*. It is relatively easy to see the reasonableness of this criterion from the satiation of the consumer property. There is also a *consistency over time* property. This simply says that the preference order over investments is unchanged if we shift all investment returns being considered by exactly the same number of periods. In making this shift, we must be careful that we do not lose any investments. The consistency over time property simply states that if **x** = $[x_0, x_1, \ldots, x_{N-1}, 0]^T$ and **y** = $[y_0, y_1, \ldots, y_{N-1}, 0]^T$ and if **x** → **y**, then we

---

*x is not preferred to **y** if **y** ≈ **x** or if **y** → **x**.

must have $\mathbf{x}' \to \mathbf{y}'$, where $\mathbf{x}' = [0, x_0, x_1, \ldots, x_{N-1}]^T$ and $\mathbf{y} = [0, y_0, y_1, \ldots, y_{N-1}]^T$. The consistency over time requirement is based on the interest or discount rate being constant throughout the investment time interval 0 to N.

If we accept these five requirements as reasonable, then it turns out that we can only accept the NPV criterion, or one that would yield the same preference ordering as the NPV criterion, as truly reasonable. It is possible to show that only preferences that satisfy these requirements are those given by the net present value criterion in which interest or discount rates, which may vary from period to period, are positive. Surely, it would be difficult to argue that any of these requirements are unreasonable. Although this discussion and the five requirements, which could be posed as axioms, appear quite abstract, it is just this sort of formulation that leads to normative, or axiomatic, theories.

Based on these requirements and the conclusion that follows from them, we should value investment $\mathbf{x}$ according to a net present value criterion, or decision rule, in which we obtain the present value of the investment component at the nth period x_n by use of the standard discounting relation

$$\text{NPV}(x_n) = x_n \prod_{j=1}^{n} (1 + i_{j-1})^{-1}$$

and then add together all of these present values over all of the interest bearing periods to obtain

$$\text{NPV}(\mathbf{x}) = \sum_{n=0}^{N} \text{NPV}(x_n)$$

or

$$\text{NPV}(\mathbf{x}) = \sum_{n=0}^{N} x_n \prod_{j=1}^{n} (1 + i_{j-1})^{-1}$$

where i_{j-1} is the interest rate that is valid in the time interval from the period $j - 1$ to the period j.

8.2.3. The Discount Rate

In our discussions thus far in this section, we have assumed that there is a single market rate of interest, which we have called the discount rate or the opportunity cost of capital. However, in reality, there is no such single rate. Several related approaches to determining a discount rate are possible: marginal interest rates, marginal time preference rates (MTPR) for individuals, corporate discount rates, minimum attractive rates of return (MARR), government borrowing rates, and the social opportunity costs of capital.

Other key distinctions are also important, such as the difference between debt and equity. Since corporate debt payments are deductible from income for tax purposes, corporate interest rates need to be doubled to be compared with public sector interest rates. Also, we are discussing hard discount rates and potentially soft discount rates. Rates that actually have to be paid with real money by government, business, or individuals are said to be *hard* discount rates. *Soft* discount rates involve speculations about social or personal discount rates. These do not necessarily involve real money and are, instead, a kind of welfare economics market research rate. A potential problem is that one may be called to pay, in hard currency or real money, for expenditures that have been justified by unrealistic *soft* rates. Yet, there is an inevitable need for trade-offs among consumption now and consumption later and for assumption of a discount rate. Many acknowledge skepticism about the notion of social discount rate. It is doubtlessly correct that many involved in public policy debates have never heard a serious political statement regarding a public sector investment that invoked the concept of social discount rate. Nevertheless, the concept does have useful value.

Corporate discount rates include a premium for risk, a markup for corporate taxes, and a profit to be returned to shareholders. If government bonds yield 10-percent interest and the corporation desires to pay a 4-percent premium to stockholders in order for the assumed stockholder risk to be reasonable, then the corporation will have to obtain almost a 28-percent annual return on investment in order to be able to provide a 14-percent return to the stockholders and pay a 50-percent tax on corporate profit. Many feel that this is quite high and argue for a capital gains tax that is lower than that on ordinary income, which is not subject to the double taxation rules extant in the United States to partially offset this difficulty.

Market interest rates for government and corporate bonds generally vary with the perceived risk that is associated with the investment. We can compute a market value of outputs, perhaps as a function of industry, and the inputs to the project, all measured in dollars and discounted in time. Then, if a return of $1,120 results from an investment of $1,000, we would say that the market interest rate, which would be the realized MARR, is 12 percent. This would need to be doubled by corporations because of the tax they must pay on profits before distributing them to shareholders.

The individual marginal time preference rate or personal discount rate is the rate at which an individual is willing to trade off present personal consumption for future personal consumption. Related to this is the concept of a personal social discount rate. The personal discount rate represents personal deferred consumption preferences. The social discount rate of an individual represents that individual's preferences in terms of future consumption versus present consumption. Many modifications to rates such as these can be made. One might believe that individuals will set too high a personal social discount rate in order to enjoy consumption today at the expense of future generations. Thus one might argue, as does the Pigouvian* discount rate, for a social discount rate that is lower than the personal discount rate of present-day individuals. Several items, many of which relate to questions of sustainable development, are of importance:

1. Evaluation of long-term public sector investments needs to consider other than strictly financial issues. Among the most important is that of generational trade-offs: How willing am I to pay taxes for multidecade public works that will benefit primarily my children and not myself? This equity issue requires that a social discount rate be used to help determine the present value of such projects. This is the case since actual discount rates will not be known long into the future. Even if they are, the effective social discount rate can be used as a figure of merit and metric from which the worth of a project may be judged.

2. Evaluation of public sector investments also needs to consider how current consumers, or taxpayers, balance their desire for current consumption (i.e., not paying taxes for public works projects) with their desire for future benefits that might be produced by the project in question. This is far less of an equity issue than it is a matter of how to accurately determine the today-versus-tomorrow preference curves of current consumers/taxpayers.

3. Evaluation of public investment must not lose sight of the practical question: Will the project in question continue to attract public financing as work on it progresses? History is replete with public works that one group of legislators or administrators consider worthwhile, perhaps through the use of a low social discount rate, but that subsequent legislators abandon even when there have been no major changes in the conditions that once supported the initial decision. Thus, when employing a social discount rate, it seems reasonable not to deviate too far from market rates. If benefits are discounted at too low a rate, an erroneous inference of ability and willingness to pay for the project in the future is obtained, particularly in the case of groups not identified as stakeholders in the current cost–benefit analysis.

4. However useful it may be to apply a social discount rate to public sector projects, there is no merit in applying such a concept to private sector projects. Discount rates for private investments should be market driven, since this is the only way to secure financing in our free market system. Even in public sector projects, assumption of a high social discount rate for expenditures made now will result in projects that will require major subsidies later in order to keep them economically viable.

---

*After Pigou, a noted nineteenth century economist.

A reasonably good illustration of item 4 concerns defined benefit retirement plans. A firm, or government agency, can justify very small present day contributions to retirement plans designed to pay a defined future benefit by the simple artifice of assuming a very large discount rate. Firms and state governments have used this artifice to justify small contributions to these plans, and even to justify taking money from them. Of course, retirement day will come, as will the day of reconciliation and atonement.

In summary, it is difficult to conclude other than that the social discount rate should be essentially the real cost of capital. However, the social discount rate is not useless as a guide to investment. It is simply a decision variable. If a social discount rate is set too high, then there needs to be some sort of bail out to rescue those projects that are funded through assumption of an artificially high rate that cannot, in practice, be obtained.

Thus, although social discount rates can be very useful, particularly in evaluating public sector projects, they must be used with considerable caution. In general, it would seem better for arguments to be made that important attributes of projects have been omitted and that the benefits of a project are greater than those identified by an analysis than to distort the analysis through use of an unreal discount rate. Thus the argument that the social discount rate should represent the opportunity cost of the project, that is to say, the returns that could be obtained from these funds if used on some other project, is a very potent one. In other words, resources must be used in the most productive way. To not acknowledge this is to forget, in effect, that there are always resource constraints on the use of capital. The task should be to define productivity in an appropriate way and not to use artificial discount rates to justify actions and projects that cannot pay their own way and that thereby cast the burden of payment, in an undefined and unexplored manner, on those that the artificial and distorted discount rate is supposed to benefit.

8.3. COST – BENEFIT AND COST – EFFECTIVENESS ANALYSIS

The broad goals of cost–benefit analysis (CBA), or perhaps more appropriately cost–benefit assessment, are to provide procedures for the estimation and evaluation of the benefits and costs associated with alternative courses of action. In many cases, it is not possible to obtain a completely economic evaluation of the benefits of proposed courses of action. In this case, the word benefit is replaced by the term effectiveness, and a multiattribute effectiveness evaluation is used. We may use cost–benefit analysis (CBA) and cost–effectiveness analysis (CEA) either to help choose among potential new projects or to evaluate existing systems for various purposes, such as identifying potential system modification needs.

There are several stages in a cost benefit analysis which correspond to steps in the systems engineering process. The iterative nature of these steps are very similar to the expanded systems engineering steps.

Systems Engineering Step	Corresponding CBA or CEA Step
Formulation of the Issue	
Problem definition	Identify needs, constraints and alterables
Value system design	Identify objectives for CBA or CEA
System synthesis	Identify alternative COA* or projects
Analysis of the Impacts of Alternatives	
Systems analysis	Identify costs, benefits (effectiveness)
Systems refinement	Eliminate dominated COA, iterate/refine
Evaluation and Interpretation of the Impacts	
Decision making	Select best alternative
Planning for action	Prepare CBA or CEA reports

*Course of action.

The process is illustrated in Figure 8.9. Let us now describe some of the majors efforts needed to accomplish these steps.

8.3.1. Benefit–Cost Ratio and Portfolio Analysis

In this section, we discuss calculation of the benefit-cost ratio.* Also, we indicate why and where it is an important criterion. To do this we need to introduce some concepts from portfolio analysis that concern decisions with a constrained budget. Generally, benefit–cost analysis ratios are defined in terms of discounted benefits and costs. We compute the benefit–cost ratio (BCR) as we discussed at the beginning of Section 8.2.2.

As is easily seen, we do not have to discount the benefits and costs to the present time to determine the BCR; any time will do. Nor do we need to do this to obtain the return on investment, ROI = BCR − 1. It is convenient, however, to discount benefits and costs to the present time, for then we can easily obtain the net present value, or net present worth, from $PB_0 - PC_0$. Ranking projects by benefit–cost ratio will generally lead to results different from those obtained by the use of net present worth. For example, if one project has present benefits of 10 and present costs of 2, and a second project has present benefit of 100 and present costs of 50, then project 2 is the best

*For some reason, cost–benefit analysis generally uses a benefit–cost ratio rather than a cost–benefit ratio. Perhaps this is because we like to evaluate items on a scale where a larger number is better than a smaller number. It is desirable for a benefit-cost ratio to be large and the larger the better. The potential confusion is unintentional.

in terms of its net present worth of 50 but project 1 is best in terms of its BCR of 5. Project 1 is also best in terms of its ROI of 4, or 400%, whereas the ROI in the second project is 100%.

8.3.2. Identification of Costs and Benefits

Identification and quantification of the benefits and costs of possible alternative courses of action is a difficult task, although generally not quite as difficult as formulating meaningful alternatives themselves. Here we use the word benefits to mean the possible overall effects that result from implementation of a project. These include both positive and negative benefits or disbenefits. We must first identify benefits and then assign quantitative values to them. Many benefits (and disbenefits) are intangible and accrue to differing groups or individuals in differing amounts, especially in public sector projects. Problems with intangibles are especially difficult to deal with in the public sector, where agencies are designed primarily to deliver services or public goods, rather than products for individual consumption. A major goal of a classic private sector organization is profit maximization and it is relatively easy to measure profit as a benefit. The benefits of a public service, such as a school, or a public good, such as a subway system, are much more difficult to define because they are intangible or indivisible (or both).

Two very different concepts are introduced here: that of intangible benefits, and that of benefits accruing differently to different individuals. It might seem that intangible benefits are inherently more likely to be associated with public goods, and tangible benefits with private sector goods. But this is not necessarily so, and to assume it may seriously confuse issues. Take the field of transportation, for example. Public transit is an example of a public service whose benefits are quite difficult to define because they are intangible or indivisible. But what about private sector airplane transportation, bus transportation, or toll roads? To take other examples, both public and private sectors provide various forms of insurance; both public and private sectors provide various forms of security services; both public and private sectors provide recreational facilities. As our economy incorporates more service economy features that enhance the value of output from production, the private sector will find itself more and more providing intangible benefits. In no way will this, or should it, prevent application of market-based pricing concepts. And, information value is primarily intangible.

One valuation philosophy that we might adopt is based on two suppositions. The first of these is that the value of a project to an individual is equal to the fully informed willingness of the individual to pay for the project. The second is that the social value of a project is the sum of the values of the project to the individual members of society. The fundamental conclusion of microeconomics, under perfect competition conditions, confirms that these two suppositions or premises should, under ideal conditions, serve as good guideposts for cost–benefit analysis.

The full information called for in the first supposition is almost never available. The use of pricing as the basic measure of value may produce the sort of equity concerns that were addressed in Section 8.1.4 on welfare economics. Measurement problems also arise in determining the value of a complex project, such as a large sewage disposal system serving residential, commercial, and industrial users, or a wide-area computer network upgrade for a giant corporation. Usually, decomposition of the benefit of such a project into attributes and assigning of numerical values to these attributes produces a more reliable indication of the value of a project than unaided intuitive judgment, especially when dealing with unfamiliar issues.

The implementation of any project alters the supply of inputs, consumed in the production process, and the supply of outputs, resulting from the project. Such changes may, in turn, affect the ultimate value of the project.* Another measurement problem with cost–benefit analysis arises in trying to foresee contingencies. CBA may be a relatively simple technique when we are evaluating operational-level projects, such as the introduction of word-processing equipment in an office to replace electric typewriters. But many projects are predicated on substantial changes in external environments, such as population growth, business expansion, or changing consumer tastes. In sustainable development situations such as these, the evaluation of alternative options must evolve over time. Useful description of situations such as these often require complex and difficult judgments.

Usually, a financial accounting of the various alternative projects results in the information that is needed to identify and quantify economic benefits and costs, perhaps in a balance or spreadsheet format. However, adjustments to the financial accounting data are often needed. Sometimes we find that financial accounting neglects some benefits and costs, particularly those of a secondary and/or intangible nature, that may be necessary for a useful analysis of the economic benefits and costs of a project. Also, it is often necessary to adjust financial information for those in which market prices either do not exist or, even if they do exist, do not reflect true economic value. Again, economic value is a multifaceted term.

Where competition is imperfect, such as with a monopoly, prices do not reflect value. Also for products made for use internal to a given firm and not marketed, there is no market price. In all cases, investment decisions are likely to be based on market prices, whether or not they reflect true economic value, assuming the investor believes such prices are likely to be reasonably steady during the term of the investment.

Information that should not be specifically included in the calculations made in a cost–benefit analysis includes the following.

---

*For example, much of the 1991 debate over such "big science" projects as the genome project and the super-conducting super-collider revolves around the appropriate use of top-level scientists. There are only a limited number of such people, and their commitment to a big-science project precludes their use elsewhere. Thus, this commitment may diminish society's ability to exploit the research findings of the genome work or the super-conducting super-collider.

1. *Transfer Payments:* Often there are intersectoral flows of benefits or costs that do not in and of themselves constitute cost. For example, there are zero costs associated with using otherwise unemployed capital or land on a project. But there may be a transfer of costs associated with using otherwise unemployed labor, capital, or land. There are other transfer payments, such as taxes and subsidies, which in certain situations should not be treated as costs or benefits.

2. *Sunk Costs:* Economic sunk costs are those that have occurred prior to the time at which a project decision is made. They represent money already spent and are costs that cannot be avoided (any more), regardless of the wisdom or judgment once used in making this resource allocation.

3. *Secondary Effects and Externalities:* Often some costs or benefits of a project cannot be directly observed or measured but may have indirect effects on performance. A good example is the impact of a new airport or the quality of life in a neighborhood. These secondary effects and externalities should be considered as part of a more general economic cost–benefit analysis, perhaps through sensitivity analysis concepts, even though they are often very difficult to identify and quantify.

4. *Contingencies and Risk:* Often a project is implemented without precise knowledge of the environment that will exist during the entire planning horizon for the project. Consequently, various contingency plans are considered to account for risks that might materialize. Generally, the costs and benefits of these risks, such as product price rises, should be included.

Although it is not difficult to state items of this sort, it is often quite difficult to fully account for them in practice. This makes it difficult to do a fully meaningful CBA and to read and understand those that have been conducted, especially with respect to whether facets like those just listed are included in the calculations.

8.3.3. Identification and Quantification of Effectiveness

Often there are a variety of reasons why people are uncomfortable with providing a strict economic measure for benefits. The word *effectiveness* is often used instead of benefit when more than a strictly economic valuation is needed or when an economic (only) analysis is not possible. When effectiveness is substituted for benefit we obtain a cost–effectiveness analysis.

In cost–effectiveness analysis, we desire to rank projects in terms of effectiveness and costs, including in our consideration attributes measured according to different standards. Certainly we would wish to eliminate obviously inferior projects, that is, projects that both are more expensive and less effective than others. Beyond this, a cost–effectiveness analysis does not specify which of the remaining projects is best. This can be accomplished if one is willing to trade off cost for effectiveness so as to obtain a scaler performance index. This can be done by considering cost as one of the attributes in the effectiveness evaluation.

The effectiveness of an alternative is the degree to which it satisfies identified objectives. The effectiveness assessment approach described here provides an explicit procedure for quantifying multiple attributes. This is accomplished by identifying and organizing the attributes into a tree-type hierarchy, attribute tree, or worth structure, and then attaching measures of effectiveness to each such attribute. Effectiveness assessment is most appropriate when we want a single, consistent approach for measuring the impacts of proposed policies and then ranking these policies.

Such an approach also results in an easily communicable picture of the effectiveness, value, or importance that individuals or groups place upon different impacts of the proposed projects. Typically, there also results a significant amount of learning by decision makers concerning their own preferences and the consistency of their evaluations and decisions regarding these preferences. Increased understanding of the decision process and of decision-maker preferences for possible outcomes are other results of this approach. Generally an effectiveness assessment study involves the following major steps which were also discussed in Chapter 7.

1. *Formulation of the Issue:* The group whose preferences and standards of measure are to be used is defined. The scope of the analysis, including objectives or attributes to be studied, is determined. The attributes should be restricted to those of the highest degree of importance. No attribute should encompass any other attributes or objectives. Each should be independent, in the sense that the decision maker is willing to trade partial satisfaction on one attribute for reduced satisfaction on another attribute. Once the high-level effectiveness attributes have been established, they must be disaggregated into lower-level attributes. Each of these is further subdivided until the decision maker feels that project effectiveness can ultimately be measured. This dividing and subdividing process results in a tree-type hierarchical structure of effectiveness attributes.

2. *Selection of Appropriate Attributes of Effectiveness Performance Measures:* Some quantifiable characteristic of performance or effectiveness for an alternative is assigned to each lowest-level attribute.

3. *Definition of the Relationship between Low-Level Attributes and Quantifiable Attribute Measures:* This relation is established by assigning an effectiveness score to all possible values of a given attribute measure. The score given a particular attribute can range from 0 to 1. Zero is the worst score, and 1 is the best score. In determining the score of each alternative on all lowest-level attributes, the following questions must be answered.

   3.1. Is the scale of attribute values continuous or discrete? Generally it can be discrete, such as a good or bad outcome.

3.2. For a continuous scale, does the attribute measure possess either a logical upper bound or a logical lower bound, or both? If this is not the case, the particular attribute in question needs to be redefined.

3.3. What values of each attribute measure are identified with alternative performance of 0 to 1?

3.4. Does the rate of change of worth or effectiveness with respect to attribute measures stay fixed, or does it increase, or decrease?

3.5. If the rate of change of worth of the attribute varies, does it always decrease (or increase) or does it first decrease (or increase) and then increase (or decrease)?

4. *Establishment of Relative Importance within Each Level of Attributes:* Some of the attributes within a given level are normally more important than others, and these differences need to be accounted for. The first step is to rank the subattributes of a particular attribute by relative importance. The most important subattribute is assigned a temporary value of 1.0. The second most important may be assigned a temporary value that reflects its relative weight, such as $1.0 \times 3/5 = 3/5$. If the third attribute is two-thirds as important as the second, it is assigned a temporary value of $3/5 \times 2/3 = 2/5$. This process is continued until all subattributes have been assigned temporary weights. The temporary weights are scaled so that the sum of attribute weights is unity.

5. *Determination of the Equivalent Weights for Each Lowest-Level Attribute:* The relative weighting process is applied to all lowest-level attributes in the decision tree and all of the attribute weights are determined.

6. *Calculation of Effectiveness:* The effectiveness of each project is calculated by multiplying the equivalent weights of each attribute by its individual worth score and summing to yield an overall effectiveness score.

7. *Sensitivity Analysis:* The sensitivity of the effectiveness scores to variation in parameters is determined by a sensitivity analysis, in which different values are assigned to the attribute worth scores, and the effectiveness scores are then recalculated.

8. *Final Results:* The final results of the analysis are used for comparison, ranking, and prioritization of alternatives according to effectiveness.

The process involved is very much like that used for multiple attribute utility assessment in our previous chapter. Any of the variants suggested for MAUT may be adapted for use here.

Any cost–effectiveness evaluation program must begin with a study of the goals of the program. In some cases, those planning the evaluation are given the goals. Some organizations have a very narrowly focused purpose. The goals may be focused on development of a particular technology and may not allow consideration of whether other technologies or approaches may address the same purpose more effectively. Often, goals are elusive.

Once the goals of the program or project to be evaluated have been identified, it is necessary to determine how attainment of these goals will be measured. This is not difficult to do in some cases but it may be impossible to do in others without ambiguity. In many cases, the difficulty in assigning a measure of performance depends on how well the goals have been defined.

If there is a single goal for a program and it is very specific, it is fairly simple to select a performance measure. If, however, there is a single goal that is very broad, there are likely no universally accepted metrics. A goal such as this provides no direction to the research program and has no functional value other than as the first step toward more specific goals. In this case, step 1 cannot easily be completed.

If there are several different groups or individuals involved in the decision, it is unlikely that all will accept the same trade-off weights or measures. In very polarized cases, an increase in a weight may enhance perceived worth for one group while diminishing it for another; or one group may be insensitive to weights of great importance to another group.

For example, suppose that we are trying to aid someone in deciding whether to spend $50 million to upgrade the local sewage system, or, alternately, to expand the local airport. For the sewage system, the top three benefits might be satisfying the needs of a potential new manufacturer coming to the area, less maintenance than associated with the present system, and less risk of rupture and subsequent health threat. The three major projected effectiveness attributes of a new airport might be fetching more business, providing quicker access for local residents, and easier and less expensive maintenance. We would then do a rough-and-ready cost–effectiveness analysis for each of these six postulated attributes, and, based on our findings, would then assign percentage weights to the various attributes. We would often need to add other perspectives to pure economic calculations in order to reflect political, environmental, and similar factors. Then we would be prepared to identify and rank the various effectiveness attributes.

Often, it is necessary to employ sophisticated forecasting techniques to obtain answers to relevant questions. This is particularly true for the costs of "long life" systems. The two sources of innovation, demand pull and technology push, commonly influence such costs. Demand pull describes innovation arising from recognition of an unsatisfied demand. The innovator, noting a need, considers ways to address that need until one or more viable approaches are found. Technology push occurs when available or achievable technology make it possible to do something that was not possible before, or to do something in a new and better way. The force behind technology push may be a new technology or it may be a new material, process, or application of existing technology. Conceivably, a new view of an existing technology can also be a source of inspiration innovation.

Three kinds of forecasting are often needed. The first is economic, which seeks to establish demand for new products, technologies, and services (both public and private) and the costs of providing these. Generally, the demand for new products is highest in expanding segments of the economy and the lowest in those that are declining. Therefore, patterns of economic growth are a principle topic for systems engineering research and development.

A sociological forecast is also usually needed. The objective here is to identify changes in the society that effect the demand for new technologies and the environment for innovation. Specific areas included in the forecast [9] should be

1. *Demographic Structures:* Includes location and makeup of future populations.
2. *Spending Priorities:* Includes concerns relating to public versus private consumption, preference shifts within product groups, acceptability of new technologies, and emphasis on various areas of expenditure.
3. *Role of Government:* Includes those activities of government that influence the demand for new technologies through both direct purchase and regulations that affect the use of existing technologies.
4. *Public and Legal Attitudes toward Business:* Includes attitudes toward monopolies and oligopolies, patent issues, and public control of product, material, and service process.
5. *International Affairs:* Includes issues relating to foreign markets, foreign political shifts, tariff barriers, and international currency stability.
6. *Labor Conditions:* Includes those conditions favorable to or adverse to the introduction of automation.
7. *Education:* Includes the number of trained scientists and engineers, level of consumer sophistication, and the propensity of university research to develop new products.

A technological forecast assesses general developments in the specific scientific community and among competitors and emerging technological needs of the customers. While identifying opportunities for technology push, this approach is much more oriented to studying demand. Certainly it is easier for a research organization to address known needs than to identify the potential in new technologies. Normally, we desire an economic forecast, a social forecast, and a technological forecast.

8.3.4. Cost Estimation

There are several approaches that we can use to estimate costs of a project under analysis.

1. Analogy.
2. Bottom up.
3. Expert opinion.
4. Parametric models.
5. Top down or design to cost.
6. Price to win.

Each is reasonable in particular circumstances. The approach we outline here is just the multiattribute utility theory approach of our previous chapter. If desired, uncertainty effects may be considered in this.

A number of approaches may be used to estimate effort or cost rates.

Direct labor rate.
Labor overhead rate or burden.
General and administrative rate.
Inflation rate.
Profit.

These are not mutually exclusive, nor are they collectively exhaustive.

There are a number of approaches to cost as pricing strategies. These include full-cost pricing, investment pricing, and promotional pricing. Another approach to costs is determining the cost required to achieve *functional worth*, that is, to fulfill all functional requirements that have been established for the system. While this is easily stated, it is not so easily measured. A major difficulty is that there are essential and primary functions that a system must fulfill, and ancillary and secondary functions that, although desirable, are not absolutely necessary for proper system functioning.

After the functional worth of a system has been established in terms of operational effectiveness, it is necessary to estimate the costs of bringing a system to operational readiness. If this cost estimate is to be useful, it must be made before a system has been produced. It is easily possible to think conceptually of three different costs:

1. *Could Cost:* The lowest reasonable cost estimate to bring all the essential functional features of a system to an operational condition.
2. *Should Cost:* The most likely cost to bring a system into a condition of operational readiness.
3. *Would Cost:* The highest cost estimated that might have to be paid for the operational system if significant difficulties ensue.

It is interesting to relate nonfunctional value-adding costs to risks. If the system is well-designed in the first place, all costs of implementation should contribute to the functional worth of the system. Any additional costs would then fall into the category of risk, that is, unanticipated expenses. Quite obviously, it is very difficult to estimate each of these costs. The should cost estimate is the most likely cost that results from meeting all essential functional requirements in a timely manner. Could cost is the cost that would result if no potential risks materialize and all nonfunctional value adding costs are avoided. Would cost is the cost that will result if risks of functional operationalization materialize and if non-value-adding costs are not avoided. There is a strong notion of uncertainty in any discussion of costs such as

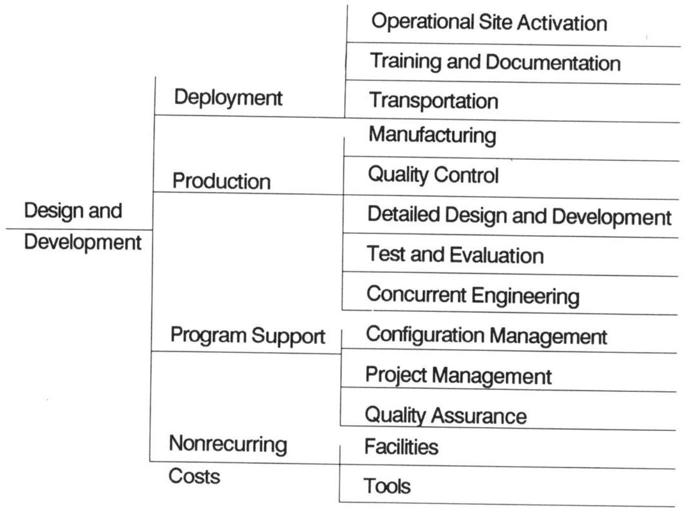

		Operational Site Activation
		Training and Documentation
	Deployment	Transportation
		Manufacturing
	Production	Quality Control
Design and		Detailed Design and Development
Development		Test and Evaluation
		Concurrent Engineering
	Program Support	Configuration Management
		Project Management
		Quality Assurance
	Nonrecurring	Facilities
	Costs	Tools

Figure 8.6. Work Breakdown Structure for System Design and Development

these, and various probabilistic notions need to be used in obtaining useful cost estimates.

As we have often noted, there are three fundamental phases in the systems development life cycle: system definition, system design and development, and system operation and maintenance. These phases may be used as the basis for a *work breakdown structure* (WBS), or *cost breakdown structure* (CBS), for depicting cost element structures. [A work break down structure (WBS) is mandated for proposals and contracts for work with the federal government in efforts that involve either, or both, of the first two phases of the systems life cycle. Military Standard STD-881-A governs this. The terms research and development, investment, and operation and support are used instead of the terminology used here.] In general, we would expect that the total acquisition cost for a system would represent the initial investment cost to the customer for the system. This would be the aggregate cost of designing, developing, manufacturing, or otherwise producing the system, and the costs of the support items that are necessary for such purposes as maintenance.

Figures 8.6, 8.7, and 8.8 illustrate some of the many components that comprise a WBS or CBS for the three-phase systems engineering life cycle. There are a number of related questions that, when answered, provide the basis for reliable cost estimation for the work breakdown structure (often called a contract work breakdown structure-CWBS-by the U.S. DoD) suggested here. One DoD publication describes these in terms of organizational questions, planning and budgeting questions, accounting questions, analysis

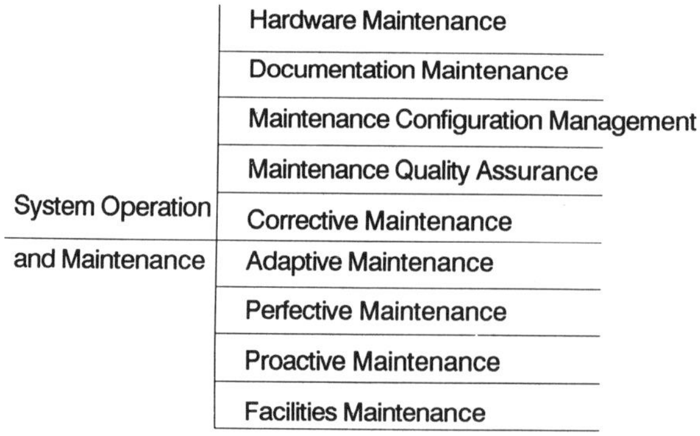

System Operation and Maintenance	Hardware Maintenance
	Documentation Maintenance
	Maintenance Configuration Management
	Maintenance Quality Assurance
	Corrective Maintenance
	Adaptive Maintenance
	Perfective Maintenance
	Proactive Maintenance
	Facilities Maintenance

Figure 8.7. Work Breakdown Structure for System Operation and Maintenance

System Definition	Requirements Identification
	Specifications Development
	Research and Advanced Development
	Design and Development Plan
	Prototype Production
	System Test and Evaluation Plan
	Configuration Management
	Operation and Maintenance Plan
	Facilities

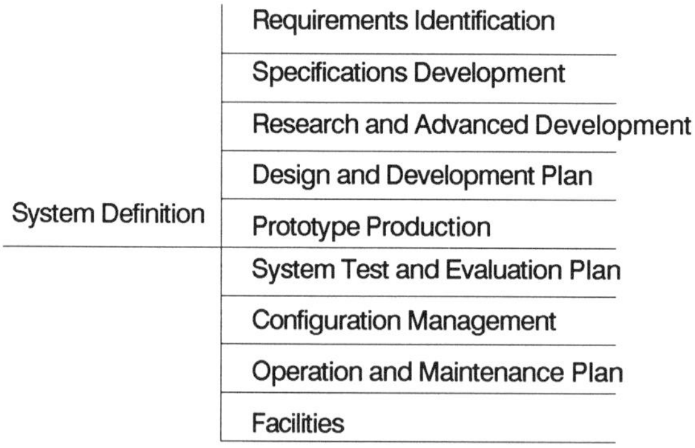

Figure 8.8. Work Breakdown Structure for System Definition

questions, and program and project revision questions. There a number of questions that can be posed [10, 11], and the responses to these provide valuable input for estimating WBS and costs.

A recent study of cost estimation [12] provides nine guidelines for cost estimation.

1. Assign the initial cost estimation task to the final system developers.
2. Delay finalizing the initial cost estimates until the end of a thorough study of the conceptual system design.
3. Anticipate and control user changes to the system functionality and purpose.
4. Carefully monitor the progress of the project under development.
5. Evaluate progress on the project under development through use of independent auditors.

6. Use cost estimates to evaluate project personnel on their performance.
7. Management should carefully study and appraise cost estimates.
8. Rely on documented facts, standards, and simple arithmetic formulas rather than guessing, intuition, personal memory, and complex formulas.
9. Do not rely on cost estimation software for an accurate estimate.

These guidelines were established specifically for information system software and there is every reason to believe that they have more general applicability. This study identifies the use of cost estimates as selecting projects for implementation, staffing projects, controlling and monitoring project implementations, scheduling projects, auditing project progress and success, and evaluating project developers and estimators. Associated with this is the need for evaluating the costs of quality, or the benefits of quality, and the costs of poor quality.

In many of our previous discussions, especially in Chapters 4, 5, and 6, we have examined notions relating to both operational and strategic quality assurance and management. These notions carry over to cost considerations as well. Although it is not unrealistic and it is relatively standard practice to attempt to place a cost on the production of high-quality producers, it is potentially more meaningful to address the *cost of poor quality* (COPQ). This has been defined by Juran [13] as the sum of all costs that would disappear if there were no quality problems.

Feigenbaum [14] identifies quality costs as including both (1) the cost of quality control to prevent defects from occurring and to appraise and correct those that do occur and (2) the costs of internal or external failure to bring about quality control. The establishment of costs for a quality management and control program involves the formulation of quality costs and of the costs of poor quality. This involves the identification of both quality cost items and poor-quality cost items. These would include the direct costs of the operational level quality assurance items we discussed in Chapter 4. Importantly also, it should include the identification of the costs that are ultimately added to fielding a large system, such as rework and additional scrap (scrap may be hardware or software), due to a low quality system development process. Identification of the costs of the strategic quality effort and the operational level costs of implementing this are also needed. This includes the costs of continuous improvement in the quality of processes and products over time.

8.3.5. A Guide to Cost–Benefit and Cost and Operational Effectiveness Analysis

Cost–benefit and cost–effectiveness analysis are methods used by systems engineers, and other professionals, to aid decision makers in the interpretation and comparison of proposed alternative plans or projects. They are

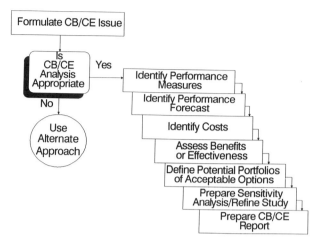

Figure 8.9. Generic Cost–Benefit and Cost–Effectiveness Assessment Process

based on the premise that a choice should be made by comparing the benefits or effectiveness of each alternative with its costs. Benefit is an economic term that is generally understood to be a monetary unit. Effectiveness is a multiattribute term used when the consequences of project implementation are not reduced to dollar terms. We will describe an approach for CBA and then indicate modifications to adapt it to COEA. Figure 8.9 illustrates the general steps involved in the process.

First, objectives for the project must be identified and alternatives generated and defined carefully. Then the costs as well as the benefits or effectiveness attributes of proposed projects are identified. These costs and benefits are next quantified and expressed in common economic units whenever this is possible. Discounting is used to compare the costs and/or benefits that accrue to a project at different times. Net present worth, or value, is the usual discount criterion of choice. Overall performance measures, such as the total costs and benefits, are computed for each alternative. In addition to this quantitative analysis, an account is made of qualitative impacts, such as social aesthetic and environmental effects. Equity considerations regarding the distribution of costs and benefits across various societal groups may be considered. The cost–benefit analysis method is based on the principal that a proposed economic condition is superior to the present state if the total benefits of a proposed project exceed the total costs.

Among the results of a cost–benefit or cost–effectiveness analysis are

1. Tables or graphs containing a detailed explanation of the costs and benefits over time of each alternative project and present value of costs or benefits of each alternative project.

2. Computation and comparison of overall performance measures, in terms of benefits and costs for each of the alternative projects. Alternatively, effectiveness and costs may be used if a cost–effectiveness analysis is desired.

3. An accounting of intangible and secondary costs, such as social or environmental or even aesthetic, and the set of multiattributed benefits or effectiveness value that is associated with each alternative project.

The following major steps are generally carried out in a cost–benefit analysis.

1. Issue Formulation
    1.1. Problem definition.
    1.2. Value system design.
    1.3. System synthesis.
    These efforts are accomplished using techniques specifically suited to issue formulation, such as identification of objectives to be achieved by projects, some bounding of the issue in terms of constraints and alterables and generation of alternative projects. The results of this formulation of the issue include a number of clearly defined alternatives, the time horizon for and scope of the study, a list of affected individuals or groups, and perhaps some general knowledge of the impacts of each alternative.

2. Issue Analysis
    2.1. *System Analysis:* Identification of costs and benefits including benefits of effectiveness measures for each alternative is accomplished here. A list is made of the costs and benefits for each project. Measures for different types of costs and benefits are specified, and, if possible, conversion factors are developed to express different types of costs or benefits in the same economic units. For example, one of the benefits of a proposed highway project might be reduced travel time between two cities. To make this comparable to monetary costs, we have to determine how many dollars per time unit are gained by the reduction of the travel time. Determination of such conversion factors can be a sensitive issue, since the worth of various attributes can be dramatically different for different stakeholders. For example, consider the difficulties involved in transforming additional safety benefits of a proposed project, measured in human lives saved, into monetary benefit units. Further complicating economic benefit evaluation are equity considerations, which may require the costs and benefits of the project to be allocated in different amounts to different groups.
    2.2. *Refinement of the Alternatives:* This is accomplished through more detailed quantitative analysis of costs and benefits. Costs and

benefits are expressed in common economic units insofar as possible. Comparisons of projects may also be made with respect to different quantified units if they can be converted to economic costs and benefits. Economic discounting is used to convert costs and benefits at various times to values at the same time, and net value analysis is then used to compare benefits and costs. The impacts that cannot easily be quantified are assessed for each alternative project. These usually include intangible and secondary or indirect effects, such as social and environmental impacts, legal considerations, safety, aesthetic aspects, sustainable development, and equity considerations.

3. Interpretation
   3.1. *Decision making:* This is accomplished through selection of a preferred project or alternative course of action. There are a variety of criteria that can be used. We may select the project that maximizes benefits for a given cost. We may select the project that has minimum costs for a given level of economic benefits. We may maximize the cost–benefit ratio. We may maximize the net benefits or benefit minus cost.
   3.2. *Planning for and Communication of Results:* This is the last effort in a CBA and often takes the form of a report on both the quantitative and the qualitative parts of the study. The report may include a ranking or prioritization of alternative projects or a recommended course of action. It is important that all assumptions made in the study are clearly stated in the report. The report should be especially clear with respect to costs and benefits that have been included in the study, costs and benefits that have been excluded from the study, approaches used to quantify costs and benefits, the discount rate that have been used, and relevant constraints and assumptions used to bound the CBA.

Cost–effectiveness analysis is accomplished by a very similar approach. Methods such as decision analysis and multiattribute utility theory can be used to evaluate effectiveness. We are able to use the resulting effectiveness indices in conjunction with cost analysis to assist in making the trade-offs between quantitative and qualitative attributes of the alternative projects. There are a number of studies available addressing this important subject [15–22].

8.3.5. Cost–Effectiveness in Systems Management of Emerging Technologies

In this section, we summarize cost–effectiveness analysis as it relates to the problem of systems management of the research and development of emerging technologies. There are four major objectives.

1. Determining an appropriate specific process to use for the identification and evaluation of potential technologies for development and/or transfer.
2. Identifying the groups that should be involved in this identification and evaluation process.
3. Identifying the criteria that will be used to determine length and type of support.
4. Identifying of appropriate criteria to determine transferability of the technology to full scale operational deployment status or termination.

Our efforts are based, in large part, upon previous research by Miller and Sage [23] and Sage [24].

One major product of research is information. Systems management of research and development must attach a value to this information in terms of the final product that might be developed. This, unfortunately, is not easy, both because the product is normally far downstream in the development process and because a single research effort, if successful, normally contributes to a variety of technologies and products.

The large amount of uncertainty regarding the ultimate value of research suggests that many parallel, generally low cost, projects be initially funded. As a result, methodologies for early research cost–effectiveness studies must be able to deal with a large number of uncertain projects. Methodologies for later life cycle cost–effectiveness analysis generally deal with a smaller number of less uncertain, but more costly, projects.

It might seem that the second task is easier than the first. But the large sums of money involved in the later stages of development make it critical to fund the best alternative. The analysis may be easier, but the penalty for error is much greater. If we accept the view that the value of research may be measured by the value of the resulting products, the focus of program evaluation becomes the forecasting and estimation of the market performance of the resulting innovations. Market penetration estimates and analysis of the expected return on investment are the principal measures of research worth. Conventional market methodologies may fail, however, when a technology is not explicitly marketed or when it has no conventional competitor. When an innovation does have a conventional competitor, the potential market for an emerging technology will be:

1. That fraction of the conventional product's market in which the emerging technology ultimately enjoys an economic advantage.
2. Minus some fraction of users of the conventional technology who are unable to convert to the new one for economic, environmental, institutional, technical, cultural, or other reasons.
3. Plus some additional market which was unable to use the conventional product but is now able to use the newly developed technology.

Neither of these three market sources are particularly easy to estimate. When the emerging technology is mostly unlike anything else that exists, it is much more difficult to address issues associated with estimating the number of potential users of an emerging technology product, how rapidly they will adopt it, and how much they are willing to pay for it. The second problem with assessing research programs by the market potential of final products is that, as we have noted, some products do not individually or directly compete in a marketplace.

There are many other related concurrent variables and it is desirable to discuss some of these. Often there will be simultaneous development of two potentially competing technologies. When the products eventually compete, assuming that both are superior to existing products, the one with the lowest cost and the highest effectiveness, will ultimately dominate the market.

If the technology which is marketed first is not the best, but is still better than the conventional alternative by enough of a margin to justify the switch to it, then it will sell to those segments of the market which:

1. Are unaware of the forthcoming, superior technology.
2. Are aware of the forthcoming technology but unwilling to wait for this better alternative.
3. Are willing to purchase the immediate alternative for use until the improved version is available.

The market shares of the competing alternatives will, therefore, not reflect their relative economics because of the different time horizons associated with product introduction. The implication of this is that strategic management and management controls reflect timing consideration when considering the effectiveness of emerging technology projects. The bottom line associated with this is that the value of an emerging technology development project completed some time in the future is usually not as great as the value of the same project completed now, except when market conditions related to supporting technologies prohibit immediate adoption of the emerging technology product under research, development, test and evaluation (RDT&E).

Some products enhance the value of other products. The basic issue with augmenting one technology to support existing technologies is that of evaluating the conditional and independent worth of the individual technologies. The value of each technology must be evaluated assuming that none of the other new ones are successful, and then reevaluated for various possible outcomes of the other research projects. This is not an easy task. One very difficult complication is that the different technologies may compete in different markets. Market penetration for products is very difficult to estimate. In the special case where two or more emerging technologies are worthless without each other, but do have an estimable value when all technologies are adopted, the group of projects can be treated as one for management purposes. This may create a very significant problem in formulating or framing the emerging technology development issue. But, the decision itself is simplified to a binary all-or-none comparison. This assumes a degree of centralization of authority over emerging technology developments that may not always be achievable.

In the situation where there are several competing technologies which augment a third, effectiveness analysis becomes especially complicated. There must be careful study of the market, the decision makers and their preference structure, and the time horizon associated with product introductions. When two projects share similar information or software or hardware elements, the RDT&E on one technology can potentially be used to support efforts relative to the other. The associated information transfer can substantially complicate the cost–effectiveness evaluation process. If concurrent development of the two projects makes it possible to eliminate or reduce work in the common areas, the cost of doing both projects is not the sum of doing either without the other project. Thus, it is not possible to use simple evaluation methods without modification to enable consideration of this dependence. This brings about some serious difficulties in modeling emerging technology project interactions. Clearly, this is the situation today with the very significant interest in such areas as computer integrated manufacturing, concurrent engineering, and reusable software development.

The appropriate approach to augmenting technologies will vary widely depending on the situation. As a consequence, it is neither feasible nor wise to try to develop a single, all embracing analytic approach. For example, imagine a small, highly specialized, poorly capitalized firm developing a specific new product, say a new kind of valve for large sewage systems. Say, further, that other research in that firm's laboratory is working on projects that could augment the value of this valve. Given these conditions, the firm is likely to allocate RDT&E funding for each project based solely on that project's clearly identified market potential for itself, without reference to synergies with other technologies.

As another extreme, let us consider investments by the U.S. government in "big science" projects, like a $30 billion space station or $10 billion atom smasher. In such instances, it is generally agreed that no one person can foresee all the possible uses to which such a project could be put, and that, in a sense, the sole purpose of such projects to augment other technologies. To take a variation on this theme, advocates of a federal technology or industrial policy often like to talk about the need to fund enabling technologies, the argument being that such technologies will be a catalyst for wide-ranging subsequent technological developments (RDT&E).

Situations in between these two extremes are also quite interesting. For example, consider a large, for-profit corporation conducting research in four interrelated areas: expert systems, reduced instruction chips, gallium arsenide substrates, and laptop computer design. Obviously, potential synergies exist among all four areas of research, and a corporation considering RDT&E

would surely like to identify these. But it would be very difficult to find a reasonably reliable methodology for identifying, quantifying, and ranking such synergies and, on the basis of these rankings, rationally apportioning RDT&E funds among them. In practice, a methodology to enable prioritization of such research is greatly needed.

It is reasonable to begin cost–effectiveness assessment by evaluation of individual candidate projects. Analysis of a single project can be viewed as a two-step process. First, the technical outcome of each project is forecasted. The objective of this is to characterize the technology that the project will ultimately produce. The second step is to relate the technical outcome of a project to its effectiveness. For example, in evaluating a project directed toward producing a commercial product, the relationship between various technical performance features and future sales volume should be forecasted.

We recommend separating the technology forecast from the analysis of effectiveness, since the information and skills needed to forecast performance of an emerging technology are generally quite different than those needed to evaluate its long-term effectiveness. In the first case, technical speciality skills are most needed, whereas, in the second, a broad spectrum of information and accompanying interpretive skills, are most needed, as we have noted in Chapter 3. The value of an emerging technology is a complex function, including such factors as its performance relative to other technologies that have the same functional end and the resources and preferences of the consumers. The approaches described in two excellent works by Porter and colleagues [17, 25] describe a number of potentially useful approaches for technology forecasting and evaluation.

After each emerging technology project is examined separately, it is necessary to consider combinations of projects, or portfolios, to determine which grouping maximizes the combined value of all projects under consideration. If the projects are independent, then the effectiveness of a portfolio is the sum of the individual scaler effectiveness, and the costs of a portfolio is the sum of the individual costs. If the projects are not independent, then there must be a third step in addition to the two discussed above. The third step is to characterize the relationship between the new technologies in their final market; that is, to develop a model of the impact of each project's technical success on the value of the other projects. The next and final step in the resource allocation methodology is to select the best portfolio.

An organization needs a general idea of how much revenue a new product will yield before it can make a wise decision concerning RDT&E investments to result in that product. Certainly, when demand is uncertain, various risk analysis approaches may be employed that might not be employed when demand is reasonably predictable. These lead to risk management strategies for the technology development in question. In the final analysis, however, it will still be the case in the future that Revenue = Price × Quantity Sold, and that is what management needs to know with some reasonable degree of precision. Profitability, as expressed in monetary terms, is the normal measure of market performance, although other special measures, such as market share or profit as percent of sales may be equally appropriate.

RDT&E cost is not directly related to either the ultimate value of the technology or the number of fielded systems. Market price must first be determined. Assuming that it is possible to develop some relationship between number of units sold and market price in the form of a product demand curve, it is potentially a fairly simple effort to determine the market price that maximizes return of investment. If return on investment (ROI) is the measure of effectiveness to the technology developer or can be related to it in an appropriate manner, then a cost–effectiveness based formulation is quite appropriate.

As development of an emerging technology moves to the final phases of the RDT&E life cycle, evaluation focuses on a smaller set of candidate projects, but with much more detailed and accurate cost and effectiveness measures. The value of a project to the developing organization can be a function of many attributes. Clearly, the best procedure for assessment depends on the measures of effectiveness applied. Any assessment of cost effectiveness must include an estimate of how well a fielded product will perform, which involves many performance attributes, and how much it will cost. The cost–effectiveness assessment may stop there if the technology is being developed for internal use, or it may be carried through a forecast of market penetration if the technology is intended for use outside the developing organization.

It is generally not possible to predict accurately how an emerging technology will perform in advance of its ultimate deployment. But forecasts do need to be made. Initially, it is appropriate to regard cost and effectiveness as uncertain parameters, with probability distributions, rather than as certain values. Given some measures of the effectiveness of an emerging technology under development, the next step in cost–effectiveness assessment is to estimate the number of units of the new technology that will come into use over time. Market penetration estimation is very important in many financial decisions. Consequently, market penetration models have been developed.

Market penetration potential may be defined as the number of units of the new technology that could be adopted in a unit of time if every potential user preferred it to all other alternatives. Market share is the fraction of the potential market that is actually captured. If the technology addresses a totally new use such that there is no conventional alternative, the market share is necessarily 100 percent. The unit of time used in a penetration analysis should be chosen on the basis of expected rate of penetration and the market life of the product.

Implicit in this discussion is the assumption that the funding organization wishes to select a limited number of projects. The presence of such a resource constraint makes it necessary to consider not only the effectiveness of a project, but also its cost in terms of money, people, equipment, and any other quantity that could be binding on the RDT&E program. Our earlier discussions of work breakdown structure and cost breakdown structure are quite relevant here. Relevant projects costs include the following.

1. *Sunk Costs:* Costs already expended. In the idealized allocation problem, time begins when the allocation decision is made. Project history up to this point is not considered, except to the extent that it influences future effectiveness and cost. Of course, organizations normally have a stock of R&D projects in progress. The costs of the projects underway from initiation to the point in time used for the allocation analysis are defined as sunk costs.

2. *Fixed Costs of the Organization:* Costs that do not vary with the size of the organization or the size and makeup of the research program and cannot be directly associated with one or more specific projects.

3. *Variable Costs of the Organization:* These costs are similarly free of any association with individual projects, but they do vary with the size of the organization and the research program. The distinction between fixed and variable costs is often ambiguous. Some costs are fixed over ranges of organization and research program size but vary when a major change in size occurs. Also, some costs are fixed for the short-term but are varied over the long-term. The definition of which costs are fixed and which are variable must be made when the timing horizon and the size of a research program are determined.

4. *Fixed Costs of a Project:* Those costs that are directly attributable to the project and do not vary with the level of effort.

5. *Variable Costs of a Project:* Those costs that are directly attributable to one or more projects and do vary with the level of effort in the associated projects. Equipment purchase would be a variable cost if equipment requirements were different in different configurations of the project.

In practice, the total cost of a project is not easily defined or determined. The definition of total project cost is a function of the purpose for which detailed cost estimates are prepared. An evaluation of the overall effectiveness of the organization should include all cost elements. The fixed costs of the organization, which by definition are not attributable to an individual project, may or may not be used in the resource allocation analysis, depending upon the objective of the cost accounting. Even though fixed and variable costs of the organization do not enter into the comparison of alternative portfolios except in determining cost constraints, they play an important role in determining how large a RDT&E program the organization should undertake.

When the constituent projects in an emerging technology research portfolio are independent in the sense that the effectiveness and costs of the portfolio equal the sum of the effectiveness and costs of the projects, evaluation is simple, and mathematical programming techniques can be used to isolate the optimal mix of projects quickly and inexpensively. When effectiveness and cost cannot be added in this fashion, evaluating a portfolio becomes substantially more difficult.

There are many reasons why the effectiveness of a group of projects may differ from the sum of the scaler effectiveness indices of its members. The projects may produce technologies that enhance each other's effectiveness, so that the effectiveness of the group is greater than the effectiveness of the technologies alone. On the other hand, the projects may produce technologies that compete with each other so that the effectiveness of the group is less than the effectiveness of the individual projects. It is rather difficult to define general classes of ways in which one product can enhance the effectiveness of another. There are numerous examples of how the development of one specific product has made another more valuable. Furthermore, the degree of the enhancement in effectiveness seems to cover a broad range depending upon the specific products under consideration. In general, however, the steps illustrated in Figure 8.10 appear appropriate for a CEA of emerging technology RDT & E issues.

As we have noted, it is dangerous to assume that one or another technology is, in an absolute sense, superior to another or cheaper than another. Sound evaluation should start with a clear concept of the intended use or intended market of the technology if one hopes to be able to come up with

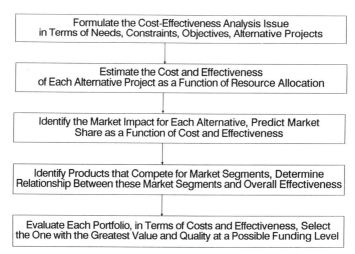

Figure 8.10. Cost–Effectiveness Analysis for Systems Management of Emerging Technology RDT & E

reasonably meaningful quantitative evaluation. Purposeful RDT&E must have a market in mind! Basic scientific or engineering research is a completely different matter, as we have discussed. There is an underlying notion of the quality of process and product in these issues also; and quality concerns should not be neglected—either the cost of quality or the costs of poor quality.

These procedures are, of course, neither mutually exclusive nor exhaustive. A rather detailed example is considered by Miller and Sage [26]. An excellent summary of the evaluation criteria we have been discussing is provided by Dutton and Crowe [27], and are identified in Chapter 3, page 113, of this book, and illustrated in Figure 3.11 in the form of a multiple attribute evaluation.

The theory surrounding the allocation of funds to research as a function of the costs and revenue requirements of the organization and the alternative options for investment is a large, complex, and heavily studied area and the interested reader will find it rewarding to consult recent studies in this area [28–30]. We could only provide a very general overview of this important area of systems management here.

8.4. SUMMARY

In this chapter, we have provided an overview of economic systems analysis approaches to costs, benefits, and effectiveness measures of systems engineering processes and products. As these approaches are strongly rooted in microeconomic analysis, we have discussed the principle topics of this subject: the theory of the consumer, the theory of the firm, economic equilibrium in a perfectly competitive economy, and normative economics. The time value of money is an important subject for economic systems analysis of projects that extend over several years, and we have examined salient features of this important subject. This provided us with a background for cost–benefit and cost–effectiveness analysis, which was our major concern in the latter portions of this chapter. We concluded with a brief discussion of criteria of importance in selecting an approach to project evaluation.

Our efforts in these subjects have blended decision analysis with classic economic cost–benefit analysis. The economics-based approaches rely primarily on net present value of costs and benefits, expressed in monetary terms. Our decision analysis approach is essentially a multiattributed cost–effectiveness analysis. The chapter concludes with an application of this approach to selection of a research and development portfolio.

There are many roles for human and human system interaction issues in cost–effectiveness analysis. The role of humans, or cognitive ergonomics, in economic analysis has been undervalued. There are many interrelations, of course, and we explore a few of these in our next chapter.

PROBLEMS

8.1. What are the optimum purchase amounts for a consumer with the following utility function and budget

$$U(\mathbf{x}) = x_1 x_2$$
$$x_1 + 2x_2 \le 40$$

8.2. What are the optimum purchase amounts for a consumer with the following utility function and budget

$$U(\mathbf{x}) = x_1 + 3x_2 + 5x_3$$
$$x_1 + 2x_2 + 3x_3 \le 40$$

8.3. Consider a consumer whose utility function reflects desire for income and desire for leisure. Suppose that the specific utility function is

$$U(I, L) = aI + bL - cL^2$$

where I is income and L is leisure. This constant marginal utility of income and decreasing marginal utility for leisure is perhaps reasonable.

a. Show that the effect of an income tax, calculated as a percentage of income, is such as to reduce the desirability of work and increase the desirability of leisure. Is there a percentage of income tax that, when increased, will actually reduce the tax paid to the government? Please explain.

b. The consumer wishes to maximize utility through purchase of a commodity bundle $\mathbf{x}$. What is the effect of the income tax on this commodity bundle? Suppose that the income tax is replaced by a sales tax. What is the difference between these two forms of taxation if the revenue to the government is to be the same in each case? What is the effect on consumer utility of the two forms of taxes?

8.4. Suppose that the utility for computers and automobiles is given by $U(\mathbf{x}) = x_1 x_2$. Suppose that the price of a car is \$20K and the price of a computer is \$5K. Suppose that income is \$100K. What is the effect of a drop in the price of computers from \$5K to \$3K. How have consumers and producers benefited?

8.5. The production of a particular firm follows a Cobb-Douglas production equation

$$q = 4x_1^{0.6} x_2^{0.3} x_1^{0.1}$$

Suppose that the costs per unit of these three factors of production are 10, 20, and 30. What are the optimum inputs to production, and what does it cost per unit product produced? What is the effect of a 50 percent increase in the price per unit of the first factor input to production?

8.6. The production of a particular firm follows a Cobb-Douglas production equation

$$q = 4x_1^{0.1}x_2^{0.4}x_1^{0.5}$$

Suppose that the costs per unit of these three factors of production are 10, 20, and 30. What are the optimum inputs to production, and what does it cost per unit product produced? What is the effect of a 50 percent increase in the price per unit of the first factor input to production? Compare this result with the result you obtained in problem 8.5. Why does this difference occur?

8.7. The price equation for a monopolistic producer is given by $p(q) = 20 - 5q$ and that the production cost of the monopolist is $PC = 10q$. What is the profit equation for the monopolist? What is the optimum production level and the price that maximizes profit?

8.8. The demand curve for a product is given by $q = 20 - p$ and the production cost to make the product is $c(q) = 3q$. The firm is operating as a monopoly. What is the equilibrium price to maximize profits, the resulting profits, and the production quantity? Suppose that price ceilings for the product are set by the government. How will this affect the behavior of the firm, and what are the profits and quantity produced under various price ceilings?

8.9. There are 100 firms in a perfectly competitive economy. Each of them has the production cost function $c_i = q_i^2$. The market demand function for this particular product is given by $q = 2,000 - 200p$. What is the supply curve for a single firm and for the entire set of 100 firms? What are the equilibrium price, quantity produced, and profit of each firm?

8.10. Suppose that the production function for a firm is given by

$$q = 5K^{0.5} + 3L^{0.5} + 2M^{0.5}$$

where the wages for the capital, labor, and resource factors are $w_K = 3$, $w_L = 1$, $w_M = 5$. What is the supply equation for the firm? If there are 1,000 of these firms, what is the resulting market supply equation?

8.11. The supply and demand functions in a perfectly competitive economic situation are $SQ = SQ_0 + \alpha p$, $DQ = DQ_0(1 - p/p_{max})$. What are the market equilibria, and what do these become for the special case where $DQ_0 = 1000$, $p_{max} = 30$, $SQ_0 = 0$, and $\alpha = 50$? What is the effect of a unit sales tax on these results?

8.12. Suppose that there are two firms in a particular economic system. Their production cost equations are

$$PC_1 = 20q_1 + 10q_1^2 - 3q_2^2$$
$$PC_2 = 40q_2 + 5q_2^2 + 3q_1^2$$

and the price of each product is \$12. What are the requirements for profit maximization of each firm? What are the requirements for Pareto optimality of each firm? How might an imposed Pareto optimality, to maximize the sum of the profits for the two firms, be obtained? What are the implications of this for each firm?

8.13. Suppose that the production cost function for two firms are given by

$$PC_1 = 20q_1 + 2q_1^2 - q_1q_2$$
$$PC_2 = 50q_2 + 2q_2^2$$

The market prices are $p_1 = 3$, $p_2 = 2$. Determine the maximum profit for each firm. That is the maximum profit if the sum of profits of the individual firms is maximized? Are there any taxes and subsidies that will result in these operating conditions with each firm maximizing their own individual profit?

8.14. A person borrows \$20,000 at a bank to buy a car. The loan arrangement calls for a 4-year loan and for a quoted interest rate of 6 percent. The total interest is computed as 4 times 0.12 times 20,000, or \$4,800. The total to be repaid is \$24,800 and, spread over 48 months, the monthly payment is \$516.67. What is the actual, or effective rate of interest in this case?

8.15. Based on the time value of money relations developed in this chapter and illustrated in Figure 8.5, derive the algorithm that expresses the relationship between annual interest rate, amount borrowed, duration in months of the loan, and the constant monthly payments needed in order to fully amortize a mortgage.

8.16. What do the results of problem 8.15 become if there is a certain amount of the principal, called the balloon amount, which is not retired over time and which becomes due upon expiration of the mortgage.

8.17. How much should you be willing to pay now for a promissory note to pay \$100,000 thirty years from now if the discount rate is 6 percent? 10 percent?

8.18. Suppose that you need to decide whether to accept investment *A* or investment *B* as the best investment. Each investment requires an initial investment of \$10,000. Investment *A* will return \$30,000 in a period of 2 years. Investment B will return \$20,000 in 1 year. Suppose that the opportunity cost of capital, or discount rate, is 20% per year. Please compare the net present value of the two investments and the internal rate of return of the two. Which project is the best according to these criteria? Suppose that the opportunity cost of capital is considered to be a variable. Is there some OCC where the two investments have the same value? Explain the implications of this.

8.19. Please indicate how you might modify the net present value criterion such that you could consider a borrowing interest rate that is different from a lending interest rate.

8.20. Suppose that the supply and demand curves for a particular product are given by $S(q) = p = q$, $100 - D(q) = p = 100 - q$. An emerging technology investment is proposed that will change the production supply curve to $S^*(q) = p = 0.75q$. Consider the new project both from the perspective of the firm and the consumer. What are their reactions to the new technology? What is the maximum cost of the emerging technology development effort that will make it marginally justified?

8.21. It is sometimes stated that the benefit cost ratio is a more useful criterion than the net present worth criterion in that it more easily allows comparison of projects of different size. Please evaluate and discuss this suggestion.

8.22. Identify the fixed costs, variable costs, and marginal costs for a firm with the production cost functions
a. $PC = 120 + 36q^2$
b. $PC = 120 + 10q + q^2$
c. $PC = 120 + 10q - q^2$

8.23. Costs and benefits can be calculated from the perspective of an individual or firm (private sector analysis), the economic system (public economic analysis), or a complete sociopolitical economic system (social cost–benefit analysis). Market prices would be used for benefits and costs in the first analysis, and shadow prices in the second. The third analysis would necessarily attempt to weigh various potentially noncommensurate factors into the analysis. Please write a brief paper on how we might go about each analysis.

8.24. Prepare a brief synopsis of the application of the cost–effectiveness analysis approach discussed here to the energy conservation program planning effort reported by Miller and Sage [23, 26].

8.25. Prepare a brief synopsis of the application of the cost–benefit and cost–effectiveness analysis approach discussed here to the information systems project evaluation suggestions in [21].

8.26. Identify relevant factors that will enable you to bring about a cost–effectiveness analysis of a make or buy decision for

  a. Database management software.
  b. A complete office automation system.
  c. A decision support system [31].

8.27. What are roles for economic systems analysis in operational quality assurance and product management?

8.28. What are roles for economic systems analysis in strategic quality assurance and process management?

8.29. Prepare for and conduct a cost–effectiveness analysis of the potential decision to purchase a personal computer. Please do this from the perspective of an individual, the organization in which the individual is employed, and society.

8.30. Two projects each cost $300,000 and are directed at technologies that will supply the same service. The decision tree associated with these projects is shown in Figure 8.11. The technology resulting from one project has a 0.6 probability of supplying the service at a cost of $1 per unit and a 0.40 probability of supplying it at a cost of $2 per unit. Comparable risks and outcomes result from project B. The worth to the project developer of being able to supply the service at $1 per unit is estimated at $2M and the worth of supplying it at $2 per unit is $1M. What is the expected worth of the two projects and which should be selected if only one project is to be developed? What is the probability that the technology produced by the nonfunded project is at least as good as that developed by the funded project? What does this say in terms of risk management?

8.31. A company wishes to explore the possibilities of implementing a computer integrated manufacturing process for its production efforts. Please prepare a cost–benefit analysis or cost–effectiveness analysis of

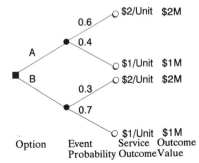

Figure 8.11. Decision Tree for Example Cost Effectiveness Evaluation

a CIM effort in terms of productivity, profitability, and competitiveness. There are tangible and intangible benefits to a CIM effort discussed elsewhere [32–33].

8.32. Using the approach suggested in Section 8.3.5 for portfolio cost–effectiveness evaluation, prepare a case study report of cost–effectiveness evaluation of strategic investments in information technology [34–35].

8.33. The most elementary approaches to cost–effectiveness analysis are intended for application to those situations where the following conditions exist.

    a. Enough is known about the technologies under development to develop credible forecasts and estimates of their performance.

    b. There is no transfer of information between projects.

    c. The performance of each product does not depend on the performance of the others.

These criteria are generally met by large-scale projects directed at the final stages of technology development but are not met by many basic research efforts. Discuss issues associated with expanding the approach such that it is applicable to each of these situations.

8.34. Please write a brief hypothetical case study of CEA that includes the cost of quality and the cost of poor quality. Illustrate the symbiotic role of operational and strategic level quality assurance and management and CEA.

8.35. How do, or should, we associate the costs and benefits of quality and the costs and benefits of poor quality with cost issues or effectiveness issues for a CEA? Please illustrate your discussion with a simple assessment scenario for an RDT&E effort.

8.36. This problem calls for some research, innovative thinking on your part, and speculation. With considerable extension, a high-quality response to the sort of query posed here could form a rather nice research paper. The problem itself concerns *strategic and process level cost and operational effectiveness analysis.* In it you are asked to do some very preliminary work concerning the interactions of such strategic level concerns as process management and total quality management on the resulting cost and operational effectiveness of the resulting operational products. The following process efforts are suggested as positive influencers of quality and effective products. These are based on Deming's 14 points for TQM.

    a. The current preoccupation with product and current outcome must be abandoned and replaced by process improvement objectives.

    b. The notion of inspections to identify defective products should be sublimated to detect deterrent strategies throughout the life cycle to enable process improvements that enhance product quality and trustworthiness.

c. Employees and human resources in general are an organization's most valuable resource. When they perform poorly, much of the blame must be shared by management.

d. Quality of product is the major product consideration. It leads directly to market share, sales volume, and profits.

e. Continuous process improvements for the purpose of maximum customer satisfaction is the most important process consideration.

f. Flexible relative standards may be invaluable in supporting excellence in all critical processes.

g. Management must lead through example, education, long-range strategic planning, understanding the essential statistical variability of key processes, and the pursuit of venturesome and well-thought out breakthroughs that lead to major increases in quality, market share, and profitability.

h. Employees must be motivated to improve work processes and work product as if they were partners in the enterprise.

i. Organizational plans are theories of action in practice documents that enable each person in the enterprise to understand what they must do in order to support the organization in pursuit of strategic objectives.

j. Education is a long-term investment leading to increased market share and profitability of the enterprise. It is not an expense. All employees receive generous directed educational support to enhance quality and productivity.

k. Selection of product suppliers based on initial price alone must be abandoned and replaced with selection based on total life-cycle cost effectiveness maximization.

l. Reward systems that strongly encourage teamwork and group effort must be encouraged and traditional reward systems must be reoriented to reflect individual contributions that are within the framework and supportive of overall unit performance.

m. Organizational structures must be redesigned to support cross-functionality and to enhance rapid communications and decision-making through efficient and effective reporting structures. All managers are employees. All employees are managers.

n. Organizational policies and procedures are not only words; they are a strategy for thinking, culture and people appraisal.

There are many more relations that can be stated. We could bring in notions of concurrent engineering, enterprise integration, just-in-time manufacturing, and configuration management. Also, we could bring in the relationships for excellence identified by Tom Peters, or the TQM approaches of John Juran and others. We could relate these efforts to the critical success factors for the Baldridge award:

Attribute	Points
Leadership	120
Information and Analysis	60
Strategic Planning	80
Human Resource Utilization	150
Quality Assurance of Product and Services	140
Quality Assurance Results	150
Customer Satisfaction	300

So there is a lot of material for you to review to be responsive to the needs of this problem statement. Clearly, you can't do it all here. Please prepare a report on your thoughts about how to bring about a meaningful tool for doing strategic and process level cost and operational effectiveness analysis.

REFERENCES

[1] Sage, A. P., *Economic Systems Analysis: Microeconomics for Systems Engineering, Engineering Management, and Project Selection*, North Holland, New York, 1983.

[2] Layard, P. R. G., and Walters, A. A., *Microeconomic Theory*, McGraw-Hill, New York, 1978.

[3] Mansfield, E., *Microeconomics: Theory and Applications*, 2nd ed., W. W. Norton, New York, 1975.

[4] Hendeson, J. M., and Quandt, R. E., *Micro-economic Theory: A Mathematical Approach*, 2nd ed., McGraw-Hill, New York, 1971.

[5] Pindyck, R. S., and Rubinfeld, D. L., *Microeconomics*, Macmillan, New York, 1989.

[6] Quirk, J., and Saposnik, R., *Introduction to General Equilibrium Theory and Welfare Economics*. McGraw-Hill, New York, 1968.

[7] Mishan, E. J., *Introduction to Normative Economics*, Oxford University Press, Oxford, 1981.

[8] Ng, Y. K., *Welfare Economics*, Halsted Press, New York, 1980.

[9] Quinee, J. B., "Long Range Planning of Industrial Research," *Harvard Business Review*, 1961, pp. 88–102.

[10] *Cost/Schedule Control Systems Criteria: Joint Implementation Guide*, DARCOM P714, AFSCP/AFLCP 173-5, NAVMAT P 5243, 1982.

[11] Michaels, J. V., and Wood, W. P., *Design to Cost*, John Wiley, New York, 1989.

[12] Lederer, A. L., and Prasad, J., "Nine Management Guidelines for Better Cost Estimating," *Communications of the Association for Computing Machinery*, Vol. 35, No. 2, 1992, p. 51–59.

[13]   Juran, J. M., *Juran on Quality: An Executive Handbook*, Free Press, New York, 1989.

[14]   Feigenbaum, A. V., *Total Quality Control*, Third Edition, McGraw Hill Book Co., New York, 1991.

[15]   Bussey, L. E., *The Economic Analysis of Industrial Projects*, Prentice-Hall, Englewood Cliffs NJ, 1978.

[16]   Mishan, E. J., *Cost–Benefit Analysis*, Praeger, New York, 1976.

[17]   Porter, A. L., Rossini, F. A., Carpenter, S. R., and Roper, A. T., 1980 *A Guidebook for Technology Assessment and Impact Analysis*, North-Holland, New York, 1980.

[18]   Sassone, P. G., and Schaffer, W. A., *Cost–Benefit Analysis—A Handbook*, Academic Press, New York, 1978.

[19]   Sugden, R., and Williams, A., *The Principles of Practical Cost–Benefit Analysis*, Oxford University Press, Oxford, 1978.

[20]   King, J. L., and Schrems, E. L., "Cost-Benefit Analysis in Information Systems Development and Operation," *Computing Surveys*, Vol. 10, No. 1, 1978, pp. 20–34.

[21]   Ewusi-Mensah, K. K., "Evaluating Information Systems Projects: A Perspective on Cost-Benefit Analysis," *Information Systems*, Vol. 14, No. 3, 1989, pp. 205–217.

[22]   Merkhofer, M. W., *Decision Science and Social Risk Management*, Reidel, Dordrecht, 1987.

[23]   Miller, C., and Sage, A. P., "A Methodology for the Evaluation of Research and Development Projects and Associated Resource Allocation," *Computers and Electrical Engineering*, Vol. 8, No. 2, 1981, pp. 123–152.

[24]   Sage, A. P., "Systems Management and Integration of Information Processing Technologies," *Proceedings United Nations University Workshop on Information Technology*, Paris France, September 1990.

[25]   Porter, A. L., Roper, A. W., Mason, T. W., Rossini, F. A., and Banks, J., *Forecasting and Management of Technology*, John Wiley, New York, 1991.

[26]   Miller, C., and Sage, A. P., "Application of a Methodology for Evaluation, Prioritization, and Resource Allocation to Energy Conservation Program Planning," *Computers and Electrical Engineering*, Vol. 8, No. 1, 1981, pp. 49–67.

[27]   Dutton, J. A., and Crowe, L., "Setting Priorities among Scientific Initiatives," *American Scientist*, Vol. 76, 1988, pp. 599–603.

[28]   Roussel, P. A., Saad, K. N., and Erickson, T. J., *Third Generation R & D*, Harvard Business School Press, Cambridge MA, 1991.

[29]   Oral, M., Kettani, O., and Lang, P., "A Methodology for Collective Evaluation and Selection of Industrial R & D Projects," *Management Science*, Vol. 37, No. 7, 1991, pp. 871–885.

[30]   Liberatore, M. L., and Titus, G. J., "Managing Industrial R & D Projects: Current Practice and Future Directions," *Journal of the Society of Research Administrators*, Vol. 18, No. 1, 1986, pp. 5–12.

[31] Gremillion, L. L., and Pyburn, P. J., "Justifying Decision Support and Office Automation Systems," *Journal of Management Information Systems*, Vol. 2, No. 1, 1985, pp. 5–17.

[32] Putrus, R., "Accounting for Intangibles in CIM Justification," *The Journal of CIM Management*, Vol. 6, No. 2, 1990, pp. 24–28.

[33] Shah, S. K., "Implementing a Manufacturing Information System," *The Journal of Information Solutions Management*, Vol. 17, No. 1, 1990, pp. 8–13.

[34] Clemons, E. K., and Weber, B. W., "Strategic Information Technology Investments: Guidelines for Decision Making," *Journal of Management Information Systems*, Vol. 7, No. 2, 1990, pp. 9–28.

[35] Kumar, K., "Post Implementation Evaluation of Computer Based Information Systems: Current Practices," *Communications of the ACM*, Vol. 33, No. 2, 1990, pp. 203–212.

Chapter **9**

Cognitive Ergonomics

The systems engineering life cycle is intended to enable evolution of high-quality, trustworthy systems that have appropriate structure and function support for identified purposeful objectives. Many major efforts called for throughout all phases of a systems engineering life cycle require human interaction with various aspects of the system.

This chapter continues our discussion of systems engineering and concludes this introductory text. The focus is on the role of the human in systems and the information support aids that should be made available to improve development and use of systems by humans. Our effort begins with a very brief overview of classic ergonomics and human factors engineering. Then we turn our attention to human problem-solving endeavors. This leads directly into a study of experientially based models for human information processing and cognition. Organizational information processing and decision related issues are of much importance also, and we study these here. Each of these leads to a systematic study of human error and approaches to ameliorate the effects of human error in systems and in organizations.

9.1. INTRODUCTION

One primary purpose in this chapter is to discuss systems engineering concerns relative to such problem-solving tasks as fault detection, diagnosis, and correction for purposes that range from planning to operations. Advances in information technology, together with the desire to improve productivity and the human condition, render physiological skills that involve only, or primarily, strength and motor abilities relatively less important than in the past. They significantly increase the importance of cognitive and intellectual skills. The need for humans to monitor and control conditions necessary for effective operation of systems is greater than ever. In many cases, primarily cognitive efforts are ultimately translated into physical control efforts. Often, systems that accomplish this are called *human–machine systems.*

A second major concern is the design of information technology-based support systems to aid human performance. The first concern, monitoring and human controller effort, has been addressed for a much longer period of time than the second. We provide some historical perspectives on it in this introductory section. Such task categories as controlling and problem-solving describe typical human activities in human–machine systems. We are particularly concerned here with system design for human interaction, with the cognitive tasks involved in problem solving, with the physiological tasks involved in controlling complex systems, and with the interfaces between humans and systems that are an integral part of this. We call this area of interest *cognitive ergonomics.*

Some of the roles for a human in a human–machine system include the following.

1. Assessing the situation at hand, in terms of system operation, to identify needs for human supervisory control, objectives to be fulfilled, and issues to be resolved.
2. Identifying task requirements to enable determination of the issues to be examined further and the issues not to be considered further.
3. Identifying alternative courses of actions, which may resolve the identified issues.
4. Identifying probable impacts of the alternative courses of action on the functioning of the system.
5. Interpreting these impacts in terms of the objectives or needs that are to be fulfilled.
6. Selecting an alternative for implementation and implement the resulting control.
7. Monitoring performance to enable determination of how well the human–machine system is performing.

Classical human–machine systems engineering, often called human factors engineering or ergonomics, was focused on training for skill-based behavior and associated physiological concerns. Contemporary efforts necessarily also emphasize the integration of physiological concerns with cognitive concerns. This has been motivated by the great deal of evidence that humans do not react to task requirements in a way capable of being easily stereotyped, but rather in a fashion that is very much a function of the task, the requirements perceived for the task, the environment into which the task is imbedded, and

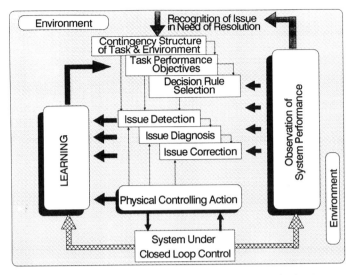

Figure 9.1. The Formal Process of Human Supervisory Control of a System Involving Perceptual, Cognitive, and Motor Processes

the experiential familiarity of the human with the task, the task requirements, and the environment into which these are imbedded—in other words, the contingency task structure.

Human tasks in a human–machine system generally involve controlling, which usually involves some physiological effort, and problem detection and diagnosis, which generally involves cognitive efforts. In addition, monitoring or feedback loops also exist in order to support learning and more precise control. Figure 9.1 indicates these generic constructs.

Issue, or fault, detection concerns the identification of a potential difficulty that impedes operation of a system. Issue or fault diagnosis is concerned with identification of a set of hypotheses concerning the likely cause of a system malfunction, and the evaluation and selection of a most likely cause. Issue or fault resolution, correction, or control is concerned with resolving issues or solving problems in actual situations. Whereas detection and diagnosis are primarily cognitive efforts, correction will often also involve physiological efforts. In the classic human–machine system, correction or controlling is accomplished with the objective of returning the overall system to a satisfactory operating state [1–3]. It is not uncommon to distinguish the cognitive effort in controlling from the physiological effort [4]. The efforts involved in detection and diagnosis [5], which are primarily cognitive, may also be called problem formulation or situation assessment. The subsequent efforts at correction may be termed solution execution. Together, the situation assessment and (re)solution execution comprise problem solving. Then, the human–machine systems problem may be regarded as one of problem finding and solution execution.

There are many other tasks in human–machine systems, such as monitoring of performance and communications. These are subtasks, or supporting tasks, for the two primary activities of problem finding and controlling. Communicating falls into two very different categories. The first is the verbal communication among the human members of the group responsible for aspects of system operation. The second involves communication between humans and the technological system that is being controlled. Often this latter form of communication is often called human–computer dialog [6–8] and the design of dialog generation and management systems is an important part of information systems engineering efforts [9].

Issues relating to human–machine systems go far back in time.* Early designers of human–machine systems relied on their skill-based experience, because system designers were often system users, to insure efficiency and effectiveness of the resulting system. Today, the system developer and the system user are two different people or groups. This, together with the need for more complex and faster responding technological systems that operate in unfamiliar environments, means that a more formal approach is often needed. Most of the early investigations of human–machine systems were concerned with physiologically based manual control tasks, often applied to aircraft piloting, later also to ship steering, car driving, and industrial process control. This work was accomplished by experimental psychologists, systems engineers, or application-oriented engineers. Overviews that now have primarily historical value have been presented by Kelley [10], Edwards and Lees [11], and Johannsen et al. [12]. Many control theoretic models have been developed to describe the behavior of the human operator in manual control and estimation tasks [13]. The most sophisticated and well-validated model seems to be the optimal control model first described by Kleinman, Baron, and Levison [14].

Even with such slow response systems as ships and industrial process plants, it has become obvious that human operator behavior is not easily explained by well-established control theoretic methods. Human controller behavior is highly nonlinear and error prone, even in these simple and slow-response cases. Human operator workload in physical task performance has been investigated for many years [15]. Although appropriate definitions and a common understanding concerning the importance of this research area exist, the reliability and validity of most workload measures was initially quite poor. A primary reason for this was that human performance modeling was primarily based on physiological considerations. Initially, there existed a neglect of human cognitive dynamics. Current efforts attempt to alleviate

---

*Plows had to be designed, for example, so that humans could control the motion of the animals pulling them or, in some cases, so that other humans could pull them.

these deficiencies; it is with these that our efforts in this chapter are concerned.

This chapter is organized as follows. First, we present a number of studies of *cognitive control* in Section 9.2. We begin with a discussion of a common physiological, or component, model of human performance. Then we will look at some of the stereotypical models in which humans are characterized as having fixed cognitive styles. More modern views suggest that these styles, which are more properly called control modes due to the dynamics involved, are very strongly influenced by such contingency task structural features as the task, the environment into which the task is imbedded, the experiential familiarity of the decision maker with the task and environment, the criticality of the task at hand, and the time available for its accomplishment. We continue our efforts with a presentation of several dynamic cognitive control mode models.

Knowledge representation is much related to the approaches humans bring to problem resolution efforts. We discuss a number of ways that can be used, by humans and by machines, to represent knowledge. Organizations of humans perform tasks. We devote our efforts in Section 9.4 to a study of a variety of rationality perspectives that indicate how individuals in organizations process information for a variety of decision-related tasks. This leads to the very important subject of human error in the development and operation of systems. The concluding parts of this chapter provide discussions of human task performance abilities and human error and its amelioration.

9.2. COGNITIVE CONTROL IN TASK PERFORMANCE

Several approaches explain how individuals and organizations acquire, represent, and use information to describe their perceptions of the world around them, as well as issues and situations that are of importance. Human information processing is a vital and crucial ingredient in effective decision making. Information processing theories of problem solving, judgment, and decision making are normally based on the assumption that individuals have the following qualities.

1. An input mechanism for acquisition of information.
2. An output mechanism for interpretation and choice making.
3. Internal processes for filtering and other analysis efforts associated with information.
4. Memories for long- and short-term storage of information.

A large number of ways represent human information processing. Many of these are described in texts in cognitive psychology such as those by

Anderson [16, 17], Estes [18], Mayer [19], Eysenck [20], and Mandler [21], as well as in applied works in such areas as consumer choice [22].

The key functions that determine how a specific problem or decision situation is cognized depend upon an interaction of the memory and higher order cognition of the problem solver with the environment, which occurs through what we denote as the contingency task structure. The various information analysis and interpretation processes of thinking, task performance objective identification, evaluation and decision rule identification, are called *higher order cognition*. This is not because they are somehow more important than the so-called *lower order cognition* efforts of information acquisition, which involve such efforts as sensation, attention, perception, and pattern recognition. They are called higher order because they occur later in time in the overall information processing effort.

Information processing and decision-making efforts intimately involve memory. Human memory is often claimed to be comprised of two major components, short-term memory and long-term memory. Short-term memory plays a key role in immediate recall of actively rehearsed limited information. Unless conscious effort is put forth in recalling information from short-term memory, this cannot be done after a lapse of 30 to 60 seconds from initial presentation. Models of a working short-term memory involve a number of mechanisms, such as an articulatory rehearsal loop that has the capacity to retain short verbal sequences. This is just one mechanism by which short-term retention is possible. There are a number of other sensory registers. It is important to note that short-term memory is an integrated network of many mechanisms and is associated with a number of processes.

There are generally acknowledged to be four components of information acquisition, as indicated in Figure 9.2:

1. *Memory:* Both short term and long term memory as we have just discussed.
2. *Sensation:* The initial experience of stimulation from the sensory modalities.
3. *Attention:* The concentration of cognitive effort on sensory stimuli.
4. *Perception:* The use of higher order cognition to interpret sensory stimuli.

In sensation, information is acquired through the five major senses (touch, taste, hearing, smell, and sight), which are environmentally activated in response to a specific array of stimulus energies. In a specific decision-making situation, the decision maker filters out bits of data believed to be irrelevant. The filtering process is based upon task characteristics, experience, motivation, as well as other features and demands of the specific decision-making situation. If such a filtering mechanism were not to exist, the decision maker

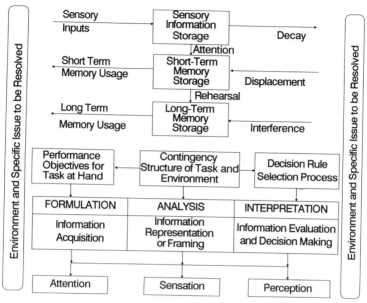

Figure 9.2. Model of Perceptual and Cognitive Process for Problem Solving

would often encounter information overload. This would generally result in a form of cognitive saturation and the inability to process sufficient information for the task at hand.

Ultimately involved in retention processes is the notion of attention. For information to be transferred from short-term to long-term memory, constant conscious attention is required. Information entering short-term memory that is not attended to through specific conscious processes is lost. Processing of information demands attending to relevant bits of incoming data and transfer of the data into long-term memory for future retrieval for purposeful use. Interferences of various types may interrupt attention and thus hinder transfer and retention of relevant stimuli into long-term memory.

The process of perception, or pattern recognition, is also inherent in the processing of information acquisition. This process generally involves two phases: extraction, and identification. A given stimulus is coded in terms of its features and may be received through any of the senses. The meaning that a stimulus conveys to a human, or the manner in which the human perceives the stimulus, is dependent upon the patterns extracted from the stimulus. In the identification phase, the sensory-perceptual system classifies the stimulus object. The way in which this is often assumed to occur is by a weighted matching of the current feature list against a likely set of prototypes in long-term memory with the input being classified according to the name of the best matching prototype. The quality or extent of the sensory information extracted determines the accuracy of identification that is thereby obtained.

Short-term and long-term memory components play key roles in the information acquisition process as the decision maker proceeds with efforts that ultimately culminate in choice. A response system couples the memory system to the sensory system and the environment. Thus it controls or activates the sensory modalities on the basis of the actions taken. Through the response system, we close the information flow feedback loop. A model of the principal components of information flow might consist of the response system, the sensory system, the memory system, and the central processor. The central processor coordinates memorizing, thinking, evaluation of information, and final decision making.

We have just described what might be regarded as a component, or physiological, model of information processing. In such a stimulus-response model, behavior is seen as being initiated by the onset of stimuli. An apparent deficiency in approaches of this sort is that there is little consideration of how information bits are aggregated to influence choice and how the decision maker goes about the process of information formulation or acquisition, analysis, and interpretation. These are needed in studies involving human and organizational information processing. Thus, we would like to be able to transition from a physiological model of human information processing to a systems engineering model of various stages of information processing, based on cognitive ergonomic considerations. We describe several such models here. The first is a group of stereotypical models, based on Jung's early work. Then we examine the genetic epistemology model of Piaget, which allows for learning and dynamic evolution, and other recent models.

9.2.1. Stereotypical Cognitive Styles

It is becoming increasingly clear that it is necessary to incorporate not only problem characteristics, but also problem solver or decision-maker characteristics, into the design of systems of all types, especially those that necessarily involve human interactions. A deficiency in some past systems designs has been the neglect of the human decision maker's role and characteristics and their effects. Essentially all available evidence suggests that problem characteristics and user characteristics influence the planning and choice strategies adopted by the decision maker.

System designs must be responsive to the observation that there are two fundamentally different thought or cognition processes. These are often and correctly associated physiologically with different halves of the brain [23–26]. One type of thought process is typically described by adjectives or attributes such as verbal, logical, sequenced, thinking, formal, rational, and analytical. The second type is described as nonverbal, intuitive, wholistic, sensing, feeling, and heuristic. Analytic processes are typically viewed as superior in engineering and natural science and intuitive processes are more often associated with the arts.

Mason and Mitroff [27] have suggested that each person possesses a particular specific psychological cognitive style or personality and that each personality type utilizes information in different ways. In their research on management information system design, they claim that an information system consists of a person of a certain psychological type who faces a problem in some organizational context for which needed evidence to arrive at a solution is made available through some mode of presentation. The person uses a fixed stereotypical cognitive style in pursuit of a solution to the problem.

There are five essential variables in the information system characterization of Mason and Mitroff. Each of these are disaggregated into subelements. Mason and Mitroff characterize the psychological-type variable according to the Jungian stereotypology. In this typology, people differ according to their preference for information acquisition and analysis and the preferred approach to information evaluation and interpretation. At extremes in the information acquisition dimension are sensing-oriented or sensation types who prefer detailed well-structured problems and who like precise routine tasks and intuitive-oriented type people who dislike precise routine structured tasks and perceive issues wholistically. At extremes in the information evaluation dimension are feeling-oriented people who rely on emotions, situational ethics, and personal values in making decisions and thinking-oriented individuals who rely on impersonal logical arguments in reaching decisions.

In the Bariff and Lusk cognitive style model [28], individuals may be categorized according to whether they are tightly bound by external referents in structuring cognitions, in which case they are called field dependent or low analytic, or whether they can make use of internal referents as well as external referents in structuring cognitions, in which case they are high analytic or field independent. In a field-dependent mode, perception is dominated by the overall organization of the field. There is limited ability to perceive discrete parts of a field, especially as distinct from a specific organized background. Field independent people have more analytic and structuring ability in comparison to field-dependent people in that they can disaggregate a whole into its component parts.

The systematic-heuristic categorization of Bariff and Lusk describes cognitive styles associated with people who either search information for causal relationships that promote algorithmic solutions, or who search information by trial-and-error hypothesis testing. Systematic individuals utilize abstract logical models and processes in their cognition efforts. Heuristic individuals utilize common sense, past experience, and intuitive feel. Systematic individuals would be able to cope with well-structured problems without difficulty and would approach unstructured problems by attempting to seek underlying structural relations, whereas heuristic individuals would attempt to cope with unstructured problems without a conscious effort to seek structural identification.

Bariff and Lusk discussed three cognitive style characteristics relevant to information system design: cognitive complexity, field dependent/independent, and systematic/heuristic. The cognitive complexity characteristic involves three structural characteristics of thinking and perception: differentiation, which refers to the number of dimensions sought or extracted and assimilated from data discrimination, the fineness of the articulation process in which stimuli are assigned to the same or different categories, and integration, which refers to the number and completeness of interconnections among rules for combining information.

McKenney and Keen [29] have done extensive work on cognitive style measurements. They conceptualize cognitive style in two dimensions: information acquisition and information processing and evaluation. The information acquisition mode consists of receptive and preceptive behavior lying at the opposite extremes of a continuum. The authors claim that preceptive decision makers use concepts or precepts to filter data, to focus on patterns of information, and to look for deviations from or conformities with their expectations. Receptive people tend to focus on data rather than patterns and derive implications from data by direct observation of it, rather than by fitting it to their own precepts.

With respect to information processing and evaluation, McKenney and Keen measured individuals on a scale, with the systematic thinker at one extreme and the intuitive* thinker at the other extreme. They have shown, using a battery of pencil and paper tests, that systematic thinkers approach a problem by structuring it in terms of some method that would lead to a solution, whereas intuitive thinkers use trial and error, intuition, and previous experience to obtain solutions. A well known Myers-Briggs personality test, based on the Jungian typology, attempts to analyze personalities on four dimensions: intuitive-sensing, thinking-judgmental (or feeling), perceiving-judging, and introvert-extrovert. The intuitive-sensing (I-S) and thinking-judgmental (T-J) dimensions are those most closely related to our discussion here.

In another effort, Driver and Mock [30] developed what they call decision-style theory, a set of four decision styles intended to relate conceptual structure of decision makers to both the amount of information they

---

*Intuition is used here to mean experientially based judgment that does not require the use of the formal analytical processes, or formal production rules, that might initially have been used to reach sound judgment. The Jungian view of intuition is a fundamentally different way of judging than explicit formal knowledge or rule based judgment. Intuition, somehow, enables reaches for understanding with an approach that is outside the formal knowledge based context of facts, logic, and sensations. I suspect that it must be concluded that I believe that only correct experiences will generally lead to correct intuitive thoughts.

tend to use and the degree of focusing that they exhibit in the use of information. A *decisive* person will use the minimum amount of information possible and will generally identify a single acceptable course of action. A *flexible* person will use minimum information but will generally identify and evaluate a number of alternatives. A *hierarchic* person will use much information and will identify a single acceptable course of action. Finally, an *integrative* person will utilize much information to identify and evaluate many potentially acceptable courses of action. This two-dimensional decision-style typology is related by Driver and Mock to the Jungian styles, as in the McKeeney and Keen effort just discussed.

Most of the research in this area assumes that there are individual differences that influence information usage, and that these involve cognitive style, personality, and demographic/situational variables. Thus, the term *cognitive style* refers to the process behavior that individuals exhibit in the formulation or acquisition, analysis, and interpretation of information or data of presumed value for decision making. Doubtlessly cognitive style is somewhat influenced by such personality variables as dogmatism, introversion, extroversion, and tolerance for ambiguity and equivocality. The demographic/situational variables involve personal characteristics such as intellectual ability, education, experience with and knowledge of specific contingency tasks, age, and the like. An important situational variable is the level of time–stress encountered by the decision maker in a specific problem situation. The level of time–stress, which results in the adoption of a coping pattern, influences the decision maker's ability in acquisition and processing of the information necessary for decision making. The subject of time–stress will be dealt with in our discussion of the Janis-Mann model.

A number of studies such as those by Taylor [31] and Simon [32] indicate that human decision makers attempt to bring order into their information processing activities when confronted with excess information or the lack of sufficient information. Many early studies assumed that fixed patterns of dealing with information were preferred by the decision maker for the process of experiencing the world, and these were referred to as cognitive style. Some early studies view cognitive style as a mode of functioning that is static and pervasive throughout all of a person's perceptive and intellectual activities and that is invariant with respect to tasks, environment into which the task is imbedded, and experiential familiarity of a person with task and environment.

We have now examined four cognitive-style characteristics in this section. Figure 9.3 summarizes the models of cognitive style that result from these efforts. We note the considerable similarity among these four constructs. There are many other related models. In a recent overview and model development effort, for example, Taggart and Valenzi [33] identify three fashions for rational information processing and three for intuitive information processing. The three fashions relate to the approach to tasks or the type of logic used, the approach to forecasting or preparation for the future, and

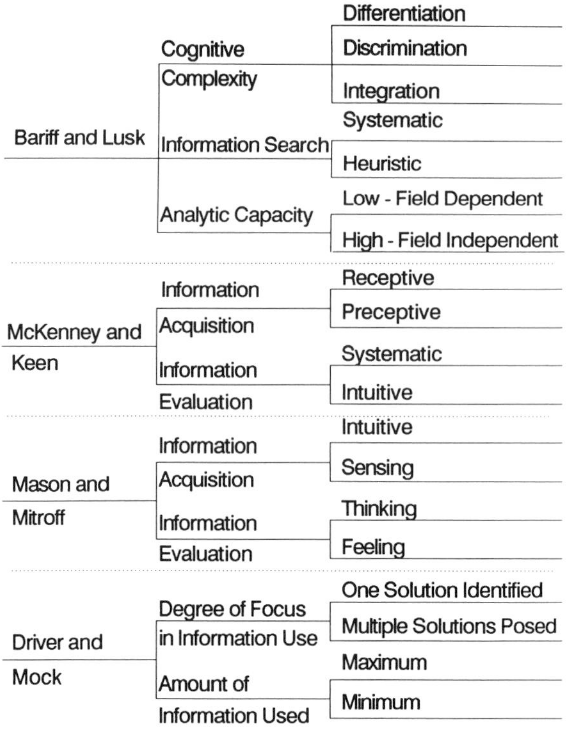

Bariff and Lusk	Cognitive Complexity	Differentiation
		Discrimination
		Integration
	Information Search	Systematic
		Heuristic
	Analytic Capacity	Low - Field Dependent
		High - Field Independent
McKenney and Keen	Information Acquisition	Receptive
		Preceptive
	Information Evaluation	Systematic
		Intuitive
Mason and Mitroff	Information Acquisition	Intuitive
		Sensing
	Information Evaluation	Thinking
		Feeling
Driver and Mock	Degree of Focus in Information Use	One Solution Identified
		Multiple Solutions Posed
	Amount of Information Used	Maximum
		Minimum

Figure 9.3. Stereotypical Cognitive Style Dimensions

ways of working with people on tasks. The fundamental differences are as follows.

Intuitive	Rational
Insight	Logic
Vision	Planning
Feeling	Ritual

These authors use this three dual-fashion metaphor to develop a sample survey approach to measurement. In general, the model is validated for the stereotypical evaluations that were conducted.

Tiedemann [34] indicates that most cognitive-style measures are at best interpreted as ability tests. He indicates that a considerable amount of the research on this subject now needs to be reinterpreted. We agree with this, and note that one of the major determinants of ability is experiential learning and familiarity with the task at hand. All of this suggests one of the several dynamic evolutionary cognitive control mode models that we study later in this section.

We conclude that the static and stereotypical viewpoint suggesting an unchanging nature of human thought processes appears unsatisfactory for our purposes. The two thought processes, intuitive and rational, are complementary and compatible. They are not necessarily competitive and incompatible in any meaningful way. One thought process be quite deficient, in our view, if it is not supported by the other. For example, a person may be taught rules and skills, or concrete operational thought relative to some specific task. They may be quite proficient as long as the task performance is occurring *naturally*. However, environmental or other changes may create the need for formal operational thought or vigilant information processing. If there is no capacity for this, major difficulties may ensue. The intuitive thought process supports the rational thought process by suggesting ideas, alternatives, and so on. The rational thought process supports the intuitive process by expressing, structuring, analyzing, and validating the creative ideas that occur in the intuitive process.

An appropriate support process must be tolerant and supportive of the cognitive thought processes of an individual or a group. These will typically vary across individuals and within the same individual as a function of the environment, the individual's previous experience with the environment, and those associated factors that introduce varying amounts of time–stress. Thus, a contingency task structural view of individuals and organizations in decision situations is needed, as contrasted with a stereotypical view in which individuals are assumed to process fixed, static, and unchanging cognitive characteristics that are uninfluenced by environmental and experiential considerations. This leads naturally into consideration of the adaptive character of thought and associated information processing efforts, a subject that has recently been explored by Anderson [35–37].

The dynamic evolutionary view suggested here encourages us to consider and assess changes over time and encourages the growth of experiential familiarity and expertise. The metamorphosis of future events over time is inherently uncertain and associated with much information imprecision. It is especially important, also, that we consider values as containing noncommensurate, ambiguous, equivocal, and uncertain components, rather than as being absolute, consistent, invariant over time, precise, and exogenous with respect to choice [38].

There have been a number of studies of the measuring instruments involved in classifying people according to these cognitive styles. Zmud's [39] findings indicate that perceptual differences can indeed be observed for specific cognitive styles and among subjects with different educational and experiential backgrounds. However, his results also indicate that there is no apparent relationship between cognitive style perceptions and actual behavior. Zmud [40] has indicated low correlation also among test scores on different instruments used to measure cognitive styles. Chervany, Senn, and

Dickson [41, 42] have expressed much concern and pessimism concerning the validity of much of the contemporary research in this area. They comment that the study of individual personality differences as predictors of human behavior and performance have been basically unsuccessful in that it has not been possible to predict performance on the basis of personality characteristics. They and others argue that the characteristics of the task in which the individual involved is a prime determinant of human behavior often overrides the effects of personality. This position appears unassailable.

Benbasat and Taylor [43] note that much cognitive complexity research deals with interpersonal perception and has limited value for modeling activities of managers in processing information and making decisions. Mischel [44], in his study of the interface between cognition and personality, is especially perceptive in discussing the potential hazards of attributions and enduring categorizations of people into fixed compartments on the basis of a few behavioral personality-based signs. The assumptions that static characterizations are sufficiently informative to enable behavior predictions in specific settings are strongly challenged. An evaluation of the uses and limitations of static trait characterization of individuals is presented and the strong interacting role of context is emphasized. Mischel is especially concerned with cognitive economics, that is, the recognition that people are easily overloaded with an abundance of information and that simplified methods of acquisition and processing of information are therefore used. He is especially concerned also with growth of self-knowledge and rules for self-regulation with maturation, topics closely related to the Piaget model we discuss below.

We concur with these views in that we believe that it is the individual's prior experience with the task at hand, blended with previous experiences, that is the primary determinant of the cognitive control mode. An individual's information processing capacity under various levels of time–stress and in different contingency task structures determines the quality of the decision making. These factors depend strongly upon experience. Thus, we support the information processing view of Simon that few characteristics of the human information processing system are invariant over the decision maker and the task. These characteristics are generally experiential and evolve over time in a dynamic fashion. They are not static and cannot be treated as static and task invariant for a given individual. This suggests that personalities evolve gradually over time, adjusting to changing circumstances and new experiences and changing perspectives. It does not deny the existence of some enduring aspects of character and preferred approaches to governance and management.

Huber [45] characterized the situation well in the title of one of his papers: "Much Ado About Nothing." Our view is that these studies and follow-on efforts do indeed suggest a cognitive style for human information processing endeavors. Rather than being static and invariant over time, it is a dynamic function of the contingency task structure. We provide and discuss additional evidence supporting a dynamic cognitive control mode characterization that incorporates the contingency task structure and the decision maker's related task experiences. In particular, we emphasize the strong need for consideration of the structure and the content of decision situations to evolve contextually meaningful support for humans. Of particular importance with respect to these dynamic cognitive control modes are relationships between the environmental complexity of contingency task structures, information processing characteristics, and how these evolve over time and with learning.

9.2.2. Piaget's Model of Cognitive Development

Insights into the nature of cognitive development, including a conceptual model of cognitive activity are contained in the works of Piaget, who founded genetic epistemology in the 1930s, and in recent accounts of this development [46–49]. According to Piaget, there are four stages of intellectual development in a human being:

1. Sensory motor.
2. Preoperational.
3. Concrete operational.
4. Formal operational.

The last two of these are of particular importance to our efforts here. In the writings of Piaget, intellectual development is seen as a function of four variables: maturation, experience, education, and self-regulation, which is a process of mental struggle with discomforting information until identification of a satisfactory mental construct allows intellectual growth or learning. In Piaget's model of intellectual development, both formal and concrete operational thinkers can deal logically with empirical data, manipulate symbols, and organize facts toward the solution of well-structured and personally familiar problems. But concrete only thinkers lack the formal thinker's capacity to reason hypothetically and to consider the effect of different variables or possibilities that are outside of their own personal experience. This suggests that concrete operational (only) thinkers may be capable of learning skills through the repeated use of rules to which they are exposed. It suggests that they will be unable to formally reason relative to such things as possible inapplicability of these rules in some situations.

Concrete operational only thinkers, for instance, often have difficulty in responding "true" or "false" to the statement, "six is not equal to three plus four." As another example: "A card has a number on one side and a letter on the other; test the hypothesis that a card with a vowel on one side will have an even number on the other side." Concrete operational thinkers will have difficulty selecting cards for bottom side examination if the top sides of four cards are a, b, 2, 3. However, failure to pick the cards with "a" and "3" on top may not indicate inability as a formal operational thinker but, rather, failure to properly diagnose the task and determine the need for formal operational thought.

We wish to illustrate model(s) of higher order cognitive processing that describes the mature adult decision maker. Such a decision maker will typically be capable of both formal and concrete operational thought. We argue that selection of a formal or concrete cognitive process depends upon the decision maker's diagnosis of need with respect to a particular task. That diagnosis depends upon a decision maker's maturity, experience, and education with respect to a particular problem. Each of these influence cognitive strain or stress, a subject that is discussed later in this section. Ordinarily, a decision maker prefers a concrete operational thought process and makes use of a formal operational thought process only when concrete operational thought is perceived inappropriate for the task at hand. In general, a concrete operational thought process involves less stress and may involve repetitive and previously learned behavioral patterns. Familiarity and experience with the issue at hand or with issues perceived to be similar or analogous play a vital role in concrete operational thought. In novel situations, which are initially unstructured and where new learning is required, formal operational thought is typically more appropriate than concrete operational thought.

It is of interest to describe some salient features of these two processes. In concrete operational thought, people use concepts that

1. Are drawn directly from their personal experiences.
2. Involve elementary classification and generalization concerning tangible and familiar objects.
3. Involve direct cause-and-effect relationships, typically in simple two-variable situations.
4. Can be taught or understood by analogy, algorithms, affect, standard operating, procedures or recipes.
5. Are closed, in the sense of not demanding exploration of possibilities outside the known environment of the person and the given data.

In formal operational thought, however, people use concepts that may

1. Be imagined, hypothetical, based on alternative scenarios, and potentially counterintuitive.
2. Be open ended, in the sense of requiring speculation about unstated possibilities.
3. Require deductive reasoning using unverified and perhaps flawed hypotheses.
4. Require definition by means of other concepts or abstractions that may have little or no obvious correlation with contemporary reality.
5. Require the identification and structuring of intermediate concepts that are not initially specified.

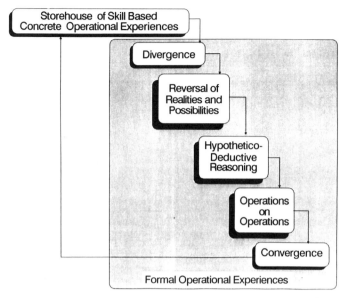

Figure 9.4. Learning Through Formal Operational Experiences (Piaget)

Figure 9.4 presents an illustration of how these processes blend together, and how concrete operational thought may evolve from a learning process that initially involves formal operational thought or from training. The formal operational thought processes are accomplished through reflective observation, abstract conceptualization, and the testing of the resulting concept implications in new situations. It is in this way that the divergence produced by discomforting new experiences allows the learning of new developments and concepts ultimately to be "stored" in memory as part of one's concrete operational experiences.

The formal operational thought process of Piaget appears not fundamentally different from the systems engineering approach that is comprised of the following.

1. Identification of hypotheses, laws, assumptions, or generalizations that may be in the form of axioms or postulates.
2. Deduction, including both formal and informal, or default logic or rules of inference.
3. Observation, experiment, and confirmation or verification of degrees of support for hypotheses.
4. Induction and abduction, using any of a variety of approaches that deal with imperfect information, such as to result in generalizations.

These processes have evolved over the history of systems engineering and very much support information processing, both in systems and organizations. They support the interpretation process of going from model to theory, and the formalization process of going from theory to model. They provide the basis for a number of axioms, postulates, inference rules, the confirmation or verification of degrees of factual support to hypotheses, and a number of inductive and deductive reasoning approaches.

9.2.3. The SHOR Model

Wohl [50] developed, in 1981, a dynamic model for tactical decision processes called SHOR, based upon the following.

1. **S**timulus arrival, suggesting a potential issue associated with some detected decision situation.
2. Generation of **H**ypotheses concerning diagnosis of the situation.
3. Generation of a number of **O**ptions or response alternatives.
4. Some approach to evaluation of the options and selection of a **R**esponse or course of action for implementation.

This model was intended to prescribe a portion of the foundation required for a theory of military command and control, including guiding principles for system development in this environment. Wohl is careful to relate this model, which appears very much as a normative model, to the descriptive realities of decision making. A number of decision assessment prescriptions have been based on this model, illustrated in Figure 9.5. There have been a number of

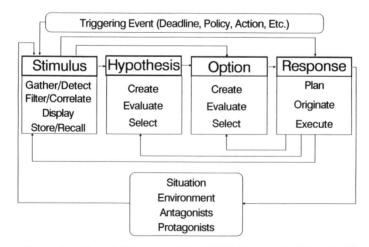

Figure 9.5. Essential Features of the SHOR Human Decision Model

applications of this model, particularly in command-and-control environments, and SHOR has led to a number of innovative designs of sensors and data fusion hardware elements. A number of SHOR applications have been documented [51].

9.2.4. The Rasmussen Model of Judgment and Choice

One particularly useful taxonomy developed by Rasmussen [52, 53] conceptualizes three distinct types of problem-solving behavior or knowledge to support reasoning:

1. Formal reasoning-based behavior.
2. Rule-based behavior.
3. Skill-based behavior.

The choice of which type of reasoning to employ is made by the problem solver on the basis of experiential familiarity with the task at hand, and the environment in which this task is imbedded. All three cognitive control modes of reasoning can exist at the same time, even though the primary control mechanisms move toward skill-based as expertise increases. Figure 9.6 presents an interpretation of the Rasmussen model, which was initially devised to describe the judgment and choice processes of process control operators. This figure does not explicitly show the dynamic learning over time that enables a person to transfer formal rule-based reasoning results to a set of rule-based judgments and then, in turn, to skill-based reasoning. There notions are present throughout many of the works of Rasmussen.

Eight stages of decision assessment are suggested: activation, observation, identification, interpretation, evaluation, goal selection, procedure selection, and activation. Rather than follow a sequenced set of steps, shortcuts may be taken and steps omitted or abbreviated, in the form of associative leaps, when these are perceived appropriate for the contingency task structure.

Rasmussen [54] indicates that human information processing can be described in terms of *data*, the mental representations on information describing the state, *models*, the mental representation of the system's anatomical or functional structure, and *strategies*, the higher level structures of the mental processes that relate goals to sets of models, data, and tactical process rules. Tactical rules describe the control of the detailed processes within a formal strategy. Rasmussen says that a diagnostic search can be performed in two basic ways: symptomatic search and topographic search.

1. *Symptomatic search* uses a set of symptomatic observations as a search template for comparison with a library of symptomatic characterizations of different conditions to find a matching set. The search can be a parallel data driven pattern recognition or a sequential decision table

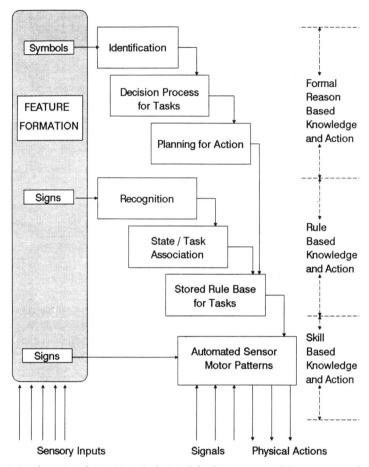

Figure 9.6. Three-Level Decision Style Model of Rasmussen (The Meta-Level Systems Management for Choice of Style is Not Shown)

search. A hypothesis and test search strategy uses reference patterns that are generated on-line by modifying a functional model to correspond to a postulated situation. This strategy is efficient if the method for generating hypotheses is efficient.

2. *Topographic search* is performed with reference to a cognitive template. This method is applied by comparing an actual situation with a template that represents a normal or desired situation. The template may depict situations at several levels of abstraction. It is important for effective topographic search that observed information be available in

forms that correspond to levels of the reference template which represent the desired situation. Here, mismatches represent situations that potentially need attention.

Rasmussen concludes that topographic search is not economic in its use of information but uses simple data processing and small short-term memory load and thus is basically unaided human search. On the other hand, search by hypothesis and test is economic in its use of information but uses complex data processing and has high memory requirements. Thus, it might be more appropriate for mechanistic or computer oriented search.

Hunt and Rouse [55] use this symptomatic–topographic search approach to develop a pattern-recognition-like approach to problem solving. Topographic search rules (T-rules) are used, as are symptomatic search (S-rules), in a search for appropriate rules that have reasonable expected utility, are recallable and available, applicable to the current situation, and are simple. Rule selection, where S-rules are sought before T-rules, is strongly influenced by both the frequency and the recency of past successful use. The model has been used in fault diagnosis tasks with success. Symptomatic search, which is a kind of inductive reasoning, and topographic search, which is a kind of deductive reasoning, are fundamentally different ways of looking at the world.

Rasmussen [56] discusses cognitive control domains in terms of skill, rule, and formal knowledge-based behavior. This characterization is used to distinguish among the psychological mechanisms behind typical categories of errors considered as occurrences of human-task mismatches. The argument is that there are three levels of cognitive control that can be identified. These depend on different kinds of knowledge about the environment and different interpretation of the available information, that is, situation assessment. A model of the three levels of control is shown in Figure 9.6.

At the *skill-based level*, sensed information is perceived as time–space signals and performance is governed by schemata (stored patterns of preprogrammed instructions) represented as analog structures in a time–space domain. At the *rule-based level*, sensed information is perceived as signs that activate stored rules or productions and performance is goal-driven. Signs cannot be used for functional reasoning, to generate new rules, or to predict the response of an environment to unfamiliar conditions. Signs can only be used to trigger or modify rules controlling the sequencing of skilled programs. The *formal knowledge-based level* is triggered by sensed information that indicates an unfamiliar situation.

Conceptual foundations have recently been developed [57] for human–machine interface design, based primarily on suporting these three cognitive levels. In general, humans use skill-based knowledge, rule-based knowledge, and formal reasoning-based knowledge in an attempt to keep processing effort at the lowest cognitive level that trustworthy performance of the task requires. The ecological interface design construct attempts to minimize the difficulty of controlling a complex system while, at the same time, supporting the entire range of activities that specific users may require.

Vicente and Rasmussen suggest [58] that the usual approach to interface design, which is generally based on a *direct manipulation interface* (DMI), for example a computer mouse, fails to consider the following.

1. Practical problem solving can take place at various levels of abstraction on a hierarchical problem domain representation.
2. The same interface can be interpreted in different ways, and the way in which information is interpreted triggers qualitatively different modes of information processing, each requiring a different type of computer support.

They indicate that human errors may be related to problems of learning and adaptation, interference among competing control structures, lack of resources to avoid error, and entrenched human variability. These four categories of human error are impacted somewhat differently as a function of whether skill-based, rule-based, or formal reasoning-based approaches to performance are used. The generic type of errors that result are depicted in Figure 9.7.

The ecological interface construct suggests that an interface design must take these factors into account if it is to be a viable aid that supports human interaction. Ecological interfaces are related to direct manipulation interfaces, direct perception interfaces, object displays, and graphics-based dis-

	Cognitive Control Mechanism		
Error Source	Formal Knowledge	Rule Based	Skill Based
Learning and Adaptation	Acts Judged as Errors	Underspecified Cues Due to Low Effort	Speed Accuracy Trade-Offs
Interference Among Control Structures	False Analogies	Functional Fixation from Adherence to Familiar Rules	Frequently Used Motor Schemata
Lack of Resources	Causal Network Reasoning Limits	Inadequate Memory for Rules	Lack of Speed or Precision
Stochastic Variability	Memory Slips	Errors in Recall of Parameters in Rules	Attention Variability and Lapses

Figure 9.7. Cognitive Control Error Taxonomy of Rasmussen and Vicente

plays. This construct is based upon *Ashby's Law of Requisite Variety* [59], which indicates that in designing a human-computer system interface system, the complexity inherent in the process cannot be reduced below some minimum level if effective control is to be achieved. To ensure that these requirements are met satisfactorily, two questions must be addressed by interface designers.

1. What is the best way of describing or representing the complexity of the domain?
2. What is the best way of communicating this information to the system operator?

These questions translate directly into more specific questions for the designer of systems, one relating to the problem side of design and one relating to user resources. They may be stated as follows:

1. *Problem Side:* What is the best way of presenting the complexity inherent in the problem or issue at hand?
2. *User Side:* What is the most effective resource that the operator has for coping with complexity and how can this be best utilized?

It is important here to comment upon the information requirements to activate skill, rule-, and formal reasoning-based behavior in the model of Rasmussen.* In general, skill-based behavior can only be activated when information is presented in the form of time–space signals. Rule-based behavior is triggered by familiar perceptual structures or signs. Formal reasoning-based behavior is activated by the presence of meaningful relational structures or symbols. Although not apparent in the preceding three sentences, the signals, signs, and symbols that represent perceived knowledge relevant to a situation are very much a function of the user experiential familiarity with the task and environment. It appears possible to design appropriate display and presentation aids for each knowledge form.

Skill-Based Behavior. To support interaction based on time–space signals, the user should be able to act directly on the display, and the structure of the displayed information should have a one-to-one correspondence to the part-whole structure of operator movements. The use of chunked or highly aggregated information is useful here and this leads, for example, to the concept of coarse or more general data for preceptive information gathering. The associated interface should be designed so that the aggregation of elementary movements into more complex routines corresponds with a

---

*It should be noted that Rasmussen uses the term *knowledge* to refer to what we call *formal reasoning-based* or *formal knowledge-based* behavior. We would prefer to associate the term knowledge with skills and rules also, and thus use a slightly different terminology.

concurrent integration of visual features into higher level cues for these routines. In this way, it is possible to represent higher level information as an aggregation of lower level information. Flexibility is maintained by allowing the user to define the desired level of chunking.

Rule Based Behavior. The major element of support here is to provide a consistent one-to-one mapping between the invariant properties of the process and the cues or signs provided by the interface. The ecological interface design attempts to overcome the usual inconsistent mapping capability by developing a unique and consistent mapping between the invariants that govern the process and the cues that the interface provides. This should decrease the frequency of errors due to improper cues since the cues are more aptly related to the fundamental characteristics of the process. The user can control the system relying on perceptual cues rather than resorting to a higher level, generally more complex form of knowledge, which we call formal reasoning.

Formal Reasoning Based Behavior. The objective here is to display the domain's relational structures in the form of an abstraction hierarchy in order to serve as an externalized mental model that will support formal reasoning-based problem solving involving issue formulation, analysis, and interpretation. Often, there is a mismatch between the problem solving behavior that a particular human might wish to use in a given situation and the behavior that a machine designed to support a human in judgment and choice activities might be programmed to emulate. Expert systems often attempt, for example, to use rule-based representations of skill-based knowledge, or alternately use rules when a very experienced person would use a more skilled-based representation of knowledge. This knowledge may initially be expressed in any of several affective or intuitive forms that have been learned by the expert on the basis of much relevant experience. Although the expert may well have learned how to be an expert through formal means that require an absorbed and alert monitoring of the task components, experts may no longer be formally aware of all of their monitoring actions, and may therefore give incomplete information concerning their activities. This simply reflects the view that expertise develops through perceptual learning and through the acquisition of a larger number of rules. Experts are not just faster and more accurate with their use of a larger number of rules, but they can perceive situations, similarities, and task objectives with considerably greater clarity. Additional conceptual discussions of human diagnostic search and associated human reasoning efforts may be found in Ref. 60.

9.2.5. The Klein Cognitive Control Model

Studies of information support for U.S. Air Force command-and-communication systems accomplished by Klein [61, 62] express a number of concerns regarding artificial intelligence and formal decision support-based ap-

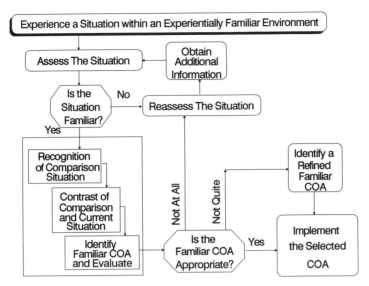

Figure 9.8. Iterative Application of Recognition Promoted Decision Making

proaches to aid decision making. Specifically, these reservations concern potential inabilities and unwillingness of experts to disaggregate situations into components and to analyze these discrete components. Klein indicates that the proficient performance of experts may be based more on reasoning by analogy than by representations in terms of step-by-step descriptions capable of traditional digital computer processing.

Further, expert and very proficient performers may not follow explicitly conscious rules. Requiring them to do so may reduce performance quality, and they will be unable to accurately describe the rules that they do follow. Klein views expertise as arising from perceptual ability to recognize analogous situations, to appreciate the significance of subtle variations in the environment, and to orient learning toward accomplishment of a goal. He presents a model for decision making based on these observations, our interpretation of which is illustrated in Figure 9.8.

1. A current decision situation is perceived in terms of objectives.
2. The decision maker's experience allows recognition of a comparable situation.
3. Similarities and differences between the comparison situation and the current situation are noted.
4. Familiar options are identified and evaluated and a selection is made.
5. The decision is adjusted, as appropriate, and new options may be generated if none of the initial ones are appropriate.

Klein encourages development of decision aids to support the recognitional capacity of the expert. These aids will assist the expert in looking at new situations in terms of analogous cases that suggest possible options. Such an aid would also keep track of options, assist in generation of new ones, and perform computations to assess the impacts of various options. Although information technology systems can contribute to such decision aids, it must be remembered that not all users of such a system will be proficient in all its aspects. Thus, there will generally also exist a need for provisions for formal operational thought-type processes for those contingency task situations that have not been sufficiently cognized such that recognition of comparison situations is feasible. This decision-making model, which is generally a model of the judgment and choice process of experts, does not include a formal process for option generation, since options are suggested on the basis of past experience.

This model is the basis for a *naturalistic decision-making model* [63] that is comprised of, and characterized by, ten characteristics:

1. Ill-structured problems.
2. Uncertainty, ambiguity, and missing data.
3. Dynamic environments.
4. Shifting, ill-defined, and competing objectives.
5. Action feedback loops.
6. Time–stress.
7. Multiple actors or players.
9. Multiple and potentially changing organizational goals and norms.
10. Experienced personnel as decision makers.

In this model, a decision event occurs when the decision maker becomes aware of and identifies, or detects, a situation that exists. Then the decision maker diagnoses the situation and, finally, controls or manages the situation. The terms *recognition*, *generation*, and *selection* are used by Klein to describe these efforts, and this gives rise to the name *recognition-primed* decision making. Figure 9.8 also illustrates the situational context for application of the strategies in the recognition-primed decision-making model.

Klein notes that there will often be a propensity for subconscious information processing, since this takes less effort than formal conscious processing. Even in critical incidents, experts rely mainly on the recognitional mode of decision making. The expert is usually able to decide on an appropriate action to take based on experience-based intuition, rather than consciously generating and evaluating a detailed set of alternatives. Thus, conscious control is guided by displays of systems and processes, whereas subconscious control is guided by one's mental model of the underlying process.

9.2.6. The Dreyfus Cognitive Control Model

Dreyfus and Dreyfus [64, 65] also argue that experienced and expert human decision makers solve new problems primarily by seeing similarities to previously experienced situations in them. They argue strongly that, since similarity-based processes actually used by experienced and expert humans lead to better performance than formal approaches practiced by beginners, decision making based on proven expertise should not be replaced by formal computer based models that do not replicate the performance of experts. In particular, they challenge the Socratic assumption that intelligence is based on knowledge principles (only), the Platonic assumption that principles are (only) expressed in the form of (production) rules, and more modern artificial intelligence based assumptions that we solve problems (only) by acquiring rules by abstracting and generalizing them from specific cases. We do not intend to demean Socrates and Plato. Plato believed in the existence of ideals, a difficult concept, rather than rules per se. Socrates was, above all else, one who questioned established perceptions; this is presumably why the Athenians made him drink his cup of hemlock.

Dreyfus and Dreyfus pose a model that contains six developmental stages through which a person passes in acquiring a skill such as to become a proficient expert or master. Their basic tenet is that people depend less and less on formal and abstract principles and more and more on concrete experience as they become increasingly proficient. These development stages influence the mental tasks and mental attributes used for problem solving. These six stages of expertise are as follows.

1. *Novice:* The novice decomposes the task environment into context-free nonsituational features that can be recognized without experience. A novice relies on rules for determining action and needs monitoring and feedback to improve rule following. A novice consciously monitors performance of the task, lacks perspective from which to judge efforts, and is detached relative to the commitment to the task.

2. *Advanced Beginner:* The process is the same as for the novice, except that the advanced beginner can view features from a situational perspective.

3. *Competent:* A competent person is able to identify significant features rather than calling attention to recurrent sets of features. Such a person is able to recognize dangerous aspects or conditions and employs knowledge of guidelines to correct these conditions. The competent person is also consciously aware of personal monitoring of performance and is beginning to develop a perspective relative to the task at hand. There is still a detached commitment relative to understanding of the situation, but the competent person has become personally involved in the outcome that follows from a decision.

4. *Proficient:* This comes with the increased practice that exposes one to a variety of situations. Aspects appear more or less important depending upon relevance to goal achievement. Contextual identification of similar features and aspects of the task is now possible, and memorized principles, called maxims, are used to determine action. There is still a detached commitment relative to deciding, but the competent person has become involved in the outcome that follows from a decision and has understanding of the features and aspects of the task. The proficient person has developed an experienced perspective relative to the task.

5. *Expert:* Vast experience characterizes an expert, such that the occurrence of a specific situation triggers an intuitively appropriate action. The expert is consciously aware of monitoring of performance and has an involved commitment to all facets of the task.

6. *Master:* The master is absorbed and no longer needs to devote constant attention to performance. There is no need for self-monitoring of performance and energy is devoted only to identifying the appropriate perspectives and appropriate alternative actions.

It should be noted here that the identification of six levels of capability is somewhat arbitrary. Clearly a larger or smaller number could be identified and the model presented here is a composite of the evolving Dreyfus models. In their view, computers can, often with ease, emulate the performance of novices, but these authors are very strong in their arguments that the higher level capabilities of the expert and the master will not be so easily emulated through machine intelligence.*

Dreyfus and Dreyfus associate the development of these six skill categories with successive transformation of six mental functions. Figure 9.9 indicates how these transformations occur with increased stages of proficiency. Although developed initially for training, this model contains much of importance with respect to information system design and general automated knowledge support as well. There are three mental tasks that occur in the judgment and choice process.

1. *Recollection of Similar Features:* Describes the way in which a person is able to sense environmental characteristics and task needs. A novice can only view these in a nonsituational way, as the novice is, by definition, unfamiliar with the task at hand and the environment into which it is embedded.

*Computer chess programs now play at the level of world masters. Computers perform calculations in the field of astronomy that match the calculational ability of Newton and Galileo. On the other hand, no computer can match the bedside manner of an experienced nurse, and perhaps cannot even match that of a nursing-student. One major point is that it makes little sense to talk about the capabilities of computers without first talking about the task at hand. A second major point is that this chess expertise is due to very rapid rule instantiation, not because of emulating the thought process of the grand master of chess.

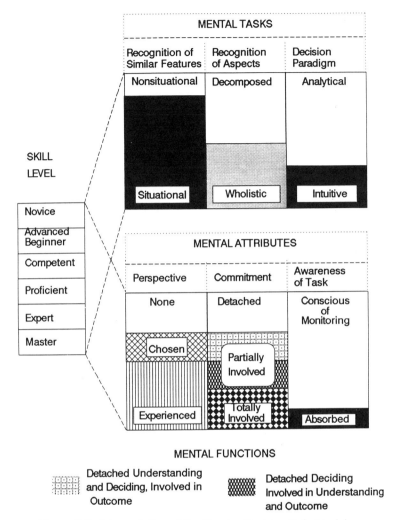

Figure 9.9. Interpretation of the Dreyfus Decision Style Model

2. *Recognition of Aspects:* Refers to the way in which the activities to resolve the task requirements are cognized. The novice and competent person must decompose these into component parts, such as the rules needed to make a left-hand turn while driving an automobile. Above the level of competence, a person is able to perceive in a wholistic or gestalt fashion, without the necessity of decomposition of activities that will constitute potential solutions to the identified problem or task.

3. *Decision Paradigm:* Refers to how one makes a decision. Up to the level of expertise, people will generally use an analytic decision style.

At and above the level of expertise, intuitive judgment will be the decision style of choice.

There are also three mental attributes that also vary across the six capability levels.

4. *Perspective* relative to the unfamiliar task at hand is lacking from the mental traits of novices and advanced beginners. A person at the competent level will have a chosen perspective, and above the level of competent, an experienced perspective is developed.

5. *Commitment* relative to all mental functions is detached for novices and advanced beginners. They can perform only by being given instructions from others. At the level of competence, a person becomes involved in the outcome to be achieved from a decision or task solution, but is detached in commitment to all other mental functions. The commitment increases with increasing mastery of the task and experts and masters are involved in all facets of a task.

6. *Monitoring of Activity*, up to the level of mastery, is such that there is an awareness of some conscious task monitoring efforts. At the level of mastery, a person becomes totally absorbed in performance of the task and is not consciously aware of monitoring efforts.

Human errors of all types, and at all levels of expertise, are possible. We will provide a detailed discussion of these, and strategies for their amelioration, in Section 9.5. This is needed as very important concerns exist with respect to possible use of poor information processing heuristics and cognitive biases in the skill-based decision making of perceived experts or masters. Questions related to the effects of changing environments upon the judgment and decision quality of masters and novices alike are also very important in all of these models. For example, intuitive experience may also face problems in rapidly changing situations because contingency task structural models are outmoded. Finally, the paralysis-through-analysis possibilities we discussed in Chapter 5 are a constant source of dangers.

In our view it is possible to become a master, but unfortunately possible to become a master of the art of self-deception as well as of a specific task. The external behavior of the two masters may well be the same–situational, wholistic, intuitive, and absorbed. What was an appropriate style for one master may well be inappropriate for another. Figure 9.10 illustrates some of the effects of these potentially erroneous contingency task structure assessments. There is nothing at all specific to these effects and the particular dynamic cognitive control-mode model discussed here. But each of the mental functions—*tasks*, including recognition of similar features of the situation, recognition of aspects of the tasks, and selected decision paradigm, and *attributes*, including perspectives relative to the task, commitment to the task and the outcome, and awareness of task monitoring activities—can be associated with errors. So, we see that there are dangers in using intuitive

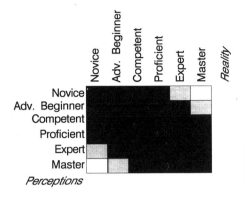

Figure 9.10. Errors in Diagnosis of Contingency Task Structure (Lighter Colors Indicate Increasing Seriousness of Errors)

cognition when analytical cognition is appropriate. There are also potential dangers, of a different character but no less important, in using analytical cognition when intuitive cognition is appropriate. We discuss a number of error-related aspects of task performance and related cognitive ergonomics issues in Section 9.5.

9.2.7. The Janis and Mann Cognitive Mode Model

Judgment and decision-making efforts are often characterized by intense emotion, stress, and conflict; especially when there are significant consequences likely to follow from decisions. As the decision maker becomes aware of various risks and uncertainties that may be associated with a course of action, this stress becomes all the more acute. Janis and Mann [66] have developed a conflict model of decision making that reflects this thinking. Here, *conflict* refers to "simultaneous and opposing tendencies within the individual to accept and reject a given course of action." The most frequent of such symptoms of conflicts are hesitation, feelings of uncertainty, vacillation, and acute emotional stress with an unpleasant feeling of distress [67]. For this reason, this model is often called a *stress-based model*. The term *time–stress-based model* would seem more appropriate. The major elements associated with the conflict model of Janis and Mann are the concept of vigilant information processing, the distinction between hot and cold cognitions, and several coping patterns associated with judgments.

Cold cognitions are defined to be those cognitions made in a calm, detached state of mind. The changes in utility possible due to different decisions are small and easy to determine. Hot cognitions are those associated with vital issues and concerns, and are associated with a high level of time–stress. Whether a cognition should be hot or cold is dependent upon the task at hand and the experiential familiarity and expertness of the decision maker with respect to the task. The symptoms of stress include feelings of apprehensiveness, a desire to escape from the distressing choice dilemma, and self-blame for having allowed oneself to get into a predicament where one is forced to choose between unsatisfactory alternatives. Janis and Mann state that *psychological stress* is used as a generic term to designate unpleasant emotional states evoked by threatening environmental events or stimuli. They define a *stressful* event as "any change in the environment that typically induces a high degree of unpleasant emotion, such as anxiety, guilt, or shame, and that affects normal patterns of information processing."

Janis and Mann describe several functional relationships between psychological time–stress and decision conflict.

1. The degree of time–stress generated by decision conflict is a function of those objectives that the decision maker expects to remain unsatisfied after implementing a decision. If implementation of a course of action is expected to produce a timely high-quality result, there is little stress associated with the decision.

2. Often a person encounters new threats or opportunities that motivate consideration of a new course of action. The degree of decision stress is a function of the degree of commitment to adhere to the present course of action.

3. When the degree of time–stress is low, and there is satisfaction with the present course of action, unconflicted adherence to the present course of action will be the chosen decision. When the time–stress is low and there is dissatisfaction with the present course of action and a single satisfactory alternative can be easily identified, the decsion maker will implement a decision involving unconflicted change to a new course of action.

4. When decision conflict is severe because all identified alternatives pose serious risks, failure to identify a better decision than the least objectionable one may lead to defensive avoidance, or undue procrastination.

5. In severe decision conflict when the decision maker anticipates having insufficient time to identify an adequate alternative that will avoid serious losses, the level of stress remains extremely high. The likelihood that the dominant pattern of response will be hypervigilance, or panic, increases.

6. A moderate degree of stress, which results when there is sufficient time to identify acceptable alternatives, in response to a challenging situation, induces a vigilant and formal effort to carefully scrutinize all identified alternative courses of action and to then select a best decision.

Based upon these relationships or propositions, Janis and Mann present five coping patterns that a decision maker would use as a function of the level of stress: unconflicted adherence or inertia, unconflicted change to a new course of action, defensive avoidance, hypervigilance or panic, and vigilance.

Figure 9.11 presents an interpretation of this conflict model of decision making in terms of the contingency models discussed here. This model points

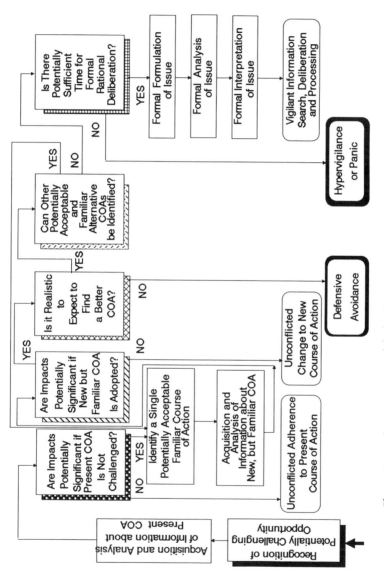

Figure 9.11. Contingency Model of Judgment and Choice due to Janis and Mann

521

to a number of markedly different tendencies that become dominant under particular conditions of time–stress. These include open-mindedness, indifference, active evasion of potentially disconfirming information, failure to assimilate new information, and other cognitive information processing biases that we commented upon in Chapter 7. There are a number of cognitive information processing preferences and decision modes potentially generated by this conflict model. Particularly evident is the striking complexity entailed by the vigilant information processing pattern in comparison to the other coping patterns. This vigilance pattern is characterized by formulation, analysis, and interpretation of a number of alternative courses of action and implementation of the most preferred course. Selection of a coping pattern may be made properly or unwisely, just as selection of a decision style may be proper or improper. We note that there are three proper coping strategies and two improper coping strategies in this figure. Unconflicted adherence to a present course of action, unconflicted change to a new course of action, and vigilant search and deliberation are all appropriate if the contingency task structure is assessed correctly. The first two of these coping patterns involve basically intuitive cognition and the latter involves analytic cognition. Defensive avoidance, or decidophobia, and hypervigilance are basically inappropriate, and humans and organizations should proactively seek to avoid the creation of conditions that lead to them.

Janis and Mann present a decision balance sheet, an adaptation of the moral algebra of Benjamin Franklin, on which to construct a profile of the identified options together with various cost and benefit attributes of possible decision outcomes. They have shown that decision regret reduction and increased adherence to the adopted decision results from use of this balance sheet. Strategies for challenging outworn decisions and improving decision quality are also developed in this seminal work.

In this time–stress-based conflict theory model, there are many activities that are associated with information processing and associated situation assessment. These include obtaining information, or additional information, to:

1. Enable recognition of a potentially challenging opportunity.
2. Enable determination of potential losses if the present course of action is continued.
3. Enable determination of potential losses if a change is made to a new, but familiar course of action.
4. Enable determination of whether it is reasonable to find a better course of action than the familiar ones already considered and initially dismissed as improper.
5. Ascertain if familiar courses of action not previously considered are acceptable.

6. Ascertain whether the remaining time until the decision must be appropriate to formal rational deliberation.
7. Support a formal formulation, analysis, and interpretation of the issues and the resulting vigilant search, processing, and deliberation.

It is of considerable interest and value to indicate the typical interactions between these basically contingency task structure models of decision style that we have discussed in this section. Each of these models addresses somewhat different aspects of task evaluation, information processing preference, and decision rule selection, in terms of contingency elements associated with the environment and the decision maker's prior experiences. Janis has examined the use of this time–stress-based model, and others, in group settings that encourage *groupthink* [68], and in prescriptive settings for leadership in policymaking and crisis management [69] to avoid this.

9.2.8. Hammond's Model of Intuitive and Analytical Cognitions

A conceptual framework for human information processing and judgment must, or at least should, be appropriate to real decisions that are made by real people in real decision situations. This is one reason why operational test and evaluation of systems engineering products is so very important as it seeks to determine whether a product is operationally functional. From our discussions thus far, it is clear that the terms judgment and choice refer to human cognition over a continuum of perspectives that range from formal analytical thought at one end of the spectrum to wholistic intuitive thought based on concrete operational experiences at the other.

The differences between analytic and perceptual thought processes is seen in the writings of many authors. Piaget's theory of intellectual development, which we have briefly described, was among the first to identify relevant variables that influence selection of judgment and choice style: maturation, experience, education, and self-regulation. Brunswick [70, 71] describes perception as an intuitive, continuous, and rapid process—as contrasted with analytic reasoning, which is typically deterministic, discontinuous, and slow.

Hammond [72] has evolved a definitive model of this cognitive continuum that is most appropriate for systems engineering efforts. He indicates that the meaning of analytic cognition in ordinary language is clear; it signifies a step-by-step, conscious, logically defensible process of problem solving. He indicates that the ordinary meaning of intuition signifies a cognitive process that somehow permits the achievement of an answer, solution, or idea without the use of a conscious, logically defensible, step-by-step process. There are numerous normative models for formal analytic cognition to which system engineers may turn. There are a considerably smaller number of fully worked-out models of intuitive cognition. Hammond [73, 74] is much concerned with developing cognitive continuum models that enable integration of the facets of intuitive and analytical cognition. Toward this end, he identifies three principal features of a theory of task systems.

1. The *task inducement principle* stimulates, but does not force, a person to employ a form of cognition that is compatible with the contingency task structure. This task-cognition congruency exists because the contingency task structure offers more support for use of one form of cognition than it does for another.

2. Conventionally, there is a separation of the representations of a system, or surface features, from the structures and functional aspects, or depth features, of a system. This simply says that the interfaces to the human controller of a system may not be appropriately matched to either the system itself, or the environment in which the system is intended to operate, or the task that the system is intended to perform.

3. The *task continuum index* is an ordering that ranges from analysis inducement to intuition inducement. It is influenced by a number of task characteristics, as indicated in Figure 9.12.

Hammond also identifies and describes six properties of intuitive and analytic cognition. These are illustrated in Figure 9.13. This figure also shows the typical position attained by analytic and intuitive adjudication relative to

	Number of Cues	Small (<5)	Large (>5)
	Measurement of Cues	Reliable	Perceptual
	Cue Value Distribution	Unknown	Continuous
	Cue Redundancy	Low	High
	Task Decomposition	High	Low
Task/Cue	Task Uncertainty	Low	High
Attribute	Cue-Criterion Relationship	Nonlinear	Linear
	Cue Weighting	Unequal	Equal
	Organizing Principle	Available	Unavailable
	Cue Display	Sequential	Simultaneous
	Time Period	Long	Short
		Analytical Cognition	Intuitive Cognition

Figure 9.12. Inducement of Analytical and Intuitive Cognition by Task and Cue Attributes

	Cognitive Control	High	Low
	Data Processing Rate	Slow	Fast
	Conscious Awareness	High	Low
Attributes of	Organizing Principle	Task Specific	Weighted Average
Cognitition	Errors	Few, But Large	Normally Distributed
	Confidence in Result	Low	High
	Confidence in Process	High	Low
		Analytical Cognition	Intuitive Cognition

Figure 9.13. Attributes of Analytical and Intuitive Cognition

these. They follow from three central features of Hammond's theory of dynamic task cognition.

1. Cognition is defined in terms of a *cognitive continuum* that moves from analysis, at one extreme of the continuum, to intuition, at the other extreme of the continuum.
2. There is a *correspondence-accuracy principle*, which maintains that judgment veridicality is greater when there is an appropriate correspondence between cognitive activity and task properties than where this does not exist.
3. *Changes in cognition* will be induced by changes in the properties of the contingency task system. The location of the type of cognition on the cognitive continuum will change, and the environmental scanning mechanism will shift from pattern seeking, or symptomatic search, to function-relation seeking, or topographic search.

This model is based on the precept that decision-making tasks are generally *dynamic decision-making tasks* in that they possess, or are described by, the following characteristics.

1. A series of decisions are required.
2. These decisions are interrelated.
3. The decision situation changes, both autonomously and as a consequence of the decsion maker's actions over time.
4. The decisions are made in real time.

Brehmer [75] has also been very concerned with dynamic decision making. He indicates that an individual or group, attempting either to design a new system or control an existing system, needs to have a model of the system to envision how the system is changing. If the system is of large scale and scope, and therefore complex, then the model that characterizes it must often also be complex and dynamic. Brehmer suggests that, in these cases, the system to be designed or controlled can be characterized by the following distinctions.

1. *Complexity:* Refers to the number of goals of the system, the number of processes that must be controlled to reach the system's goals, and the number and descriptions of the means available by which the processes can be controlled.

2. *Processing Delays:* Generally result from information system response delays or delays in the command system.

3. *Characteristics of the Processes and the Means of Control:* Can affect the controllability of the processes.

4. *Rate of Change:* The process to be controlled and the means for achieving this control may affect the controllability of the system.

5. *Delegation of Decision-making Power:* Occurs as a centralized system migrates to a decentralized system, or vice versa, and a hierarchical or distributed organizational control structure results.

6. *Feedback Quality:* Affects the ability of decision makers to perceive situations clearly.

7. *Information Imperfections:* Result from uncertain aspects of a situation or, more generally, from incomplete, imprecise, and/or inconsistent information.

Dynamic decision tasks differ from static tasks in that a series of interdependent and interrelated decisions are required to achieve objectives. They differ from sequential decision tasks in that the time aspect is very important. These tasks require a different conception of decision making from the usual formulation of discrete choice options prevalent in traditional research on decision making. Brehmer suggests that a new conception of decision tasks is required and that the traditional uncertainty constructs, which are based upon gambles in which decision nodes are connected to outcomes event nodes by means of probabilities, will not suffice.

Brehmer has also suggested [76] that decision making in dynamic tasks should be seen as a matter of trying to achieve control, instead of only an attempt to resolve discrete choice dilemmas. He proposes a conceptualization of dynamic decision tasks in terms of six basic characteristics: complexity, feedback quality, feedback delay, possibilities for decentralization, rate of change, and prerequisites for control. These six characteristics are indeed present in many empirical studies on dynamic decision making and are, clearly, very important for any study of cognitive ergonomics. This is especially the case when the dynamic nature of appropriate cognitive continuum adaptation is also considered.

9.2.9. Interpretation of Human Information Processing or Cognitive Control Models

Those who use concrete operational thought do not necessarily have limited abilities to process or integrate information in a formal sense. It may be that they are experts, or masters, of the situation at hand. They fully understand what they are doing and employ skills to implement an action alternative that involves unconflicted change or continuation of the present course of action. In a similar way, the formal operational thinker is not necessarily averse to use of concrete operational thought, which we envision to be of a rule-based or skill-based nature. Our contingency task structural model for the mature, perhaps expert, adult decision maker suggests use of either formal or concrete operational thought, based primarily on diagnosis of the contingency structure of the decision situation, and the time–stress perceived to be associated with the decision situation. The election of a formal or concrete operational mode of thought may be appropriate or inappropriate. Figure 9.10 illustrates some potential for correct action and errors in diagnosis of the contingency task structure. These errors generally are of a very different and quite varied character. We have a good bit more to say about human error, and its amelioration, later.

A support system should be designed such that it assists the system user in minimizing errors between perceived knowledge level relative to a particular task, the environment into which the task requirements are embedded, and actual knowledge level relative to the task and environment. When both perceived and actual knowledge level are at the level of master, for example, then skill-based knowledge is generally appropriate for judgment and choice tasks. When the perceived and actual knowledge level is that of novice, then the knowledge-support system user is generally aware of the need for support. When the perceived knowledge level is that of master and the actual knowledge is that of novice, perhaps due to an unrecognized change in environment, then it is very likely that acts of judgment and choice will reflect self-deception. This is what we attempt to illustrate in Figure 9.10.

A major conclusion from this is that it would be desirable to develop predictable interrelations among human cognitive processes for human information processing, associated judgment and choice, and environmental characteristics. Although these depend upon the decision task, surrounding environment, and decision-maker experiential familiarity with these, it should be expected that most computer-based decision support processes tend to encourage use of formal analytic thought processes by both novices and experts [77]. This suggests a prescriptive approach based on matching the cognitive control mode encouraged by the system with the cognitive control mode appropriate for the task. Also, people tend to adjust their initial hypotheses when provided with relevant information feedback [78]. Without information feedback, it is possible for alternatives to be selected without feeling a need to justify this selection. Thus, the cognitive cue presentations, including feedback of cognitive cues, is very important.

A general underpinning for this is that we normally learn through experience. We should be rather cautious however in the apparently reasonable inference that we *always* learn correctly from experience. Important studies by Brehmer [79, 80] have shown that by no means do people always improve their judgment and decision-making ability on the basis of experience. Biases, such as the tendency to use or to associate the most importance with potentially confirming evidence and to neglect potentially disconfirming evidence, are the key culprits. Brehmer indicates how these biases can be understood in terms of available information. He concludes that truth is not manifest. It needs to be inferred in order to extract from experience information components that truly lead to better judgments and decisions.

9.3. INFORMATION AND KNOWLEDGE REPRESENTATION

The importance of information and knowledge in problem-solving tasks is well recognized. As our previous section has indicated, experiential familiarity with situations and the broad milieu of elements that comprises the contingency task structure strongly influences the method of information processing, knowledge representation, and associated human communications.

Human information and knowledge representation is important in systems engineering from a variety of perspectives. It is important for internal representation and design of data, information, and knowledge bases, such that interoperability across existing and future systems is a reality. It is important for use in detection and diagnosis of errors. For example, expert judgment may be such that a very small number of critical attributes are used in a very simple manner, one which is often inconsistent and which varies from case to case, and expert to expert, and which cannot be explicitly described by the people exercising the judgment [81]. Thus, knowledge representation is important, from an external or user perspective, in that many systems will be designed for humans with very different levels of familiarity with the system, and the environment in which it is used. In brief, we wish to design systems that can easily be integrated with other systems. We wish to be able to predict the effect of human input, including human error, on system functioning. Also, we wish to be able to design better systems. In this section, we present a description and interpretation of several approaches to data, information, and knowledge representation and aggregation. Our discussion is based upon and follows portions of [82].

A knowledge representation is a set of symbols used for illustration, a method for arranging them, and a reasoning mechanism for using the arranged symbols to hold and convey semantic knowledge. Sound approaches to knowledge representation are needed as the foundation for contemporary system design and development efforts, especially in the human–machine systems area, regardless of whether the application efforts involve fault diagnosis and repair, planning, language understanding, or any of a large number of areas.

It is a significant understatement to remark that our understanding of the cognitive processes that transform data into information and then into knowledge and action remains incomplete. For this reason, any discussion of information and knowledge representation by humans is bound to be incomplete. This is the subject of considerable ongoing research by professionals in a wide variety of disciplines. Much of the research is directed toward understanding what appears to be two antithetical types of information processing. These are accorded different names by different researchers, as we have seen. Piaget used the terms formal operational thought and concrete operational thought. Hammond uses the terms analytical and intuitive cognition. Rasmussen uses the terms knowledge-based reasoning and skill-based reasoning. He includes a third form of reasoning, called rule-based reasoning, that lies between knowledge-based and skill-based reasoning.

Contemporary approaches to knowledge representation typically utilize an information processing approach that allows consideration of operational components which may be either human or machine. These operational components typically include [83]:

1. *Perceptors:* Used to receive information.
2. *Effectors:* Used to perform such actions as communications.
3. *Representation Scheme:* Used to interpret and identify information.
4. *Control:* Used to monitor and regulate the actions to be performed.
5. *Decision-Making Effort:* Used to allocate cognitive resources.
6. *Domain Specific Knowledge:* Describes the represented "facts" that describe a situation.
7. *Meta-Knowledge:* Higher level information about knowledge representation that indicates how we know what we know and how we decide how to decide.
8. *Memory:* A physical embodiment of the knowledge base.
9. *World Model:* A model of the environment, or those elements outside of the boundaries within which the system itself is modeled.

These are illustrated, very conceptually, in Figure 9.14. All of these components are essential for intelligent interaction among humans and within

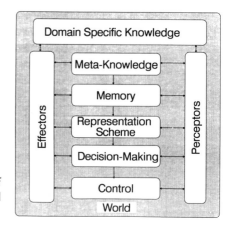

Figure 9.14. Operational Components of an Intelligent System (After Simon and Newell)

human–machine systems. Obviously, no claim can be made that a system will necessarily contain all of these features. But if a human–machine system does not possess these characteristics, there will generally be major potential problems with human–system operation. No claim is made that this model is physiologically correct, but rather that each of these components is necessary for a substantive representation of intelligence and knowledge.

Approaches that enable effective knowledge representation in large knowledge bases have been the subject of investigation by many researchers. A definitive artificial intelligence handbook [84] describes seven representation schemes: logic, procedural representations, semantic networks, production systems, direct or analogical representations, semantic primitives, and frames and scripts. There are, of course, others. There are a number of representations that may be used to describe the four different types of factual knowledge elements that may be captured in a knowledge base: objects, events, performance, and meta-knowledge. These representations will also assist in identification of the values that need to be associated with facts to enable judgment and choice. Finally, they may be used to describe knowledge sensing, retrieving, and reasoning, of new knowledge and relating this to already known knowledge. Here, we will describe seven of the more often used representations.

Production Systems. Various mechanisms can be used to represent the organization of declarative and procedural knowledge. The most common and simplest representation is that in which knowledge is structured as a set of facts, with each fact related to one or more facts in a causal type of inferential relationship. This modular cause–effect structuring of knowledge into the form of a production rule, which was first developed by Newell and Simon [83], has seen many applications including strategic planning, policy analysis, decision making and other areas. The basic idea behind a production system is that there exists a set of productions, or rules, in the form of various condition-action pairs, generally in the form of *if–then* combinations. Initially these were exclusively explicit rules, although there is much current interest in incorporating fuzziness and imprecision into production rule concepts for automated reasoning [85]. Generally, production rules are heuristic and symbolic in nature, and may be appropriate or inappropriate to the task at hand.

Semantic Networks. The problem of representing knowledge in terms of *nodes*, that represent objects concepts or events, and *links* between the nodes to represent their interrelations, is one of the forms of knowledge representation that has been of continuous interest in the field of artificial intelligence. Much of the initial research concerning semantic networks, as these representations are called, stems from the desire to model human associative memory. A semantic network is a convenient computational representation that is readily implementable on computers and used to represent knowledge that is expressed in natural language. The extent to which this type of knowledge representation can be effectively utilized for inference and decision assessment processes depends on the existence, or lack thereof, of a well-formulated set of rules. Semantic networks are capable of representing both the physical descriptions of actions as well as the purposeful aspects of these actions. Thus, a semantic network representational system must include the goals or objectives that can be obtained by actions, the scripts that describe scenarios in simple stereotyped situations, the plans that allow for flexible description of potentially appropriate action-impact pairs, and the themes that allow for such environmental descriptions as the occupations and aspirations of actors involved in the issue under consideration. In this way, a semantic network may be a descriptive as well as a prescriptive mechanism.

Unfortunately, semantic networks do have a number of defects. Defects often occur when the size of the semantic network, in terms of the associated database, becomes sufficiently large such that it can represent a nontrivial amount of knowledge. The computational difficulties in processing the network and the cognitive difficulties in coping with the associated complexity may be overwhelming. This leads to the need for network aggregation in order to obtain summary representations that are efficient and effective. These aggregate networks are called *frames*. Thus the concept of a frame, originally due to Minsky [86], as a chunk of slots and their contents, does eliminate many of the defects of semantic networks. There are also difficulties associated with representing time, as structural models are basically static devices. There are difficulties associated with maintaining a distinction between facts and values, and in incorporating concepts of uncertainty, fuzziness, and imprecision. Situations often arise, for example, in which reasoning must be accomplished using incomplete, inconsistent, and perhaps even contradictory data. These may arise when collecting data from information summaries or from imperfect and/or distributed sources.

Cognitive Maps. A much simpler network-based representation is one in which knowledge is structured as a set of concepts—with each concept related to one or more facts by a single causal type of relationship. Needed to accomplish a structural model of this sort is a theory of *psychologic* [87–89] or pulsed digraphs in which relations may have enhancing $(+1)$, inhibiting (-1), or neutral (0) causal influences on other relations. A number of applications of the resulting cognitive maps, as the resulting structural models are often called, have been reported by Eden and others [90, 91].

Unlike semantic networks, which are typically multirelational structures and as such require a sophisticated set of production rules or grammars for representation and interpretation, a cognitive map is based on a single specific contextual relation, such that any given element will have enhancing, inhibiting, or neutral causal impacts on each other element in the cognitive map. Thus the representation and analysis of a cognitive map is usually simpler than is the case for semantic networks. The simple contextuality of the cognitive map may make it difficult for such a map to replicate a complex belief structure. However, the elements that represent concepts in a cognitive map are variables that can take on different values. The linkage, or contextual relation, among concepts may represent causal assertions and perceptions concerning how one concept variable affects another concept variable.

Schemas. The schema theory of judgment suggests that people have images or schemas that they use for comparison purposes. Some of these schemas describe individual goals or objectives, some describe an individual's view of the present and the future. It is these schemas, including their structure and interrelatedness, that determine the judgments that an individual will make. When choosing among alternatives, an individual views relevant schemas associated with the alternatives and their possible impacts, and accepts those alternatives whose schemas are congruent with desirable outcome schemas. Those incongruent with schemas, or images of desirable impacts, are rejected. If there is more than one alternative with congruent schemas, then any of several strategies for schema adjustment are made, some sort of adjustment that leads to a single dominant schema is found, and the alternative associated with this is then selected.

Fundamentally, a schema is a data structure for representing generic concepts as stored in memory [92]. Schemata, or schemas as the plural of schema is often called, represent knowledge about concepts. The purposeful definition of a schema is simply that it is a framework to enable comprehension and understanding of acquired information. Of course, the schemas that a person has in memory will, at least in part, determine the way they go about acquiring information. Schemas do exist in memory and describe what a person knows. Every schema is structurally organized about some theme. Schemas are said to contain slots that are filled by specific information about some concept.

Four generically different types of schemas have been identified.

1. *Self-Schema:* The objectives and goals to which one aspires.
2. *Normative Trajectory Schema:* The images of the paths that an individual would like to be on.
3. *Nominal Trajectory Schema:* The images of the paths that will evolve if the current course of action is not changed.
4. *Action Alternative Schema:* A set of alternative courses of action and the trajectory images that will follow if the action alternatives are implemented.

The schema-based theory of judgment suggests that an individual first compares the nominal trajectory schema with the normative trajectory schema. If the incongruence between these two schemas is below some critical threshold level, then unconflicted adherence to the original course of action, the nominal trajectory schema, results. When the incongruence between these two is above some threshold, an examination is made of the action alternative schemas to see if there is a single familiar schema that has a sufficiently high concordance or congruence with the normative trajectory schema, such that unconflicted change to a new course of action is possible. We see that schemas are both knowledge sources and preprogrammed guides for routing operational judgments. They are closely related to two related constructs, frames and scripts, which we discuss below.

Also cases exist in which the nominal trajectory schema is unacceptable and where none of the impacts of the readily identified action alternatives, or action alternative schema, is acceptable, in the sense of being congruent with the normative trajectory schema. This incongruence will lead to decidophobia when the stakes are small and to hypervigilance or panic when they are large and where there does not appear to be adequate time to identify potentially suitable alternatives.

When there is time for vigilant search and deliberation, when the stakes are sufficiently high, and where the nominal trajectory schemas and familiar action alternative schemas are unacceptable, then the individual will synthesize new alternative action schemas and identify utility functions for evaluating identified alternative courses of action. The switch from a satisficing or sufficient paradigm of choice, in which expert intuitive judgment is the basis for action, to a utility maximizing paradigm, that is based on formal analytic reasoning, is apparent.

Seen in this is also a multilayer approach to judgment. First an anticipative judgment is made with respect to whether there is a need for a change in the present course of action. This judgment is based on detection of incongruity between nominal trajectory schemas and normative trajectory schemas. When the decision maker has considerable familiarity with the situation, unconflicted change to a new course of action is fairly easy. When this familiarity is not present, decidophobia may result when the stakes are small. Formal knowledge-based judgment results when the stakes are high and there is sufficient time for thorough search and deliberation. This results in vigilant construction of new schemas. Hypervigilance will generally result when this is not the case.

There are major questions of flexibility, and particularly of imperfect schemas in all of the foregoing. Often, schemas will contain uncertainties, incomplete information, imprecise information, and inconsistent information. There are several ways to test this theory. One would be a simulation, in which an expert system model would be used to construct the schema-based theory. This would force the identification of areas where empirical research might be most productive. For example, we do not know how the congruence and/or incongruence thresholds are set, and it might be possible to obtain answers to questions such as these. There are a number of other potentially useful questions that might be asked. Although they are posed specifically for schema theory here, they could be rephrased for other forms of knowledge representation, including the following:

1. Can individuals describe the four types of schemas just identified in either, or both, a prospective and retrospective way?

2. Do individuals have reasoning forward or reasoning backward type schemas that they can use for inductive and deductive formal-operational reasoning?

3. How do individuals cope with imperfect schemas, especially when the imperfection leads to inconsistencies or conflicts among schemas?

4. How do people go about identifying new alternative courses of action and describing them as alternative action schemas?

5. Why do some people identify different alternative action schemas? Why do some individuals continue to pursue new action alternatives, whereas others will reverse their initial judgments because of any of several decision conflicts?

6. How is the cognitive continuum, representing the contingency task structure that determines appropriate strategies for information and judgment, to be accommodated?

7. What support processes can be used to aid people in any of these tasks, and how would this be evaluated?

8. What are the implications of all of this for decision making and for the design, development, and fielding of systems?

Noble [93] is among those who have been especially concerned with using schema representations to assist in memory organization and human and machine information processing support for situation assessment to support recognition primed decision making.

Frame Representations. Minsky [86] also presents a rather different approach to semantic networks that eliminates some of their defects. He advocates the use of local procedures within a frame in order to represent structured knowledge. A frame is, as we have noted, a chunk of knowledge for representing a stereotyped situation. Attached to each frame are several kinds of information. Some of this information concerns how to use the frame and some of the information may concern what can happen next. A frame can then be represented as a hierarchical network of nodes and relations. The top-level element in the frame will represent facts that are always assumed to be true about the generic situation at hand. The lower levels of the frame will have many empty terminal slots that must be filled by the specific context-dependent information about the frame and the situation at hand.

Although this frame-based theory attempts to address identification of a general system whose purpose is to represent knowledge as a collection of separate and simple fragments, there seems to exist many technical gaps at present concerning how to design an operational system that makes best use of this theory of frames. Automated procedures to organize knowledge in such a hierarchically structured framework appear necessary to make this theory functional. The problem of efficient search involving a very large knowledge base produces serious problems both in terms of time and in terms of the possible combinatorial explosion of knowledge that can occur.

Scripts. The script concept [94] is very closely related to that of frames and schema. Each of these consists of organized structures of stereotypical knowledge about general concept. Each results from extraction and synthesis of common related elements in a series of events, situations, or actions. There are three related notions.

1. The way in which an issue is represented in memory strongly influences understanding of the issue.
2. The frame through which an issue is understood influences how alternatives are formulated and how issues are resolved.
3. Previous experience with similar issues influences the frames, scripts, or schema that are stored in memory and that strongly influence representation of new information.

Like schemas and frames, a script is a knowledge structure about a stereotypical sequence of frequently performed actions. They capture the action–event relations in situations that are so frequently encountered that the need for formal methods of problem solving seldom arises. They are self-contained knowledge chunks. As a consequence, it is difficult to transfer knowledge from one script to another. Also scripts may lack understanding. A person, according to the script representational construct, will follow the prescriptions of the script without necessarily understanding the reasons for behavior.

Clearly, there are instances where such behavior is appropriate and others where it may be inappropriate.

Analogous Representations. As we have noted previously, analogies and analogous inference play a very important role in human judgment. Philosophers of science often claim that reasoning by analogy is the basis for hypothesis formation and identification in science. Often analogous reasoning is used when there exists uncertainty and imprecision associated with the judgment task at hand. When there is a lack of explicit, certain, and concrete knowledge about a specific issue, reasoning by analogy is often used. In such cases, one searches for similarities between the extant task situation and a previously experienced and familiar situation. When the situations are sufficiently analogous that one can see parallels between elements in one situation and elements in the other, then reasoning by analogy becomes possible [95, 96].

Silverman [97], in an extensive review of research concerning analogous inference in systems management, identifies a taxonomy that facilitates the development of procedures and protocols for the identification and correction of pitfalls in this form of reasoning. The form, or frame, of knowledge representation that a person uses is very much a function of the perspective that the person has with respect to the particular issue under consideration. This suggests that the contingency task structural approach, described in our last section, is very important. As we have often indicated, it is the particular task at hand, the environment into which this task is imbedded, and the familiarity of the decision maker with the task and environment that determines the information acquisition and analysis strategy that is adopted as a precursor to judgment and choice. Thus we need to be aware of a variety of knowledge representations and the way in which meta-level knowledge leads to a knowledge representation in terms of the information requirements determined for a particular task, the method of analyzing the acquired information, and the way in which associated facts and values are aggregated to enable judgment formation.

There is no available theory comparing the different types of knowledge representation schemes that is capable of precisely and incontrovertibly indicating which will be the most useful in any particular application. We have, however, indicated many of the considerations affecting this choice. Among the various criteria for choosing among knowledge representation schemes, four seem particularly important.

1. *Cognitive Complexity:* The representation scheme used should be capable of being adapted to the cognitive continuum requirements of specific users of the system.
2. *Task Adequacy:* Sufficient knowledge should be present and capable of being captured by the knowledge representation scheme used such that the task requirements can, in principle, be met.

3. *Heuristic Adequacy:* A knowledge representation approach should offer ways of avoiding or greatly reducing information search complexity.

4. *Extendability:* The knowledge base should be designed so as to minimize the difficulty of associating and linking new information to existing information.

It is especially necessary that the knowledge representation scheme be capable of coping with the types of expertise and the reasoning perspectives that can reasonably be expected to exist among the users of a support system in the operational environment under consideration. In this way, we are much better able to accomplish needed activities that involve learning and discovery, such as the diagnosis of faulty theories, the proper assimilation of new knowledge, and posing questions in an understandable and interpretable way. There are many additional studies of these issues [98–100].

In many cases, it is necessary to acquire, represent, use and/or communicate knowledge that is imperfect. There are a great number of approaches that can potentially be used to aid in this. In describing the structure of the beliefs and the statements that people make about issues that are of importance to them, the nature of the environment that surrounds them, as well as the ways in which people reason and draw conclusions about the environment and issues that are imbedded into the environment, especially when there are conflicting pieces of information and opinions concerning these, people often attempt to use one or more of the forms of logical reasoning. There are many important works that deal with this subject. Of particular interest here is the work of Toulmin [101], who has described an explicit model of logical reasoning that is subject to analytic inquiry and computer implementation. The model is sufficiently general that it can be used to represent logical reasoning in a number of application areas.

Toulmin assumes that whenever we make a claim there must be some ground in which to base our conclusion. He states that our thoughts are generally directed, in an inductive manner, from the grounds to the claim, each of which are statements that may be used to express both facts and values. As a means of explaining observed patterns of stating a claim, there must be some reason, one that can be identified, with which to connect the grounds and the claim. This connection is called the warrant, and it is the warrant that gives to the grounds–claim connection its logical validity.

We say that the grounds support the claim on the basis of the existence of a warrant that explains the connection between the grounds and the claim. It is easy to relate the structure of these basic elements with the process of inference, whether inductive or deductive, in classical logic. The warrants are the set of rules of inference, and the grounds and claim are the set of well-defined propositions or statements. Only the sequence and procedures, as used to formulate the three basic elements and their structure in a logical fashion, determine the type of inference that is used.

Sometimes, in the course of reasoning about an issue, it is not enough that the warrant is the absolute reason to believe the claim on the basis of the grounds. For that, Toulmin allows for further backing, which, in his representation, supports the warrant. It is the backing that provides for the reliability, in terms of truth, associated with the use of the warrant. The relationship here is analogous to the way in which the grounds support the claim. An argument will be valid and will give the claim solid support only if the warrant is relied upon and is relevant to the particular case under examination. The concept of logical validity of an argument seems to imply that we can only make a claim when both the warrant and the grounds are certain. However, imprecision and uncertainty in the form of exceptions to the rules or low degree of certainty in both the grounds and the warrant does not prevent us on occasions from making a *hedge* or a vague claim. Commonly, we must arrive at conclusions on the basis of something less than perfect evidence, and we put those claims forward not with absolute and irrefutable truth but rather with some doubt or degree of speculation.

To allow for these cases, Toulmin adds modal qualifiers and possible rebuttals to his framework for logical reasoning. Modal qualifiers refer to the strength or weakness with which a claim is made. In essence every argument has a certain modality. Its place in the structure presented so far must reflect the generality of the warrants in connecting the grounds to the claim. Possible rebuttals, on the other hand, are exceptions to the rules. Although modal qualifiers serve the purpose of weakening or strengthening the validity of a claim, there may still be conditions that invalidate either the grounds or the warrants, and this will result in deactivating the link between the claim and the grounds. These cases are represented by the possible rebuttals.

The resulting structure of logical reasoning provides a very useful framework for the study of human information processing activities. It is beneficial to stress at this point that the order in which the six elements of logical reasoning have been presented serves only the purpose of illustrating their function and interdependence in the structure of an argument about a specific issue. It does not represent any normative pattern of argument formation. In fact, due to the dynamic nature of human reasoning, the concept formation and framing that results in a particular structure may occur in different ways. The six element model of logical reasoning is shown in Figure 9.15.

Most implementations of Figure 9.15 assume a Bayesian inferential framework for processing information. Frameworks for Bayesian inference require probability values as primary inputs. Since most events of interest are unique or little is known about their relative frequencies of occurrence, the assessments of probability values usually requires human judgment. Substantial psychological research has shown that people are unable to elicit probability values consistent with the rules of probabilities and that they are unable to process information as they should, according to the laws of probability, in revising probability assessment when new information is obtained. For example, when people have both causal and diagnostic implications, they should

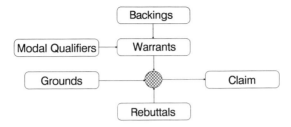

Figure 9.15. The Six Element Model of Logical Reasoning

weigh the causal and diagnostic impacts of the evidence. More often, however, unaided humans reach judgments that suggest they apparently assess conditional probabilities primarily in terms of the direct causal effect of the impacts. If A is perceived to be the cause of B, for example, people usually associate higher probabilities with $P(B|A)$ than with $P(A|B)$.

Studies concerning the elicitation of probability values often report that individuals found it easier and showed more confidence in assessing $P(A|B)$ if B is causal to A. This strongly suggests that the choice of which form of inference to invoke depends more on the level of familiarity of the observer with the task at hand and the frame adopted to initially represent the knowledge. Again, this supports the wisdom of appropriate structuring of decision situations. A decision assessment system based on this concept has been described [102, 103] and summarized [104]. Laskey, Cohen, and Martin [105] are especially concerned with another appropriate application of the Toulmin logic argument structure. Smith, Benson, and Curley [106] provide a cognitive analysis of subjective probability judgments as assessments of belief processing activities, which differentiates reasoning and judgment. They pose a model in which reasoning is used to translate data into conclusions and in which judgment qualifies the conclusions with degrees of belief. Alternative to Bayesian probabilistic representations [99] are a very important topic but are, sadly, beyond the boundaries of the present effort.

9.4. HUMAN INFORMATION PROCESSING IN ORGANIZATIONAL SETTINGS

Organizations can be viewed from a closed system perspective, which views an organization as an instrument designed to enable pursuit of well-defined specified objectives. In this view, an organization will be concerned primarily with four objectives [107]: efficiency, effectiveness, flexibility or adaptability to external environmental influences, and job satisfaction. Four organizational activities typically follow from this: complexity and specialization of tasks, centralization or hierarchy of authority, formalization or standardization of jobs, and stratification of employment levels. In this view, everything is functional and tuned, such that all resource inputs are optimum and the associated responses fit into a well-defined master plan.

March and Simon [108] are among many who discuss the inherent short-comings that are associated with this closed system model of humans as machines. Not only is the human-as-machine view inappropriate, there are pitfalls associated with viewing environmental influences as only noise. Cyert and Simon [109] describe *behavioral rules* as modes of deportment that an individual or an organization develop as guidelines for decision making in complex environments characterized by uncertainty and incomplete informa-tion. These rules incorporate the decision maker's assumptions about both the nature of the internal environment of the organization and the external environment of the world surrounding the environment. In this model, judgment is capable of being decomposed into a set of behavioral rules that change and move closer to those described by simple known situations as uncertainties give way to certainty and as knowledge increases.

In the open systems view of an organization, concern is not only with objectives but with appropriate responses to a number of internal and external influences. Weick [110, 111] describes organizational activities of enactment, selection, and retention that assist in the processing of ambiguous information that results from an organization's interactions with ecological changes in the environment. The overall result of this process is the mini-mization of information equivocality so that the organization is able to

1. Understand its environment.
2. Recognize problems.
3. Diagnose their causes.
4. Identify policies to potentially resolve problems.
5. Evaluate efficacy of these policies.
6. Select a priority order for problem resolution.

These are the primary steps in the systems engineering formulation, analysis, and interpretation efforts we have discussed throughout this text.

The result of the enactment activities of the organization is the enacted environment of the organization. This enacted environment contains an external part, which represents the activities of the organization in product markets, and an internal part which is the result of organizing people into a structure to achieve organizational goals. Each of these environments is subject to uncontrollable economic, social, and other influences. Selection activities allow perception framing, editing, and interpretation of the effects of the organization's actions upon the external and internal environments, such as to enable selection of a set of relationships believed of importance. Retention activities allow admission, rejection, modification of the set of selected knowledge in accordance with existing knowledge, and integration of previously acquired organizational knowledge with new knowledge. A potentially large number of cycles may be associated with enactment, selection, and retention. These cycles generally minimize informational equivocality and allow for organizational learning, so that the organization is able to cope with complex and changing environments.

A very important feature of many models of organizations is that of *organizational learning*. Much of the organizational learning that occurs in practice is not necessarily beneficial or appropriate in either a descriptive or normative sense. For example, there is much literature that shows that organizations and individuals use improperly simplified and often distorted models of causal and diagnostic inferences, and improperly simplified and distorted models of the contingency structure of the environment and task in which these realities are embedded.

This surely occurs in group and organizational situations, as well as in individual information processing and judgment situations. Individuals often join groups to enhance survival possibilities and to enable pursuit of career and other objectives. These coalitions of like-minded people pursue interests that result in emotional and intellectual fulfillment and pleasure. The activities that are perceived to result in need fulfillment become objectives for the group. Group cohesion, conformity, and reinforcing beliefs often lead to what has been called *groupthink* [112]. This is an information acquisition and analysis structure that enables processing only in accordance with the belief structure of the group. The resulting selective perceptions and neglect of potentially disconfirming information preclude change of beliefs.

Organizational learning results when members of the organization react to changes in the internal or external environment of the organization by detection and correction of errors [113–115]. An error is a feature of knowledge that makes action ineffective; detection and correction of error produces learning. Individuals in an organization are agents of organizational action and organizational learning. In the studies just referred to, Argyris cites two information-related factors that inhibit organizational learning:

1. The degree to which information is distorted such that its value in influencing quality decisions is lessened.
2. Lack of receptivity to corrective feedback.

Two types of organizational learning are defined. *Single-loop learning*, illustrated in Figure 9.16, is learning that does not question the fundamental objectives or actions of an organization. It is essential to the quick action often needed. Members of the organization discover sources of error and identify new strategic activities that might correct the error. The activities are analyzed and evaluated, and one or more is selected for implementation. Environmental control and self-protection through control over others, primarily by imposition of power, are typical strategies. The consequences of

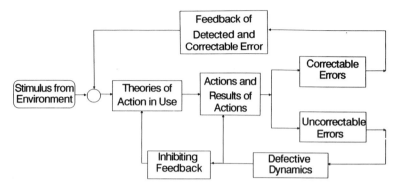

Figure 9.16. Single-Loop Learning and Associated Error Detection, Diagnosis, and Correction

this approach may include defensive group dynamics and low production of valid information.

This lack of information does not result in disturbances to prevailing values. The resulting inefficiencies in decision making encourage frustration and an increase in secrecy and loyalty demands from decision makers. All of this is mutually self-reinforcing. It results in a stable autocratic state and a self-fulfilling prophecy with respect to the need for organizational control. So, although there are many desirable features associated with single-loop learning, there are a number of potentially debilitating aspects as well.

Double-loop learning, shown in Figure 9.17, involves identification of potential changes in organizational goals and of the particular approach to inquiry that allows confrontation with and resolution of conflict, rather than continued pursuit of incompatible objectives leading to intergroup conflict. Not all conflict resolution is the result of double-loop learning, however. Good examples of this are conflicts settled through imposition of power rather than inquiry. Double-loop learning is the result of organizational inquiry, which resolves incompatible organizational objectives through the setting of new priorities and objectives. New understandings are developed that result in updated cognitive maps and scripts of organizational behavior. Studies show that poor performance organizations learn primarily on the basis of single-loop learning and rarely engage in double-loop learning.

Double-loop learning is particularly useful in the case when people's espoused theories of action, which are the official theories that people claim as a basis for action, conflict with their theories in use, which are the descriptive theories of action that may be inferred from actual behavior. Although people are often adept at identifying discrepancies in other people between espoused theories of action and theories in use, they are not equally capable of self diagnosis. The dictates of tactfulness normally prevent us from

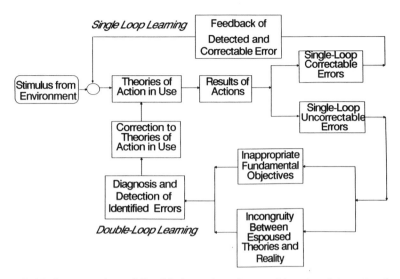

Figure 9.17. Interpretation of Double-Loop Learning, or Meta-Level Learning for Improved Theories of Action in Use and Conformance to Espoused Theories

calling to the attention of others the inconsistencies between their espoused and actual theories of action, and thereby inhibiting double-loop learning.

There are several potential dilemmas associated with the results of this theory of action building. Among these nonmutually exclusive dilemmas, which aggregate together to result in conflicting and intolerable pressures, are the following.

1. Incongruity between espoused theory and theory in use which are often recognized but are usually not corrected.
2. Inconsistency between theories in use.
3. Ineffectiveness as objectives associated with theories in use become less and less achievable over time.
4. Disutility as theories in use become less valued over time.
5. Unobservability as theories in use result in suppression of information by others such that evaluation of effectiveness becomes impossible.

Detection and correction of inappropriate espoused theories of action and theories in use is suggested as potentially leading to a reduction in those factors that inhibit double-loop learning. Of course, single-loop learning often will be appropriate and is encouraged. The result of double-loop learning, however, is a new set of goals and standard operating policies that become part of the organization's knowledge base. It is double-loop learning that leads to meta-level improvements in performance. Inability to accommodate double-loop learning is a flaw. Ability to successfully integrate and utilize the appropriate blend of single- and double-loop learning is called deutero, or dialectic, learning. Ideally, decision support systems and expert systems, or knowledge-based systems, should support double-loop learning as well as single-loop learning.

In more or less classical terminology, an expert system is constructed with expert knowledge, usually in the form of production rules, for use by novices and advanced beginners. A classic decision support system makes formal rational thought procedures more amenable to use. Many of our discussions indicate the pitfalls potentially associated with use of systems this narrowly confined on the cognitive continuum.

Huber [116, 117] and Vroom and Yetton [118] have indicated a number of potential advantages and disadvantages to group or organizational participation in decision making. Since a group has more information and knowledge potentially available to it than any individual in the group, it should be capable of making a better decision than an individual. Group decisions are often more easily implemented than individual decisions since participation will generally increase decision acceptance as well as understanding of the decision. Also group participation increases the skills and information that members may need in making future organizational decisions. On the other hand, there are potential disadvantages to groups. They consume more time in making decisions than individuals do, however. Their decisions may not fully support higher organizational goals. Group participation may lead to unrealistic anticipations of involvement in future decisions and resentment by individuals toward subsequent decisions in which they have not participated. Finally, there is no guarantee that the group will converge on a decision alternative.

Huber asks four primary questions, the answers to which determine guidelines for selection of group decision making strategies. These include

Involving others.
Encouraging group activities.
Delegating authority to the group.
Including the leader in the group.

The responses to these questions determine an appropriate form of group decision making. There are a number of subsidiary questions concerned with each of the primary questions. For example, we may determine whether or not to involve others by posing questions involving decision quality, understanding and acceptance, personnel development and relationships, and time required for decision processing and decision making.

Vroom and Yetton have been also concerned with leadership and decision making. Their primary concern is with effective decision behaviors. They develop a number of clearly articulated normative models of leadership style for individual and group decisions. These should be of use to those attempting to structure normative or prescriptive models of the leadership-style portion of decision situations, which are capable of operational implementation. It is the apparent goal of Vroom and Yetton to come to grips with, and use explicitly, leadership behavior and situational variables in order to enhance organizational effectiveness. This is, of course, a primary purpose of a leadership model.

Keen [119] acknowledges four causes of inertia relative to organizational information systems:

1. Information is only a small component task.
2. Human information processing is experiential and relies on simplification.
3. Organizational change is incremental and evolutionary, with large changes being avoided.
4. Data are a political resource to particular groups as well as an intellectual commodity.

Each of these factors suggests problems in determining how information is processed by organizations.

Tushman and Nadler [120] have developed a number of propositions, based on their own and other research that reflects various aspects of information processing in organizations. The general conclusion of these studies is that in an effort to enhance efficiency, organizational information processing typically requires selective routing of messages and summarization of messages. In the classical normative theory of decision making, it is easily shown that information about the consequences of alternative courses of action should be purchased only if the benefits of the information, in terms of precision, relevance, reliability and other qualities exceed the cost. Feldman and March [121] present an alternate point of view in their description of information use in organizations. Their discussions of information incentives indicate systematic bias in estimating the benefits and costs of information due to the fact that the costs and benefits do not occur at the same place and at the same time such that one group has responsibility for information use whereas another has responsibility for information availability. Also, people are prone to obtain more information than is needed since, under uncertainty conditions the postoutcome probabilities of events that do occur will be judged higher than the prior probabilities of these events. This suggests that less information was obtained than should have been obtained and, typically, leads to incentives to obtain too much information.

Feldman and March also indicate that much of the information that is obtained is obtained for surveillance purposes to uncover potential surprises rather than to directly clarify uncertainties for decision making. Strategic misrepresentation of information, due to interpersonal conflicts and power struggles, is a third factor suggested as decoupling information gathering from decision making. In this case, information must be suspected of bias. Finally, information is a symbol that indicates a commitment to rationality; there are incentives to displaying the symbol even if it is not used.

Identification of other variables that influence information processing in organizations would be useful. To determine how these information processing variables are influenced by the information processing biases of individuals is especially desirable, both in its own right and in terms of the likely usefulness of the results and the need for an expanded theory of the information processing biases we discussed in Section 7.3.3. There appear to have been only limited results obtained in the area of cognitive information processing biases and use of inferior heuristics on the part of groups. The development of appropriate normative strategies for group information requirements determination are especially important toward these ends. Also important is the development of modeling strategies that will enable effective and efficient representation of organizational information structures. The efforts of Levis [122] are particularly significant in this regard. Organizational information processing is related to rationality perspectives and environments for organizational decision making, which is our next topic of discussion.

9.5. RATIONALITY PERSPECTIVES AND ENVIRONMENTS FOR INDIVIDUAL AND ORGANIZATIONAL INFORMATION PROCESSING

In this section, we describe some rationality perspectives concerning individual and organizational information processing. To do this, we assume some rather stereotypical perspectives and environments in which people acquire, represent, and use information. These perspectives are not, in any sense, mutually exclusive. In the concluding portion of the section, we attempt to relate these to the important subject of dialog and interface design.

The research of Diesing [123] and Steinbrunner [124] and many others have dealt with *rationality* as it pertains to human information processing and other judgment and choice activities. The various forms of rationality are very helpful in providing at least a partial explanation of why people seek the information they do. Understanding these will help support system design that results in more effective ways to determine information needs. Although a variety of rationality forms may be defined, the following appear to be the most common ones.

9.5.1. Economic Rationality

Maximum goal achievement with respect to technical production of a single product, subject to a production cost constraint, is the typical desired end of classic microeconomic rationality. Economic rationality extends this concept to a number of products. It seeks to maximize the overall worth, in an

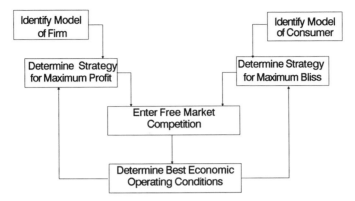

Figure 9.18. Conceptual Illustration of Economic Rationality

economic sense, of a number of investments [125]. This is possible if desired goals are well defined and measurable, the techniques employed to attain these goals are not limited in scope or hindered in application, if supply and demand operates in a stable manner, and if the interrelationships of supply and demand are known and available to all. In other words, the requirements for a perfectly competitive economy are satisfied. Using this model of rationality, it is possible to maximize goal achievement should there be any constraints placed on the above requirements. Some goals can typically be achieved and this results in enhanced economic progress. Achievement of some goals become the means toward the achievement of other societally desirable goals. This process continues and the continuation of economic progress itself becomes the top level goal to be achieved. Over the long term and only as long as decisions do not adversely affect society as a whole will these decisions be acceptable. From an economic rationality perspective, those items of information that do not provide a basis for increasing the profit goals of an organization are to be avoided. Thus, this is a useful but very incomplete form of rationality. Figure 9.18 illustrates some salient features of pure economic rationality. It is focused primarily on the production side. A consumption side of economic rationality is also possible. Such a perspective would be close to technical rationality. The pure economic rationality model is generalized somewhat by technical, or techno-economic, rationality.

9.5.2. Technical Rationality

The activities of an individual are determined in such a way as to maximize the return, or benefit, to the individual from the investment cost of that activity. In a similar way, we can define technical rationality of an organization. All activities within an organization are formed in a manner so as to reach the goals set forth by the organization. Most traditional engineering and organizational analysis has presumed, at least implicitly, technical rationality. Systems are presumed to be designed to achieve optimal attainment of objectives. The presence of multidimensional and noncommensurate objectives often prevents this sort of optimization from being easily accomplished. The need for coordination and communication among people in modern decentralized organizations also makes technical rationality very difficult to attain. There are a number of other reasons as well. Implementation of technical rationality results in what is often called the rational actor model.

Most formal decision analytic efforts are based on the *technically rational actor model*. In this model, the decision maker becomes aware of a problem, studies it, carefully weighs alternative means to a solution and makes a choice or decision based on an objective set of values. At first glance, the rational actor model appears to contain much of value. It is especially well matched to the decision theory schools of thought in Chapter 7. However, we must be aware that it is a normative substantive model. There may be any number of descriptive process realities that make it infeasible to realize. This form of rationality is almost exclusively concerned with efficiency and effectiveness. It may result in the sublimation of fundamental values and objectives as objectives measures. When a value becomes sublimated as an objectives measure, pursuit of the objectives measure may not achieve the fundamental objective.

In rational planning or decision making, following techno-economic rationality, the following steps are typically performed.

1. The decision maker is confronted with an issue that can be meaningfully isolated from other issues.
2. Objectives, which will result in need satisfaction, are identified.
3. Possible alternative activities to resolve needs are identified.
4. The impacts of action alternatives are determined.
5. The utility of each alternative is evaluated in terms of its impacts upon needs.
6. The utilities of all alternatives are compared and the policy or activity with the highest utility is selected for action implementation.

From the perspective of a technical rational actor model, the decision maker becomes aware of a problem, structures the problem space, gathers information, identifies the impacts of alternatives, and implements the best alternative based on a set of values.* Figure 9.19 illustrates some of the central

---

*From a technically rational perspective, the effects of alternatives that are proposed for implementation, including the effects they have on the environment in which they are to be utilized are determined in detail using the technologies of systems engineering. Cost/benefit analyses are determined, statistics are used, computer models are constructed, and other systems science and operations research methods are employed to ascertain the effects and results of a

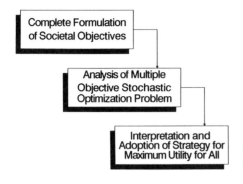

Figure 9.19. Technical, or Techno-Economic, Rationality

features of technical rationality. Since a complete identification of all needs, alterables, objectives, and so on is not usually possible, one cannot be completely rational in the purest unconstrained sense. This observation and the more important observation that humans often do not attempt to follow the normative implications of the rational actor model, due to cognitive limitations, led Simon to develop the satisficing or bounded rationality framework.

9.5.3. Satisficing or Bounded Rationality

Satisficing decision implementations are based on a minimum set of requirements that provide a degree of acceptable achievement over the short term. The decision maker does not attempt to extremize an objective function, not even in a substantive way, but rather attempts to achieve some aspiration level. The aspiration level may possibly change due to the difficulty in searching for a solution. It may be lowered in this case, or raised if the goal or aspiration level is too easily achieved.

---

particular alternative that is being considered for implementation. That this perspective is very important is shown by the extensive use of information and statistics by virtually all organizations. Of course, there are dangers in this. Often the quantity that is being measured is not a quantity that is fundamentally of interest, but a surrogate for this. A danger with overuse, and improper use, of the technological perspective is that sublimation of objectives measures for objectives may easily occur. For example, the objective "to obtain a high school education that is a stepping stone to college and an appropriate career" may be replaced by the instrumental objective "to obtain a high score on the Scholastic Achievement Test." There is ample evidence that this sublimation of values and objectives for objectives measures does occur. This is not a tautology. An objectives measure, which many would say is a very good one, of preparation for post-secondary education is performance on standardized achievement and aptitude tests. This may become interpreted in such a way that students are specifically trained to take and do well on standardized tests. While this may well enhance the probability of acceptance into post-secondary educational programs, it does not necessarily prepare students for success in post-secondary education or in a subsequent career.

Simon [126–129] was perhaps the first to make use of the observation about two decades ago that unaided decision makers may not be able to make complete substantive use of the economic and technically rational actor model. In these situations, the concepts of bounded rationality and satisficing represent more realistic substantive models of actual decision rules and practices. According to the satisficing or bounded rationality model, the decision maker looks for a course of action that is basically good enough to meet a minimum set of requirements. The goal, from an organizational perspective, is "do not shake the system" or "play it safe" by making decisions primarily on the basis of short-term acceptability rather than seeking a long-term optimum. Simon introduced the concept of satisficing or bounded rationality as an effort to " ... replace the global rationality of economic man with a kind of rational behavior that is compatible with the access to information and computational capabilities that are actually possessed by organisms, including man, in the kinds of environments in which such organisms exist." He suggested that decision makers compensate for their limited abilities by constructing a simplified representation of the problem and then behaving rationally within the constraints imposed by this simplified model. We may satisfice by finding either optimum solutions in a simplified world or satisfactory solutions in a more realistic world and, as Simon also says, "neither approach dominates the other."

Satisficing is actually searching for a good enough choice. Simon suggested that the threshold for satisfaction, or aspiration level, may change according to the ease or difficulty of search. If many satisfactory alternatives can be found, the conclusion is reached that the aspiration level is too low and needs to be increased. The converse is true if no satisfactory alternatives can be found. This may lead to a unique solution through iteration. The principle of bounded rationality and the resulting satisficing model suggest that simple heuristics may well be adequate for complex problem-solving situations. While satisficing strategies may be excellent for repetitive problems by encouraging one to do what we did last time if it worked last time and the opposite if it didn't, they may also lead to premature choices that result in unforeseen disastrous consequences, consequences that could have been foreseen by more careful analysis. Figure 9.20 illustrates central features of a satisficing model.

9.5.4. Social Rationality

A potentially simplistic or idealistic look suggests that society functions as a unit seeking betterment for itself. All its energy is directed toward the realization of this goal. The social system is cohesive in that all its activities reinforce achievement of the desired goal. Present decisions are related to those of the past and are projected into the future. Although these actions and decisions are usually not efficient and sometimes not even effective, the cohesiveness of society provides continuity for the system. The roles and

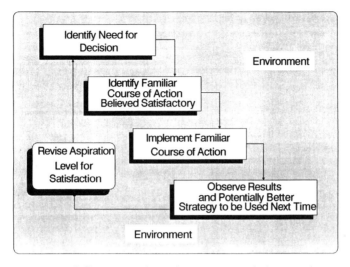

Figure 9.20. Conceptual Illustration of Satisficing, or Bounded Rationality, Perspective

structure of society are reinforced from previous results, both good and bad, lending credence to the fact that a social system is rather intractable. It maintains a conservative appearance and avoids risk. That it ought to be adaptive to change perhaps can be shown by sudden changes in the morality or consciousness of the members of the society through a violent opposition to the status quo, such as opposition to or blatant disregard of the law by some leaders of the society. Figure 9.21 illustrates some of the features of social rationality. Again, this model is very simple and somewhat stereotypical and needs to be modified to represent the many other influences that would augment pure social rationality in a realistic setting. Social rationality would, for example, not exist independently of some facets of political and legal rationality, or economic and technical rationality. These are the five rationality perspectives delineated in Diesing's treatise.

9.5.5. Political Rationality

The decision-making structure is assumed to be influenced by embedded beliefs, values, and interpersonal relationships, the interaction of which define roles under which actions and decisions are based. The three characteristics of this rationality are that all actors remain independent regardless of the pressures to be dependent on one another, the workload is distributed among all members to balance and moderate actions of the group, and future decisions are chosen in such a way that the impacts of these decisions will act to bind the group further together and increase participation. Figure 9.22 provides an interpretation of political rationality.

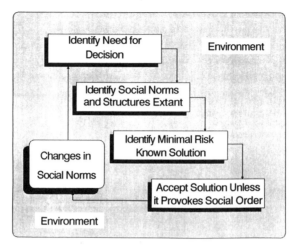

Figure 9.21. Conceptual Illustration of Social Rationality

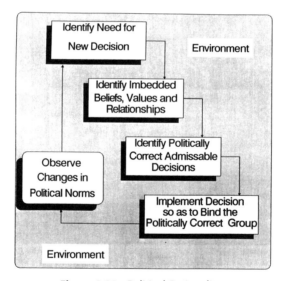

Figure 9.22. Political Rationality

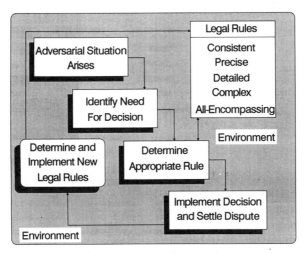

Figure 9.23. Legal Rationality

9.5.6. Legal Rationality

A system exhibiting this form of rationality operates on the basis of rules that are complex, consistent, precise, and detailed. As a result, no ambiguous conflict can occur. It is effective in preventing disputes even though the results of this system apply differently, to some extent, to each person. The prevention of disputes is accomplished through a legal framework that provides a means for settlement of disputes and sets precedents to guide members of society. Figure 9.23 illustrates salient facets of legal rationality.

Other approaches could be used to characterize rationality. Simon [130] has developed a two-element categorization of rationality as a function of whether an act is rational from an input–output perspective, or whether the internal components of the process used are rational. This leads to a description of substantive rationality and procedural rationality.

9.5.7. Substantive Rationality

This is the classic, input–output, or means–ends rationality of economics and is illustrated in Figure 9.24. This form of rationality is outcome-oriented, in that behavior is considered acceptable when given goals are achieved. Given a set of goals, rational behavior is determined by the characteristics of the environment in which it takes place. For example, use of the methods of optimum systems control will result in a system that achieves a goal (minimum cost perhaps) while satisfying a set of constraint equations that governs the behavior of the physical system concerned. That there are several possible mathematical representations of a given input–output behavior is

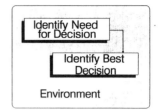

Figure 9.24. Substantive Rationality

immaterial. Substantive rationality is not concerned with this internal structural aspect of rationality. It is very closely related to economic and technical rationality. Highly product oriented, it could also be called functional rationality.

9.5.8. Procedural Rationality

This is the prevalent rationality of descriptive decision making in which any decision-making process must necessarily correspond to the capabilities of the user. It must allow a person to make use of those knowledge components that make maximum use of that person's abilities (reasoning ability, managerial ability, etc.) and minimize use of those knowledge components concerning areas in which the decision maker is not able to perform effectively. Behavior is rational, in a procedural sense, when a person effectively uses existing cognitive powers to choose actions to alleviate some problem. It is the process of selecting procedures for resolution of issues that is the basis for and the justification of rationality, rather than the outcome of the decision. Procedural rationality is, therefore, the method of searching for information for solutions to problems. Figure 9.25 provides an illustration of procedural, or process, rationality.

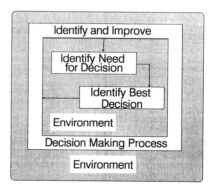

Figure 9.25. Procedural, and Substantive, Rationality

9.5.9. Bureaucratic Politics, Incrementalism or Muddling-Through Rationality

In 1959, Lindblom postulated the approach called *incrementalism*, or *muddling through* [131–133], to cope with perceived limitations in the economically rational model. From this perspective, bold changes to existing policies are avoided and decisions are based on a rather limited set of alternatives, which basically are minor perturbations of existing policies. Long-range side effects are not dealt with, but rather are left to future decision makers, who must try to ameliorate these side effects with other policies.

The bureaucratic politics, incrementalism, or muddling-through model represents another attempt to characterize individual and organizational behavior. After problems arise that require a change of policy, policy makers typically consider only a very narrow range of alternatives differing to a small degree from the existing policy. One alternative is selected and tried with unforeseen consequences left to be discovered and treated by subsequent incremental policies. This is the incremental view, a form of political rationality, and is partially illustrated in Figure 9.26.

Marginal values of change only are considered, and these for only a few dimensions of value, whereas the rational approach calls for exhaustive analysis of each identified alternative along all identified dimensions of value. A number of authors have shown incrementalism to be the typical, common, and currently practiced decision process of groups in pluralistic societies. Coalitions of special interest groups make cumulative decisions and arrive at workable compromise through a give and take process that Lindblom calls

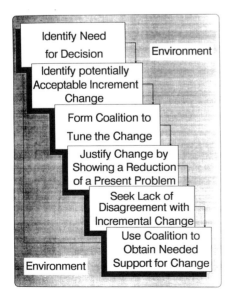

Figure 9.26. Incremental, or Muddling-Through, Rationality

partisan mutual adjustment. He indicates that ideological and other value differences do not influence marginal decisions as much as they influence major changes, and that, in fact, considering marginal values subject to practical constraints will lead to agreement on programs. Furthermore, Lindblom indicates that incrementalism can result in agreement on decisions and plans even by those who are in fundamental disagreement on values.

The main features of the model proposed by Lindblom, and illustrated conceptually in Figure 9.26, are as follows. Ends and means are not viewed as distinct. Consequently means–ends analysis is often considered inappropriate. Identification of values and goals is not distinct from the analysis of alternative actions. Rather, the two processes are confounded. The test for a good policy is typically that various decision makers, or analysts, agree on a policy as appropriate without necessarily agreeing that it is the most appropriate means to an end. Analysis is drastically limited, important policy options are neglected, and important outcomes are not considered. By proceeding incrementally and comparing the results of each new policy with the old, decision makers reduce or eliminate reliance on theory. There is a greater preoccupation with ills to be remedied rather than positive goals to be sought. Incremental analysis is a good description of political decision making and is sometimes referred to as the *political rationality* or *political process* model. There have been a number of studies that indicate incrementalism to be an often used approach in practice. Without doubt, this is a realistic process-oriented descriptive model of judgment and choice, especially in political environments, despite our simplistic interpretation.

9.5.10. Organizational Processes Rationality

Again, we adopt a simple view of a classic organization. In its purest form, everyone in the organization is aware of how the organization functions, as this is based on a well-communicated set of standard operating procedures. Decisions to be made are structured around and evaluated in terms of these procedures. Information needs are determined through discovery of how these standard operating policies or rules interact with existing problems.

From the perspective of *the organizational process model* or *organizational rationality*, plans and decisions are the result of interpretation of standard operating procedures. Improvements are obtained by careful identification of existing standard operating procedures and associated organizational structures and determination of improvements in these procedures and structures. The organizational process model, originally due to Cyert and March [134], functions by relying on standard operating procedures, which constitute the memory or intelligence bank of the organization. Only if the standard operating procedures fail will the organization attempt to develop new standard operating procedures.

The organizational process model, illustrated in Figure 9.27, may be viewed as an extension of the concept of bounded rationality to organizations. There are four main concepts of the behavioral theory of the firm

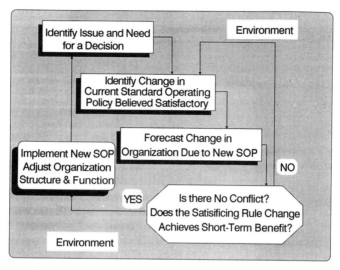

Figure 9.27. Conceptual Illustration of Organizational Process Rationality

which are suggested as descriptive models of actual choice making in organizations achieved through organizational process rationality:

1. *Quasiresolution of Conflict:* Decision makers avoid conflicts arising from noncommensurate and conflicting goals. Major problems are disaggregated and each subproblem is attacked locally by a department. An acceptable conflict resolution between the efforts of different departments is reached through sequential attention to departmental goals and through the formulation of coalitions that seek power and status. When resources are scarce and then there are unsatisfied objectives, decisions concerning allocations are met largely on political grounds.

2. *Uncertainty Avoidance:* This is achieved by reacting to external feedback, by emphasizing short-term choices, and by advocating negotiated futures. Generally there exist uncertainties about the future—uncertainties associated with future impacts of alternatives and uncertainties associated with future preferences. Generally, deficient information processing heuristics and cognitive biases are used to avoid uncertainties. The effects are, of course, suboptimal.

3. *Problem Search:* The organization is stimulated by encountering issues and does nothing before issues have surfaced. A form of "satisficing" is used as a decision rule. Search in the neighborhood of the status quo only is attempted and only incremental solutions are considered.

4. *Organization Learning:* Organizations adapt on the basis of experience. They often pay considerable attention to one part of their environment at the expense of another.

The organizational process model may be viewed as suggesting that decisions at time t may be forecasted with almost complete certainty, from knowledge of decisions at time $t - T$, where T is the planning or forecasting period. Standard operating procedures or programs, education motivation, and experience or programming of management are the critical determinants of behavior for the organizational process model. Cohen and March recommend a strategy of management leadership to cope with organizational process realities. Managers are encouraged to be intimately involved in organizations such that they are able to strongly influence decisions; to become widely informed such that they will be highly valued in the information-poor organization; to be extraordinarily persistent in their quest for change; to encourage those with opposing views to participate; and to overload organizational systems, potentially through the introduction of complex and highly bureaucratic procedures, in order to make themselves more necessary. In this view, the descriptive characteristics of the organization are seen as performance inhibiting factors. They are factors not to be overcome but to be understood and used to the advantage of the manager.

9.5.11. Garbage Can Rationality

Of course, an organization is also a mechanism for problem solving and decision making. When the realities of ambiguity are associated with organizational problem solving and decision making, the result is what is termed a *garbage can model* of organizational choice [135]. In this model, which has generated much recent interest, there are five fundamental elements:

> Issues or problems.
> Organizational structure.
> Participants, actors, or agents.
> Choice opportunities and actions.
> Solutions or products of the choice process.

The problems, solutions, and choice opportunities are assumed to be quasiindependent, exogenous streams that are linked in a fashion that is determined by organizational structure constraints. There are several of these. The most important are *access structure*, or the access of problems to choice opportunities, *decision structure*, or the access of choice opportunities to solutions, and *energy structure*, which evolves in a dynamic fashion in terms of the number of problems or solutions that are linked to choice opportunities at a particular time.

The participants in the process can also be regarded as variables since they come and go over time, and devote varying amounts of time and energy to problems, solutions, and choice opportunities due to other competing demands on their time. This relatively new organizational decision model views organizational decision making as resulting from four variables: problems, solutions, choice opportunities, and people.

Decisions result from the interaction of four factors:

Solutions looking for problems.

Problems looking for solutions.

Decision opportunities.

Participants in the problem-solving process.

The model allows for these variables being selected more or less at random from a garbage can. The interaction of these variables provides the opportunity for decision making. Generally these interactions are not controlled. Rather, they occur in an almost random fashion due to the vexing equivocality associated with problematic preferences, unclear procedures, and fluid participants. The major reason for the garbage can approach is the chronic ambiguity and considerable equivocality that is present in the environment.

This mixture of decision factors is not necessarily associated with poor management. Doubtlessly this is often the case, but it is not necessarily the case that garbage can decisions are poor decisions or that garbage can environments are poor environments. They may simply be unavoidable from the perspective of a given human in a given situation. We can easily envision numerous instances of well-run organizations in which these four factors would come together to drive the decision-making process. For example, many of the high-technology defense contractors trying to expand into commercial markets find their decision making driven by these four factors. The existing environment may simply suggest this approach to decision making.

In a garbage can environment, decisions generally occur through rational problem solving, through ignoring the problem until it goes away or resolves itself (sometimes called *decision flight*), or through the oversight of having the problem solved inadvertently by having another related problem solved. Doubtlessly, this is a realistic descriptive model. It has been used to analyze a number of existing decision situations, especially in a university decision context [136]. Examination of garbage can decision environments has led to conclusions concerning normative strategies to maximize decision effectiveness [137]. It has been shown, for example, that a chief executive officer can maximize success through using a personal hands-off policy, and by hiring very conservative people to manage high criticality projects, and very liberal people to manage very low criticality projects.

Three areas of equivocality are generally present in a garbage can environment:

1. *Problematic Preferences:* Different decision-making units have different objectives, and these generally evolve over time in an imprecise and unpredictable manner.

2. *Unclear Procedures for Making Decisions:* Responsibility and authority are usually separated and fragmented.
3. *Fluid Decision Participation:* Members of the decision-making units change over time, often in an unpredictable manner.

In the garbage can model, the problems, solutions, and choice opportunities are mixed together as if in garbage cans. The division of human effort among problems, solutions, and choices is fuzzy and not fixed in any highly organized way. Problems, solutions, and choice opportunities may not coalesce in the right way at the right time to lead to a "rational" solution to a problem.

There are many unanswered questions relative to descriptive and normative use of this interesting model, and there is much study to progress at this time concerning extensions of this model. A recent book edited by March and Weissinger-Baylon [138] describes application of the garbage can model to military decision making. There have been a considerable number of earlier studies in which university decision making, especially that of faculty search committees, is represented by a garbage can model. Among the necessary and appropriate questions are the following.

1. How are the number of garbage cans determined and what choice situations do they represent?
2. Who participates and how do they participate in the various choices?
3. What structures and flows represent the various problems, solutions, and choice opportunities?
4. How does the relative structure of the problems, solutions, choice opportunities, and participants evolve over time?
5. How is the interaction and information among the various garbage cans determined?
6. What influences situation interpretation and variables that represent situation interpretation in the various garbage cans?
7. What is the appropriate role for models, such as this, in the normative design of information systems that support enhanced organizational efficiency and effectiveness?
8. How can such designs be evaluated in an operational context, and how can the resulting information be used to enhance system designs and organizational effectiveness?

One of the most potentially useful facets of the garbage can model or garbage can rationality is that it is a definitive approach for relating social structure to cognitive structure. Not only does it have this descriptive appeal, but it is a potentially useful organizational model that can cope with such potential crisis situations as breakdown in organizational communications [139, 140]. Figure 9.28 illustrates some central features of the garbage can model of organizational choice.

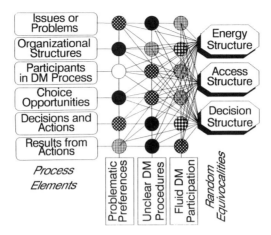

Figure 9.28. Linkages Among Problems, Solutions and Choice Opportunities in Garbage Can Perspective

Distributed decision making often occurs in garbage can modeling situations [141]. Fundamentally, distributed decision making is the process of organizational decision making that occurs when the information and the responsibility for the decision are distributed across time, space, and the various decision-making entities. A decision-making unit can be a single human, a group of humans, a machine, or a collection of humans and machines; each of these are responsible for some part of an overall decision. From this perspective, organizational decisions are a product of human information processing and associated decision making in organizational units.

Coordination takes place across the various information processing and decision making units. Each unit generally takes part in this coordination process. Each unit has access to somewhat different information. It has somewhat different standards for acquiring, processing, and evaluating information. The information processing factors affecting decision making in an organization depend upon the structure of the organization, the environment,* and the information processing and decision-making capabilities of the organizational units. In a recent effort, this is extended to issues of organizational learning and personnel turnover [142]. Thus, this is much more of an environment for organizational decision making than it is a rationality perspective. Nevertheless, the model is sufficiently robust that economic, legal, social, political, legal, and other rationality perspectives fit within the model. Thus, it is an organizational model within an environmental context.

---

*Carley calls the operational environment of the decision-making units the *event theatre*.

9.5.12. Comparison of Rationality Perspectives

These rationality perspectives are all related to one another. For example, economic rationality is necessary for and is also part of technical rationality. Social and political rationality are each concerned with internal processes and procedures. Generally, one would exhibit substantive rationality in an environment of technical or economic rationality. In the environments of social or political rationality we would expect procedural rationality to be the dominant form. Legal rationality would typically be a hybrid of the two in that the initial development of codes would be based on procedural concerns, but the actual functioning of an established code is substantive.

All of these models (or frameworks or perspectives) for rational decision making have both desirable and undesirable characteristics and any of them may be relevant in specific circumstances for individuals or for organizations. However, it seems generally true that most decision makers use a variety of methods to select among alternatives for action implementation, these methods are frequently suboptimal, and most decision makers desire to enhance their decision-making efficiency and effectiveness. Given this, we must conclude that there is much more motivation and need for research and ultimate design and development of systems that incorporate and support various rationality perspectives and environments.

These rationality perspectives and models make it very clear that improved decision-making efficiency and effectiveness, and support and training aids to this end, can only be accomplished if we understand human and organizational decision making as it is, as well as how it might be, and allow

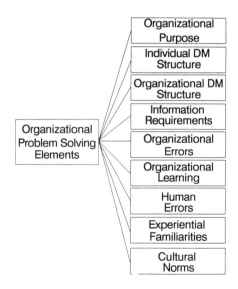

Figure 9.29. Some Elements of Large-Scale Problem Solving in Organizations

for incorporation of this understanding in the design of systems. One of the requirements imposed is relevance to the individual and organizational decision-making structure. Another requirement is relevance to the information requirements of the decision maker. Especially important also is accommodation of organizational learning. These lead to a model of the elements for large-scale problem solving and decision making, a model that has relevance for system design to support error minimization. A conceptual model of this is illustrated in Figure 9.29.

We might also attempt to describe human information seeking and decision behavior in terms of the type of rationality that a person employs to guide this search and evaluation. In general terms, a person may be said to be rational if he or she acts in a way that is sensible in terms of the existing environment, the task requirements, and the experiential familiarity or state of knowledge of the person with respect to these. According to this very informal definition, rationality is an instrumental way of thinking and a person may be said to be rational if they choose those activities most likely to lead to achievement of their goals. We immediately see that this is a multidimensional or multiobjective problem, and that there is no simple unique determination of what is "rational" that is independent of the contingencies of the particular decision situation, especially since the dimensions that are involved will include technical, economic, organizational, social, legal, and political, and perhaps even those of the garbage can.

Lyles and Thomas [143] review a number of models and provide some indication of potential bias difficulties associated with use in strategic situations. Etzioni [144] provides a caution concerning thoughtless rationality and difficulties associated with the automatization of rules of thumb as unchanging guides to behavior. He also discusses a number of unifying themes in this area of inquiry, which is important for systems management and society.

9.6. HUMAN ABILITIES, HUMAN ERROR, AND THE AMELIORATION OF HUMAN ERROR

Humans are able to do many things. Usually these are done in a quite correct and appropriate manner. Occasionally, errors are committed. More often than not, the errors are small. Committing errors is not necessarily bad, if no major harm is done and if humans accomplish single- and double-loop learning through error commission to avoid future errors. Occasionally, errors can be catastrophic. Generally, we wish to design systems that assist in amelioration of the effects of human error, and ultimately of eliminating most possibilities for human error. Of course, we also wish to support human efforts in such a way that errors, to the extent possible, do not occur.

A number of authors have identified taxonomies of human abilities. In a very authoritative work, Fleishman and Quaintance [145] identify 50 human performance abilities. These are:

arm-hand steadiness	gross body equilibrium	reaction time
auditory attention	inductive reasoning	response orientation
category flexibility	information ordering	selective attention
control precision	manual dexterity	sound localization
deductive reasoning	mathematical reasoning	spatial orientation
depth perception	memorization	speed of closure
dynamic flexibility	multilimb coordination	speed of limb movement
dynamic strength	near vision	stamina
explosive strength	night vision	static strength
extent flexibility	number facility	time sharing
far vision	oral comprehension	trunk strength
finger dexterity	oral expression	visual color discrimination
flexibility of closure	originality	visualization
fluency of ideas	perceptual speed	wrist-finger speed
general hearing	peripheral vision	written comprehension
glare sensitivity	problem sensitivity	written expression
gross body coordination	rate control	

These 50 performance abilities involve cognitive, perceptual-motor, sensory, and purely physical performance realms. Clearly, many of these performance abilities involve both cognitive and physiological control capabilities. It is very interesting to note that many of them are almost totally cognitive in nature. Most of these are self-descriptive.*

Many of these are mostly cognitive performance ability facets. Some, such as flexibility of closure, the ability to detect a known pattern imbedded in background information, involve both cognitive and physiological abilities but are primarily cognitive. The design of systems such that humans can more effectively interact with them is very important to the success of information processing in humans and organizations. *Human factors engineering*, as indicated by Tom Sheridan [146] and Ron Hess [147], is the science and art of interfacing people with the machines they operate and interact with, such as the displays, controls, and computer hardware, which send, receive, and process information. These are each illustrations of machine–machine interfaces, or human–machine interfaces. Computer software and administrative procedures that determine how people allocate their effort, make decisions,

---

*Most of the non-self-descriptive terms relate to physical performance. For example, category flexibility is the ability to identify many rules that enable the categorization of a set of elements in different ways according to some contextual relation. A classical example might be to take the parts of the body and categorize them in a hierarchy according to what it is connected to. Control precision is the ability to physically move control actuation devices to desired positions. Trunk strength is the ability to move stomach and lower back muscles within reasonable fatigue limits. Dynamic flexibility is the ability to move the body repeatedly and with reasonable speed. Stamina is the ability to exert one's body while maintaining acceptable lung and blood system performance. Fleishman and Quaintance provide very adequate definition and discussion of each of these 50 terms.

and take actions is another type of interface. A third relevant interface is between people and the seating, temperature, humidity, lighting, sound, radiation, vibration, and other environmental factors that determine their ability to function. Finally, there is the selection and training of people for their machine interactive tasks. Because human factors engineering includes human and machines in a common system, it emphasizes system performance over and above performance of either person or machine by itself. It is this integrated performance of people and machines that is of interest here.

Human factors research on the interactions between users and information systems are possible. With careful attention to the human factors in information presentation on computers, it is possible to build appropriate systems for use by both experts and novices. The human–machine interface can be significantly improved by screen formats, data entry and display methods, menu structures, and graphical interfaces that reflect the cognitive abilities of humans and how humans are influenced by the contingency task structure. Here, we examine many of these concerns.

9.6.1. Problem Formulation, Issue Detection and Diagnosis, and Situation Assessment

There have been many discussions of problem formulation [148]. In one model of problem identification, Cowan [149] identifies three stages: gestation, categorization, and diagnosis. The three stages are described as follows.

1. *Gestation:* The period when conditions in the environment are changing and building toward identification and latency. There is potential for interpreting conditions as a problem, which have not yet been perceived as such. There is a single process variable for this stage, *scanning*, which involves attending to the environmental stimuli.
2. *Categorization:* A process whereby the decision maker becomes aware that a problem exists. There are three process variables for this stage: *arousal*, which is increased readiness to respond, *clarification*, which is an attempt to understand or verify, and *classification*, which is the attaching of a descriptive label to the situation. Also, there are three process determinants for this stage:
   2.1. *Cue discrepancy*, which is the deviation between the observed state and the expected or desired state.
   2.2. *Perceived urgency to respond*, which is the decision maker's temporal need to respond.
   2.3. *Persistence or accumulation of discrepancies*, which is the manner in which discrepancies can be considered important. For example, these could include a single significant discrepancy lasting for some period of time, or accumulated small discrepancies.
3. *Diagnosis:* The attempt to achieve greater certainty about a problem description, such as searching for more information to confirm or discon-

firm the current hypothesis. There are three process variables for this stage: *information search*, which is the gathering of additional evidence about a problem, *inference*, which is the drawing of conclusions from the accumulated information, and *problem description*, which is the classifying of the specific type and nature of the problem. There are also three process determinants for this stage:

3.1. *Familiarity*, which is how well known a particular problem is to the decision maker or to the support system in the case of aided support in task performance.

3.2. *Priority*, which is the precedence assigned to a particular problem relative to other problems.

3.3. *Availability of information*, that is, the capability to acquire the needed information in the allotted time.

It appears possible to aggregate gestation and categorization and denote the combination as issue detection. Determining whether a given situation and the associated decision is important or not and determining the extent of experiential familiarity with a potentially challenging situation are each accomplished by assessing the situation. Since every decision begins, or at least should begin, with an assessment of the situation, it is an important fundamental characteristic of the decision process to possess the capability to properly assess the situation in an accurate and timely manner. Situation assessment [150] should be a trustworthy process that can be relied upon for proper detection and diagnosis of problems that may confront any particular user, in any particular environmental context.

It is tempting to say that the situation assessment process should be a well-defined and well-structured one. The recent emphasis in decision assessment on the behavioral aspects of human decision making recognizes the importance of past experiences, tasks to be accomplished, cues, and the evolving nature of cognitive control in developing effective systems. It suggests that, just like appropriate prescriptive decision processes in general, situation assessments are not always well defined and well structured. The initial identification, including detection and diagnosis or definition of a problem, depends on the observations of pertinent parameters or salient features in the environment. Such observations, whether objectively or subjectively arrived at, are usually deficient or imperfect in some manner, involving measurement errors, judgmental or bias errors, and other errors of omission or commission. There are three primary categories of errors;

1. Errors in detection that there is a problem.
2. Errors in diagnosis of the cause of problem.
3. Errors in correction of the problem, or errors in execution.

We wish to discuss each of these errors, as well as errors in planning for problem solving. These are especially important in situations involving poor information, which implies the need for good information requirements and an appropriate information acquisition system to supply higher quality information. They are also important where mistakes are costly. This implies the need for very small type 3 errors and good decision options and selection processes. Finally, they are important where recovery from missed opportunities is difficult. This brings about the need for very small error in opportunity categorization.

The approaches to the study of problem identification and solution generally fit into one of the following categories.

1. *Error Analysis:* Characterized by analyzing, generally through hindsight, errors that could have produced disorders.

2. *Process Analysis:* Generally characterized by an attempt to disaggregate problem solution into finer components that can be more easily subject to detailed scrutiny. This is the formal systems engineering and scientific approach to problem solution. If not handled well, however, this can simply result in the replacement of one set of vague constructs with another. Also, failure to consider contingency task situational contingencies can result in many difficulties, as we have noted.

3. *Task Analysis:* A process to characterize human behavior by an analysis of the interaction of individual goals or inner psychological environment with the outer environment. Task analysis considers the outer, or task, environment in explaining individual adaptation to task demands.

The categories are not at all mutually exclusive, however. Each approach or category is important, and an integrated approach would appear better than one that considers only one category or type of analysis.

9.6.2. Human–Machine or Human–System Interactions

Norman has described seven stages of user cognitive activity in performing a task [151]:

1. Establishing goals.

2. Forming intentions, or internal mental characterizations of goals.

3. Specifying or selecting the action sequence, generally through review and evaluation of all identified actions.

4. Executing actions.

5. Perceiving the resultant system state.

6. Interpreting this with respect to the initially specified goals.

7. Evaluating whether to continue action or to accept the task as now accomplished.

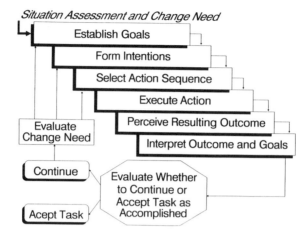

Figure 9.30. Norman's Activity Steps in Task Performance

In this model, the problem solver first forms conceptual intent, reformulates this into action semantics, and ultimately produces a selected action. Errors sometimes occur in the process as intents are reformulated to lower and lower levels. Figure 9.30 illustrates our interpretation of these activities and thus provides a basis for discussion of human–machine interactions and errors.

Norman associates a *gulf of execution* and a *gulf of evaluation* with this model of problem-solving activity. The gulf of execution is reduced by making system commands and mechanisms match the thoughts and goals of the problem solver as much as possible. The gulf of evaluation is reduced by providing conceptual models of the situation that are easily perceived and evaluated. Norman's model indicates seven phases of action:

Framing the goal.
Framing the intention.
Identifying an action.
Executing the action.
Perceiving the resultant system state.
Interpreting the system state.
Finally evaluating the outcome.

These are not unlike the analogous phases in the systems life cycle or a generic approach to problem solving.

There are a number of other suggested life cycles for human–computer interface design. For example, Polson [152] has developed a four-phase methodology.

1. *System Definition:* The phase in which the user requirements, task environment, and basic structure needs for the user interface are identified. Functionality is defined in terms of the diversity of applications to be supported by the resulting system.

2. *Task Analysis:* The phase in which the top and mid levels of the goal structure of the user are identified through decomposition of the system requirements. The interface design team develops methods that accomplish these decomposed tasks.

3. *Detailed Design:* Details of the user interface are specified. This phase includes development of detailed specifications for such elements as menus, command languages, and so on. A *test suite*, which specifies the tasks and associated characteristics that are to be utilized in design evaluation, is identified.

4. *Evaluation:* The final phase in the human–computer interaction effort. Here, the design is evaluated, generally using simulation approaches. Rapid prototyping is suggested as an alternative to simulation-based evaluation.

These efforts are, of course, easily found in any of the myriad of systems life cycles that we discussed in Chapters 2 and 3.

Many efforts in this area have been applied to human–computer interface design issues and associated evaluation issues [153–155]. Given the importance of this area, it is not surprising that there have been a number of recent approaches to human–computer interaction, and associated interface design issues. Most of these identify objectives and design life cycles. Shneiderman, for example [156], identifies eight primary objectives, which he calls *golden rules for dialog design*.

1. Strive for consistency of terminology, menus, prompts, commands, and help screens.

2. Enable frequent users to use shortcuts that take advantage of their familiarity with the computer system.

3. Offer informative feedback for every operator action that is proportional to the significance of the action.

4. Design dialogs to yield closure, such that the system user is aware that specific actions have been concluded and that planning for the next set of activities may now take place.

5. Offer simple error handling, such that, to the extent possible, the user is unable to make a mistake. Even when mistakes are made, the user should not have to, for example, retype an entire command entry line but, rather, edit only the portion that is incorrect.

6. Permit easy reversal of action, such that the user is able to interrupt and then cancel wrong commands rather than having to wait for them to be fully executed.

7. Support internal locus of control, such that users are always the initiators of actions rather than the reactors to computer actions.

8. Reduce short-term memory load, such that users are able to master the dialog activities that they perform.

Clearly, all of these have specific interpretations in different systems engineering environments and must be capable of extension into a variety of environments. They appear to be generally applicable as high-level prescriptions for system design for effective human interaction.

In another significant effort, Moran [157] identified six levels in a framework of user interface design, denoted as the *Command Language Grammar* (CLG): task level, semantic level, syntactic level, lexical level, spatial layout level, and device level. These are clearly related in a hierarchical manner. They also relate to the information and knowledge representation efforts we discussed in Section 9.3. Additional levels can easily be defined. For example, we can distinguish between high-level and low-level tasks.

Other human problem-solving representations attempt approximate predictive models that are potentially useful for design purposes. For example, Card, Moran, and Newell [158] have evolved a specific information processing architecture called the *Model Human Information Processor*. Sensory, cognitive, and effector processors are included in the model, as well as a long-term memory and a working memory. Each are assigned such appropriate parameters as storage capacity and cycle processing time. A general-purpose model such as this might well be capable of describing the way one goes about such low-level lexical functions as keystroke entry. But, it may not be as relevant to modeling as how one goes about syntactic or semantic tasks.

These same authors recognize this, and propose [159] a *Goals, Operators, Methods, and Selection* model that is intended to represent higher level information processing perspectives. In this model, *goals* provide a representation of the user's ultimate purposeful intent in performing a task. The model is based on user decomposition of a task into subtasks. *Operations* provide the system user's representation of elementary physiological actions. *Methods* are operational sequences that are designed to accomplish some intermediate level objective. *Selection* rules are used to specify conditions that must be fulfilled to execute a method. Usually these draw upon complex pieces of experiential knowledge concerning the computer system being used.

These authors identify principles, or prescriptions, for design of human–computer interfaces. These provide an important linkage between the reactive and interactive efforts involved in evaluation of existing interface designs, and the proactive effort that should be involved in the design of interfaces.

1. Consider the psychology of both the user and the designer of the user interface.
2. Identify interface performance requirements.
3. Identify the user group.
4. Identify generic applications to be performed and associated user interface tasks.
5. Specify appropriate methods with which to accomplish the tasks.
6. Match the methods to the task and the interface performance requirements.
7. Design the interface such that it can be used appropriately by people with greatly differing familiarity with the computer system.
8. Design a set of alternative interface use strategies, and provide clear guidance concerning how to select an appropriate interface.
9. Identify, design, and implement a set of error detection, diagnosis, and correction methods.
10. Determine the sensitivity of the performance of the interface to variations in design assumptions.

One word that has appeared frequently in this chapter is *consistency*. This is Shneiderman's first *"Golden Rule of Dialog Design,"* and many other authors advocate this as well. Grudin [160] notes that issues associated with consistency should be placed in a very broad context. Three types of consistency are defined.

1. *Internal Consistency:* Consistencies across physical applications and graphic layout of the computer system, command naming and use, dialog forms, and similar features.
2. *External Consistency:* If an interface has features consistent with those of another interface with which the user is familiar, it is said to be *externally consistent*.
3. *Applications Consistency:* If the user interface uses metaphors or analogous representations of objects that correspond to those of the real-world application, then the interface may be said to correspond to experientially familiar features of the world and to be applications consistent.

Grudin makes several related observations. Ease of learning can conflict with ease of use, especially as experiential familiarity with the interface grows. Also consistency can work against both ease of use and learning. On the basis of some experiments illustrating these hypotheses, Grudin establishes some appropriate dimensions for consistency. These are generally based on the six levels for interface design of Moran, which we discussed earlier. The general conclusion is that the higher level task considerations can override consistency considerations at lower levels. Thus, the system designer is encouraged to consider global aspects of the user work environment, rather than to focus on narrowly defined consistency, a definition in and of itself and solely for the sake of consistency, as a design requirement. Thus, this work represents an enhancement to the consistency notions of Schneiderman.

There are a number of desirable characteristics for user interfaces. Roberts and Moran [161] identify functionality of the editor, learning time required, time required to perform tasks, and errors committed, as the most important attributes of text editors. To this might be added the cost of evaluation. Harrison and Hix [162] identify usability, completeness, extensibility, escapability, integration, locality of definition, structured guidance, and direct manipulation as well in their more general study of user interfaces. They also note a number of tools useful for interface development, as does Lee [163].

Evaluating usability of human computer interfaces is the subject of a recently published monograph by Ravden and Johnson [164]. Nine top-level attributes or criteria, are identified: visual clarity, consistency, compatibility, informative feedback, explicitness, appropriate functionality, flexibility and control, error prevention and correction, and user guidance and support. These are each disaggregated into a number of more measurable attributes. These attributes can be used as part of the evaluation strategy, utilizing the decision assessment or cost effectiveness approaches described in Chapters 7 and 8.

Although attributes of this sort are potentially very useful for evaluation of existing interfaces, they do not provide a proactive guide to system design. We have discussed some design guidelines here. Guidelines are needed for all phases of the research, development, test, and evaluation life cycle.

A major claim in recent efforts in interface design is that human errors are just simply not probabilistic random events that might be removed either through improved operator training or through better interface design alone [165]. Supporting this is the observation that errors arise from one of two important sources of error.

1. Errors represent systematic interference and incongruities among models, rules, and procedures. This apparently could represent a deficiency in single-loop learning.

2. Errors represent some disfunctionality of the effects of adaptive, or double-loop, learning mechanisms.

From this perspective, trustworthy human-system interactions are achieved through the design of interfaces that minimize and correct for these difficulties that cannot be eliminated through error recovery or correction approaches.

In this work, four categories of human error are identified and appropriate design guidelines are developed. These four error categories are also associated with the type of skill-based, rule-based, or formal knowledge-based cognitive control mode that is used. As we know from our previous discussions in Section 9.2, this is a function of user experiential familiarity with task and environment and the criticality of the situation at hand.

1. *Errors Related to Learning and Interaction:* These cannot and should not be removed totally in that occasional errors may play an important role in skill-based developments. Four basic design guidelines are postulated.

   1.1. Experiments are necessary to enable interface users to optimize their skills and so that appropriate interface design should make acceptable performance limits visible while their effects are still both observable and reversible.

   1.2. System operators should be provided with feedback on the effects of actions potentially taken such that they are able to cope with necessary time delays between execution of an action and the observation of its effect. This expedites error recovery through the support provided to functional understanding and knowledge based monitoring of rules and constraints associated with the applicability of rules, particularly when these constraints are originated by another person, such as another decision maker or the system developer.

   1.3. Displays should be designed such that the rule-based mappings, from the signs that define cues for action and from the symbols that describe how processes function, are consistent and unique.

   1.4. System users should be provided with the ability to perform formal reasoning-based simulation experiments, such as to test hypotheses, without the necessity to apply the associated decisions or controls on a possibly irreversible real system or process. This simulate and evaluate capability is especially needed in initially unforeseen situations.

2. *Interference among Competing Cognitive Control Structures:* These should be reduced to the extent possible. This relates to the fact that experiential familiarity with a task will change the type of cognitive control used for task performance. Interference may exist due to improper judgment at the meta-level of decisions concerning how to decide. Three design guidelines are conjectured to deal with this interference.

   2.1. Provide overview-of-the-entire-situation-type displays in order to support skill-based reasoning.

   2.2. Use integrated patterns for symbolic representations and functional monitoring of performance that are based on defined rules, such as to avoid inadvertent activation of familiar task actions when these are inappropriate.

   2.3. Support formal knowledge-based reason by formalizing mental models to avoid undesired interference among models.

3. *Lack of Cognitive and Physiological Resources:* This deficiency should be ameliorated through automation of human information processing tasks, whenever appropriate. Two guidelines are developed to support this.

   3.1. Enable effective data and information integration by using available data to develop consistent information conversion notions such as to support design of an interface that will allow information presentation at the cognitive control level that is most appropriate for decision making in a specific situation.

   3.2. Present information in structured representations that are effective for the level of cognitive control adopted. Since the type of cognitive control mode used depends upon the contingency task structure, the information presentation framework should be adaptable to this.

4. *Intrinsic Human Variability:* Cope with this by attempting to utilize this random behavior in application of skills, and occasionally rules, in an adaptive manner that is associated with and supports learning. The nine guidelines just mentioned are applicable here, also. One additional guideline is identified.

   4.1. Memory aids, and appropriate representations, are useful in coping with intrinsic human variability.

On the basis of this, Vicente and Rasmussen [57] propose a theoretical framework, based on the Rasmussen cognitive control mode representation of Section 9.2.4, for designing interfaces. The purpose of the design is reduction of human error through improvement in systematic interactions among models, and improvements in adaptive learning. The two primary objectives of *Ecological Interface Design* (EID) are support for all of the levels of cognitive control and not forcing human information processing to a higher level, in terms of not being more cognitively demanding, than required for the task at hand and associated contingency structure. In principle therefore, an ecological interface design does not contribute to overall task difficulty. The ecological interface design concept is viewed as extending and enhancing, not replacing, the earlier direct manipulation interface thinking.

Three principles are viewed as being of primary importance in interface design to avoid errors associated with skill-based, rule-based, or formal knowledge-based cognitive control.

1. To support interaction through the use of skill-based and time–space signals, the interface user should be able to directly act on the information that is provided by the display itself. The structure of the displayed information should be commensurate with a wholistic or gestalt structure of the situation and should allow potential views of the situation from several hierarchical levels of abstraction. It is postulated that the use of a mouse is much to be preferred to a command language here as it is easier to retain spatial and temporal aspects of a situation.

2. A consistent and unique transformation should be provided between the invariant properties of the situation and the resulting cues or signs that are provided to the user by the system interface in order to support situation understanding through rule selection. This support for cognitive control at the level of rules helps avoid the selection and application of inappropriate rules.

3. A formal knowledge-based display of the structure of the evolving situation should be provided to the interface user. This provides the necessary support in structuring initially unstructured or semistructured problems so that formal knowledge-based problem solving strategies can be more effectively utilized in those situations where they are needed.

A theoretical foundation such as this provides a framework for designing human interfaces to complex systems. This ecological interface design construct supports the three cognitive control levels of knowledge: skill-based, rule-based, and formal reasoning-based. The EID design is intended to encourage information processing and cognitive control at the lowest cognitive level that the task, and associated experiential familiarity with the task and environment, indicates as appropriate.

In this section, we have identified some design guidelines for the systems engineering of human–machine system interfaces, a simplified architecture for which is shown in Figure 9.31. One of the fundamental notions in most contemporary studies of information technology and its use to improve

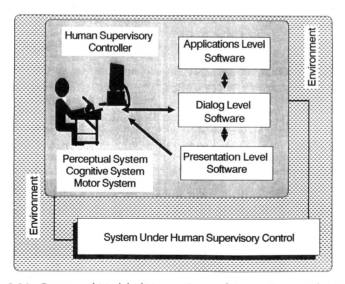

Figure 9.31. Conceptual Model of Human System (Human Computer) Interaction

performance is that there is an intimate association between human intent and human error. Realistic efforts to discuss human error and to design systems for human interaction that cope with human error possibilities will, therefore, consider the different types of human intentions and associated errors. In a very insightful work, Reason [166] indicates the importance of knowing whether human actions are directed by conscious intent, whether human actions proceeded as planned, and whether they achieved the desired result. Six types of actions result from this observation, as indicated in Figure 9.32. Successful interface and dialog designs will necessarily seek to create systems that encourage human intent and associated actions that result in successful act performance. There is only one bottom line box in Figure 9.32 that indicates successful act performance. Involuntary and spontaneous acts may or may not accomplish desired objectives. These result from unintended behavior and it is unusual that they accomplish a desired objective. Three of the four types of actions based on existing cognitive intention are erroneous. In the next section, we comment further on this important subject of human error and its amelioration.

9.6.3. Human Error and Its Amelioration

We have just introduced some elementary notions of human error. Figure 9.32 illustrates the variety of action types that are possible. This representation of human actions and associated errors assumes that the human is attempting to act in a supportive manner to aid in achieving overall objectives. The figure needs to be modified somewhat to consider an uncooperative human who is attempting not to achieve overall objectives, and perhaps to sabotage them. However, we will not consider this here.

We may act in an unintended manner, either spontaneously or involuntarily and without prior cognitive intent in either case. The following are illustrative unintended acts. A burglar pushes a gun in your back. It exacerbates a very sore boil, causing you to violently turn around while you are carrying a sledge hammer. The sledge hammer strikes the burglar in the face, causing severe injury. There is no prior cognitive intent. The act itself is a spontaneous intentional act. You are running down the hallway very rapidly to get to an appointment. I open a door in order to leave a room. You bump into the door and fall to the floor. I have, presumably, no prior intent to knock you to the floor. I have just performed a nonintentional or involuntary act, that of knocking you to the floor, even through I did have a cognitive intent to the leave the room. You are carrying a very valuable antique Satsuma vase. In falling to the floor, you drop the vase and it shatters. Your action in breaking the vase surely has no prior cognitive intent. There was neither prior cognitive intention to break the vase nor was this your intent in the action of falling to the floor. Thus, this vase breakage is a nonintentional and involuntary act. It was totally subsidiary to falling to the floor.

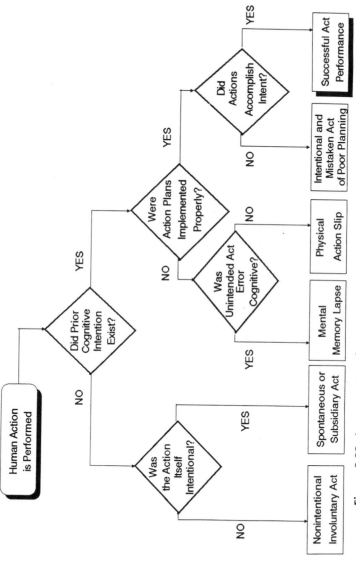

Figure 9.32. Interpretation of Reason's Taxonomy of Human Actions and Associated Types of Human Error

Our concern here with human actions that are performed without prior cognitive intent is one of providing support aids for situation assessment such that humans do form prior cognitive intents when it is appropriate that this be done. In the example just cited, my intention in leaving the room by opening the door without looking to see who was coming down the hall, was a prior intent. It is the basic cause of the shattering of the Satsuma vase. My prior intent in leaving the room should have been a bit more tempered with safety concerns such that my action, without prior intent, of causing you to fall to the floor, would not have taken place.

Thus, our concerns are, mostly, directed at actions for which a prior cognitive intent exists. There are two fundamental types of errors in situations where a prior cognitive intent exists. There may exist planning errors or detection and diagnosis mistakes. These represent failure to either identify suitable objectives or a suitable course of action to obtain it. Alternately, there may exist execution or controlling errors. An execution error, which results from actions that do not accomplish their intent, is a lapse when there is an unintended memory error that leads to poor action, and a slip when the physical action taken is itself improper and not congruous with the prior cognitive intent.

In other words, actions based on inappropriate intentions are mistakes, and inappropriate actions that are based on proper or improper plans are lapses or slips [167, 168]. Figure 9.33 illustrates how the various cognitive intent errors may exist at several cognitive control modes of behavior. There may be errors associated with either detection or diagnosis, or control. Generally, slips and lapses are associated with control, and mistakes are associated with detection and/or diagnosis. Specific error categories, representing potential errors of omission and commission, may be identified for a

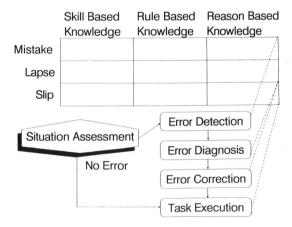

Figure 9.33. Error Possibilities for Various Cognitive Control Modes

given application domain and the matrix of Figure 9.33 completed. Generally, this would consider inherent human limitations and inherent system limitations. It would also consider specific contributing conditions and events that are particular to the application domain. It would involve [169]:

1. *Error Identification:* Histories of user behavior and system response are correlated with operational procedures and scripts such as to enable us to detect deviations between observed behavior and nominally expected behavior.
2. *Error Classification:* The causal factors leading to errors and the consequences of various errors are tabulated, structured, organized, and perhaps prioritized.
3. *Error Remediation:* Various monitoring, feedback, and control possibilities are delineated and the most suitable procedures for remediation of errors of various types are determined.

This can be accomplished from a structural, functional, or purposeful perspective. All three horizons are desirable to explore the detailed form of issues and solutions, to conceptualize issues and solutions in terms of input–output relations, and to explore issue and solution needs and requirements. Often, it would also be desirable to be able to allocate these identification, classification, and remediation efforts across humans and machines. We discussed error mechanisms for skill rule, and formal knowledge-based cognitive control earlier and this led to Figure 9.7. Any of the other cognitive control mode representations could, of course be used.

Mistakes, of failures of detection and diagnosis, may be further categorized as due to failures of expertise or lack of expertise. In the first case, a skill-based reasoning approach to situation assessment is applied inadequately. In the second case, lack of skill may force cognitive control at the level of formal reasoning based knowledge and this may turn out to result in a poor solution. Alternately, the lack of sufficient experiential familiarity needed to use skill-based reasoning may be unrecognized. A skill-, or perhaps rule-, based knowledge approach may be improperly used and this may lead to significant error. Of course, any cognitive control mechanism may be applied properly or improperly, as we have noted in our earlier discussions and illustrated in Figure 9.10.

There has been much recent interest in this subject. Event tree models often result from the sort of analysis suggested here. These may be used to both analyze and identify support mechanisms to aid human reliability [170] and safety [171]. At an organizational level, these concepts support the integration of systems and organizations [172]. Cooperative work efforts [173] and skill development through information technology [174] are each supported strongly by these developments. Clearly, a unified theory of cognition [175] would need to consider tasks, errors in task performance, and related issues. There are a number of very interesting current speculations relative to the adaptive character of thought [35–38] that are appropriate here. Sadly, we are not able to examine these issues in any detail here.

9.7. SUMMARY

In this last chapter, we have illustrated a number of models for human cognitive control. Also, we considered a number of models of rationality that considered a variety of perspectives relative to judgment and choice situations. Finally, we examined some human error taxonomies. One substantial effort that we did not consider is that the rationality perspectives and cognitive control modes are not at all mutually exclusive concepts. Each of these exist simultaneously for any real situation. Furthermore, human error may exist in either of these cells. This two-dimensional representation could be expanded to a third by considering the phases of effort needed to develop a system. Complexity is added also by the need to implement and preserve quality, in all meanings of the term, at each facet of effort in developing and using systems to support human endeavor. So, we see that we have addressed a truly very large problem. We have not solved it. *That may be your job.* Hopefully, we have set forth some very useful guidelines for solution in this text. That was my job. I hope it was done well and that you find these guidelines for systems engineering useful.

PROBLEMS

9.1. Contrast and compare the Janis Mann model of decision making, or cognitive control, with the Dreyfus model. Contrast and compare the Klein model of decision making with the Dreyfus model. Please contrast and compare the Rasmussen model of decision making with the Dreyfus model and the Janis and Mann model. In what ways are these models similar and dissimilar.

9.2. Please contrast and compare the various models for cognitive control with the rationality perspectives that we discussed. In what ways are models and rationality perspective approaches similar and dissimilar.

9.3. Stress is a major driver of the Janis-Mann model. We have interpreted this to mean time–stress. A complimentary way of viewing stress is to suggest that it is the difference between the coherency of the task and cognitive control mode for the human, or organization, charged with task performance. Illustrate the compatibility of these thoughts.

9.4. Does our discussion of cognitive style indicate that the various stereotypical models of cognitive style are incorrect or that they need to have some adaptive mechanism to enable them to track the contingency task structure? How does this relate to the use of various personality tests for prediction of human performance?

9.5. Identify one human information processing bias from Chapter 7 that appears to be of interest and write a case study scenario of how a human might process or misprocess information if afflicted with this bias. How does this affliction appear on the cognitive control mode scales we have discussed here? How would you represent this bias using the various rationality perspectives that we have discussed here?

9.6. Prepare an evaluation methodology, using approaches developed in Chapters 7 and 8, for human–computer interaction and interface evaluation. Use this to evaluate a human–computer interface, or dialog, with which you are familiar.

9.7. How would you expect the human error discussions to vary across the phases of the systems engineering life cycle. In particular, please write a report describing human error in identification of user requirements, in detailed system design, in operational test and evaluation of a system, and in maintenance and modification of a system.

9.8. Please write a summary of Section 9.6.3 that is applicable for the special case of intentional human performance violations or sabotage.

9.9. Please write a report indicating how you might go about determining appropriateness of assignment of various aspects of task performance to humans, or to automated machines. How do knowledge representations relate to this?

9.10. Please write a report indicating interrelationships between total quality management and cognitive ergonomics.

9.11. Please relate and discuss Figure 9.29 in connection with the total quality management objectives described in Chapter 5.

9.12. On page 564, we noted 50 performance abilities that involve cognitive, perceptual–motor, sensory, and purely physical performance realms. Please develop a structural model of these abilities with these four realms at the first level of your hierarchy.

9.13. A major goal of the systems engineering approach is to assist in research, development, test, and evaluation of systems, based upon knowledge of human roles, capabilities, and tasks performed; such that these systems are designed for human interaction. A framework for system design for human interaction should assist systems engineers in determining:

a. the implications of cognitive limitations on the requirements specifications for systems;

b. efficient techniques for evaluation of cognitive issues;

c. approaches for incorporating human interaction concerns into the analysis, design, and evaluation procedures for systems; and

d. methods for structuring and presenting this framework to system developers and system users.

Each of these serves a very appropriate purpose. It will be appropriate and desirable also to cope with a number of related issues that also support these purposes.

a. There are a large variety of environments into which various fielded system activities are imbedded. These include a number of perspectives such as economic, technical, legal, political, and social. This results in a number of "natural" architectures or environments for system use. The garbage can architecture is one of these that has been recently used for modeling decision assessment environments.

b. The experiential familiarity of the decision-making individual or group with the tasks at hand and the environment into which tasks are imbedded with strongly influence the time–stress and resulting cognitive control mode chosen for such information processing tasks as issue detection, issue diagnosis and issue resolution. Support systems for command and control in organizations must be accommodating to these cognitive control mode realities in order to enhance the resulting decision perspectives, rather than to inhibit them.

c. Errors can occur in formulation, analysis, or interpretation (or detection, diagnosis, or correction) efforts. There is need for an error taxonomy that relates human, human–machine, error to environment, cognitive–control mode, and other facets.

d. A computer-based support system to improve human functioning will necessarily have to interface and be interactive with humans, and hardware and software, in the command and control environment. This brings about the needs for both system design for human interaction, and open systems architectures that enable the fielded system to be successfully maintained and proactively adapted to new environments. Doing this well results in operational level quality assurance.

e. The systems engineering process itself must be accommodating to these four needs, and others. This leads to the need for strategic, or process, level quality assurance and management in systems engineering.

These five assertions are posed as a major integrating theme that provides linkages among the systems engineering facets highlighted on the dust jacket for this book: methodology, design, management, quality, decisions, life cycles. If accepted, they strongly indicate the wholistic nature of the entire systems engineering effort. Comment, discuss, and find reasons for agreement and disagreement. Why is this an important concluding problem for this text?

REFERENCES

[1] Pelsma, K. H. (Ed.), *Ergonomics Sourcebook: A Guide to Human Factors Information*, Ergosyst, Lawrence KS, 1987.

[2] Sheridan, T. B., and Ferrell, W. R., *Man-Machine Systems: Information, Control, and Decision Models of Human Performance*, MIT Press, Cambridge MA, 1974.

[3] Sheridan, T. B., and Johannsen, G. (Eds.), *Monitoring Behavior and Supervisory Control*, Plenum Press, New York, 1976.

[4] Johannsen, G., Rijnsdorp, J. E., and Sage, A. P., "Human Interface Concerns in Support Systems Design," *Automatica*, Vol. 19, No. 6, 1983, pp. 595–603.

[5] Rasmussen, J., and Rouse, W. B. (Eds.), *Human Detection and Diagnosis of System Failures*, Plenum Press, New York, 1981.

[6] Card, S. K., Moran, T. P., and Newell, A., *The Psychology of Human-Computer Interaction*, Lawrence Erlbaum, Hillsdale NJ, 1983.

[7] Shneiderman, B., *Software Psychology: Human Factors in Computer and Information Systems*, Winthrop, Cambridge MA, 1980.

[8] Shneiderman, B., *Designing the User Interface: Strategies for Effective Human-Computer Interaction*, Addison Wesley, Reading MA, 1987.

[9] Sage, A. P., *Decision Support Systems Engineering*, John Wiley, New York, 1991.

[10] Kelley, C. R., *Manual and Automatic Control*, John Wiley, New York, 1968.

[11] Edwards, E. and Lees, F. P. (Eds.), *The Human Operator in Process Control*, Taylor and Francis, London, 1974.

[12] Johannsen, G., Boller, H. E., Donges, E., and Stein, W., *Der Mensch im Regelkreis—Lineare Modelle*, Oldenbourg Verlag, Munich, 1977.

[13] Rouse, W. B., *Systems Engineering Models of Human-Machine Interaction*, Elsevier North-Holland, New York, 1980.

[14] Kleinman, D. L., Baron, S., and Levison, W. H., "An Optimal Control Model of Human Response, Part 1: Theory and Validation," *Automatica*, Vol. 6, 1970, pp. 357–369.

[15] Moray, N. (Ed.), *Mental Workload: Its Theory and Measurement*, Plenum Press, New York, 1979.

[16] Anderson, J. R., *Cognitive Psychology and Its Implications*, W. H. Freeman, San Francisco CA, 1980.

[17] Anderson, J. R., *The Architecture of Cognition*, Harvard University Press, Cambridge MA, 1983.

[18] Estes, W. K., *Handbook of Learning and Cognitive Processes*, Volumes 1–6, Lawrence Erlbaum, Hillsdale NJ, 1975–1979.

[19] Mayer, R. E., *Thinking, Problem Solving, Cognition*, W. H. Freeman, San Francisco, 1983.

[20] Eysenck, M. W., *A Handbook of Cognitive Psychology*, Lawrence Erlbaum, Hillsdale NJ, 1984.

[21] Mandler, G., *Cognitive Psychology*, Lawrence Erlbaum, Hillsdale NJ, 1985.

[22] Bettman, J. R., *An Information Processing Theory of Consumer Choice*, Addison Wesley, Reading MA, 1979.

[23] Blakeslee, T. R. *The Right Brain*, Anchor Press, New York, 1980.

[24] Craik, F. I. M., "Human Memory," *Annual Review of Psychology*, Vol. 30, 1979, pp. 63–102.

[25] Fishbein, M., and Azjin, I., *Belief, Attitude, Intention, and Behavior*, Addison-Wesley, Reading MA, 1975.

[26] Gregory, R. L. (Ed.), *The Oxford Companion to the Mind*, Oxford University Press, New York, 1987.

[27] Mason, R. O., and Mitroff, I. I., "A Program for Research on Management Information Systems," *Management Science*, Vol. 19, No. 5, 1973, pp. 475–485.

[28] Bariff, M. L., and Lusk, E. J., "Cognitive and Personality Tests for the Design of Management Information Systems," *Management Science*, Vol. 23, No. 8, 1977, pp. 820–829.

[29] McKeeney, J. L., and Keen, P. G. W., "How Managers Minds Work," *Harvard Business Review*, Vol. 52, No. 3, 1974, pp. 79–90.

[30] Driver, M. J., and Mock, T. J., "Human Information Processing, Decision Theory Style, and Accounting Information Systems," *Accounting Review*, Vol. 50, 1975, pp. 490–508.

[31] Taylor R. N., and Dunnette, M. D., "Relative Contribution of Decision-Maker Attributes to Decision Processes," *Organizational Behavior and Human Performance*, Vol. 12, 1974, pp. 286–298.

[32] Simon, H. A., "Information Processing Models of Cognition," *Annual Review of Psychology*, Vol. 30, 1979, pp. 363–396.

[33] Taggart, W., and Valenzi, E., "Assessing Rational and Intuitive Styles: A Human Information Processing Metaphor," *Journal of Management Studies*, Vol. 27, No. 2, 1990, pp. 150–169.

[34] Tiedemann, J., "Measures of Cognitive Styles: A Critical Review," *Educational Psychologist*, Vol. 24, No. 3, 1989, pp. 261–275.

[35] Anderson, J. R., *The Adaptive Character of Thought*, Lawrence Erlbaum, Hillsdale NJ, 1990.

[36] Anderson, J. R., and Milson, R. M., "Human Memory: An Adaptive Perspective," *Psychological Review*, Vol. 96, No. 4, 1989, pp. 703–719.

[37] Anderson, J. R., "Is Human Cognition Adaptive," *Behavioral and Brain Sciences*, Vol. 14, 1991, pp. 471–517.

[38] March, J. G., "Bounded Rationality, Ambiguity, and the Engineering of Choice," *The Bell Journal of Economics*, Vol. 10, 1978, pp. 587–608.

[39] Zmud, R. W., "Individual Differences and MIS Success: A Review of the Empirical Literature," *Management Science*, Vol. 25, No. 10, 1979, pp. 966–979.

[40] Zmud, R. W., "On the Validity of the Analytic-Heuristic Instrument Utilized in the Minnesota Experiments," *Management Science*, Vol. 24, 1978, pp. 1088–1090.

[41] Dickson, G. W., Senn, J. A., and Chervany, J. J., "Research in Management Information Systems: The Minnestoa Experiments," *Management Science*, Vol. 23, 1977, pp. 913–923.

[42] Chervany, N. L., and Dickson, G. W., "On the Validity of the Analytic-Heuristic Instrument Utilized in the Minnesota Experiments—A Reply," *Management Science*, Vol. 24, 1978, pp. 1091–1092.

[43] Benbasat, I., and Taylor, R. N., "The Impact of Cognitive Styles on Information System Design," *Management Information Systems Quarterly*, Vol. 2, 1978, pp. 43–54.

[44] Mischel, W., "On the Interface of Cognition and Personality: Beyond the Person-Situation Debate," *American Psychologist*, Vol. 34, 1979, pp. 740–754.

[45] Huber, G. P., "Cognitive Style as a Basis For MIS and DSS Designs: Much Ado About Nothing," *Management Science*, Vol. 29, No. 5, May 1983, pp. 567–579.

[46] Brainerd, C. J., *Piaget's Theory of Intelligence*, Prentice-Hall, Englewood Cliffs NJ, 1978.

[47] Flavell, J. H., *Cognitive Development*, Prentice-Hall, Englewood Cliffs NJ, 1977.

[48] Ginsburg, H., and Opper, S., *Piaget's Theory of Intellectual Development*, Prentice Hall, Englewood Cliffs NJ, 1979.

[49] Stone, C. A., and Day, M. C., "Competence and Performance Models and the Characterization of Formal Operational Skills," *Human Development*, Vol. 28, 1980, pp. 323–353.

[50] Wohl, J. G. "Force Management Decision Requirements for Air Force Tactical Command and Control," *IEEE Transactions on Systems, Man, and Cybernetics*, Vol. 11, No. 9, 1981, pp. 618–639.

[51] Andriole, S. J., and Halpin, S. M. (Eds.), *Information Technology for Command and Control: Methods and Tools for System Development and Evaluation*, IEEE Press, New York, 1991.

[52] Rasmussen, J., "Skills, Rules, Knowledge; Signals, Signs, and Symbols; and Other Distinctions in Human Performance Models," *IEEE Transactions on Systems, Man, and Cybernetics*, Vol. SMC 13, No. 3, 1983, pp. 257–266.

[53] Rasmussen, J., *On Information Processing and Human-Machine Interaction: An Approach to Cognitive Engineering*, North Holland, New York, 1986.

[54] Rasmussen, J., "Models of Mental Strategies in Process Plant Diagnosis," in Rasmussen, J., and Rouse, W. B. (Eds.), *Human Detection and Diagnosis of System Failures*, Plenum Press, New York, 1980, pp. 241–258.

[55] Hunt, R. M., and Rouse, W. B., "A Fuzzy Rule Based Model of Human Problem Solving," *IEEE Transactions on Systems, Man, and Cybernetics*, Vol. 14, No. 1, 1984, pp. 112–120.

[56] Rasmussen, J., "Use of Computer Games to Test Experimental Techniques and Cognitive Models," in Rasmussen, J. and Zunde, P. (Eds.), *Empirical Foundations of Information and Software Science III*, Plenum Press, New York, 1987, pp. 187–195.

[57] Vicente K. J., and Rasmussen, J., "Ecological Interface Design: Theoretical Foundations," *IEEE Transactions on Systems, Man, and Cybernetics*, Vol. 22, No. 3, 1992.

[58] Rasmussen, J., and Vicente, K. J., "Coping with Human Errors through System Design: Implications for Ecological Interface Design," *International Journal of Man–Machine Studies*, Vol. 31, 1989, pp. 517–534.

[59] Ashby, W. R., *Introduction to Cybernetics*, John Wiley, New York, 1956.

[60] Rasmussen, J., "Diagnostic Reasoning in Action," *IEEE Transactions on Systems, Man, and Cybernetics*, Vol. 22, No. 6, 1992, in press.

[61] Klein, G. A. and Weitzenfeld, J., "Improvement of Skills for Solving Ill-Defined Problems," *Educational Psychology*, Vol. 13, 1978, pp. 31–41.

[62] Klein, G. A. "Automated Aids for the Proficient Decision Maker," *Proceedings 1980 IEEE Systems, Man, and Cybernetics Conference*, October 1980, pp. 301–304.

[63] Klein, G. A., Orasanu, J., Calderwood, R., and Zsambok, C. E. (Eds.), *Decision Making in Action: Models and Methods*, Ablex Publishers, New York 1992.

[64] Dreyfus, H. L., and Dreyfus, S. E., *Mind Over Machine: The Power of Human Intuition and Expertise in the Era of the Computer*, Free Press, New York, 1986.

[65] Dreyfus, H. L., "The Socratic and Platonic Basis of Cognitivism," *AI and Society*, Vol. 2, 1988, pp. 99–112.

[66] Janis, I. L., and Mann, L., *Decision Making: A Psychological Analysis of Conflict, Choice, and Commitment*, Free Press, New York, 1977.

[67] Broadbent, D. W., *Decision and Stress*, Academic, London, 1971.

[68] Janis, I. J., *Groupthink: Psychological Studies of Policy Decisions and Fiascoes*, Houghton Mifflin, Boston MA, 1982.

[69] Janis, I. J., *Crucial Decisions: Leadership in Policy Making and Crisis Management*, Free Press, New York, 1989.

[70] Brunswick, E., "The Conceptual Framework of Psychology," in *International Encyclopedia of Unified Science*, Vol. 1, No. 10, University of Chicago Press, Chicago, 1952.

[71] Brunswick, E., *Perception and the Representative Design of Psychological Experiments*, University of California Press, 1956.

[72] Hammond, K. R., "Intuitive and Analytical Cognition: Information Models," in A. P. Sage (Ed.), *Concise Encyclopedia of Information Processing in Systems and Organizations*, Pergamon Press, Oxford, 1990, pp. 306–312.

[73] Hammond, K. R., Hamm, R. M., Grassia, J., and Pearson, T., "Direct Comparison of the Efficacy of Intuitive and Analytical Cognition in Expert Judgment," *IEEE Transactions on Systems, Man, and Cybernetics*, Vol. 17, No. 5, 1987, pp. 753–770.

[74] Hammond, K. R., "Judgment and Decision Making in Dynamic Tasks," *Information and Decision Technologies*, Vol. 14, No. 1, 1988, pp. 3–14.

[75] Brehmer, B., "Systems Design and the Psychology of Complex Systems," in Rasmussen, J. and Zunde, P. (Eds.), *Empirical Foundations of Information and Software Science III*, Plenum Press, New York, 1987, pp. 21–32.

[76] Brehmer, B., "Dynamic Decision Making," in Sage, A. P., *Concise Encyclopedia of Information Processing in Systems and Organizations*, Pergamon Press, Oxford UK, 1990, pp. 144–149.

[77] Hagafors, R., and Brehmer, B., "Does Having to Justify One's Judgments Change the Nature of the Judgment Process," *Organizational Behavior and Human Performance*, Vol. 31, 1983, pp. 223–232.

[78] Crocker, J., Fiske, S. T., and Taylor, S. E., "Schematic Basis for Belief Change," in Eiser, J. R. (Ed.), *Additudional Judgment*, Springer Verlag, New York, 1984.

[79] Brehmer, B., "Response Consistency in Probabilistic Inference Tasks," *Organizational Behavior and Human Performance*, Vol. 22, 1978, pp. 103–115.

[80] Brehmer, B., "In One Word: Not from Experience," *Acta Psychologica.*, Vol. 45, 1980, pp. 223–241.

[81] Brehmer, B., "Models of Diagnostic Judgment," in Rasmussen, J., and Rouse, W. B. (Eds.), *Human Detection and Diagnosis of System Failures*, Plenum Press, New York, 1981, pp. 231–241.

[82] Sage, A. P., and Lagomasino, L., "Knowledge Representation and Man-Machine Dialog," in Rouse, W. B. (Ed.), *Advances in Man Machine Systems Research*, Vol. 1, JAI Press, Greenwich CT, 1984, pp. 223–260.

[83] Newell, A., and Simon, A. H., *Human Problem Solving*, Prentice-Hall, Englewood Cliffs NJ, 1972.

[84] Barr, A., Cohen, P. R., and Feigenbaum, E. A., (Eds.), *Handbook of Artificial Intelligence*, Vol. I Kaufman, Los Angeles CA, 1981.

[85] Post, S., and Sage, A. P., "An Overview of Automated Reasoning," *IEEE Transactions on Systems, Man, and Cybernetics*, Vol. 20, No. 1, 1990, pp. 202–224.

[86] Minsky, M., "A Framework for Representing Knowledge," in Winston, P. A. (Ed.), *The Psychology of Computer Vision*, McGraw Hill, New York, 1975.

[87] Abelson, R. P., "The Structure of Belief Systems," in Schank, R. C., and Colby, K. M. (Eds.), *Computer Models of Thought and Language*, Freeman, San Francisco, 1973, pp. 287–339.

[88] Axelrod, R. M. (Ed.), *Structure of Decision: The Cognitive Maps of Political Elites*, Princeton University Press, 1976.

[89] Axelrod, R. M., "Framework for a General Theory of Cognition and Choice," Institute of International Studies, University of California at Berkeley, 1972.

[90] Eden, C., Jones, S., and Sims, D., *Thinking in Organizations*, MacMillan Press, London, 1979.

[91] Eden, C., "Using Cognitive Mapping for Strategic Options Development and Analysis," in Rosenhead, J. (Ed.), *Rational Analysis for a Problematic World*, Wiley, Chichester, 1989.

[92] Holland, J. H., Holyoak, K. J., Nisbett, R. E., and Thagard, P. R., *Induction: Process of Inference, Learning, and Discovery*, MIT Press, Cambridge MA, 1987.

[93] Noble, D. F., "Schema-Based Knowledge Elicitation for Planning and Situation Assessment Aids," *IEEE Transactions on Systems, Man, and Cybernetics*, Vol. 19, No. 3, 1989, pp. 473–482.

[94] Abelson, R. P., "The Psychological Status of the Script Concept," *American Psychologist*, Vol. 36, 1981, pp. 715–729.

[95] Carbonnel, J. G., "Learning by Analogy: Formulating and Generalizing Plans from Past Experience," Chapter 5 in Michalski, R. S., et al. (Eds.), *Machine Learning: an Artificial Intelligence Approach*, Tioga, San Francisco, 1983, pp. 137–162.

[96] Gick, M. L., and Holyoak, K. J., "Analogical Problem Solving," *Cognitive Psychology*, Vol. 12, 1980, pp. 306–355.

[97] Silverman, B. G., "Analogy in Systems Management: A Theoretical Inquiry," *IEEE Transactions on Systems, Man, and Cybernetics*, Vol. 13, No. 6, 1983, pp. 235–261.

[98] Shapiro, S. C., (Ed.), *Encyclopedia of Artificial Intelligence*, John Wiley, New York, 1987.

[99] Garcia, O. N., and Chien, Y. T. (Eds.), *Knowledge Based Systems: Fundamentals and Tools*, IEEE Computer Society Press, Los Alamitos CA, 1991.

[100] Sage, A. P. (Ed.), *Concise Encyclopedia on Information Processing in Systems and Organizations*, Pergamon Press, Oxford, 1990.

[101] Toulmin, S., Rieke, R., and Janik, A., *An Introduction to Reasoning*, Macmillan, New York, 1979.

[102] Lagomasino, A., and Sage, A. P., "Representation and Interpretation of Information for Decision Support with Imperfect Knowledge," *Large Scale Systems*, Vol. 9, No. 2, 1985, pp. 169–181.

[103] Lagomasino, A., and Sage, A. P., "An Interactive Inquiry System," *Large Scale Systems*, Vol. 9, No. 3, 1985, pp. 231–244.

[104] Sage, A. P., "On the Processing of Imperfect Information Using Structured Frameworks," Chapter 7 in Kandel, A. (Ed.), *Fuzzy Expert Systems*, CRC Press, New York, 1991, pp. 99–112.

[105] Laskey, K. B., Chen, M. S., and Martin, A. W., "Representing and Eliciting Knowledge about Uncertain Evidence and its Implications, *IEEE Transactions on Systems, Man, and Cybernetics*, Vol. 19, No. 3, 1989, pp. 536–545.

[106] Smith, G. F., Benson, P. G., and Curley, S. P., "Belief, Knowledge and Uncertainty: A Cognitive Perspective on Subjective Probability," *Organizational Behavior and Human Decision Processes*, 1992.

[107] Haye, J., "An Axiomatic Theory of Organizations," *Administrative Science Quarterly*, Vol. 10, No. 3, 1965, pp. 289–320.

[108] March, J. G., and Simon, H. A., *Organizations*, Wiley, New York, 1958.

[109] Cyert, R. M., and Simon, H. A., "The Behavioral Approach: With Emphasis on Economics," *Behavioral Science*, Vol. 28, 1983, pp. 95–108.

[110] Weick, K. E., *The Social Psychology of Organizing*, Addison-Wesley, Reading MA, 1979.

[111] Weick, K. E., "Cosmos versus Chaos: Sense and Nonsense in Electronic Contexts," *Organizational Dynamics*, Vol. 14, No. 3, 1985, pp. 50–64.

[112] Janis, I. L., *Groupthink*, Free Press, New York, 1982.

[113] Argyris, C., and Schon, D. A., *Organizational Learning: A Theory of Action Perspective*, Addison Wesley, Reading MA, 1978.

[114]   Argyris, C., and Schon, D. A., *Theory in Practice: Increasing Professional Effectiveness*, Jossey Bass, San Francisco, 1974.

[115]   Argyris, C., *Reasoning, Learning, and Action: Individual and Organizational*, Jossey-Bass, San Francisco, 1982.

[116]   Huber, G. P., "Organizational Decision Making and the Design of Decision Support Systems," *Management Information Systems Quarterly*, Vol. 5, No. 2, 1981.

[117]   Huber, G. P., "Organizational Information Systems: Determinants of their Performance and Behavior," *Management Science*, Vol. 28, No. 2, 1982, pp. 138–155.

[118]   Vroom, V. H., and Yetton, P. W., *Leadership and Decision Making*, Pittsburgh, PA: University of Pittsburgh, 1973.

[119]   Keen, P. G. W., "Information Systems and Organizational Change," *Communications of the Association for Computing Machinery*, Vol. 24, No. 1, 1981, pp. 24–33.

[120]   Tushman, M. L., and Nadler, D. A., "Information Processing as an Integrating Concept in Organizational Design," *Academy of Management Review*, Vol. 3, No. 3, 1978, pp. 613–624.

[121]   Feldman, M. S., and March, J. G., "Information in Organizations as Signal and Symbol," *Administrative Science Quarterly*, Vol. 26, 1981, pp. 171–186.

[122]   Levis, A. H., "Organizational Information Structures: Quantitative Models," in Sage, A. P. (Ed.), *Concise Encyclopedia of Information Processing in Systems and Organizations*, Pergamon Press, Oxford UK, 1990, pp. 368–375.

[123]   Diesing, P., *Reason in Society*, University of Illinois Press, Urbana IL, 1962.

[124]   Steinbruner, J. D., *The Cybernetic Theory of Decision*, Princeton University Press, Princeton NJ, 1974.

[125]   Sage, A. P., *Economic Systems Analysis: Microeconomics for Systems Engineering, Engineering Management, and Project Selection*, Elsevier North-Holland, New York, 1983.

[126]   Simon, H. A., "From Substantive to Procedural Rationality," in Latiss, S. J. (Ed.), *Method and Appraisal in Economics*, Cambridge University Press, New York, 1976, pp. 129–148.

[127]   Simon, H. A., *Models of Thought*, Yale University, New Haven, 1979.

[128]   Simon, H. A., "Rational Decision Making in Business Organization," *American Economic Review*, Vol. 69, No. 4, 1979, pp. 493–513.

[129]   Simon, H. A., "The Behavioral and Social Sciences," *Science*, Vol. 209, July 1980, pp. 72–78.

[130]   Simon, H. A., "Rationality as Process and as Product of Thought," *American Economic Review*, Vol. 68, 1978, 1–16.

[131]   Lindblom, C. E., "The Science of 'Muddling Through,'" *Public Administrative Review*, Vol. 19, 1959, pp. 155–169.

[132]   Lindblom, C. E., *The Intelligence of Democracy: Decision Making Through Mutual Adjustment*, Free Press, New York, 1965.

[133]   Lindblom, C. E., *The Policy Making Process*, Prentice-Hall, Englewood Cliffs NJ, 1980.

[134] Cyert, R. M., and March, J. G., *A Behavioral Theory of the Firm*, Prentice-Hall, Englewood Cliffs NJ, 1963.

[135] Cohen, M. D., March, J. B., and Olsen, J. P., "A Garbage Can Model of Organizational Choice," *Administrative Science Quarterly*, Vol. 17, No. 1, 1972, pp. 1–25.

[136] Cohen, M. D., and March, J. G., *Leadership and Ambiguity*, McGraw Hill, New York, 1974.

[137] Padgett, J., "Managing Garbage Can Hierarchies," *Administrative Science Quarterly*, Vol. 25, No. 4, 1980, pp. 583–604.

[138] March, J., and Wessinger-Baylon, T. (Eds.), *Ambiguity and Command: Organizational Perspectives on Military Decisionmaking*, Pitman, Boston MA, 1986.

[139] Carley, K., "An Approach to Relating Social Structure to Cognitive Structure," *Journal of Mathematical Sociology*, Vol. 12, No. 2, 1986, pp. 137–189.

[140] Carley, K., "Designing Organizational Structures to Cope with Communications Breakdowns: A Simulation Model," working paper, September 1990.

[141] Carley, K., "Distributed Information and Organizational Decision Making Models," in Sage, A. P. (Ed.), *Concise Encyclopedia of Information Processing in Systems and Organizations*, Pergamon Press, Oxford, 1990, pp. 137–144.

[142] Carley, K., "Organizational Learning and Personnel Turnover," *Organizational Science*, Vol. 3, No. 1, 1992, pp. 20–46.

[143] Lyles, M. A., and Thomas, H., "Strategic Problem Formulation: Biases and Assumptions Embedded in Alternative Decision-Making Models," *Journal of Management Studies*, Vol. 25, No. 2, 1988, pp. 131–145.

[144] Etzioni, A., *The Moral Dimension: Toward a New Economics*, Free Press, New York, 1988.

[145] Fleishman, E. A., and M. K., Quaintance, *Taxonomies of Human Performance: The Description of Human Tasks*, Academic Press, San Diego CA, 1984.

[146] Sheridan, T. B., "Human Factors Engineering," in Sage, A. P. (Ed.), *Concise Encyclopedia of Information Processing in Systems and Organizations*, Pergamon Press, Oxford UK, 1990, pp. 209–217.

[147] Hess, R., "Human Factors Engineering: Information Processing Concerns," in Sage, A. P. (Ed.), *Concise Encyclopedia of Information Processing in Systems and Organizations*, Pergamon Press, Oxford, 1990, pp. 217–223.

[148] R. J. Volkema, "Problem Formulation," in A. P. Sage (Ed.), *Concise Encyclopedia of Information Processing in Systems and Organizations*, Pergamon Press, Oxford, 1990, pp. 377–382.

[149] Cowan, D., "Developing a Process Model of Problem Recognition," *Academy of Management Review*, Vol. 11, No. 4, 1986, pp. 763–776.

[150] Smith, C. L., and Sage, A. P., "A Theory of Situation Assessment for Decision Support," *Information and Decision Technologies*, Vol. 17, 1991, pp. 91–124.

[151] Norman, D. A., "Cognitive Engineering," in Norman, D. A., and Draper, S. W. (Eds.), *User Centered System Design*, Lawrence Erlbaum, Hillsdale NJ, 1986, pp. 32–61.

[152] Polson, P. G., "A Quantitative Theory of Human-Computer Interaction," in Carroll, J. M. (Ed.), *Interfacing Thought: Cognitive Aspects of Human–Computer Interaction*, MIT Press, Cambridge MA, 1987, pp. 184–235.

[153] Harrison, H. R., and Hix, D., "Human Computer Interface Development: Concepts and Systems for Its Management," *ACM Computing Surveys*, Vol. 21, No. 1, 1989, pp. 5–92.

[154] Hix, D., and Schulman, R. S. "Human Computer Interface Development Tools: A Methodology for their Evaluation," *Communications of the ACM*, Vol. 34, No. 3, 1991, pp. 74–87.

[155] Thimberly, H., *User Interface Design*, Addison Wesley, Reading MA, 1990.

[156] Shneiderman, B., *Designing the User Interface: Strategies for Effective Human Computer Interaction*, Addison Wesley, Reading MA, 1987.

[157] Moran, T. P., "The Command Language Grammar: A Representation of the User Interface of Interactive Computer Systems," *International Journal of Man-Machine Studies*, Volume 15, 1981, pp. 3–50.

[158] Card, S. K., Moran, T. P., and Newell, A., *The Psychology of Human Computer Interaction*, Lawrence Erlbaum, Hillsdale NJ, 1983.

[159] Card, S., Moran, T., and Newell, A., *Applied Information Processing Psychology*, Lawrence Erlbaum, Hillsdale NJ, 1983.

[160] Grudin, J., "The Case Against User Interface Consistency," *Communications of the ACM*, Vol. 32, No. 10, October 1989, pp. 1164–1173.

[161] Roberts, T. L., and Moran, T. P., "A Methodology for Evaluating Text Editors," in Curtis, B. (Ed.), *Proceedings of the IEEE Conference on Human Factors in Software Development*, Gaithersburg MD, 1982.

[162] Harrison, H. R., and Hix, D., "Human-Computer Interface Development: Concepts and Systems for Its Management," *ACM Computing Surveys*, Vol. 21, No. 1, 1989, pp. 5–92.

[163] Lee, E., "User-Interface Development Tools," *IEEE Software*, Vol. 7, No. 3, 1990, pp. 31–36.

[164] Ravden, S., and Johnson, G., *Evaluating Usability of Human-Computer Interfaces: A Practical Method*, John Wiley, Chichester, 1989.

[165] Rasmussen, J., Duncan, K., and Leplat, J. (Eds.), *New Technology and Human Error*, John Wiley, Chichester, 1987.

[166] Reason, J., *Human Error*, Cambridge University Press, Cambridge UK, 1990.

[167] Norman, D. A., "Categorization of Action Slips," *Psychological Review*, Vol. 88, No. 1, 1981, pp. 1–15.

[168] Norman, D. A., *The Psychology of Everyday Things*, Basic Books, New York, 1988.

[169] Rouse, W. B., *Design for Success: A Human Centered Approach to Designing Successful Products and Systems*, John Wiley, New York, 1991.

[170] Dhillon, B. S., *Human Reliability*, Pergamon Press, Oxford, 1986.

[171] Heslingal, G., and Stassen, H. G., The Prediction of Human Performance Safety with Event Trees, *IEEE Transactions on Systems, Man, and Cybernetics*, Vol. 22, No. 4, 1992, in press.

[172]   Booher, H. R., *Manprint: An Approach to Systems Integration*, Van Nostrand Reinhold, New York, 1990.

[173]   Rasmussen, J., Brehmer, B., and Leplat, J. (Eds.), *Distributed Decision Making: Cognitive Models for Cooperative Work*, John Wiley, Chichester UK, 1991.

[174]   Bainbridge, L., and Ruiz-Quintanilla, S. A. (Eds.), *Developing Skills with Information Technology*, John Wiley, Chichester UK, 1989.

[175]   Newell, A., *Unified Theories of Cognition*, Harvard University Press, Cambridge, MA, 1990.

Index

Abelson, R. P., 532, 535, 587, 588
Abilities, human, 563
Absolute weight, 359
Acceptance testing, 36
Access structure, 558
Accredited organization (AO) standard, 165
Accredited sponsor (AS) standard, 165
Acquisition:
  cost risk, 250, 254
  schedule risk, 250, 254
  systems, 19, 173
Activity:
  antecedent and descedent, 267, 268
  matrix, 46
  predecessor, 267
  successor, 268
  trees, 287
Activity-on-arrow, 266
Activity-on-node, 266
Acyclic graph, directed, 298
Adelman, L., 323, 347, 385, 404, 418, 419, 421
Adequacy:
  heuristic, 537
  task, 536
Adjacency matrix, 283
Advanced beginner, 515
Advanced development, 111
Akao, Y., 190, 228
Allais Paradox, 370
Allocated baseline, 139
Amelioration of human error, 563
American Management Systems, 46, 48
American National Standards Institute (ANSI), 165
American Society of Mechanical Engineers, 165
Analogous representations, 536
Analysis, 4, 11, 13, 45, 55, 63, 81, 320, 456, 469, 487, 567

Analysis:
  cost-benefit, 425
  cost-effectiveness, 425
Analytical:
  cognition, 496, 524
  hierarchy process, 409
  new-product decision model, 111
Anchoring, 382
Anderson, D. F., 404, 423
Anderson, J. R., 494, 501, 580, 583, 584
Andriole, S. J., 507, 585
Anthony, R. N., 79, 82, 125
Archibald, R., 265, 310
Architecture, integration, 175
Architectures, conceptual, 35
Argyris, C., 77, 125, 540, 588, 589
ARIADNE, 406
Arkes, H. R., 406, 423
Arrow, K., 341
Arrow diagramming, 268
Arthur Young Information Technology Group, 97, 125
Ashby, W. R., 511, 586
Ashby's Law of Requisite Variety, 511, 586
Aspiration level, 549
Assessment:
  decision, 314
  risk, 259, 263
  through decision observation, 346
Association Francaise de Normalization (AFNOR), 166
Assumption:
  Platonic, 515
  Socratic, 515
Assurance:
  quality, 131, 135
  strategic quality, 185
Atherton, G., 64, 72